Lecture Notes in Physics

Volume 847

For further volumes:
http://www.springer.com/series/5304

The Lecture Notes in Physics

The series Lecture Notes in Physics (LNP), founded in 1969, reports new developments in physics research and teaching—quickly and informally, but with a high quality and the explicit aim to summarize and communicate current knowledge in an accessible way. Books published in this series are conceived as bridging material between advanced graduate textbooks and the forefront of research and to serve three purposes:

- to be a compact and modern up-to-date source of reference on a well-defined topic
- to serve as an accessible introduction to the field to postgraduate students and nonspecialist researchers from related areas
- to be a source of advanced teaching material for specialized seminars, courses and schools

Both monographs and multi-author volumes will be considered for publication. Edited volumes should, however, consist of a very limited number of contributions only. Proceedings will not be considered for LNP.

Volumes published in LNP are disseminated both in print and in electronic formats, the electronic archive being available at springerlink.com. The series content is indexed, abstracted and referenced by many abstracting and information services, bibliographic networks, subscription agencies, library networks, and consortia.

Proposals should be sent to a member of the Editorial Board, or directly to the managing editor at Springer:

Christian Caron
Springer Heidelberg
Physics Editorial Department I
Tiergartenstrasse 17
69121 Heidelberg/Germany
christian.caron@springer.com

Ernst Bauer · Manfred Sigrist
Editors

Non-centrosymmetric Superconductors

Introduction and Overview

Editors
Ernst Bauer
Institute of Solid State Physics, Vienna
University of Technology, Wiedner
Haupstrasse 8-10
1040 Wien, Austria
e-mail: bauer@ifp.tuwien.ac.at

Prof. Manfred Sigrist
ETH Zürich
Wolfgang-Pauli-Str. 27
8093 Zürich, Switzerland
e-mail: sigrist@itp.phys.ethz.ch

ISSN 0075-8450
ISBN 978-3-642-24623-4
DOI 10.1007/978-3-642-24624-1
Springer Heidelberg Dordrecht London New York

e-ISSN 1616-6361
e-ISBN 978-3-642-24624-1

Library of Congress Control Number: 2011941704

Cover design: eStudio Calamar, Berlin/Figueres

Printed on acid-free paper

Springer is part of Springer Science+Business Media (www.springer.com)

Preface

Symmetry plays an important role in superconductivity and influences many of its properties in a profound way. Ever since the discovery of the first unconventional superconductors roughly 30 years ago, the search for the symmetry of Cooper pairs has been among the most important tasks to be addressed, when studying new superconducting materials. Two symmetries are particularly important for superconductivity—time reversal and inversion symmetry. If at least one of the two is absent in the normal state already, Cooper pairing appears in non-standard forms. Ferromagnetic superconductors, lacking time reversal symmetry, form most likely pairs of electrons of the same spin, if the pairing mechanism permits. Missing inversion symmetry in so-called noncentrosymmetric crystals gives rise to mixed-parity pairing.

Non-centrosymmetric superconductors are known since a long time, but received little attention until recently. Their rise to a prominent topic of research occurred actually 2004 with the discovery of the first heavy fermion superconductor without inversion symmetry, $CePt_3Si$. This example was followed swiftly by the synthesis of other superconductors in similar classes, such as $CeIrSi_3$ and $CeRhSi_3$, as well as several others also outside the heavy fermion family. These superconductors received special attention due to the expectation of unconventional pairing due to non-standard pairing mechanisms, most likely driven by magnetic fluctuations.

The symmetry properties are intriguing for many reasons connecting these unconventional superconductivity with several other modern research fields in condensed matter physics, such as multi-ferroics, spintronics or topological insulators. Many symmetry-related properties have been observed in experiment and others are predicted by theory, displaying most intriguing features of a superconducting phase.

This book provides an introduction to and an overview on many aspects of non-centrosymmetric superconductivity, written by several scientists who are most active in this field. We are most grateful to all authors for their contributions. In addition, we are very grateful to Prof. H. von Löhneysen (KIT Karlsruhe, Germany) for critically reading the manuscripts, suggesting many useful improvements. Finally, the editors wish to thank *Springer* for making this book project possible.

Vienna and Zürich, August 2011

Ernst Bauer
Manfred Sigrist

Contents

Part I
Basic Features of Non-centrosymmetric Superconductors

Chapter 1
Non-centrosymmetric Superconductors: Strong vs. Weak Electronic Correlations

E. Bauer and P. Rogl

Abstract Superconductivity in materials without inversion symmetry displays intriguing properties due to a strong modification of their band structures caused by antisymmetric spin-orbit coupling. This is most dramatically seen in several recently discovered heavy fermion superconductors such as $CePt_3Si$ or $CeTSi_3$ (T = Rh and Ir). These systems are interesting in view of the involvement of magnetic fluctuations in the pairing mechanism yielding dominant unconventional Cooper pairing in a so-called mixed parity form. However, also other non-centrosymmetric superconductors with weakly correlated electrons are in many respects interesting and will be reviewed here.

1.1 Introduction

The discovery of superconductivity (SC) 100 years ago has triggered countless developments relevant to practical applications as well as the fundamental understanding of this most intriguing macroscopically coherent state of electrons. In the course of time, many groups of materials exhibiting SC were found and the underlying physics was unravelled. The microscopic understanding of SC to the present day is based on the seminal work of Bardeen, Cooper and Schrieffer [1] which explains SC as the formation of a macroscopic coherent state of Cooper pairs formed by conduction electrons. In the traditional BCS theory the interaction allowing the electrons to pair results from electron-phonon coupling and has due to screening effects the character of a contact interaction which through retardation effects (slow ion motion) is able to

---

E. Bauer (✉)
Institute of Solid State Physics, Vienna University of Technology, 1040 Wien, Austria
e-mail: bauer@ifp.tuwien.ac.at

P. Rogl
Institute of Physical Chemistry, University Vienna, 1090 Wien, Austria
e-mail: peter.franz.rogl@univie.ac.at

E. Bauer and M. Sigrist (eds.), *Non-centrosymmetric Superconductors*,
Lecture Notes in Physics 847, DOI: 10.1007/978-3-642-24624-1_1,

circumvent the strong (instantaneous) Coulomb repulsion which is essentially also a contact interaction. A contact interaction requires the electrons to form Cooper pairs in their most symmetric channel, generally called "s-wave" channel. The corresponding pair wavefunction has even parity and, in order to obtain a totally antisymmetric wavefunction, spin-singlet spin configuration (odd under exchange of the two electrons). A superconductor based on such a pairing state is termed "conventional". It exhibits a quasiparticle spectrum with a nodeless gap. This gives rise to activated behaviour in the temperature dependence of many quantities at very low temperatures, such as the specific heat, NMR relaxation rates and the London penetration depth.

Leaving the realm of electron-phonon coupling opens the possibility for alternative pairing mechanisms driven by electron-electron coupling, e.g. spin fluctuation exchange. These interactions are generally longer-ranged and allow for other pairing channels with higher angular momentum. Since, in general, the corresponding pair wavefunctions have nodes, the electrons do not approach each other closely and can in this way diminish the adverse effect of the (contact) Coulomb repulsion. Superconductors relying on Cooper pairing in a channel different from "s-wave" are called "unconventional" [2]. In general, pairing states may be distinguished by even and odd parity. The Pauli principle requiring a totally antisymmetric Cooper pair wavefunction imposes the condition that even parity is tied to the spin-singlet and odd parity to the spin-triplet configuration. The pair wavefunctions or gap functions in spin space may then be written in the following way [2]:

$$\hat{\Delta}(\boldsymbol{k}) = \begin{pmatrix} \Delta_{\uparrow\uparrow}(\boldsymbol{k}) & \Delta_{\uparrow\downarrow}(\boldsymbol{k}) \\ \Delta_{\downarrow\uparrow}(\boldsymbol{k}) & \Delta_{\downarrow\downarrow}(\boldsymbol{k}) \end{pmatrix}. \tag{1.1}$$

For even-parity states we parametrize

$$\hat{\Delta}(\boldsymbol{k}) = i\hat{\sigma}_y \psi(\boldsymbol{k}) \quad \text{with} \quad \psi(\boldsymbol{k}) = \psi(-\boldsymbol{k}), \tag{1.2}$$

and for odd-parity states

$$\hat{\Delta}(\boldsymbol{k}) = i\boldsymbol{d}(\boldsymbol{k}) \cdot \hat{\boldsymbol{\sigma}}\hat{\sigma}_y \quad \text{with} \quad \boldsymbol{d}(\boldsymbol{k}) = -\boldsymbol{d}(-\boldsymbol{k}). \tag{1.3}$$

This classification relies on the presence of an inversion center in the crystal structure, as to have parity as a proper quantum number. The lack of inversion symmetry introduces a special form of spin-orbit coupling, so called anti-symmetric spin-orbit (ASOC) coupling among which Rashba-type and Dresselhaus-type of spin-orbit coupling are the best-known examples, and can be represented as

$$\mathcal{H}_{\text{ASOC}} = \sum_{\boldsymbol{k}} \sum_{s,s'} \boldsymbol{g}_{\boldsymbol{k}} \cdot \boldsymbol{\sigma}_{ss'} c^{\dagger}_{\boldsymbol{k}s} c_{\boldsymbol{k}s'} \tag{1.4}$$

where the characteristic vector $\boldsymbol{g}_{\boldsymbol{k}}$ is an odd function: $\boldsymbol{g}_{-\boldsymbol{k}} = -\boldsymbol{g}_{\boldsymbol{k}}$. It has been shown that ASOC acts detrimental on spin-triplet pairing states [3], in general, with the exception of states satisfying the condition $\boldsymbol{d}(\boldsymbol{k}) \parallel \boldsymbol{g}_{\boldsymbol{k}}$ [4]. For spin singlet states the influence of ASOC is minor.

The energy scale introduced by ASOC is $E_{ASOC} \sim 10 - 100$ meV which is much larger than the energy scale of the superconducting phase. ASOC has the structure of a Zeeman term in Eq. (1.4) with $\boldsymbol{k}$-dependent "magnetic field" such that we observe a spin splitting of the energy bands with spin-split Fermi surfaces. The effect of this splitting is most important for the pairing symmetry as it introduces a mixing of the pairing parity and we talk about "mixed-parity" pairing states in non-centrosymmetric superconductors. Consequently, there is neither separation according to parity nor spin-singlet and -triplet pairing.

1.2 Superconductivity of Strongly Correlated Electron Systems Without Inversion Symmetry

In the following sections an overview will be given on prominent physical properties of SC without inversion symmetry possessing at the same time strong electronic correlations. The examples in mind are Ce-based heavy fermion superconductors. Here, the Kondo effect in competition with the RKKY interaction and crystalline electric field (CEF) effects dictate the ground state of such materials many of which show a magnetic quantum phase transition upon changing parameters such as pressure or composition. In this family of compounds SC emerges in the proximity of such quantum phase transitions leading also to coexistence of magnetic order with SC. Thus spin-fluctuations will likely be the principal ingredient of SC and properties characterizing the NCS state above T_c may relate to the mechanism of SC.

Takimoto and Thalmeier [5] have recently shown that the lack of inversion symmetry leads to novel spin fluctuations which tend to mix spin-singlet and spin-triplet parts. Both components of such mixed-parity state display non-trivial momentum dependencies, which may give rise to accidental line nodes in the gap function. The characteristic $\boldsymbol{q}$-dependence of the anomalous (NCS) spin fluctuations responsible for this mixing originates from ASOC.

1.2.1 Ternary $CePt_3Si$

$CePt_3Si$ crystallizes in tetragonal symmetry $P4mm$ (No. 99), isotypic with the ternary boride $CePt_3B$ [6] (see Fig. 1.1). Crystallographic data (standardized) are: $a = 0.4072(1)$ nm and $c = 0.5442(1)$ nm; Ce is in site 1(b) at $\left(\frac{1}{2}, \frac{1}{2}, 0.1468(6)\right)$; Pt(1) in 2(c) at $\left(\frac{1}{2}, 0, 0.6504(6)\right)$, Pt(2) in 1(a) at $(0, 0, 0)$ (fixed) and Si in site 1(a) at $(0, 0, 0.412(3))$ [7]. $CePt_3Si$ derives from hypothetical $CePt_3$ with cubic $AuCu_3$ structure by filling the void with Si, which causes a tetragonal distortion of the unit cell to $c/a = 1.336$.

$CePt_3Si$ is a heavy fermion compound with a substantial Sommerfeld constant of the specific heat in the normal state region ($\gamma \approx 400$ mJ/molK^2) that orders

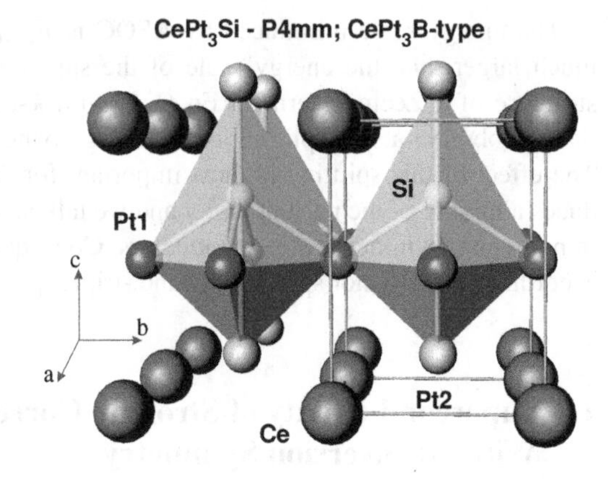

Fig. 1.1 Crystal structure of $CePt_3Si$. The coordination figure around the Si atoms $Si[Pt1_4Pt2_1]$ plus one remote Pt2 atom are outlined as capped tetragonal pyramids (*shaded*). *Note* origin shifted by $(\frac{1}{2}, \frac{1}{2}, 0.8532)$ in order to match setting of the parent $AuCu_3$-type structure

antiferromagnetically below $T_N = 2.25$ K [7]. Antiferromagnetism occurs with a simple propagation vector $\boldsymbol{Q} = (0, 0, 1/2)$, i.e. with ferromagnetic layers in the basal plane [8]. The ordered moment is small ($\mu_s \approx 0.16\ \mu_B$), originating from a crystalline electric field (CEF) doublet as ground state in relation to Kondo type interaction.

The ground state of $CePt_3Si$ results from the lifting of the 6-fold degeneracy of the $J = 5/2$ total angular momentum due to CEF effects in tetragonal symmetry. Neutron studies and polarized soft x-ray experiments reveal the first and the second excited doublet at ≈ 15 and ≈ 19 meV above the ground state, respectively. Thus, physics at low temperatures is governed by the ground-state doublet only. In standard notation the CEF ground state is given by $\Gamma_0 = 0.46| \pm 5/2\rangle + 0.89| \mp 3/2\rangle$ [10].

SC in $CePt_3Si$ occurs in high-quality polycrystalline samples below ≈ 0.75 K [7]. Quite unexpectedly, single-crystalline materials show bulk SC around 0.45 K [11]. An explanatory statement for this discrepancy is given below. A muon spin rotation study carried out as a function of temperature and field [12] gives clear evidence of a complete spatial coexistence of both long-range magnetic order and SC.

Various intriguing features of the superconducting state have been observed from macroscopic and microscopic measurements: i) the width of the SC transition is unusually large; ii) the upper critical magnetic field exceeds the paramagnetic limit H_P [7]; iii) a non-exponential temperature dependence of the NMR relaxation rate $1/T_1$ is observed below T_c, together with an unexpected Hebel-Slichter peak in some of the samples right at T_c [13]; iv) the Knight shift does not change from the normal to the superconducting state neither in the basal plane nor along the c-axis [14].

The principal conclusion drawn from such observations is that the superconducting gap likely has line nodes and the spin susceptibility is at most only modestly affected by the superconducting phase. Both features fit well into the present understanding of NCS in $CePt_3Si$. The polar symmetry of the crystal lattice suggests a shape for $\boldsymbol{g_k} = \hat{x}k_y - \hat{y}k_x$ in lowest order expansion in $\boldsymbol{k}$, i.e. the standard Rashba-type of ASOC [15]. The mixed-parity state yields different gap structures on the two spin-split Fermi surfaces which can in general have line nodes [16]. A mixed-parity state with the full lattice symmetry, $\hat{\Delta}(\boldsymbol{k}) = i\{\Delta_1 + \Delta_2 \hat{g}_{\boldsymbol{k}} \cdot \hat{\boldsymbol{\sigma}}\}\hat{\sigma}_y$, would yield line nodes, if the odd-parity is larger than the even-parity part ($|\Delta_2| > |\Delta_1|$ with $|\hat{g}_{\boldsymbol{k}}| = 1$). Moreover, such a state would contribute to the NMR signal with a finite coherence factor, explaining the NMR Hebel-Slichter peak [17].

The upper critical field exceeding H_P may have various reasons and can occur for spin-singlet pairing if i) a reduction of the Lande g-factor occurs as it was observed e.g. in URu_2Si_2 [18] or ii) a strong-coupling effect yields a large SC gap Δ ($\Delta \gg \Delta_{BCS} \approx 3.52k_BT_c$). As a consequence of both, the limiting field H_P rises above $H_P^{BCS} = \sqrt{2}\Delta/(g\mu_B)$. For NCS systems, two additional mechanisms have been proposed, weakening the Pauli-limiting effect: iii) the reduced pair-breaking effect of spin polarization due to ASOC [19, 20] and iv) the realization of a helical vortex state [21]. Frigeri et al. [20] demonstrated that the spin susceptibility of spin-singlet states behaves approximately as that of a spin-triplet state with $\boldsymbol{d}(\boldsymbol{k}) \parallel \boldsymbol{g}_{\boldsymbol{k}}$, if the ASOC is strong (i.e., the energy difference of split bands $E_{ASOC} \gg k_BT_c$). For $CePt_3Si$, the spin susceptibility for fields along [001] remains basically unchanged entering the superconducting phase, while it reduces to an intermediate size for fields in the basal plane of the tetragonal crystal lattices. Concomitantly, paramagnetic limiting is absent for fields along the [001] direction and moderate in the basal plane.

A helical vortex state has been proposed to appear in $CePt_3Si$ for fields perpendicular to the c-axis, because such a magnetic field would modify the Fermi surfaces in a way as to shift the centers by a wave vector $\boldsymbol{q}$ ($\boldsymbol{q} \propto (\hat{z} \times \boldsymbol{H})$) [21]. This introduces an additional phase factor $\exp(i\boldsymbol{q}\boldsymbol{R})$ for the order parameter without inducing, however, a net current flow along $\boldsymbol{q}$ due to gauge invariance. This helical order coincides with an increase of the upper critical field, $T_c(\boldsymbol{H}) = T_c - aH + b(\boldsymbol{n} \times \boldsymbol{H})^2$ [21], where a and b are positive constants. The in-plane enhancement of H_{c2} can be substantial and might thus explain the extraordinarily small anisotropy of the upper critical fields found for $H//c$ and $H \perp c$ in $CePt_3Si$ [22].

Microscopic studies carried out on $CePt_3Si$ by NMR experiments on high quality single crystals revealed unconventional strong-coupling SC with a line node gap below $T_c \approx 0.45$ K [23]. However, disordered domains in both single- as well as polycrystalline samples might give rise to the occurrence of a more conventional s-type superconducting state below 0.75 K [23]. Evidence for line nodes in the gap structure is also derived from London penetration depth studies, exhibiting a linear temperature dependence for $T << T_c$ [24]. Such a conclusion is drawn from thermal conductivity data as well [25].

The phase diagram of $CePt_3(Si, Ge)$ is displayed in Fig. 1.2 [26, 27, 28]. The characteristic temperatures of the system, i.e., T_N and T_c are represented as a function of the reduced volume, V/V_0, where V_0 refers to the volume of $CePt_3Si$ at 1 bar. Stoichiometric $CePt_3Si$ at ambient pressure exhibits the largest SC transition temperature. The most important features derived upon the application of pressure are the suppression of long-range magnetic order at a critical pressure $p_{cr}^{AFM} \approx 6$ to 8 kbar and of the superconducting transition temperature at $p_{cr}^{T_c} \approx 14$ to 16 kbar [27, 29, 30].

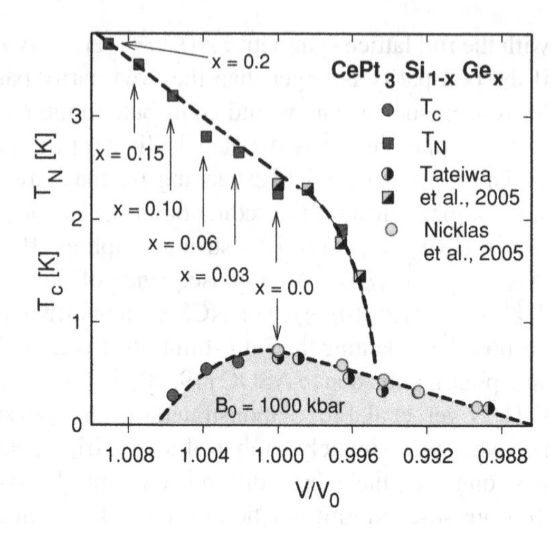

Fig. 1.2 Low-temperature phase diagram of $CePt_3Si_{1-x}Ge_x$ as a function of the reduced unit-cell volume (V/V_0), with V_0 the unit-cell volume of $CePt_3Si$. (Figure taken from Ref. [28])

Substitution of Si by Ge in $CePt_3Si$ provides the possibility to expand the unit-cell volume without substantially changing the electronic structure. In order to compare both pressure and volume effects, a bulk modulus of typical intermetallic compounds, i.e., $B_0 = 1000$ kbar is assumed. Si substitution by Ge causes an increase of T_N but a decrease of T_c. The increasing unit-cell volume releases pressure from the Ce ion; as a consequence, there is a loss of hybridization and the $4f^1$ electronic configuration becomes more localized, thus magnetism is stabilized. Additionally, pair breaking by non-magnetic impurities, appearing through the Si/Ge substitution, reduces the SC transition temperature on the side where the volume increases ($V/V_0 > 1$). The latter, however, appears to be the more relevant mechanism as was found from a pressure study carried out on $CePt_3Si_{0.94}Ge_{0.06}$ [30].

The superconducting "dome", i.e., the broad T_c maximum as a function of V/V_0 is found only in part below the regime with long-range magnetic order, while for pressure values $p > 8$ kbar, SC survives in a nonmagnetic environment. The coexistence of long-range magnetic order and SC obviously hints at magnetic fluctuations being a necessary ingredient for Cooper pairing, while due to the heavy quasi-particles formed by the Kondo interaction in $CePt_3Si$, the retardation effect with respect to phonons becomes weaker and thus, Cooper pairs mediated by phonons are rather unlikely.

1.2.2 Ternary $CeRhSi_3$

Shortly after the discovery of heavy fermion SC in $CePt_3Si$, the ternary tetragonal non-centrosymmetric compound $CeRhSi_3$ was reported by N. Kimura [31] to show SC as well, although only at elevated pressure. $CeRhSi_3$ crystallizes in the $BaNiSn_3$ structure, which belongs to the space group $I4mm$ (No. 107) with Ce in site 2a (0, 0,

Table 1.1 Normal-state and superconducting properties of correlated materials without inversion symmetry

Compound	$CePt_3Si$	$CeRhSi_3$	$CeIrSi_3$
structure type	$CePt_3B$	$BaNiSn_3$	$BaNiSn_3$
space group	$P4mm$	$I4mm$	$I4mm$
lattice parameter [Å]	$a = 4.072$, $c = 5.442$	$a = 4.269$, $c = 9.738$	$a = 4.252$, $c = 9.715$
T_c^{max} [K]	0.75	1.05	1.6
H_{c0} [mT]	26	37	54
$H_{c2}(0)$ [T]	≈4	≈7	≈11
dH_{c2}/dT [T/K]	−8.5	−12	−11.4
γ_n [mJ/molK^2]	390	110	100
ξ_0 [Å]	81	66	57
$\lambda(0)$ [Å]	11000	10100	8300
κ	≈140	≈140	≈135

Some of the superconducting properties are calculated in terms of the BCS theory using the free electron model [9]. T_c^{max} is the SC transition temperature, H_{c0} is the thermodynamic critical field at $T = 0$, θ_D is the Debye temperature, $H_{c2}(0)$ is the upper critical field, dH_{c2}/dT is the slope of the upper critical field, γ_n is the Sommerfeld value of the normal state, $\xi(0)$ is the coherence length at $T = 0$, $\lambda(0)$ is the London penetration depth at $T = 0$ and κ is the Ginzburg-Landau parameter. Data (in general the largest one reported for a certain material and a certain physical property) are taken from references of section 2.

0.5759), Pt in 2a (0, 0, 0.2313), Si1 in 2a (0, 0, 0.0) and Si2 in 4b (0, $\frac{1}{2}$, 0.3253). The $BaNiSn_3$-type structure is an ordered variant within the large family of compounds based on the parent type $BaAl_4$. The latter exhibits inversion symmetry, while compounds belonging to the $BaNiSn_3$ type, however, lack a center of inversion rendering positive and negative [001] directions inequivalent (Fig. 1.9). Crystallographic data are summarized in Table 1.1 [32].

At ambient pressure, $CeRhSi_3$ orders antiferromagnetically at $T_N = 1.6$ K. Paramagnetic properties are characterized by the coaction of crystalline electric field (CEF) effects and a Kondo-type interaction. The tetragonal crystal symmetry causes a lifting of the $2J + 1 = 6$-fold degenerate ground state into 3 doublets located at 0, 220 and 270 K, respectively. Applying the standard CEF Hamiltonian one obtains the CEF parameters as $B_2^0 = -1.75$ K, $B_4^0 = 0.381$ K and $B_4^4 = 4.74$ K [34].

Susceptibility measurements up to room temperature revealed an effective magnetic moment $\mu_{eff} = 2.65\mu_B$, fairly well corresponding to the $J = 5/2$ free ion value of Ce^{3+}. The paramagnetic Curie temperature, $\theta_p = -128$ K is indicative of antiferromagnetic interactions among Ce ions. The Kondo interaction is deduced from the large Sommerfeld constant $\gamma = 110$ mJ/molK^2, equivalent to a Kondo temperature $T_K \approx 50$ K [32]. The latter is connected to the (negative) paramagnetic Curie temperature of θ_p, since $T_K \propto |\theta_p|$.

At $T = 1.6$ K $CeRhSi_3$ exhibits an AFM instability characterized by an incommensurate propagation vector $\boldsymbol{Q} = (\pm 0.215, 0, 0.5)$ [35]. Resistivity data below 1.6 K suggest a gap $\Delta^{SW} \approx 2$ K in the antiferromagnetic spin-wave dispersion relation.

Table 1.2 Normal-state and superconducting properties of correlated materials without inversion symmetry

Material	Structure	Space Group	Lattice Parameter [Å]	T_c^{max} [K]	$2\Delta/k_BT_c$	$\theta_D[K]$	λ_{el-ph}	H_{c2} [T]	dH_{c2} /dT [T]	γ_n [mJ/molK^2]	ξ_0 [Å]	$\lambda(0)$ [Å]	κ
Y_2C_3	Pu_2C_3	$I\bar{4}3d$	$a = 8.226$	14.7	$2\frac{\Delta_1}{k_BT_c} = 4.9$ $\frac{\Delta_2}{k_BT} = 1.1$	530	0.6	26.8	−1.8	6.3	36	4600	127
La_2C_3	Pu_2C_3	$I\bar{4}3d$	$a = 8.818$	13.2	$2\frac{\Delta_1}{k_BT_c} = 5.6$ $2\frac{\Delta_2}{k_BT} = 1.3$	350	0.84	19	−1.8	10.6	43	3800	90
$Cd_2Re_2O_7$	$Ca_2Nb_2O_7$ $Cd_2Re_2O_7$	$Fd\bar{3}m$ (RT) $I\bar{4}m2$ (13 K)	$a = 10.2257$ $a = 7.2326$ $c = 10.2183$	0.97	3.5	458	0.38	0.29	−0.42	30.2	340	4600	14
Li_2Pd_3B	filled β-Mn	$P4_332$	$a = 6.7534$	7.6	3.9	221	1.1	6.2	−0.84	9	95	1900	20
Li_2Pt_3B	filled β-Mn	$P4_332$	$a = 6.7552$	2.6	3.53	228	0.48	1.9	−0.63	7	145	3640	25
β'-Mg_2Al_3	β'-Mg_2Al_3	$R3m$	$a = 19.968$ $c = 48.911$	0.87	3.55	373	0.42	0.14	−0.23	6.6	485	5300	13
$BaPtSi_3$	$BaNiSn_3$	$I4mm$	$a = 4.4009$ $c = 10.013$	2.25	3.5	345	0.5	0.053	−0.1	5.7	790	2600	2
Rh_2Ga_9	Rh_2Ga_9	Pc	$a = 6.4008$ $b = 6.3944$ $c = 8.7519$ $\beta = 93.48°$	1.95	3.5	324	–	–	–	7.6	790	2600	type I
Ir_2Ga_9	Rh_2Ga_9	Pc	$a = 6.4185$ $b = 6.3918$ $c = 8.7761$ $\beta = 95.46°$	2.25	3.54	325	0.17	0.025	–	6.9	1000	–	1.1

Some of the superconducting properties are calculated in terms of the BCS theory using the free electron model [10]. T_c^{max} is the SC transition temperature, Δ is the gap width, θ_D is the Debye temperature, λ_{el-ph} is the electron – phonon coupling constant, $H_{c2}(0)$ is the upper critical field, dH_{c2} /dT is the slope of the upper critical field, γ_n is the Sommerfeld value of the normal state, $\xi(0)$ is the coherence length at $T = 0$, $\lambda(0)$ is the London penetration depth at $T = 0$ and κ is the Ginzburg Landau parameter. Data (in general the largest one reported for a certain material and a certain physical property) are taken from references of section 3.

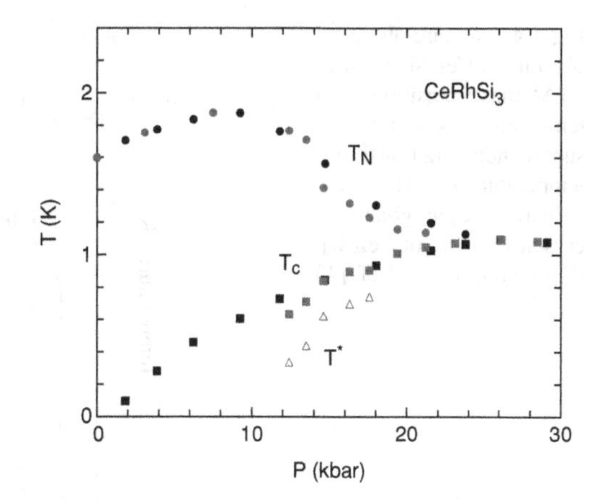

Fig. 1.3 Temperature - pressure (*TP*) phase diagram of $CeRhSi_3$ based on resistivity measurements. (Figure taken from Ref. [34])

At very low temperatures, Fermi-liquid behavior dominates, putting $CeRhSi_3$ onto the standard Kadowaki-Woods relation [33].

The application of pressure initially increases T_N; above 9 kbar the phase transition temperature starts to drop before merging with the SC phase-line above 20 kbar (Fig, 1.3). In parallel, the SC transition temperature T_c increases up to 1.05 K for $p \approx 30$ kbar. For a narrow pressure range from 12 to 18 kbar, a further characteristic temperature scale T^* is deduced [34], denoting the temperature where a distinctly different slope in the electrical resistivity occurs. The most important SC feature of $CeRhSi_3$ is an extraordinarily large upper critical field. $H_{c2}(T)$ behaves concavely as function of temperature and exceeds the paramagnetic limit by far. Specifically, for $p = 26$ kbar and $H \parallel c$, $\mu_0 H_{c2} \approx 16$ T at 0.4 K, with $\mu_0 H_{c2}' = -23$ T/K [36]. As in the case of $CePt_3Si$, involvement of spin-triplet Cooper pairs might be, at least partly, responsible for this behavior.

1.2.3 Ternary $CeIrSi_3$

Similar to $CeRhSi_3$, isotypic $CeIrSi_3$ crystallizes in the tetragonal $BaNiSn_3$ structure. Ce atoms occupy the corners and the center of the tetragonal unit cell. The arrangement of Si and Ir, however, lacks inversion symmetry along the [001] direction (see Fig. 1.9 for a sketch of the crystal structure). For crystallographic data see Table 1.1 [32].

At ambient pressure, $CeIrSi_3$ is an antiferromagnet below $T_N = 5$ K. For $T > T_N$, the magnetic susceptibility is highly anisotropic; while the effective magnetic moment in both directions is close to the value of Ce^{3+}, the paramagnetic Curie temperature θ_p ranges from -186 K ($H \parallel [100]$) to -109 K ($H \parallel [001]$) [37]. CEF effects cause a splitting of the 6-fold degenerate ground state, revealing doublets at 0, 160 and 501 K, respectively. The corresponding CEF parameters are $B_2^0 = 9$K, $B_4^0 = 0.1$ K and $B_4^4 = 9$ K [37]. Below T_N antiferromagnetic order develops; the

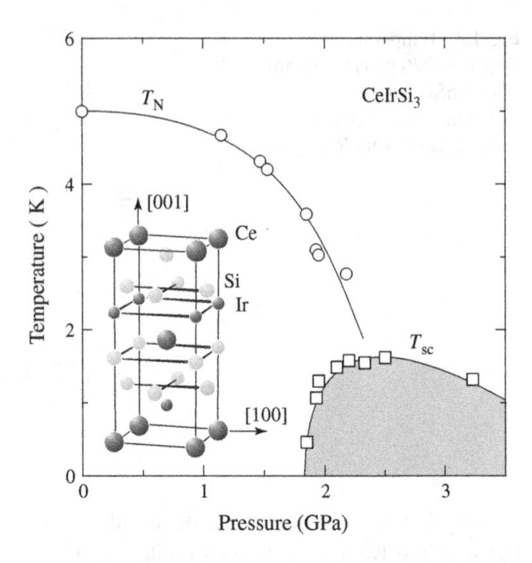

Fig. 1.4 Pressure phase diagram of CeIrSi3 for the AFM phase transition temperature T_N and the superconducting transition temperature T_{sc}. The inset indicates the tetragonal crystal structure of CeIrSi3 (Figure taken from Ref. [43])

ordered moment, however, appears to be very small, well below 0.25 μ_B [38]. Antiferromagnetism occurs with an anisotropic gap in the spin-wave dispersion relation, with $\Delta^{SW} = 2.1$ K ($J \parallel [110]$) and $\Delta^{SW} = 3.3$ K ($J \parallel [001]$). Heat capacity data also show evidence for long-range magnetic order at 5 K; the Sommerfeld value $\gamma \approx 105$ mJ/molK^2 characterizes CeIrSi3 as a heavy fermion material due to Kondo interaction.

Pressurizing CeIrSi3 causes a suppression of the antiferromagnetic order, which vanishes at about 25 kbar (see Fig. 1.4) [37]. Simultaneously, SC develops above 18 kbar, reaching a maximal value of $T_c = 1.6$ K at 25 kbar. At this optimal pressure, the upper critical field H_{c2} for $H \parallel [110]$ is extraordinarily large, extrapolating to 10 T for $T \to 0$. The initial slope, H'_{c2} is about -11.4 T/K; $H_{c2}(T)$ simply follows theory developed by Werthamer, Helfand, and Hohenberg (WHH theory) [39]. Again, the paramagnetic limit ($H_P \sim 3$ T) is exceeded by far. Measurements for $H \parallel [001]$, however, do not comply with the WHH behavior and $\mu_0 H_{c2}(0)$ extrapolates to fields well above 20 T. The heat capacity data inside the SC pressure regime reveals a very sharp transition, while $\Delta C/C_n(T = T_c, p = 22.5\text{kbar}) \approx 5.7 \pm 0.1$ substantially exceeds the BCS value of 1.43 and turns out to be one of the largest values found among SCs [40]. In general, such a behavior points to strong coupling SC. Strong-coupling might explain, at least partly, the concave upturn behaviour of $H_{c2}(T)$. Besides strong coupling, the very large value observed for $H \parallel [001]$ might be a direct consequence of the absence of inversion symmetry. Frigeri et al. [20] have demonstrated that in this case the paramagnetic limiting becomes weak for $H \parallel [100]$ and is almost absent for $H \parallel [001]$.

A microscopic study of the pressure-induced superconducting state of CeIrSi3 using ^{29} Si NMR exhibits a T^3 nuclear-spin relaxation rate $1/T_1$ below T_c without any coherence peak right at T_c. These facts provide evidence for a superconducting quasiparticle gap characterized by line nodes. For $T > T_c$, $1/T_1$ follows a $\sqrt{T}$

dependence [41], as a signature that SC emerges under non-Fermi liquid conditions around a quantum critical point. The rather high SC transition temperature, compared to other Ce-based materials might be a result of strong AFM spin fluctuations, as it is also the case in $CeCoIn_5$ [42]. Note also that the SC dome extends to very high values of pressure; the maximum of T_c roughly corresponds to that pressure where $T_N(p)$ seems to cross the SC phase line.

Further members in this class of compounds are $CeCoGe_3$ crystallizing in the $BaNiSn_3$ structure and UIr which shows the monoclinic PdBi type of structure (space group $P2_1$).

$CeCoGe_3$ is an antiferromagnet that orders at $T_{N1} = 21$ K [44] determined by a resistivity anomaly for currents along [100]; further magnetic phase transitions have been observed for $J \parallel [001]$ at 12 and 8 K, respectively. The T^2 term in the electrical resistivity would correspond to a Sommerfeld value $\gamma \approx 34$ mJ/molK^2 [45] within the Kadowaki-Woods [33] scheme, indicating only a moderate mass enhancement of charge carriers. Applying pressure causes a continuous reduction of T_{N1}, with $T_{N1} \to 0$ for $p_c \approx 65$ kbar. Concomitantly with the suppression of long-range magnetic order, SC occurs in a pressure range from about 54 kbar to 75 kbar, with a maximum $T_c = 0.69$ K at 65 kbar [46]. For fields along [001] the slope of the upper critical field amounts to -20T/K [45], referring to a huge upper critical field, comparable to $CeRhSi_3$ and $CeIrSi_3$.

At ambient pressure, monoclinic UIr is a ferromagnet, with $T_{C1} = 46$ K [47]. The ferromagnetic moments orient along the $[10\bar{1}]$ direction in the (010) plane, with a saturation value of $0.5\,\mu_B$/U (FM1 phase). The Sommerfeld value $\gamma = 49$ mJ/molK^2 refers to weakly enhanced effective electron masses.

Upon the application of hydrostatic pressure, the FM1 phase of UIr is suppressed, vanishing presumably at $p_{c1} \approx 17$ kbar [48]. Above about 10 kbar and well below 30 K, two further ferromagnetic phases (FM2 and FM3) develop above $p_{c3} \approx 28$ kbar. A superconducting phase within a narrow pressure range is embedded in the FM3 phase, below p_{c3}. The largest T_c observed is about 140 mK [48, 49]. For $p = 26.1$ kbar, $\mu_0 H_{c2} \approx 26$ mT revealing a coherence length $\xi = 1100$ Å. A $\rho \propto T^{1.6}$ dependence above T_c refers to non-Fermi liquid behaviour. Thus, SC may originate from ferromagnetic spin fluctuations. Unconventional SC can be concluded as well from a significant pair breaking effect by non-magnetic disorder [49]. The small upper critical field is well below H_P; thus the superconducting condensate may consist primarily from spin-singlet components. Besides ferromagnetic order of UIr, the small value of $H_{c2}(0)$ and the large coherence length is in strong contrast to all other SCs described in this chapter.

1.3 Superconductivity of Materials Without Inversion Symmetry and Electronic Correlations

A central issue that arises when filing physical properties of NCS SCs pertains to the role of strong correlations among electrons. Such a distinct knowledge allows disentangling the physics related to the antisymmetric spin-orbit coupling owing to the absence of inversion symmetry on the one hand, and the role of electronic correlations on the other. The absence of substantial electronic correlations provides a possibility to carry out band structure calculations, in order to derive rather accurate information on the splitting of electronic bands due to the ASOC.

In the remaining part of this chapter, a number of materials will be described in some detail, which are characterized by NCS in their crystal structure but also by the absence of *f* or *d* electronic configurations responsible for strong electronic correlations.

1.3.1 Sesquicarbides R_2C_{3-y}

The family of sesquicarbides R_2C_{3-y} ($R=$ rare earth) is known for their relatively high superconducting transition temperatures (up to 18 K) [50, 51]. Renewed interest in carbide-based SCs was stimulated by the fact that the crystal structure of the sesquicarbides does not possess inversion symmetry. Light-atomic-mass elements like B or C in BCS SCs favor a coupling of parts of the Fermi surface with high-frequency phonon modes, resulting in SCs with substantial transition temperatures.

Rare earth sesquicarbides crystallize in the bcc Pu_2C_3 type with space group $I\bar{4}3d$ (No. 220). The structure is built by eight bcc subunits and, thus, may be regarded as a partially filled superstructure of a distorted bcc unit (a $=$ 0.81350(2) nm; Pu in 16c (0.0492(5), x, x) and C in 24d (0.2896(5), 0, $\frac{1}{4}$)). A sketch of the crystal structure of isotypic La_2C_3 (a $=$ 0.8818(4) nm) is shown in Fig. 1.5, from which one can recognize carbon-dumbbells embedded in a polyhedron formed by 8 metal atoms fused from two severely distorted octahedral units of C[La_5C] sharing a common edge of two La-atoms. Concomitant with the strong distortions is a spread of La-La distances $0.360 \leq d_{La-La} \leq 0.404$ nm, some of which being closer than next-nearest neighbors in β-La (0.375 nm). Whereas the C-C distances derived from X-ray data [52, 53], $d_{C-C} = 0.151$ nm in Pu_2C_3 and 0.153 nm in Y_2C_3 (prepared at high pressure), seem to be long, neutron powder data for La_2C_3 with $d_{C-C} = 0.132(3)$ nm and for Pu_2C_3 with $d_{C-C} = 0.139(3)$ nm indicate a double bond [54, 55].

$I\bar{4}3d$ defines the tetrahedral crystal class T_d lacking a center of inversion. Initial studies evidenced superconducting transition temperatures around 11 K for La_2C_3 and Y_2C_3. Doping with Th even drives T_c up to 17 K [56]. Y_2C_3, however, turns out to be metastable. High pressure synthesis was successful in providing samples with $T_c \approx 18$ K and $\mu_0 H_{c2}(0) > 30$ T [57, 58]. The specific carbon content turns out to be the principal parameter determining superconducting properties like T_c or H_{c2}. In order to obtain R_2C_3 with optimal T_c, it is of importance compensating carbon loss during preparation [59].

The electronic structure of La_2C_3 was derived from LAPW calculations in terms of a generalized gradient approximation [60]. Bands crossing the Fermi surface are the hybridized La-*d* and the anti-bonding C-C states. The bonding C-C bands are separated by a gap of about 2.5 eV from the Fermi energy.

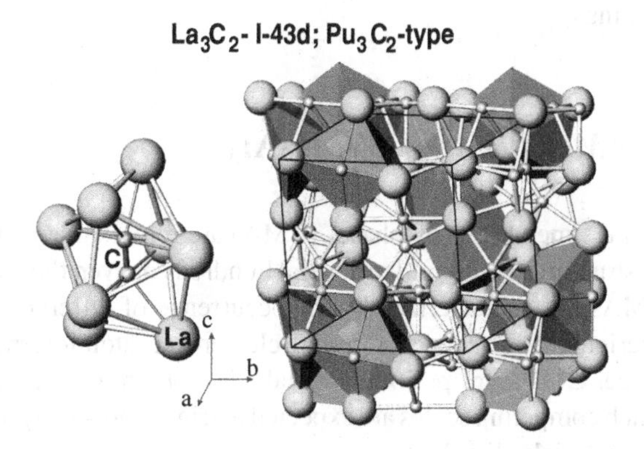

Fig. 1.5 Crystal structure of La_3C_2. Right panel shows connectivity of distorted octahedra $C[La_5C]$. Left panel outlines the La-atom coordination around each C_2-dumbbell

The electronic structures calculated with and without spin-orbit coupling are quite similar. Due to the lack of inversion symmetry, however, ASOC causes lifting of the spin degeneracy. As a consequence, bands become spin split as obvious from details of the LAPW calculations around the Fermi energy.

The DOS at the Fermi energy depends sensitively on off-stoichiometries of the system: slight C-deficiencies result in a substantial decrease of the DOS at E_F. A 2% C-deficient sample has a DOS reduced by about 25% and even 30% when taking into account the calculation based on spin-orbit coupling. These remarkable results may straightforwardly explain substantial differences of transition temperatures and upper critical fields deduced for the various sesquicarbides.

Nuclear magnetic resonance, high-resolution photoemission studies and muon spin rotation [61, 62, 63] carried out on high-quality samples revealed multigap SC for both Y_2C_3 and La_2C_3, similar to MgB_2. The gaps turn out to have no nodes with $2\Delta_1/k_BT \approx 5$ in both cases. These large gaps in the dominating Fermi surface exceed the BCS value, $2\Delta_1/k_BT = 3.5$ by far, classifying both sesquicarbides as SCs in the strong coupling limit. In fact, specific heat data reveal $\lambda_{el-ph} \approx 1.4$ for La_2C_3, corroborating the above conclusions. Similar conclusions may be drawn from measurements of the upper critical field. Although $\mu_0 H_{c2}(0) \approx 20$ T for La_2C_3 is rather large, the Pauli limiting field ($\mu_0 H_P \approx 25$ T) is even above this value. Consequently, orbital currents constitute the principal depairing mechanism restricting the upper critical field. Interestingly, $H_{c2}(T)$ deviates at low temperatures from the standard WHH behaviour. Rather, $H_{c2}(T)$ behaves almost linearly in the entire range studied. Beside localization effects and anisotropy of the Fermi surface, such a behaviour might be attributed to strong electron – phonon coupling, fully in agreement with the microscopic studies. Since $H_{c2}(0) < H_P$, spin-triplet pairing is rather unlikely.

Symmetry considerations of the NCS point group T_d led Sergienko [64] to propose for Y_2C_3 and La_2C_3 two-band SC from the fact that there are certain directions where the spin-orbit-split bands must touch each other. This conclusion fits well with the experimental findings.

1.3.2 Complex Metallic Alloy β'-Mg_2Al_3

Structurally complex metallic alloy phases (CMA) are remarkable metallic systems based on crystal structures composed of several hundreds or even thousands of atoms per unit cell. CMAs are characterized by the occurrence of different length scales, with a lattice periodicity of several nm and cluster-like atomic arrangements on a nm-scale. Hence, CMAs are periodic crystals, but on an atomic scale resemble quasicrystals. Such competing scales are expected to trigger novel physical properties of these specific materials (Fig. 1.6) .

β-Mg_2Al_3 was studied for the first time by Samson [65] and was classified as a cubic system with centrosymmetric space group $Fd\bar{3}m$ (No. 227) and lattice parameter $a = 2.8239(1)$ nm comprising 1168 atoms in the unit cell. The coordination polyhedra consist of icosahedra, Friauf polyhedra (a truncated tetrahedron where each of the four hexagonal faces are capped by one additional atom, all together forming a CN-16 polyhedron), and other irregular polyhedra of various ligancies. Of particular interest is the intrinsic disorder due to mismatch of the various adjacent polyhedra. This causes displacement- and substitutional disorder as well as fractional site occupation. The investigation of the Al-Mg phase diagram [66] assigned the β-phase to the composition $Mg_{38.5}Al_{61.5}$. The redetermination of the crystal structure via *in-situ* X-ray single crystal diffraction at 400°C essentially confirmed the structure model of Samson [65] and in consistency arrived at 1168 atoms per unit cell with $a = 2.8490(2)$ nm. Out of the 23 independent crystallographic sites, eleven positions show partial and/or random atom occupation or split sites to comply with the atom disorder. β-Mg_2Al_3 was found to undergo a structural phase transition at $T = 214$°C from the high temperature cubic phase to a trigonal low temperature modification β'-Mg_2Al_3 with space group $R3m$ (No. 160). In the hexagonal setting, the lattice parameters deduced are $a = 1.9968(1)$ nm and $c = 4.89114(8)$ nm, assembling a total of 925 atoms per unit cell [66], corresponding well to the content of the cubic high temperature cell ($1168/4 = 292$ and $292 \times 3 = 876$ atoms). It should be noted that all crystallographic sites in β'-Mg_2Al_3 are fully occupied but a considerable number of sites still exhibit random Mg/Al disorder. Although a crystallographic group - subgroup relation exists, the phase transformation was found to be of first order.

A particular feature of this crystal structure is the absence of inversion symmetry.

Intrinsic disorder in β'-Mg_2Al_3 is rendered from resistivity measurements [68] revealing a residual resistivity ratio RRR of about 1.16 only, as displayed in Fig. 1.7(a); SC appears below 0.9 K. Magnetic fields suppress SC at $\mu_0 H_{c2}(0) \approx 0.14$ T (inset, Fig. 1.7(a)). The strongly curved $\rho(T)$ data above $T = T_c$ remind us of A15 SCs and might be described in terms of a parallel resistor model.

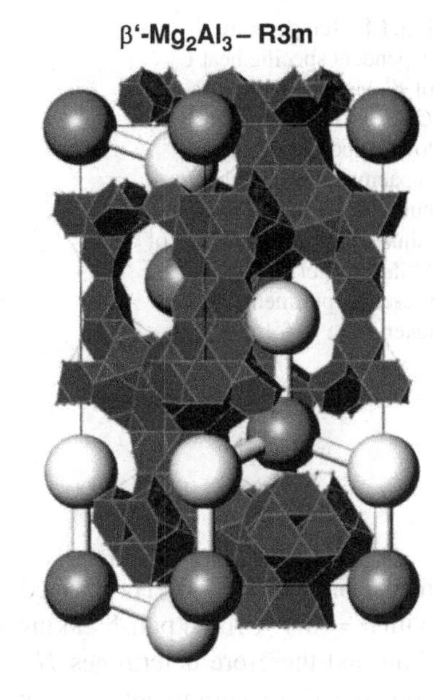

Fig. 1.6 Crystal structure of β'Mg$_2$Al$_3$ showing a three-dimensional framework built by face-connected truncated tetrahedra enclosing clusters ("*spheres*") packed in form of a diamond network. (Figure taken from Ref. [66])

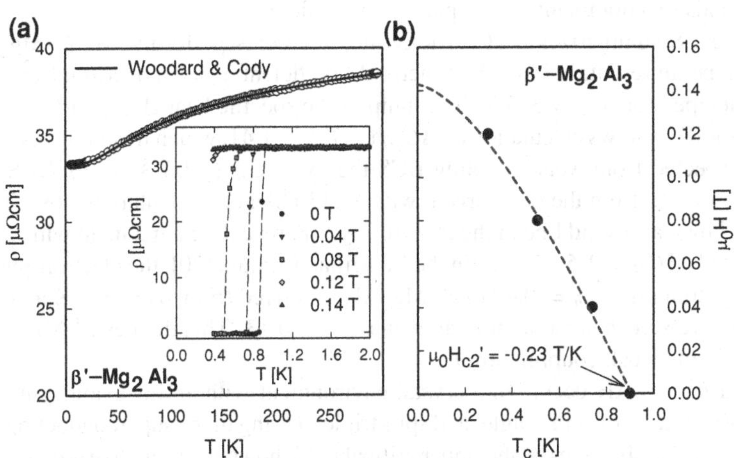

Fig. 1.7 a Temperature dependent electrical resistivity ρ of β'-Mg$_2$Al$_3$. The *solid line* is a least squares fit according to a model of Woodard and Cody [67]. The inset shows the onset of a superconducting transition at $T_c^{mid} = 0.9$ K. The application of magnetic fields suppresses SC. **b** Temperature dependent *upper* critical field H_{c2}. The *dashed line* corresponds to $H_{c2}(T)$ derived in terms of the model of Werthamer et al. [39] for $\alpha = \lambda_{so} = 0$ revealing $\mu_0 H'_{c2} \approx -0.23$ T/K. (Figure taken from Ref. [68])

The upper critical field H_{c2} of β'-Mg$_2$Al$_3$ is displayed in Fig. 1.7(b). A theoretical description is possible in terms of the WHH model [39], taking into account orbital pair-breaking, including the effect of Pauli spin paramagnetism and spin-orbit scattering. Two parameters essentially define H_{c2} : the Maki parameter α and the spin-orbit scattering parameter λ_{so}. Because of the presence of light ele-

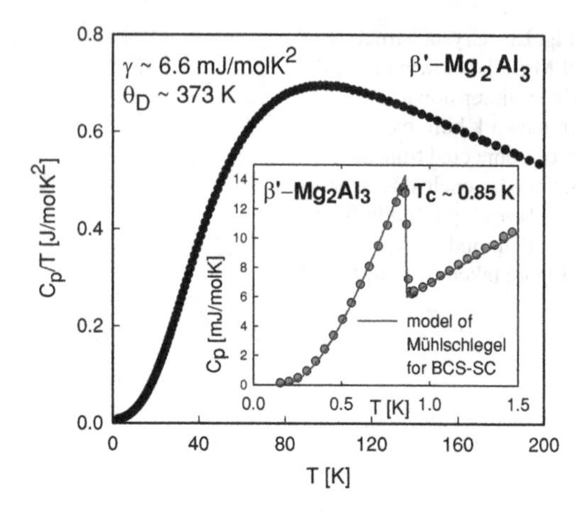

Fig. 1.8 Temperature dependent specific heat C_p of β'-Mg_2Al_3 plotted as C_p/T vs. T. The inset shows low temperature heat capacity data, evidencing bulk SC. The *solid line* adjusts the numerical data of Mühlschlegel [69] to the present experiment. (Figure taken from Ref. [68])

ments only, $\lambda_{so} \approx 0$. The dashed line in Fig. 1.7(b) represents the WHH model with $\alpha = 0.1$. Orbital pair-breaking is the most relevant mechanism in the low-field limit and therefore determines H'_{c2}. The Maki parameter $\alpha \approx 0.1$ of β'-Mg_2Al_3 corresponds to a dominant orbital pair breaking field.

Figure 1.8 summarizes heat-capacity measurements performed on β'-Mg_2Al_3. Low temperature data yield a Sommerfeld coefficient $\gamma = 6.6$ mJ/molK^2 and a Debye temperature $\theta_D = 373$ K. The jump of the specific heat $\Delta C_p/T(T = T_c) \approx$ 8.5 mJ/molK^2, allows calculation of $\Delta C_p/(\gamma_n T_c) \approx 1.41$, which matches closely the figure expected from weak coupling BCS theory ($\Delta C_p/(\gamma T_c) \approx 1.43$). BCS-type SC follows also from the comparison with Mühlschlegel's calculations [69]. These fits are shown as a solid line in the inset of Fig. 1.8, revealing agreement with a fully gapped SC ($\Delta(0) = 1.5$ K). Within the McMillan formula [70], the electron-phonon coupling strength $\lambda_{e,ph} = 0.42$ of β'-Mg_2Al_3 is derived from $\theta_D = 373$ K and from the repulsive screened Coulomb parameter $\mu^* \approx 0.13$. This defines β'-Mg_2Al_3 to be a SC in the weak coupling limit.

SC in β'-Mg_2Al_3 occurs in a crystal environment without inversion symmetry, favoring a mixture of spin-singlet and spin-triplet pairing in the superconducting condensate. The small values of the upper critical field, however, seem to exclude a substantial portion of spin-triplet pairs in the condensate. Moreover, the light elements Al and Mg may be responsible for only a minimal spin-orbit coupling in β'-Mg_2Al_3, hence, the spin-singlet condensate dominates. Additionally, the very complex crystal structure is supposed to smooth the effect of missing inversion symmetry.

1.3.3 Ternary $BaPtSi_3$

In a detailed investigation of the phases of Ba-Pt-Si at 900°C [71], the compound $BaPtSi_3$ was identified as a representative of the NCS $BaNiSn_3$ type, exhibiting a SC phase transition below 2 K. The tetragonal crystal structure corresponds to the space group $I4mm$, (as shown in Fig. 1.9), being an ordered variant of the cen-

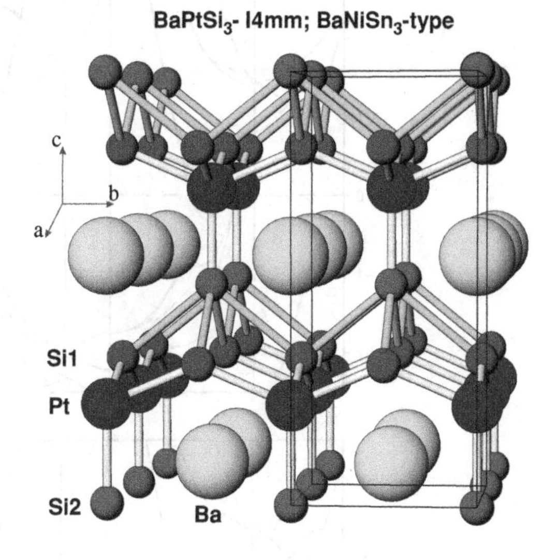

Fig. 1.9 Crystal structure of $BaPtSi_3$ clearly revealing non-centrosymmetry due to the missing mirror plane perpendicular to the c-axis. (Figure taken from Ref. [72])

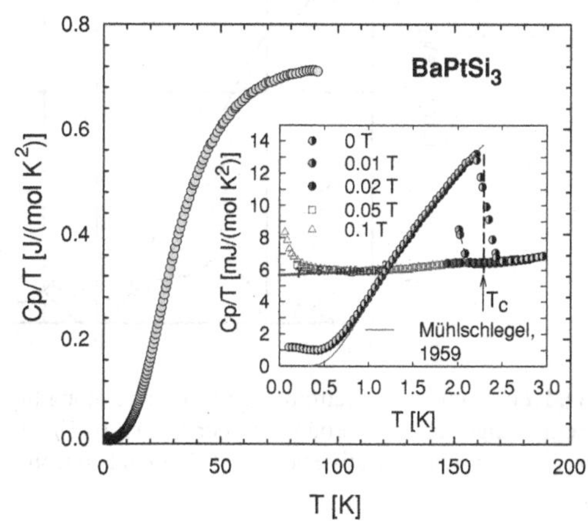

Fig. 1.10 Temperature dependent specific heat C_p of $BaPtSi_3$, plotted as C_p/T versus. T. The inset shows low temperature details of the superconducting transition for various values of externally applied magnetic fields. The *solid line* represents the temperature dependent specific heat of a spin-singlet fully gapped SC according to the model of Mühlschlegel [69]. (Figure taken from ref. [72])

trosymmetric $ThCr_2Si_2$ structure type. The standardized crystallographic data are: $a = 0.44094(2)$ nm and $c = 1.0013(3)$ nm; Ba is at the 2(a) site with coordinates (0, 0, 0.6022(8)); Pt at 2(a) with (0, 0, 0.2502(7)), Si2 at 2(a) with (0, 0, 0) and Si1 at 4(b) with (0, $\frac{1}{2}$, 0.3608(8)).

Resistivity data show SC below $T_c \approx 2$ K [72]. The normal metal state is well described by a standard Bloch-Grüneisen behaviour. Least squares fits reveal a Debye temperature of $\theta_D = 345$ K. Both, magnetic fields and pressure, rapidly suppress SC. An estimation of the electron-phonon interaction strength [70] considering $\theta_D = 345$K and $\mu^* \approx 0.13$ yields $\lambda_{e-ph} \approx 0.5$, characterizing $BaPtSi_3$ as a weak-coupling SC. Heat capacity data of $BaPtSi_3$ yield a Sommerfeld value of $\gamma \approx 5.7$ mJ/molK^2. Details of the superconducting transition under a magnetic field are compiled in the inset of Fig. 1.10.

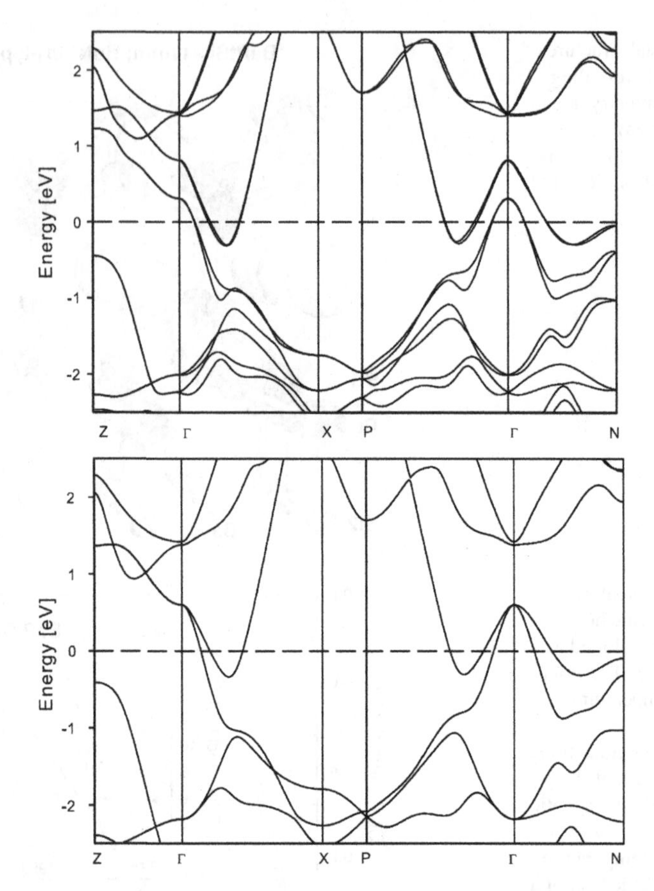

Fig. 1.11 Section of electronic band structure along high symmetry directions for $BaPtSi_3$ in the energy range ±2.5 eV around the Fermi energy E_F. *Upper panel*: spin-orbit coupling included; *lower panel*: scalar relativistic only. (Figure taken from ref. [72])

BCS-type SC follows from both the specific heat jump $\Delta C_p/(\gamma_n T_c) \approx 1.38$ as well as from the temperature dependence of the specific heat $C_p(T)$ in the superconducting state. The solid line in the inset of Fig. 1.10 shows $C_p(T)$ by applying the generalized BCS model by Mühlschlegel [69], corresponding to spin-singlet pairing and an isotropic quasiparticle gap.

The application of external fields rapidly suppresses bulk SC. Accordingly, a field of 0.01 T shifts T_c by about 0.25 K to lower temperatures, while fields of the order of 0.05 T already inhibit SC. The initial slope of the upper critical field $\mu_0 H'_{c2}(T) = -0.033$ T/K is substantially lower than the value deduced from the resistivity data ($\mu_0 H'_{c2}(T) \approx -0.1$ T/K). The Maki parameter $\alpha = 0.018$ [72], clearly indicates that orbital depairing is the essential mechanism which limits the upper critical field.

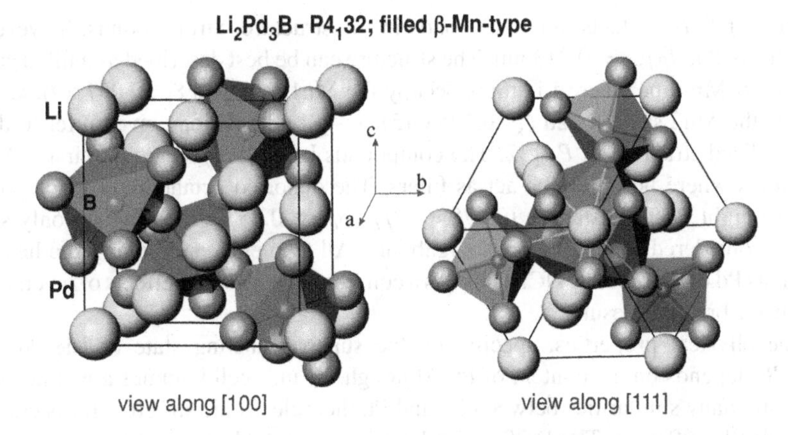

Fig. 1.12 Crystal structure of Li_2Pd_3B with corner-connected distorted Pd-octahedra (*shaded*); *left panel*: view along [100]; *right panel*: view along [111] highlighting the icosahedral coordination figure (distorted) around Li-atoms: Li[Li_3Pd_9]

The DOS of $BaPtSi_3$ is shown in Fig. 1.11, for both non-relativistic (upper panel) and relativistic (including spin orbit coupling, lower panel) band structure calculations. The main difference between these calculations occurs at about -1.5eV below the Fermi energy, at which the DOS splits into two prominent peaks of mixed Pt-d and Si-p character. The spin-orbit coupling most strongly affects the Pt-like states, which due to hybridization transfer the relativistic effect to the Si-like states. At the Fermi level one essentially finds Si-p and Pt-5d, as well as a smaller contribution from Ba-p states. Close to the Fermi energy, E_F, both calculations produce very similar and rather smooth DOS features. The values of the total DOS at E_F are $N(E_F) = 1.64$ and 1.60 states/eV for the non-relativistic and the relativistic calculation, respectively. The Sommerfeld coefficient of the electronic specific heat, $\gamma = \frac{\pi^2}{3} k_B^2 N(E_F)$ allows direct comparison with the calculated electronic density of states and suggests a phonon enhancement factor of $\lambda = 0.5$, reasonable for a weak-coupling BCS SC and perfectly in agreement with the McMillan model [70].

1.3.4 Ternary $Li_2(Pd, Pt)_3B$ *and* Mo_3Al_2C

The solid solution $Li_2(Pd, Pt)_3B$ crystallizes in a perovskite-like cubic structure (space group $P4_332$; No. 212) with a lattice constant $a = 0.67534(3)$ nm in the case of Pd and $a = 0.67552(5)$ nm in the case of the Pt compound [73, 74]. The simple cubic cell contains four considerably distorted octahedra $B[T_6]$ ($T =$ Pd, Pt) which sharing vertices form a three-dimensional framework enclose icosahedrally coordinated Li atoms. As filler atoms of the octahedra in site 4a ($\frac{3}{8}$, $\frac{3}{8}$, $\frac{3}{8}$), there are neither B-B contacts nor Li-B bonds in the structure. Strong bonds, however, exist for B-Pd: $d_{Pd-B} = 0.213$ nm. The structure can be best described as a filled and ordered β-Mn-type where Li atoms occupy the Mn1-atoms in 8c (0.0572(9), x, x) and Pd the Mn2 sites in 12d ($\frac{1}{8}$, 0.19583(5), $\frac{1}{4}$ +y); all positions given refer to the standardized structure in $P4_132$. The compounds $Li_2(Pd, Pt)_3B$ are the first β-Mn structures where boron atoms act as fillers. The distorted octahedra cover a wide span of eight Pd-Pd distances ($0.278 \leq d_{Pd-Pd} \leq 0.353$ nm) of which only six can be considered as next nearest neighbours. All elements, particularly the heavy elements Pd (or Pt) occupy NCS sites. As a consequence, stronger effects of inversion symmetry breaking result.

The physical properties, specifically the superconducting state of $Li_2(Pd_{1-x}Pt_x)_3B$, depends on the content of Pt. Although the unit-cell volumes are alike and there are many similarities between Pd and Pt, their electronic structure turns out to be distinctly different. The DOS at the Fermi energy, $N(E_F)$, increases from 2.24 states/eV in the case of Pd to 2.9 states/eV in the Pt case. Additionally, there are many more bands in a region within 1 eV below E_F as a consequence of a wider bandwidth, 7.6 eV in case of Pt and 6.7 eV in the case of Pd [75]. The band splitting due to ASOC is as large as 200 meV in Li_2Pt_3B and about 30 meV for Li_2Pd_3B.

Both ternaries Li_2Pd_3B and Li_2Pt_3B exhibit SC below 7 and 2.7 K, respectively [76, 77, 74, 78]. The distinct differences due to the much larger band splitting in the case of the Pt-based compound is obvious from the $1/T_1$ relaxation times taken from ^{11}B and ^{195}Pt NMR experiments derived by Nishiyama et al. [79]. In the case of Pd, a Hebel-Slichter peak appears around $T = T_c$ as a signature of a fully gapped DOS in the superconducting state, while such a peak is missing for the Pt-based compound. Moreover, a T^3 dependence is observed below T_c, pointing to line nodes of the superconducting order parameter as it is the case in various heavy fermion SCs (compare also Chapt. 2). An exponential, BCS-like temperature dependence of $1/T_1$ is observed for Li_2Pd_3B, resulting from a gap in the DOS around E_F without any nodes.

Unconventional SC of Li_2Pt_3B is also evidenced from ^{195}Pt NMR Knight shift data [79]. In conventional s-wave SCs the spin susceptibility decreases below T_c following the Yoshida function, attaining zero at zero temperatures. The Knight shift of Li_2Pt_3B is temperature independent for $T < T_c$ suggesting that we encounter here not simple spin-singlet superconductivity. Indeed, this behaviour is a strong indication of unconventional Cooper pairing. The striking difference between Li_2Pt_3B and Li_2Pd_3B may be attributed to substantial differences in the spin-orbit coupling, since Pt has a larger atomic number than Pd. Equally striking is the temperature dependent London penetration depth λ, behaving BCS-like ($T << T_c$) in the case of Li_2Pd_3B, but has a linear temperature dependence in Li_2Pt_3B [80]. While the former is a signature of a fully gapped SC state, the latter might be a consequence of line nodes in the SC gap.

Carbides based on Mo comprise a large body of refractory compounds, where carbon atoms (in trigonal prismatic or octahedral Mo_6C subunits) occupy a fraction of the interstitial sites either in an ordered or in a random manner. Among Mo-based carbides for which superconductivity was reported (α-MoC at $T_c = 9.95$

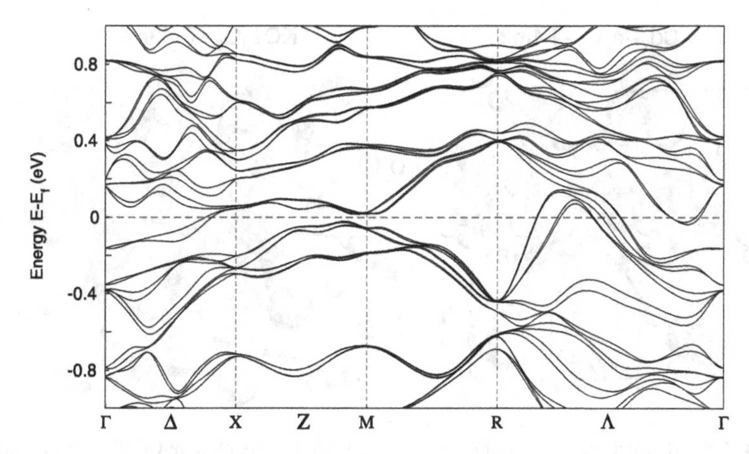

Fig. 1.13 Relativistic electronic band structure along high symmetry directions for Mo_3Al_2C in the energy range $\pm 1.$ eV around the Fermi energy E_F. (Figure taken from ref. [82])

K, ηMoC at 7.57 K, Mo_2BC at 6.33 K and Mo_3Al_2C at 9.05 K) the crystal structure of Mo_3Al_2C is outstanding, since the respective β-Mn type does not possess a center of inversion [81].

Electrical resistivity, specific heat and NMR measurements classify non-centrosymmetric Mo_3Al_2C (filled β-Mn type, space group $P4_132$) as a strong-coupled superconductor with $T_c \approx 9$ K deviating notably from BCS-like behaviour. The absence of a Hebbel-Slichter peak, a power-law behaviour of the spin-lattice relaxation rate (from ^{27}Al NMR), an electronic specific heat strongly deviating from BCS model and a pressure enhanced T_c suggest unconventional superconductivity with possibly a nodal structure of the superconducting gap [82, 83]. Relativistic DFT calculations [82] reveal a splitting of degenerate electronic bands due to the asymmetric spin-orbit coupling (compare Fig. 1.13), favouring a mix of spin-singlet and spin-triplet components in the superconducting condensate, in absence of strong correlations among electrons. In fact, both $Li_2(Pd, Pt)_3B$ and Mo_3Al_2C crystallize in the very same β-Mn structure, which lacks inversion symmetry along all principal axes. As a result, ASOC is particularly pronounced presumably driving superconductivity unconventional.

1.3.5 Pyrochlores $A_2B_2O_7$ *and* AB_2O_6

The class of pyrochlores comprises a series of compounds characterized by geometrically frustrated magnetism/spin-ice due to three-dimensional triangular lattices. The crystal chemical formula for α-pyrochlore is $A_2B_2O_6O'$ generally abbreviated as $A_2B_2O_7$. The face centered cubic lattice with space group $Fd\bar{3}m$ (No. 227) accommodates four inequivalent lattice sites, where A occupies the 16d site, B the 16c

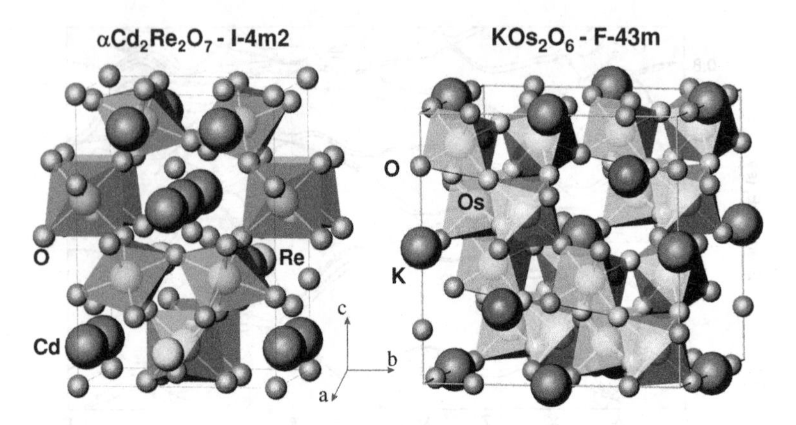

Fig. 1.14 Crystal structures in three-dimensional view for α-pyrochlore α-$Cd_2Re_2O_7$ (*left panel*) and β-pyrochlore (*right panel*) KOs_2O_6; the corner connected octahedra ReO_6 and OsO_6 are outlined

site, and oxygen atoms O are at $48f$ and O' at $8b$. The elements A are electropositive from the first, second or third main group whereas B belongs to transition elements. β-pyrochlores, AOs_2O_6 with space group $Fd\bar{3}m$, are formed for A^{1+} elements when O' is replaced by K, Cs or Rb, B = Os, and the $16d$ sites stay vacant. The defect structure of β-pyrochlore is then given by the ideal formula $\Box_2B_2O_6A'$, where $\Box$ is a vacancy. The B atom (Os) is octahedrally co-ordinated by six O atoms such that OsO_6 octahedra sharing vertices form channels, which host the A atoms. For a summary of the structural chemistry of the pyrochlore family, see the review by Subramanian et al. [84]. High-resolution neutron powder diffraction carried out on α-pyrochlore $Cd_2Re_2O_7$ [85] revealed symmetry violations below 180 K. Best agreement with the experimental data is achieved for a tetragonal structure model ($a = a_0/\sqrt{2}, c = c_0$) with space group $I\bar{4}m2$ (No. 119) which lacks a center of symmetry. Below 120 K the structure of $Cd_2Re_2O_7$ was indicated to become more complex; although the small structural distortions gave no clear hints for a specific space-group symmetry, the low-temperature structure seems to retain non-centrosymmetry in any case [85]. A closer inspection of the crystal structure of a single crystal of KOs_2O_6 by Schuck et al. [86] revealed low intensity x-ray reflections, incompatible with the space group symmetry $Fd\bar{3}m$. The best description of the single crystal X-ray diffraction pattern was claimed to be the cubic NCS space group $F\bar{4}3m$ (No. 216). Thus, SC in some of the pyrochlores may occur in a NCS environment. A very recent structure study on KOs_2O_6 confirmed, however, the centrosymmetric $Fd\bar{3}m$ space group [87]. Symmetry reductions have been reported in other studies on pyrochlores as well [88, 89, 90, 91, 92, 93] (Fig. 1.14).

Normal-state properties of the pyrochlores [94] are characterised by strong curvatures in the temperature dependence of the resistivity, particularly pronounced in case of KOs_2O_6, even at low temperatures. This points towards an extraordinary electron-phonon interaction, possibly as a precursor of SC. Except for the K-based compound, a T^2-dependence of $\rho(T)$ refers to Fermi-liquid behaviour.

Specific-heat measurements reveal enhanced Sommerfeld values γ in the range from 30 to 70 mJ/molK^2. The respective values deduced from bandstructure calculations, however, are constraint to about 10 mJ/molK^2 [94].

The coefficient A of the T^2-dependence of the electrical resistivity, together with the Sommerfeld value for the various pyrochlores, fit perfectly into the Kadowaki-Woods plot [33], grouping primarily in the lower left part of this plot where s-wave SCs (either isotropic or anisotropic) aggregate. In contrast, SCs with nodal gaps (e.g., heavy fermion compounds) are predominantly found in the upper right corner of the Kadowaki-Woods plot [95].

SC in pyrochlores was discovered for the first time in α-pyrochlore $Cd_2Re_2O_7$ with $T_c = 1.0$ K [96, 97]. Larger transition temperatures have been found in β-pyrochlore AOs_2O_6 with $T_c = 3.3$, 6.3, and 9.6 K for A = Cs [98], Rb [99, 100, 101], and K [102], respectively. This increase is concomitant with a substantial decrease of the lattice parameter a in the respective compounds [94] and might be a consequence of a significant enhancement of the electron-phonon parameter λ_{el-ph} together with a rattling mode of the A atom inside the cage-forming structure [103]. Since the β-pyrochlores are, most likely, centrosymmetric in their crystal structure, the brief discussion of SC is constrained to α-pyrochlore $Cd_2Re_2O_7$. The electrical resistivity of $Cd_2Re_2O_7$ vanishes at ≈ 1 K and the application of magnetic fields suppress T_c at a critical value of about 0.29 T.

Heat-capacity data taken at low temperatures for $Cd_2Re_2O_7$ result in a Sommerfeld value $\gamma = 30$ mJ/molK^2 and a Debye temperature $\theta_D = 485$ K [104]. A distinct jump of C_p at $T_c = 0.97$ K evidences bulk SC. However, $\Delta C_p/(\gamma T_c) \approx 1.15$ is well below the BCS value of 1.43. $C_p(T)$ below T_c follows an exponential behaviour, referring to a node-less behaviour of the superconducting gap. A least-squares fit reveals $2\Delta/(k_B T_c) = 3.6$, very close to the weak-coupling BCS value. This agrees with $\lambda_{el-ph} \approx 0.42$, derived in terms of the McMillan formula. A fully gapped BCS-like superconducting state can also be concluded from the ^{187}Re NQR $1/T_1$ relaxation rate, exhibiting a Hebbel-Slichter coherence peak at $T = T_c$ [105].

The temperature-dependent upper critical field H_{c2} follows the WHH curve and extrapolates to $\mu_0 H_{c2}(0) \approx 0.29$ T. This value is well below the Pauli limiting field $\mu_0 H_P \approx 2$ T. Orbital pair breaking is thus the most reliable scenario constraining H_{c2}. Consequently, spin-singlet pairs dominate SC of $Cd_2Re_2O_7$. The electronic mean free path in $Cd_2Re_2O_7$ is much larger than the correlation length $\xi = 34$ nm; hence SC occurs within the clean limit [104].

1.3.6 Binary T_2Ga_9, T = Rh, Ir

Binary T_2Ga_9 with T = Rh, Ir have recently been shown to exhibit SC below 1.95 and 2.25 K, respectively [106].

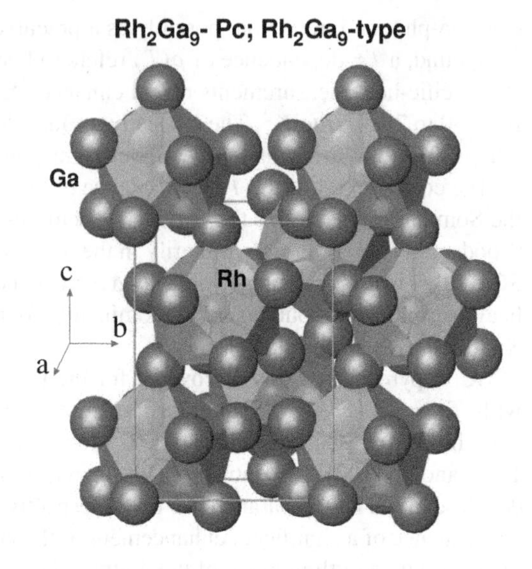

Fig. 1.15 Crystal structure of Rh_2Ga_9 with a three dimensional network of corner connected mono-capped Archimedian antiprisms $Rh[Ga_9]$

Although the centrosymmetric monoclinic crystal structure of Co_2Al_9 (space group $P2_1/c$; No. 14) is the parent structure type represented by e.g., Co_2Al_9 and Rh_2Al_9 [107], the gallides based on Rh and Ir, as a result of differences in interatomic interactions, show small distortions from centrosymmetry best described in space group *Pc* (No. 7). This structure is characterized by single-capped square antiprismatic coordination polyhedra around the transition metals Rh or Ir (see Fig. 1.15). Corner-connected polyhedra form zig-zag strands parallel to the c-axis. In the low-symmetry case of gallides, each of the zig-zag strands contains polyhedra of only one of the two crystallographic independent transition metal atoms [108]. The bond-angle of 165.8° for *T*-Ga-*T* in the case of Ir_2Ga_9 and 164.5° for Rh_2Ga_9 are much smaller than 180°, but comparable, e.g., with β-Mn type Li_2Pt_3B. These substantial deviations from 180° are the reason for the lack of inversion symmetry. Furthermore, the heavy elements Rh and Ir give rise to strong ASOC.

Electronic structure calculations were carried out for centrosymmetric Ir_2Al_9 and NCS Ir_2Ga_9. Results of the DOS show very similar features in both cases. Relatively small DOS values are found at the Fermi energy E_F, about 5 states/eV, independent of the material and also independent of the crystal structure assigned [108]. Interestingly, the electronic states at E_F originate primarily from Al respectively Ga, while the *5d* states of Ir play only a minor role. Slightly above E_F the DOS exhibits a pronounced local minimum.

SC of Rh_2Ga_9 and Ir_2Ga_9 is observed in resistivity, specific-heat and magnetisation measurements. The temperature dependent resistivity is metallic-like and can be accounted for in terms of a parallel resistance model. The application of external magnetic fields rapidly diminishes T_c, yielding critical magnetic fields of about 200 and 300 Oe for the Rh- and Ir-based compounds, respectively. The field response, however, is distinctly different in both cases. Field dependent heat capacity data show evidence for a second order phase transition in Ir_2Ga_9, while the application of a magnetic field changes the type of transition from second to first order in the case of Rh_2Ga_9. This demonstrates type-II SC for the former, but type-I SC for the latter. In fact, the critical magnetic field for Rh_2Ga_9 coincides perfectly with the thermodynamic critical field $H_c(T)$ [106]. Improved sample quality in the case of Ir_2Ga_9 [109] also reveals type-I SC, rather than type-II.

Specific-heat data at zero field are indicative of a weak-coupling s-wave BCS SC. Support for this conclusion follows from Ga-nuclear-quadrupole-resonance (NQR) measurements [110]. A Hebel-Slichter peak right at T_c and an exponential decrease of the $1/T_1$ relaxation rate reveal an isotropic gap with $2\Delta(0)/k_B T_c = 4.4$, hallmarks of BCS SC. Despite the absence of inversion symmetry of the crystal structure, there is no evidence of any feature in the physical properties characterising unconventional SC.

1.3.7 Ternary $LaNiC_2$

RNiC$_2$ compounds (R = rare earth) crystallize in the NCS orthorhombic $CeNiC_2$ structure type, space group $Amm2$ (No. 38). R-atoms in the crystal structure (Ce in site 2b ($\frac{1}{2}$, 0, 0.3857) form face-connected triangular prisms R_6 which are alternatively centered by either a Ni-atom (in site 2a (0, 0, 0.0) or by a dumbell of C-atoms in site 4d (0, 0.3447, 0.1775) (see Fig. 1.16). The structure of $CeNiC_2$ can also be conceived as a distorted AlB_2-type with alternating Ni and C_2 constituting the boron net. Whereas R-C, R-Ni and C-Ni distances are close to the sum of metal radii, C-C distances, $d_{C-C} = 0.141(3)$nm, indicate strong bonding (although confirmation by neutron diffraction is desirable) [111]. Depending on the specific rare earth element, antiferromagnetic ordering with a maximum transition temperature of 25 K ($TbNiC_2$) is observed [112].

Evidence for bulk SC of metallic $LaNiC_2$ is obtained from a variety of measurements at $T_c \approx 2.75$ K. Upon Ni/Cu substitution, T_c unexpectedly increases [113], a fact that has been interpreted in terms of a volume enlargement, accompanied by a change of the density of states and electron-phonon coupling strength. Early specific-heat measurements revealed a T^3 behaviour for $T < T_c$ rather than an exponential temperature dependence [114]. However, improved measurements carried out in a broader temperature range showed a standard exponential BCS behaviour, with $2\Delta/(k_B T_c) = 4.62$. BCS-type SC was concluded from ^{139}La-NQR studies as well from both, a Hebel-Slichter peak around T_c and an exponential temperature dependence of the $1/T_1$ rate. A fit to $1/T_1$ for $T < T_c$ reveals $2\Delta/(k_B T_c) = 3.34$ [115].

Very recently, however, Hillier et al [116] observed from muon spin relaxation studies the appearance of spontaneous intrinsic magnetization in the SC phase starting at T_c. This may be interpreted as a superconducting state with broken time-reversal symmetry (TRS). In a SC with broken TRS, spontaneous fields arise in regions where the order parameter is inhomogeneous, such as domain walls and grain boundaries.

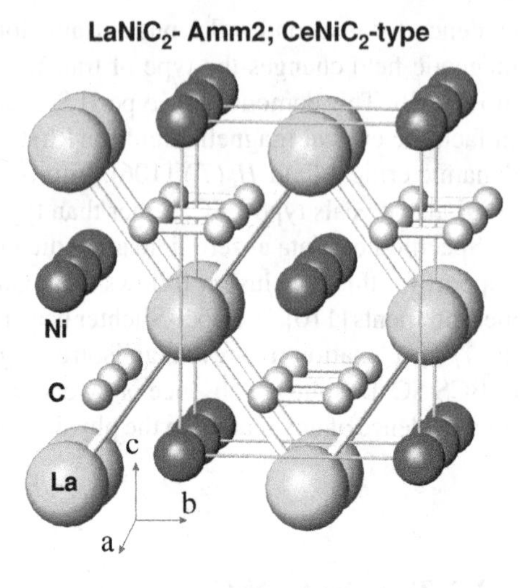

Fig. 1.16 Crystal structure of $LaNiC_2$ where La-atoms form triangular prisms which are alternatively centered by either a Ni-atom or by a dumbell of C-atoms

It is unclear so far how to explain this finding within the frame work of the symmetry classification of pairing states in this material.

Other recently discovered non-centrosymmetric superconducting compounds in this group are $Mg_{10}Ir_{19}B_{16}$, WRe_3, the low charge carrier compound LaBiPt ($T_c \approx 0.9$ K) [117] and $YNiGe_3$ ($T_c = 0.46$ K) [118], as well as numerous amorphous SCs like Zr_xCu_{1-x} [119].

$\mathbf{Mg_{10}Ir_{19}B_{16}}$ crystallizes in a body centered cubic structure with a lattice parameter $a = 1.0568$ nm and the space group $I\bar{4}3m$. Each of the elements in the compound occupies a NCS lattice site. $Mg_{10}Ir_{19}B_{16}$ behaves metallic below room temperature exhibiting a SC transition between 4 and 5 K, depending on the actual stoichiometry [120]. The extrapolated upper critical field does not exceed 0.8 T; hence it is well below the standard paramagnetic limit. As a consequence, spin-singlet pairs should dominate the type-II SC ($\kappa \approx 20$) [121]. Such a conclusion can be drawn from a BCS-like temperature dependent heat capacity in the SC temperature range as well. The relatively large Sommerfeld value $\gamma \approx 52$ mJ/molK^2 might be a result of the large number of atoms in the unit cell counting 90. A modified McMillan formula [122] reveals $\lambda_{el-ph} = 0.66$, thus classifying $Mg_{10}Ir_{19}B_{16}$ as a moderate-coupling SC. The value of the coherence length is $\xi = 206$ Å and the London penetration depth is calculated as $\lambda = 4040$ Å[121]. Penetration-depth studies [123] of $Mg_{10}Ir_{19}B_{16}$ denote a further transition or crossover around 0.8 K and an overall low-temperature response that possibly indicates a two-gap SC, ruling out an isotropic spin-singlet gap structure, as suggested by specific-heat measurements.

$\mathbf{WRe_3}$ is a cubic compound that crystallises in the α-Mn crystal structure (space group $I4\bar{3}m$) with a lattice constant $a = 9.596$ Å. SC in intermetallic WRe_3 was first studied in the 1960s. The SC transition temperature was found to be $T_c \approx 9$ K [124, 125]. A study of the low-temperature susceptibility and heat capacity by Jing et al. [126] hints at two SC phase transitions, at $T_{c1} = 9$ K and $T_{c2} = 7$ K. The Sommerfeld value $\gamma = 17$ mJ/molK^2 and $\theta_D = 348$ K, revealing $\lambda_{el-ph} = 0.74$, a value slightly beyond the weak coupling limit. The temperature dependent specific heat for $T < T_c$ follows the BCS behaviour, suggesting a fully-gapped s-wave SC state. The latter is entirely supported by temperature dependent penetration depth studies [127]. The coherence length for WRe_3 is estimated as $\xi \approx 2000$ Å [127].

1.4 Summary

Non-centrosymmetric superconductors have been identified as an intriguing family of intermetallics which challenge both theory and experiment. The materials explored so far can be sub-grouped in classes where strong correlations occur between electrons (evidenced, e.g., by a substantially large Sommerfeld value) and those which behave like simple metals. While the latter in each case, except $LiPt_2B_3$ and Mo_3Al_2C, appear to be textbook-like BCS SCs, the former have attracted much interest because of a variety of unconventional SC features, such as line nodes in the SC gap function or the involvement of spin-triplet pairs in the SC condensate.

A generally accepted description relies on the Rashba-type antisymmetric spin-orbit coupling (ASOC) which lifts the spin degeneracy of electronic bands. If the ASOC splitting is large, a genuine mixing of spin-singlet and spin-triplet components occurs. In fact, the possibility of sizable spin-triplet pairing components in non-centrosymmetric superconductors is suggested by spin-fluctuation theory, where NCS spin fluctuations enable spin-triplet paring, while the centrosymmetric fluctuations contribute only to spin-singlet components. Strong electronic correlations, i.e., a substantially large Coulomb correlation parameter U, favour such a scenario.

Acknowledgment Work supported by the Austrian FWF P22295 and by COST P16.

References

1. Bardeen, J., Cooper, L., Schrieffer, J.: J. Phys. Rev. **108**, 1157 (1957)
2. Mineev, V.P., Samokhin, K.V.: Introduction to Unconventional Superconductivity. Gordon and Breach Science Publishers, Amsterdam (1999) ISBN: 90-5699-209-0
3. Anderson, P.W.: Phys. Rev. B **30**, 4000 (1984)
4. Frigeri, P.A., Agterberg, D.F., Koga, A., Sigrist, M.: Phys. Rev. Lett. **92**, 097001 (2004)
5. Takimoto, T., Thalmeier, P.: J. Phys. Soc. Jpn. **78**, 103703 (2009)
6. Tursina, A.I., Gribanov, A.V., Noël, H., Rogl, P., Seropegin, Y.D., Bodak, O.I.: J. Alloys Compounds **383**, 239 (2004)
7. Bauer, E., Hilscher, G., Michor, H., Paul Ch., Scheidt, E.W., Gribanov, A., Seropegin Yu., Noël, H., Sigrist, M., Rogl, P.: Phys. Rev. Lett. **92**, 027003 (2004)
8. Metoki, N., Kaneko, K., Matsuda, T.D., Galatanu, A., Takeuchi, T., Hashimoto, S., Ueda, T., Settai, R., Onuki, Y., Bernhoeft, N.: J. Phys.: Cond. Mat., **16**, L207 (2004)
9. Tinkham, M.: Introduction to Superconductivity. McGraw-Hill, New York (1975)
10. Willers, T., Fak, B., Hollmann, N., Körner, P.O., Hu, Z., Tanaka, A., Schmitz, D., Enderle, M., Lapertot, G., Tjeng, L.H., Severing, A.: Phys. Rev. B **80**, 115106 (2009)

11. Takeuchi, T., Yasuda, T., Tsujino, M., Shishido, H., Settai, R., Harima, H., Onuki, Y.: J. Phys. Soc. Jpn. **76**, 014702 (2007)
12. Amato, A., Bauer, E., Baines, C.: Phys. Rev. B **71**, 092501 (2005)
13. Yogi, M., Kitaoka, Y., Hashimoto, S., Yasuda, T., Settai, R., Matsuda, T.D., Haga, Y., Onuki, Y., Rogl, P., Bauer, E.: Phys. Rev. Lett. **93**, 027003 (2004)
14. Yogi, M., Mukuda, H., Kitaoka, Y., Hashimoto, S., Yasuda, T., Settai, R., Matsuda, T.D., Haga, Y., Onuki, Y., Rogl, P., Bauer, E.: J. Phys. Soc. Japan **75**, 013709 (2006)
15. Rashba, E.I.: Sov. Phys. Solid State **2**, 1109 (1960)
16. Frigeri, P.A., Agterberg, D.F., Milat, I., Sigrist, M.: Eur. Phys. J. B **54**, 435 (2006)
17. Hayashi, N., Wakabayashi, K., Frigeri, P.A., Sigrist, M.: Phys. Rev. B **73**, 092508 (2006)
18. Brison, J.P., Keller, N., Verniere, A., Lejay, P., Schmidt, L., Buzdin, A., Flouquet, J., Julian, S.R., Lonzarich, G.G.: Physica C **250**, 128 (1995)
19. Gorkov, L.P., Rashba, E.I.: Phys. Rev. Lett. **87**, 037004 (2001)
20. Frigeri, P.A., Agterberg, D.F., Sigrist, M.: New J. Phys. **6**, 115 (2004)
21. Kaur, R.P., Agterberg, D.F., Sigrist, M.: Phys. Rev. Lett. **94**, 137002 (2005)
22. Yasuda, T., Shishido, H., Ueda, T., Hashimoto, S., Settai, R., Takeuchi, T., Matsuda, T.D., Haga, Y., Onuki, Y.: J. Phys. Soc. Jpn. **73**, 1657 (2004)
23. Mukuda, H., Nishide, S., Harada, A., Iwasaki, K., Yogi, M., Yashima, M., Kitaoka, Y., Tsujino, M., Takeuchi, T., Settai, R., Onuki, Y., Bauer, E., Itoh, K.M., Haller, E.E.: J. Phys. Soc. Jpn. **78**, 014705 (2009)
24. Bonalde, I., Brämer-Escamilla, W., Bauer, E.: Phys. Rev. Lett. **94**, 207002 (2005)
25. Izawa, K., Kasaharal, Y., Matsudal, Y., Behnia, K., Yasuda, T., Settai, R., Onuki, Y.: Phys. Rev. Lett. **94**, 197002 (2005)
26. Bauer, E., Hilscher, G., Michor, H., Sieberer, M., Scheidt, E., Gribanov, A., Seropegin, Y., Rogl, P., Amato, A., Song, W.Y., Park, J.-G., Adroja, D.T., Nicklas, M., Sparn, G., Yogi, M., Kitaoka, Y.: Physica B, 359–361, 360 (2005)
27. Nicklas, M., Sparn, G., Lackner, R., Bauer, E., Steglich, F.: Physica B 359–361, 386 (2005)
28. Bauer, E., Kaldarar, H., Prokofiev, A., Royanian, E., Amato, A., Sereni, J.G., Brämer-Escamilla, W., Bonalde, I.: J. Phys. Soc. Jpn. **76**, 051009 (2007)
29. Tateiwa, N., Haga, Y., Matsuda, T.D., Ikeda, S., Yasuda, T., Takeuchi, T., Settai, R., Onuki, Y.: J. Phys. Soc. Jpn. **74**, 1903 (2005)
30. Nicklas, M., Steglich, F., Knolle, J., Eremin, I., Lackner, R., Bauer, E.: Phys. Rev. B **81**, 180511 (2010)
31. Kimura, N., Ito, K., Saitoh, K., Umeda, Y., Aoki, H., Terashima, T.: Phys. Rev. Lett. **95**, 247004 (2005)
32. Muro, Y., Eom, D., Takeda, N., Ishikawa, M.: J. Phys. Soc. Jpn. **67**, 3601 (1998)
33. Kadowaki, K., Woods, S.B.: Solid State Commun. **58**, 507 (1986). Universal relationship of the resistivity and specific heat in heavy-Fermion compounds K. Kadowaki and S. B. Woods
34. Kimura, N., Muro, Y., Aoki, H.: J. Phys. Soc. Jpn. **76**, 051010 (2007)
35. Aso, N., Miyano, H., Yoshizawa, H., Kimura, N., Komatsubara, T., Aoki, H.: J. Magn. Magn. Mater. **310**, 602 (2007)
36. Kimura, N., Ito, K., Aoki, H.: Phys. Rev. Lett. **98**, 197001 (2007)
37. Sugitani, I., Okuda, Y., Shishido, H., Yamada, T., Thamizhavel, A., Yamamoto, E., Matsuda, T.D., Haga, Y., Takeuchi, T., Settai, R., Onuki, Y.: J. Phys. Soc. Jpn. **75**, 043703 (2006)
38. Krimmel, A., Reehuis, M., Loidl, A.: Appl. Phys. A **74**, S695 (2002)
39. Werthamer, N.R., Hefland, E., Hohenberg, P.C.: Phys. Rev. **147**, 295 (1966)
40. Tateiwa, N., Haga, Y., Matsuda, T.D., Ikeda, S., Yamamoto, E., Okuda, Y., Miyauchi, Y., Settai, R., Onuki, Y.: J. Phys. Soc. Jpn. **76**, 083706 (2007)
41. Mukuda, H., Fujii, T., Ohara, T., Harada, A., Yashima, M., Kitaoka, Y., Okuda, Y., Settai, R., Onuki, Y.: Phys. Rev. Lett. **100**, 107003 (2008)
42. Hegger, H., Petrovic, C., Moshopoulou, E.G., Hundley, M.F., Sarrao, J.L., Fisk, Z., Thompson, J.D.: Phys. Rev. Lett. **84**, 4986 (2000)

43. Okuda, Y., Miyauchi, Y., Ida, Y., Takeda, Y., Tonohiro, C., Oduchi, Y., Yamada, T., Duc Dung, N., Matsuda, T.D., Haga, Y., Takeuchi, T., Hagiwara, M., Kindo, K., Harima, H., Sugiyama, K., Settai, R., Onuki, Y.: J. Phys. Soc. Jpn. **76**, 044708 (2007)
44. Thamizhavel, A., Takeuchi, T., Matsuda, T.D., Haga, Y., Sugiyama, K., Settai, R., Onuki, Y.: J. Phys. Soc. Jpn. **74**, 1858 (2005)
45. Kawai, T., Muranaka, H., Measson, M.-A., Shimoda, T., Doi, Y., Matsuda, T.D., Haga, Y., Knebel, G., Lapertot, G., Aoki, D., Flouquet, J., Takeuchi, T., Settai, R., Onuki, Y.: J. Phys. Soc. Jpn. **77**, 064716 (2008)
46. Settai, R., Okuda, Y., Sugitani, I., Onuki, Y., Matsuda, T.D., Haga, Y., Harima, H.: Int. J. Mod. Phys. B **21**, 3238 (2007)
47. Dommann, A., Hulliger, F., Sigrist, T., Fischer, P.: J. Magn. Magn. Mater. **67**, 323 (1987)
48. Akazawa, T., Hidaka, H., Fujiwara, T., Kobayashi, T.C., Yamamoto, E., Haga, Y., Settai, R., Onuki, Y.: J. Phys.: Condens. Matter **16**, L29 (2004)
49. Kobayashi, T.C., Hori, A., Fukushima, S., Hidaka, H., Kotegawa, H., Akazawa, T., Takeda, K., Ohishi, Y., Yamamoto, E.: J. Phys. Soc. Jpn. **76**, 051007 (2007)
50. Krupka, C., Giorgi, A.L., Krikorian, N.H., Szklarz, E.G.: J. Less Common Met. **19**, 113 (1969)
51. Giorgi, A.L., Sklarz, E.G., Krikorian, N.H., Krupka, M.C.: J. Less Common Met. **2**, 131 (1970)
52. Atoji, M., Gschneidner, K. Jr., Daane, A.H., Rundle, R.F., Spedding, F.H.: Acta Crystallogr. **80**, 1804 (1958)
53. Novokshonov, V.I.: Russ. J. Inorganic Chemistry **25**, 375–378 (1980)
54. Atoji, M., Williams, D.E.: J. Chem. Physics, **35**, 1960 (1961)
55. Green, J.L., Arnold, G.P., Leary, J.A., Nereson, N.G.: J. Nucl. Mater. **34**, 281–289 (1970)
56. Stewart, G., Giorgi, A.L., Krupka, M.C.: Solid State Commun. **27**, 413 (1978)
57. Amano, G., Akutagawa, S., Muranaka, T., Zenitani, Y., Akimitsu, J.: J. Phys. Soc. Jpn. **73**, 530 (2004)
58. Nakane, T., Mochiku, T., Kito, H., Itoh, J., Nagao, M., Kumakura, H., Takano, Y.: Appl. Phys. Lett. **84**, 2859 (2004)
59. Kim, J.S., Kremer, R.K., Jepsen, O., Simon, A.: Curr. Appl. Phys. **6**, 897 (2006)
60. Kim, J.S., Xie, W., Kremer, R.K., Babizhetskyy, V., Jepsen, O., Simon, A., Ahn, K.S., Raquet, B., Rakoto, H., Broto, J.-M., Ouladdiaf, B.: Phys. Rev. B **76**, 014516 (2007)
61. Harada, A., Akutagawa, S., Miyamichi, Y., Mukuda, H., Kitaoka, Y., Akimitsu, J.: J. Phys. Soc. Jpn. **76**, 023704 (2007)
62. Kuroiwa, S., Saura, Y., Akimitsu, J., Hiraishi, M., Miyazaki, M., Satoh, K. H., Takeshita, S., Kadono, R. R.: Phys. Rev. Lett. **100**, 097002 (2008)
63. Sugawara, K., Sato, T., Souma, S., Takahashi, T., Ochiai, A.: Phys. Rev. B**76**, 132512 (2007)
64. Sergienko, I.A.: Physica B **359**(361), 581 (2005)
65. Samson, S.: Acta Crystallogr. **19**, 401 (1965)
66. Feuerbacher, M., Thomas, C., Makongo, J.P.A., Hoffmann, S., Carrillo-Cabrera, W., Cardoso, R., Grin Yu., Kreiner, G., Joubert, J.-M., Schenk Th., Gastaldi, J., Nguyen-Thi, H., Mangelinck-Noel, N., Billia, B., Donnadieu, P., Czyrska-Filemonowicz, A., Zielinska-Lipiec, A., Dubiel, B., Weber Th., Schaub, P., Krauss, G., Gramlich, V., Christensen, J., Lidin, S., Fredrickson, D., Mihalkovic, M., Sikora, W., Malinowski, J., Brühne, S., Proffen Th., Assmus, W., de Boissieu, M., Bley, F., Schreuer, J., Steurer, W.: Z. Kristallogr. **222**, 259 (2007)
67. Woodard, W., Cody, G.D.: Phys. Rev. **136**, A166 (1964)
68. Bauer, E., Kaldarar, H., Lackner, R., Michor, H., Steiner, W., Scheidt, E.-W., Galatanu, A., Marabelli, F., Wazumi, T., Kumagai, K., Feuerbacher, M.: Phys. Rev. B **76**, 014528 (2007)
69. Mühlschlegel, B.: Zf. Physik **155**, 313 (1959)
70. McMillan, W.L.: Phys. Rev. **167**, 331 (1968)
71. Melnychenko-Koblyuk, N., Grytsiv, A., Rogl, P., Bauer, E., Lackner, R., Royanian, E., Rotter, M., Giester, G.: J. Phys. Soc. Jpn. **77**(Suppl. A), 54 (2008)

72. Bauer, E., Khan, R.T., Michor, H., Royanian, E., Grytsiv, A., Melnychenko-Koblyuk, N., Rogl, P., Reith, D., Podloucky, R., Scheidt, E.-W., Wolf, W., Marsman, M.: Phys. Rev. B **80**, 064504 (2009)
73. Eibenstein, U., Jung, W.: J. Solid State Chemistry **133**, 21 (1997)
74. Badica, P., Kondo, T., Togano, K.: J. Phys. Soc. Jpn. **74**, 1014 (2005)
75. Lee, K.-W., Picket, W.E.: Phys. Rev. B **72**, 174505 (2005)
76. Togano, K., Badica, P., Nakamori, Y., Orimo, S., Takeya, H., Hirata, K.: Phys. Rev. Lett. **93**, 247004 (2004)
77. Badica, P., Kondo, T., Kudo, T., Nakamori, Y., Orimo, S., Togano, K.: Appl. Phys. Lett. **85**, 4433 (2004)
78. Takeya, H., Hirata, K., Yamaura, K., Togano, K., El Massalami, M., Rapp, R., Chaves, F.A., Ouladdiaf, B.: Phys. Rev. B **72**, 104506 (2005)
79. Nishiyama, M., Inada, Y., Zheng, Guo.-.qing.: Phys. Rev. Lett. **98**, 047002 (2007)
80. Yuan, H.Q., Agterberg, D.F., Hayashi, N., Badica, P., Vandervelde, D., Togano, K., Sigrist, M., Salamon, M.B.: Phys. Rev. Lett. **97**, 017006 (2006)
81. Toth, L.E., Zbasnik, J.: Acta Met. **16**, 1177 (1968)
82. Bauer, E., Rogl, G., Chen, Xing.-.Qiu., Khan, R.T., Michor, H., Hilscher, G., Royanian, E., Kumagai, K., Li, D.Z., Li, Y.Y., Podloucky, R., Rogl, P.: Phys. Rev. B **82**, 064511 (2010)
83. Karki, A.B., Xiong, Y.M., Vekhter, I., Browne, D., Adams, P.W., Young, D.P., Thomas, K.R., Julia, Y. Chan, Kim, H., Prozorov, R.: Phys. Rev. B **82**, 064512 (2010)
84. Subramanian, M.A., Aravamudan, G., Subba Rao, G.V.: Prog. Solid State Chem. **15**, 55 (1983)
85. Weller, M.T., Hughes, R.W., Rooke, J., Knee, Ch.S., Reading, J.: Dalton Trans., **2004**, 3032
86. Schuck, G., Kazakov, S.M., Rogacki, K., Zhigadlo, N.D., Karpinski, J.: Phys. Rev. B **73**, 144506 (2006)
87. Yamaura, J.I., Hiroi, Z., Tsuda, K., Izawa, K., Ohishi, Y., Tsutsui, S.: Solid State Comm. **149**, 31 (2009)
88. Beyerlein, R.A., Horowitz, H.S., Longo, J.M., Leonowicz, M.E.: J. Solid State Chem. **51**, 253 (1984)
89. Rouse, R.C., Dunn, P.J., Peacor, D.R., Wang, L.: J. Solid State Chem. **141**, 562 (1998)
90. Sleight, A.W., Zumsteg, F.C., Barkley, J.R., Gulley, J.E.: Mater.Res.Bull. **13**, 1247 (1978)
91. Kennedy, B.J. J.: Solid State Chem. **123**, 14 (1996)
92. Kobayashi, H., Kanno, R., Kawamoto, Y., Kamiyama, T., Izumi, F., Sleight, A.W.: J. Solid State Chem. **114**, 15 (1995)
93. Michel, C., Groult, D., Raveau, B.: J. Inorg. Nucl. Chem. **37**, 247 (1975)
94. Hiroi, Z., Yamaura, J.I., Yonezawa, S., Harima, H.: Physica C **460**(462), 20 (2007)
95. Kasahara, Y., Shimono, Y., Shibauchi, T., Matsuda, Y., Yonezawa, S., Muraoka, Y., Hiroi, Z.: Phys. Rev. Lett. **96**, 247004 (2006)
96. Sakai, H., Yoshimura, K., Ohno, H., Kato, H., Kambe, S., Walstedt, R.E., Matsuda, T.D., Haga, Y., Onuki, Y. J.: Phys.: Condens. Matter **13**, L785 (2001)
97. Jin, R., He, J., McCall, S., Alexander, C.S., Drymiotis, F., Mandrus, D.: Phys. Rev. B **64**, 180503 (2001)
98. Yonezawa, S., Muraoka, Y., Hiroi, Z.: J. Phys. Soc. Jpn. **73**, 1655 (2004)
99. Yonezawa, S., Muraoka, Y., Matsushita, Y., Hiroi, Z.: J. Phys. Soc. Jpn. **73**, 819 (2004)
100. Kazakov, S.M., Zhigadlo, N.D., Brühwiler, M., Batlogg, B., Karpinski, J.: Supercond. Sci. Technol. **17**, 1169 (2004)
101. Brühwiler, M., Kazakov, S.M., Zhigadlo, N.D., Karpinski, J., Batlogg, B.: Phys. Rev. B **70**, 020503 (2004)
102. Yonezawa, S., Muraoka, Y., Matsushita, Y., Hiroi, Z.: J. Phys.: Condens. Matter **6**, L9 (2004)
103. Nagao, Y., Yamaura, J.I., Ogusu, H., Okamoto, Y., Hiroi, Z.: J. Phys. Soc. Jpn. **78**, 064702 (2009)
104. Hiroi, Z., Hanawa, M.: J. Phys. Chem. Solids **63**, 1021 (2002)

105. Vyaselev, O., Arai, K., Kobayashi, K., Yamazaki, J., Kodama, K., Takigawa, M., Hanawa, M., Hiroi, Z.: Phys. Rev. Lett. **89**, 017001 (2002)
106. Shibayama, T., Nohara, M., Katori, H.A., Okamoto, Y., Hiroi, Z., Takagi, H.: J. Phys. Soc. Jpn. **76**, 073708 (2007)
107. Swenson, D., Chang, Y.Y.: Mat. Sci. Engineering B **39**, 52 (1996)
108. Boström, M., Rosner, H., Prots, Y., Burkhardt, U., Grin, Y.: Z. Anorg. Allg. Chem. **631**, 534 (2005)
109. Wakui, K., Akutagawa, S., Kase, N., Kawashima, K., Muranaka, T., Iwahori, Y., Abe, J., Akimitsu, J.: J. Phys. Soc. Jpn. **78**, 034710 (2009)
110. Harada, A., Tamura, N., Mukuda, H., Kitaoka, Y., Wakui, K., Akutagawa, S., Akimitsu, J.: J. Phys. Soc. Jpn. **78**, 025003 (2009)
111. Bodak, O.I., Marusin, E.P.: Dopov. Akad. Nauk Ukr. RSR (Ser. A), 1048 (1979)
112. Kotsanidis, P., Jakinthos, J.K., Gamari-Seale, E.: J. Less-Common Met. **152**, 287 (1989)
113. Sung, H.H., Chou, S.Y., Syu, K.Y., Lee, W.H.: J. Phys.: Cond. Mat. **20**, 165207 (2008)
114. Lee, W.H., Zeng, H.K., Yao, Y.D., Chen, Y.Y.: Physica C **266**, 138 (1996)
115. Iwamoto, Y., Iwasaki, Y., Ueda, K., Kohara, T.: Phys. Lett. A **250**, 439 (1998)
116. Hillier, A.D., Quintanilla, J., Cywinski, R.: Phys. Rev. Lett. **102**, 117007 (2009)
117. Goll, G., Marz, M., Hamann, A., Tomanic, T., Grube, K., Yoshino, T., Takabatake, T.: Physica B **403**, 1065 (2008)
118. Pikul, A., Gnida, D.: Solid State Commun. **151**, 778 (2011)
119. Samwer, K., Löhneysen, H.v.: Phys. Rev. B **26**, 107 (1982)
120. Klimczuk, T., Xu, Q., Morosan, E., Thompson, J.D., Zandbergen, H.W., Cava, R.J.: Phys. Rev. B **74**, 220502 (2006)
121. Klimczuk, T., Ronning, F., Sidorov, V., Cava, R.J., Thompson, J.D.: Phys. Rev. Lett. **99**, 257004 (2007)
122. Carbotte, J.P.: Rev. Mod. Phys. **62**, 1027 (1990)
123. Bonalde, I., Ribeiro, R.L., Brämer-Escamilla, W., Mu, G., Wen, H. H.: Phys. Rev. B **79**, 052506 (2009)
124. Hulm, J.K., Blaugher, R.D.: J. Phys. Chem. Solids **19**, 134 (1961)
125. Blaugher, D., Taylor, A., Hulm, J.K.: IBM J. Res. Dev. **6**, 116 (1962)
126. Jing, Yan., Lei, Shan., Qiang, Luo., Wei-Hua, Wang., Hai-Hu, Wen.: Chinese Physics B **18**, 704 (2009)
127. Zuev, Y.L., Kuznetsova, V.A., Prozorov, R., Vannette, M.D., Lobanov, M.V., Christen, D.K., Thompson, J.R.: Phys. Rev. B **76**, 132508 (2007)

Chapter 2
Non-centrosymmetric Heavy-Fermion Superconductors

N. Kimura and I. Bonalde

Abstract In this chapter we discuss the physical properties of a particular family of non-centrosymmetric superconductors belonging to the class heavy-fermion compounds. This group includes the ferromagnet UIr and the antiferromagnets $CeRhSi_3$, $CeIrSi_3$, $CeCoGe_3$, $CeIrGe_3$ and $CePt_3Si$, of which all but $CePt_3Si$ become superconducting only under pressure. Each of these superconductors has intriguing and interesting properties. We first analyze $CePt_3Si$, then review $CeRhSi_3$, $CeIrSi_3$, $CeCoGe_3$ and $CeIrGe_3$, which are very similar to each other in their magnetic and electrical properties, and finally discuss UIr. For each material we discuss the crystal structure, magnetic order, occurrence of superconductivity, phase diagram, characteristic parameters, superconducting properties and pairing states. We present an overview of the similarities and differences between all these six compounds at the end.

2.1 $CePt_3Si$

The enormous interest in superconductors without inversion symmetry started with the discovery of superconductivity in the heavy-fermion compound $CePt_3Si$ [1], which exhibits long-range antiferromagnetic (AFM) order below the Neel temperature $T_N = 2.2\,\mathrm{K}$ and becomes superconducting at the critical temperature $T_c = 0.75\,\mathrm{K}$. $CePt_3Si$ is the only known heavy-fermion (HF) compound without

N. Kimura (✉)
Center for Low Temperature Science,
Tohoku University,
Sendai Miyagi 980-8578, Japan
email: kimura@mail.clts.tohoku.ac.jp

I. Bonalde
Centro de Física, Instituto Venezolano de Investigaciones Científicas,
1020-A, Caracas, 20632 Apartado, Venezuela
email: ijbonalde@gmail.com

E. Bauer and M. Sigrist (eds.), *Non-centrosymmetric Superconductors*,
Lecture Notes in Physics 847, DOI: 10.1007/978-3-642-24624-1_2,

Table 2.1 Normal and superconducting parameters of $CePt_3Si$

Crystal structure	Tetragonal
Space group	$P4mm$
Lattice parameters	$a = 4.072$ Å $c = 5.442$ Å
Sommerfeld value of specific heat at T_c	$\gamma_n = 300 - 400$ mJ/mol K^2
Effective electron mass (Fermi sheet α)	$m^* \sim 11 - 23\ m_0$
Mean free path	$l = 1200 - 2700$ Å
Antiferromagnetic transition temperature	$T_N = 2.2$ K
Antiferromagnetic propagation vector	$\mathbf{q} = (0, 0, 1/2)$
Staggered magnetic moment $\mathbf{m}_Q$ along	[100]
Ordered moment per Ce atom	$\mu_s = 0.16\ \mu_B$
Superconducting transition temperature	$T_c = 0.75$ K (or 0.5 K ?)
Specific heat jump at T_c	$\Delta C/\gamma_n T_c \approx 0.25$
Upper critical field (small anisotropy)	$H_{c2}(0) \sim 3$ T
Thermodynamic critical field	$H_c(0) = 26$ mT
Ginzburg-Landau coherence length	$\xi(0) \sim 104$ Å
Ginzburg-Landau parameter	$\kappa = 82$
London penetration depth	$\lambda(0) \approx 0.86\ \mu$m
Nodal structure	Line nodes

inversion symmetry that superconducts at ambient pressure, as opposed to the cases of $CeIrSi_3$, $CeRhSi_3$, $CeCoGe_3$, $CeIrGe_3$ and UIr. The heavy-fermion character has been established from the large Sommerfeld coefficient $\gamma_n \approx 0.39$ J/K^2 mol [1]. The coexistence of antiferromagnetism and superconductivity has been proved by zero-field muon-spin relaxation [2] and neutron scattering [3]. Although $CePt_3Si$ has been intensively studied both theoretically and experimentally many questions regarding its superconducting state remain unresolved, in part because the observed properties have some sample dependence. We will focus here on the superconducting properties of $CePt_3Si$, giving only a brief review on normal state and magnetic behaviors (a detailed review of these can be found in Refs. [4, 5]).

2.1.1 Crystal Structure, Sample Growth and Characteristic Parameters

$CePt_3Si$ crystallizes in a tetragonal crystal structure with space group $P4mm$(No. 99) without inversion symmetry [1]. The lattice parameters are listed in Table 2.1. The unit cell has one formula unit with one Ce, one Si and two Pt inequivalent sites. The absence of inversion symmetry comes from the missing mirror plane $(0, 0, \frac{1}{2})$ (see Fig. 2.1). The antiferromagnetic lattice has an ordered wave vector

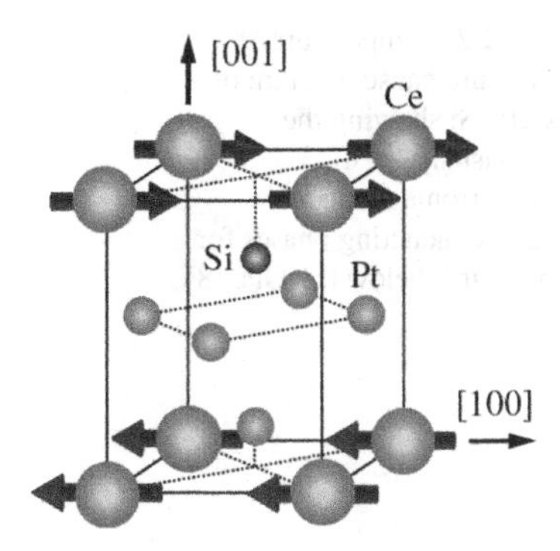

Fig. 2.1 Crystal and magnetic structures of $CePt_3Si$

(0,0,1/2), with a magnetic moment oriented ferromagnetically along the axis [100] and antiferromagnetically along the axis [001], as indicated in Fig. 2.1.

The melting temperature of $CePt_3Si$ is about 1390 °C [6]. An isothermal section of the Ce-Pt-Si phase diagram at 600 °C was presented by Gribanov et al. [6], who indicated that the interaction of Ce, Pt and Si leads to the formation of at least nine stable ternary phases. Seven of these ternary phases have a fixed composition. Polycrystalline samples of $CePt_3Si$ have been grown by argon arc melting and high-frequency melting and single crystals by the Bridgman and high-frequency techniques. These samples are usually annealed under high vacuum around 900 °C for 2–3 weeks. Interestingly, the annealing process has been linked to a Si excess [7]. Growing very high quality single crystals of $CePt_3Si$ has taken a long path that has led to the resolution of most of the problems in identifying the true superconducting properties of this compound.

2.1.2 Normal State

2.1.2.1 Phase Diagram and Magnetic Properties

Figure 2.2 shows the temperature-pressure phase diagram of $CePt_3Si$ determined by specific heat, resistivity and ac magnetic susceptibility measurements [8]. The antiferromagnetic T_N and superconducting T_c transition temperatures decrease with increasing pressure and become zero around $P_{AF} = 0.7$ GPa and $P_c = 1.6$ GPa, respectively.

The phase diagram indicates that there are two distinct superconducting phases: one below P_{AF} coexisting with the AFM phase and another above P_{AF} being presumably the only ordered phase. The coexistence of the superconducting and AFM phases was confirmed by neutron-scattering measurements that clearly show two superlattice peaks below and above the superconducting critical temperature (see Fig. 2.3) [3]. The observed peaks (0 0 1/2) and (1 0 1/2) correspond to an AFM vector $Q_0 = (0\ 0\ 1/2)$. The magnetic structure consists of ferromagnetic sheets of rather small Ce moments of 0.16 μ_B/Ce stacked antiferromagnetically along the

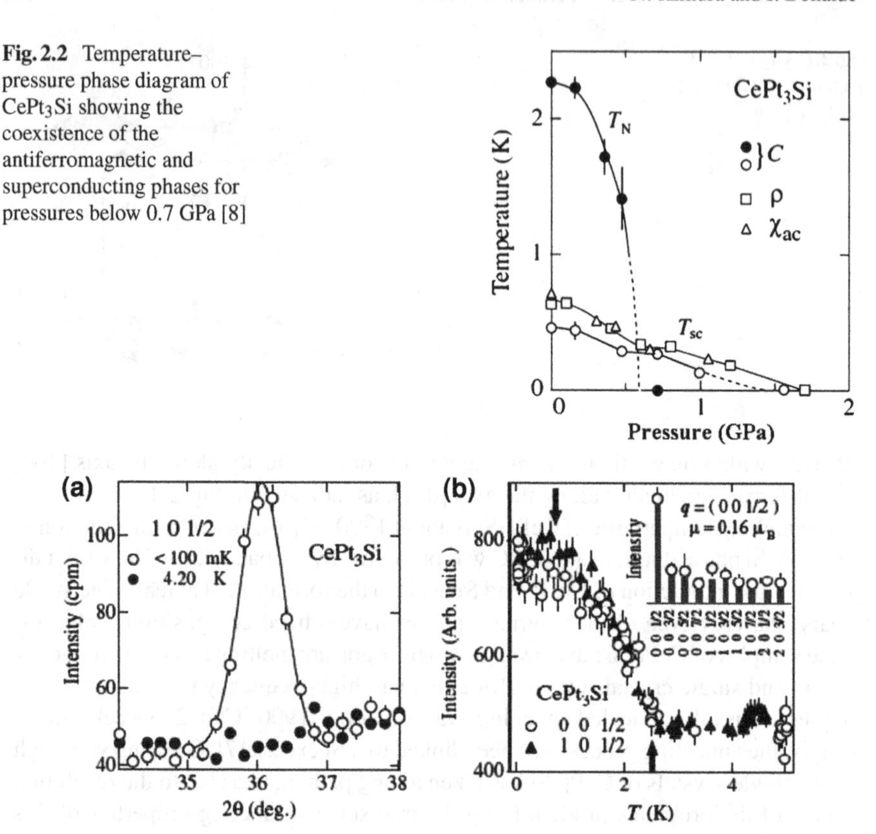

Fig. 2.2 Temperature–pressure phase diagram of $CePt_3Si$ showing the coexistence of the antiferromagnetic and superconducting phases for pressures below 0.7 GPa [8]

Fig. 2.3 **a** (101/2) AFM Bragg reflection observed below 0.1 K (*open circles*) and the background measured at 4.2 K (*solid circles*). **b** The intensity of (001/2) and (101/2) magnetic reflections as a function of temperature, shown *by open circles* and *solid triangles*, respectively. Up and down pointing arrows indicate T_N and T_c, respectively [3]

c axis (see Fig. 2.1). Here, μ_B is the Bohr magneton. The small value of the moment relative to 2.54 μ_B/Ce of Ce^{3+} may partially be explained through the itinerant character of Ce 4f-electrons involved in the formation of the heavy quasiparticles. In parts the reduction of the moment is also due to the Kondo screening effect viewing these electrons as almost localized moments [1]. In general, the magnetic response of $CePt_3Si$ in the normal state involves also the interplay of crystal electric field splitting of the 4f-orbitals and Kondo interaction (see, for example, Ref. [5]).

2.1.3 Superconducting State

At ambient pressure superconductivity in $CePt_3Si$ appears within a Fermi-liquid state, as evidenced by quantum oscillations [9] and resistivity measurements [1, 10].

However, $CePt_3Si$ becomes a non-Fermi-liquid under pressure, as indicated by the linear temperature behavior of the resistivity above 0.4 GPa [10].

Most of the unusual superconducting properties found initially in $CePt_3Si$ have been clarified by now, but others have appeared. First results, like second anomalies or a small peak just below the superconducting transition in the NMR $1/T_1T$, are not observed in the latest measurements carried out on new single crystals. These early results were associated with sample dependence: sample preparation, off-stoichiometry, impurity phases and/or annealing conditions. The new puzzling feature is a transition temperature that falls sometimes below 0.5 K [11–14].

Several theoretical approaches have been developed to try to understand this superconductor [15–19]. The models take into account the splitting of the spin-degenerate bands caused by the absence of inversion symmetry. Antiferromagnetic order effects in the heavy-fermion superconductors are also considered [18, 19]. Even though much experimental and theoretical efforts have been dedicated to clarify its physics, $CePt_3Si$ continues to be the most interesting and challenging of all superconductors without spatial inversion symmetry.

2.1.3.1 Probing the Pairing Symmetry

Several techniques have been employed to test the Cooper pairing state in non-centrosymmetric $CePt_3Si$. We will review what has been done thus far.

Spin State

One of the most direct probes of the spin state of the pairing is the ratio of the superconducting to the normal electron-spin paramagnetic susceptibility, χ_s and χ_n, respectively, which for a spin-singlet pairing state may be written as

$$\frac{\chi_s}{\chi_n} = -2\int_0^\infty d\xi \left\langle \frac{\partial f(E)}{\partial E} \right\rangle_{\hat{k}}. \tag{2.1}$$

The integration is over ξ, the energy of the free electrons relative to the Fermi level, f is the Fermi distribution function, and $E = \sqrt{\xi^2 + \Delta(\hat{k})^2}$ is the energy of the quasiparticles. The bracket $\langle \cdots \rangle_{\hat{k}}$ denotes the angular average. Here, $\Delta(\hat{k})$ is the energy gap that in general depends on the momentum direction $\hat{k}$ and temperature. For $T << T_c$ the ratio χ_s/χ_n goes as $(1/\sqrt{T})\exp(-\Delta_0/k_BT)$ for an s-wave pairing state with an isotropic gap and as T^2 for a d-wave pairing state where the gap has line nodes (Fig. 2.4(a)). Δ_0 is the zero-temperature value of Δ and k_B is the Boltzmann constant.

For spin-triplet pairing the ratio χ_s/χ_n is more complicated and depends on the field orientation. In the most simple cases, the susceptibility does not change across the transition if the field is perpendicular to the **d** vector denoting the spin-triplet gap function, but decreases continuously down to zero at $T = 0$ if the applied field is parallel to the **d** vector (Fig. 2.4(b)).

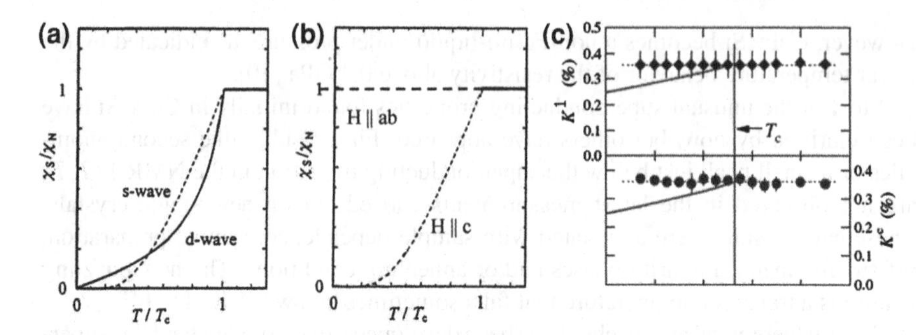

Fig. 2.4 Electronic spin susceptibility expected in **a** spin-singlet states *s*-wave and *d*-wave and **b** spin-triplet states. **c** Experimental electronic spin susceptibility of $CePt_3Si$ showing no change across the superconducting transition for all orientations of the applied magnetic field (*upper panel*: $\mathbf{H} \perp \hat{z}$ and *lower panel* $\mathbf{H} \parallel \hat{z}$) [20]

In $CePt_3Si$ the experimental result of χ_s/χ_n is puzzling: χ_s/χ_n does not change at all in the superconducting phase for any orientation of the field as shown in Fig. 2.4(c) [20]. Model calculations including a sizable ASOC characteristic for the non-centrosymmetric $CePt_3Si$ predict that for both spin-singlet and spin-triplet states $\chi_s(0)/\chi_n \rightarrow 1$ for fields parallel to the z axis and $\chi_s(0)/\chi_n \rightarrow 1/2$ for fields perpendicular to z [21, 22]. On the other hand, if electron correlations are included, χ_s/χ_n can be calculated to be constant across the transition independently of the field orientation [18, 19]. Thus, in $CePt_3Si$ spin-susceptibility measurements are not quite useful to distinguish between spin-singlet and spin-triplet pairings.

The upper critical field H_{c2} can also be used to get information about the spin configuration of a pairing state. A magnetic field induces pair breaking via paramagnetic and orbital mechanisms. The Pauli paramagnetic limiting field H_P can be estimated by comparing the (zero-field) superconducting condensation energy with the Zeeman energy

$$\frac{1}{2}(\chi_n - \chi_s)H_P^2 = \frac{1}{2}N_0\Delta_0^2, \tag{2.2}$$

where N_0 is the density of states at the Fermi energy. The Pauli susceptibility χ_n is given by $\chi_n = (g\mu_B)^2 N_0/2$, where g is the gyromagnetic ratio. H_P is then derived as

$$H_P = \frac{\sqrt{2}\Delta_0}{g\mu_B\sqrt{1-\chi_s/\chi_n}}. \tag{2.3}$$

As discussed above, for spin-singlet superconductors χ_s goes to zero as $T \longrightarrow 0$. Then, using the BCS value $\Delta_0 = 1.76k_B T_c$ for a weak-coupling superconductor and $g = 2$ for free electrons, we obtain the well-known estimate

$$H_P^{BCS} = H_P(0) = 1.86T_c[\mathrm{T/K}]. \tag{2.4}$$

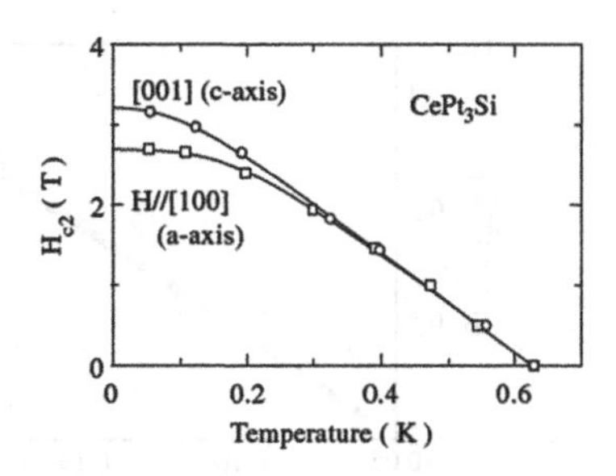

Fig. 2.5 Temperature dependence of the upper critical field in $CePt_3Si$ showing a weakly anisotropic behavior [10]

This expression is also valid for spin-triplet superconductors if the field is applied parallel to the **d** vector. On the other hand, if the field is applied perpendicular to the **d** vector $\chi_s = \chi_n$ as $T \to 0$ and Eq. (2.3) yields $H_P \longrightarrow \infty$. This would imply the absence of the Pauli paramagnetic limiting effect for fields along the *ab* plane.

Measurements on high-quality single crystals of $CePt_3Si$ show a weak anisotropy for the upper critical field $H_{c2}(0)$ with a value around 3 T (Fig. 2.5) [10] that exceeds the standard BCS weak-coupling paramagnetic limit $H_P(0) \approx 1$ T.

From Eq. (2.3) and the predictions for χ_s/χ_n in the non-centrosymmetric superconductors, no limiting behavior is expected for fields parallel to the *c* axis, whereas $H_P(0) = \Delta_0/\mu_B \approx 1.4$ T for fields in the *ab* plane. As mentioned above, modifications of χ_s due to correlation effects as well as magnetic ordering could eliminate the paramagnetic limiting for all field directions [18, 19]. It was also suggested that the realization of a so-called helical phase for fields perpendicular to the *c* axis would strongly reduce paramagnetic limiting effects [23, 24].

The orbital limiting field H_{orb} is expressed by

$$H_{orb}(T) = \frac{\Phi_0}{2\pi\xi^2(T)}, \tag{2.5}$$

where Φ_0 is the flux quantum. $H_{orb}(T=0)$ can be in principle obtained by using the BCS expression $\xi(0) = 0.18\hbar v_F/(k_B T_c)$, where v_F is the Fermi velocity. However, $H_{orb}(T=0)$ is usually estimated from the formula [25]

$$H_{orb}(T) = h(T) H'_{c2} T_c. \tag{2.6}$$

Here, $H'_{c2} \equiv -dH_{c2}/dT|_{T=T_c}$. $h(0) = 0.727$ for weak-coupling BCS superconductors in the clean limit. Using the data of Fig. 2.5 we estimate $H'_{c2} \approx -6.8$ T/K and from Eq. (2.6) $H^{BCS}_{orb} \approx 3.7$ T. This is about the value of $H_{c2}(0)$ for both field orientations, which suggests that $CePt_3Si$ may be restricted by the orbital depairing limit. In such a case the spin-triplet state should be favorable. Both paramagnetic and orbital mechanisms will be discussed in more detail in Sect. 2.2.

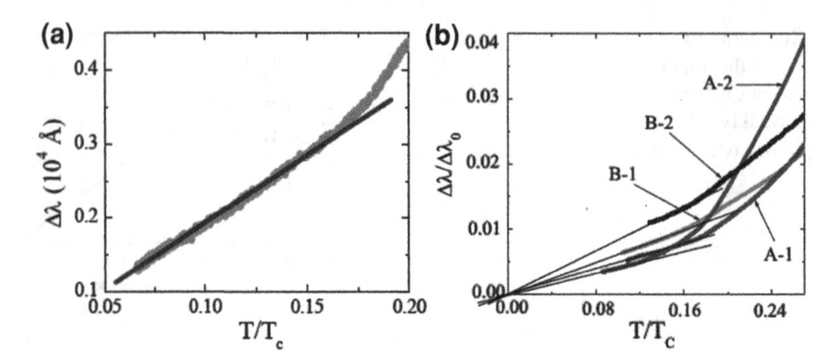

Fig. 2.6 Penetration depth in **a** a polycrystalline sample and **b** single crystals of $CePt_3Si$.The linear temperature behavior indicates line nodes in the energy gap. The single crystals used in the penetration depth measurements shown in **b** have different defect concentrations, which suggests that the linear behavior is unaffected by disorder [13]

Nodal Structure

The structure of the energy gap is directly related to the symmetry of the Cooper pairing. The energy gap is isotropic for s-wave spin-singlet superconductors and usually has zeroes (nodes) for other symmetries. Thus, the determination of the presence of nodes in the energy gap is crucial to establish pairing with symmetries lower than the s-wave. The existence of nodes in the energy gap leads to low-temperature power laws (T^n) in several superconducting properties, instead of the BCS exponential temperature response observed for an isotropically gapped excitation spectrum. In $CePt_3Si$ magnetic penetration-depth, thermal-conductivity and specific-heat measurements show power-law behaviors indicative of line nodes in the energy gap.

A linear temperature dependence of the magnetic penetration depth $\lambda(T)$ below 0.16 T_c was first found in a polycrystalline sample of $CePt_3Si$ (Fig. 2.6(a)) [26], and later also in single crystals [13]. In the local limit of the electrodynamics of superconductors, the magnetic penetration depth is given by

$$\left[\frac{\lambda^2(0)}{\lambda^2(T)}\right]_{ij} = \frac{n^s_{ij}(T)}{n} = 3\left\langle \hat{k}_i\hat{k}_j\left[1 - \int d\xi\left(\frac{-\partial f}{\partial E_{\mathbf{k}}}\right)\right]\right\rangle_{\hat{k}}. \tag{2.7}$$

In superconductors with inversion symmetry $\Delta\lambda(T) \propto T$ is expected in the low-temperature limit for an energy gap with line nodes [27]. Thus, the linear response in $CePt_3Si$ was taken as evidence for line nodes. Surprisingly, the temperature response in the penetration depth is not affected by the sample quality, as opposed to what occurs in unconventional superconductors with line nodes in the gap [13]. Figure 2.6(b) displays the low-temperature region of the penetration depth of several single crystals of different quality. A clear linear behavior is observed in all of them. We note that the superfluid density $\rho_s(T)$ shows a small anisotropy [28], in agreement with the upper critical field result.

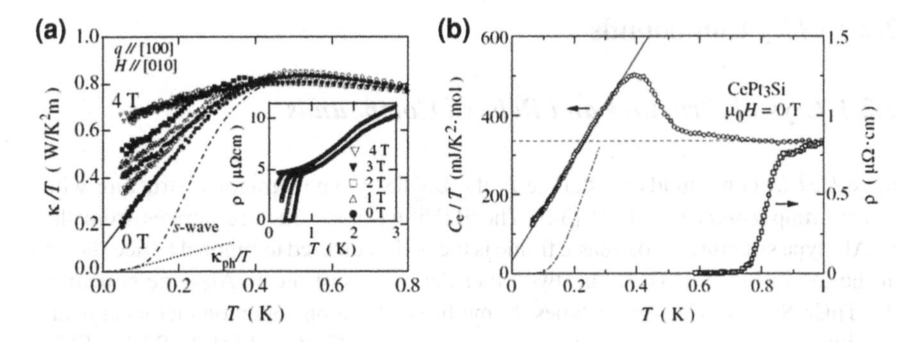

Fig. 2.7 Temperature dependence of **a** thermal conductivity [29] and **b** specific heat [30] of $CePt_3Si$. The linear temperature response suggests line nodes in the energy gap

Thermal transport measurements also suggest the presence of line nodes by showing a residual linear term in $\kappa(T)/T$ as $T \rightarrow 0$ (Fig. 2.7(a)). In superconductors with inversion symmetry such a linear term is expected when the energy gap has nodes, and is due to impurity scattering. The quasiparticle thermal conductivity has universal components in the low-temperature limit ($k_B T < \gamma$)

$$\kappa_{ii} = \frac{\pi^2}{3} N_0 v_F^2 T \left\langle \hat{k}_i \hat{k}_i \frac{\gamma^2}{(\gamma^2 + \Delta_{\mathbf{k}}^2)^{3/2}} \right\rangle_{\hat{k}} \tag{2.8}$$

that are linear functions of temperature at low T and whose proportionality constants depend on the specific form of the order parameter [27]. γ is the quasiparticle decay rate. In $CePt_3Si$ the experimental $\kappa(T)/T = 0.1$ W/(K^2· m) is in good agreement with the calculated universal conductivity limit 0.09 W/(K^2· m) [29]. Moreover, the H dependence of κ follows the prediction by a theory of Doppler-shifted quasiparticles in a superconductor with line nodes [29, 31].

In the low-temperature limit the electronic specific heat of $CePt_3Si$ has been found to follow the expression $C_{el}/T = A + BT$, with $A = 34.1$ mJ/(K^2· mol) and $B = 1290$ mJ/(K^3· mol) (Fig. 2.7(b)) [30]. In general, the electronic specific heat is given by [27]

$$C_{el} = 2N_0 \int_{-\infty}^{\infty} d\xi \left\langle E_{\mathbf{k}} \frac{\partial f}{\partial E_{\mathbf{k}}} \right\rangle_{\hat{k}}. \tag{2.9}$$

For superconductors with inversion symmetry, in the low-temperature limit $C_{el} \propto T^2$ for a gap with line nodes. The residual linear term in C_{el}/T of $CePt_3Si$ is considered to be caused by impurities or by electrons on part of the Fermi surface that do not participate in superconducting. Thus, the observed behavior of the electronic specific heat is taken as evidence for line nodes in the energy gap [30].

For a theoretical discussion on line nodes in non-centrosymmetric superconductors, see other chapters in this book.

2.2 CeTX_3 Compounds

2.2.1 Crystal Structure and Related Compounds

Most CeTX_3 compounds crystallize in the $BaNiSn_3$-type tetragonal structure with space group $I4mm$ (No.107) [32]. The $BaNiSn_3$-type structure derives from the $BaAl_4$-type structure whose basic frame is the body-centered tetragonal lattice shown at the top in Fig. 2.8. There are two other derivatives of the $BaAl_4$-type structure, the $ThCr_2Si_2$ and $CaBe_2Ge_2$ types. Some heavy-fermion superconductors crystallize into the former structure: e.g. $CeCu_2Si_2$ [33], $CeCu_2Ge_2$ [34], $CePd_2Si_2$ [35], $CeRh_2Si_2$ [36] and URu_2Si_2 [37]. The latter structure is often found in RPt_2X_2 compounds [38], where R denotes a rare-earth element. The $ThCr_2Si_2$– and $CaBe_2Ge_2$–type structures have an inversion center, while the $BaNiSn_3$-type structure does not. Fig. 2.8 displays the $BaAl_4$-type crystal lattice and its three derivatives.

The atomic framework of the $BaNiSn_3$-type structure can be alternatively displayed as a sequence of planes of the same atoms $R - T - X(1) - X(2) - R - T - X(1) - X(2) - R$ along the c axis of the tetragonal structure, where T and X denote a transition metal and Si/Ge, respectively. The point group of the $BaNiSn_3$-type structure is C_{4v}, which lacks the mirror plane and a two-fold axis normal to the c axis (z axis). Therefore, a Rashba-like spin-orbit coupling exists in this system, as it does in $CePt_3Si$ that also belongs to C_{4v} [1].

Of the CeTX_3 compounds, the series CeTSi$_3$ (T = Co, Ru, Rh, Pd, Os, Ir and Pt) and CeTGe$_3$ (T = Fe, Co, Rh and Ir) are known to crystallize in the $BaNiSn_3$-type structure. Among these, $CeRhSi_3$ [39], $CeIrSi_3$ [40], $CeCoGe_3$ [41] and $CeIrGe_3$ [42] have been found to be pressure-induced superconductors. Single crystals of $CeRhSi_3$ and $CeIrSi_3$, as well as of $CeRuSi_3$ [43], can be obtained by the Czochralski pulling method in a tetra-arc furnace using a pulling speed of 10 mm/h or 15 mm/h. In these compounds annealing treatments at 900 °C, in vacuum, for a week are usually very effective. Notably, for $CeRhSi_3$ the use of stoichiometric amounts of the components sometimes yields a different crystal; e.g., $CeRhSi_2$. To obtain a crystal of $CeRhSi_3$ an off-stoichiometric composition, typically Ce:Rh:Si = 1:1:3.3, works better. Single crystals of $CeCoGe_3$ are obtained by the Bi-flux method (the Czochralski method is unsuitable) [44]. In this procedure, arc-melt-prepared ingots of $CeCoGe_3$ and Bi are placed in an alumina crucible and heated up to 1050 °C in an argon atmosphere. After keeping this temperature for a day, the crucible is cooled down to 650 °C over a period of two weeks and then down to room temperature rapidly. Single crystals of $CeRhGe_3$ can also be obtained by the Bi-flux technique [43]. Single crystals of $CeIrGe_3$ are grown by the Bi- and Sn-flux methods [42, 43].

Other CeTX_3 compounds with $BaNiSn_3$-type structure are CeTAl$_3$ (T = Cu, Au) and $CeAuGa_3$. They order magnetically at low temperatures [45, 46, 47, 48, 49]. In particular, $CeCuAl_3$ and $CeAuAl_3$ are suggested to be HF compounds with AFM ground states [46] and have the potential to become non-centrosymmetric HF superconductors. There are at least two CeTX_3 compounds that do not have the $BaNiSn_3$-type crystal lattice: $CeNiGe_3$, which has an orthorhombic structure and

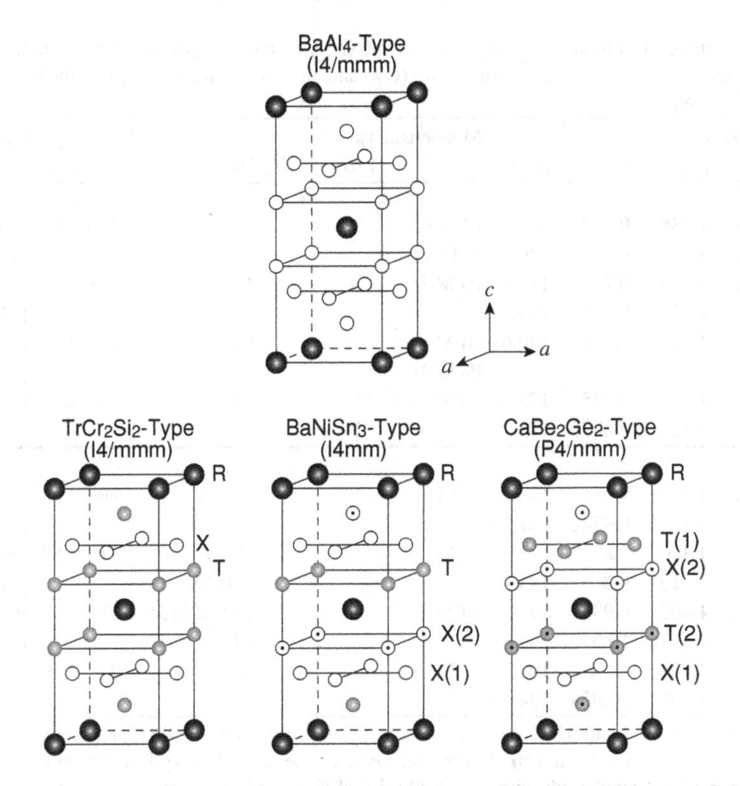

Fig. 2.8 $BaAl_4$-type crystal structure and its three derivatives $TrCr_2Si_2$, $BaNiSn_3$ and $CaBe_2Ge_2$. Only the $BaNiSn_3$-type structure does not have an inversion center

exhibits superconductivity under pressure [50], and $CeRuGe_3$, which is reported to have either orthorhombic [51] or cubic structure [52].

On the other hand, no uranium-based compound (UTX_3) has been fully confirmed to have the $BaNiSn_3$-type structure. Only $UNiGa_3$, that exhibits AFM order at 39 K, seems to have this structure [53].

Last, we briefly comment on non-heavy-fermion $LaTX_3$ compounds. Some of them, like $LaRhSi_3$, $LaIrSi_3$ and $LaPdSi_3$, also have $BaNiSn_3$-type structures and show superconductivity at the critical temperature 1.9, 0.9 and 2.6 K, respectively [32, 54, 55]. In these materials, however, no evidence for unconventional behavior has been observed [54].

2.2.2 Normal State

2.2.2.1 Magnetic Properties

The $CeTX_3$ compounds exhibit various magnetic ground states as summarized in Table 2.2. The magnetic ground states vary from AFM to intermediate valence

Table 2.2 Unit-cell volume (V), magnetic ground state, electronic specific-heat coefficient γ_n, ordering temperature (T_N), Weiss temperature (Θ_p) and effective moment (μ_{eff}) for the $BaNiSn_3$-type $CeTX_3$ compounds

Compound	a [Å]	c [Å]	V [Å^3]	Magnetism	γ_n [mJ/mol·K^2]	T_N [K]	Θ_p [K]	μ_{eff} [μ_B]	Ref.
*$CeTSi_3$									
$CeCoSi_3$	4.135	9.567	163.6	PM(IV)	37	–	−840	2.80	[59]
$CeRuSi_3$	4.21577	9.9271	176.43	PM(IV)		–			[43]
$CeRhSi_3$	4.269	9.738	177.5	AFM(HF)	110	1.6	−128	2.65	[60, 55]
	4.237	9.785	175.7						[61]
$CePdSi_3$	4.33	9.631	180.6	AFM	57	5.2/3	−26	2.56	[62, 55]
$CeOsSi_3$				PM(IV)					[63]
$CeIrSi_3$	4.252	9.715	175.6	AFM(HF)	120	5.0	−142	2.48	[60, 55]
$CePtSi_3$	4.3215	9.6075	179.42	AFM	29	4.8/2.4			[64]
*$CeTGe_3$									
$CeFeGe_3$	4.332	9.955	186.8	PM(HF)	150	–	−90	2.6	[65]
	4.3371	9.9542	187.24						[66]
$CeCoGe_3$	4.320	9.835	183.5	AFM	32	21/19	−51	2.54	[59]
	4.319	9.829	183.3			21/12/8			[44]
$CeRhGe_3$	4.402	9.993	193.6	AFM	40	14.6/10/0.55	−28	2.53	[60]
	4.3976	10.0322	194.01			14.9/8.2			[43]
$CeIrGe_3$	4.409	10.032	195.0	AFM	80	8.7/4.7/0.7	−21	2.39	[60]
	4.401	10.024	194.2			8.7/4.8			[43]

The abbreviations PM and AFM denote paramagnetic and antiferromagnetic ground states, respectively. The abbreviations IV and HF denote intermediate-valence and heavy-fermion states, respectively. We consider compounds with $\gamma_n > 100$ mJ/mol·K^2 to be heavy-fermion systems

(IV) through HF states with decreasing unit-cell volume V. For example, $CeCoGe_3$ ($V = 183.3$ Å^3) displays an AFM ground state [44, 56, 57], while $CeCoSi_3$ ($V = 163.6$ Å^3) is thought to be an IV compound [58].

Figure 2.9 shows the Néel temperature T_N and the electronic specific-heat coefficient γ_n as a function of unit-cell volume for $CeTX_3$ in which T belongs to the Group 9 (Co, Rh and Ir) in the Periodic Table [43]. T_N approximately follows a simple curve which peaks at 186 Å^3. This behavior supports the Doniach model in which the on-site Kondo effect dominates over the inter-site RKKY interaction with the coupling constant J being effectively enhanced relative to the kinetic energy with decreasing unit-cell volume [43]. γ_n is also described by a simple curve which peaks at the unit-cell volume 176 Å^3 at which T_N goes to zero, suggesting that the γ_n value is enhanced by the magnetic fluctuation arising at the corresponding volume.

Besides the Group 9 (Co, Rh, Ir) compounds, one can consider $CeFeGe_3$ to be a potential superconductor because its γ_n is comparable to those of $CeRhSi_3$ and $CeIrSi_3$. However, in $CeFeGe_3$ superconductivity has not been observed down to 0.05 K [65].

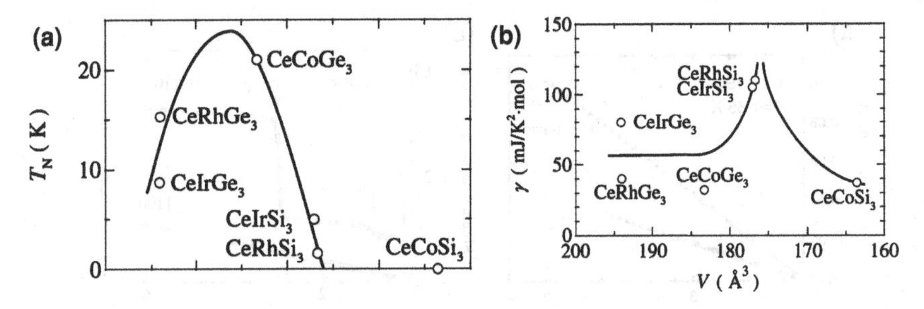

Fig. 2.9 Unit-cell volume dependence of **a** the Néel temperature and **b** the γ_n value in Ce*T*Si$_3$ and Ce*T*Ge$_3$ (*T*: Co, Rh, Ir) [43]

CeRhSi$_3$

As shown in Figs. 2.10(a) and 2.11(d), the magnetic properties of CeRhSi$_3$ are anisotropic especially at low temperatures [55, 67]. The induced magnetization along the easy axis [100] has a quite small value of 0.1 μ_B at 7 T. The magnetic susceptibility curves for H parallel to the a and c axes show a strong anisotropy at low temperatures, while they obey the Curie-Weiss law above about 150 K. The effective moments μ_{eff} for both field directions are 2.65μ_B, which is close to the expected value for the Ce^{3+} ion. The Weiss temperatures are negative and very large (−112 and − 160 K for H parallel to the a and c axes, respectively), as often found for IV compounds. The susceptibility for $H \parallel c$ has a broad peak around 50 K that is characteristic of HF compounds.

From the jump of the specific heat at T_N, the magnetic entropy gain is estimated to be only 12% of $R \ln 2$ [60]. The AFM state is robust against a magnetic field (Fig. 2.12(a)) and survives up to 8 T, although such a strong field should be sufficient to suppress an AFM state with a low T_N of the order of 1 K. The temperature and height of the specific-heat peak decrease with increasing field along the easy axis ($H \parallel a$). The magnetic contribution to the specific heat when the field is aligned with the hard axis ($H \parallel c$) does not change even at 8 T [67]. The tiny entropy gain and the insensitivity to a magnetic field are attributed to the strong Kondo screening of the 4f electron. The Kondo temperature is estimated to be about 50 K [60].

The magnetic structure at ambient pressure is revealed by neutron experiments to be a longitudinal spin-density-wave (LSDW) type characterized by the propagation vectors $Q = (\pm 0.215, 0, 0.5)$ with polarization along the a^* axis [69]. The magnetic structure is shown in Fig. 2.13. The staggered moment 0.13 μ_B is quite small [70]. The incommensurate LSDW structure with such a strongly suppressed moment suggests that itinerant-electron magnetism is realized in CeRhSi$_3$. The 4f electrons are expected to be strongly hybridized with the conduction electrons through the Kondo effect, leading to the formation of the SDW state caused by nesting of the Fermi surface.

It is not obvious whether this magnetic structure persists under pressure. In the pressure-dependent specific-heat curve, a shoulder-like transition is seen below T_N at a pressure of 0.55 GPa [71]. The origin of the transition is unclear at present.

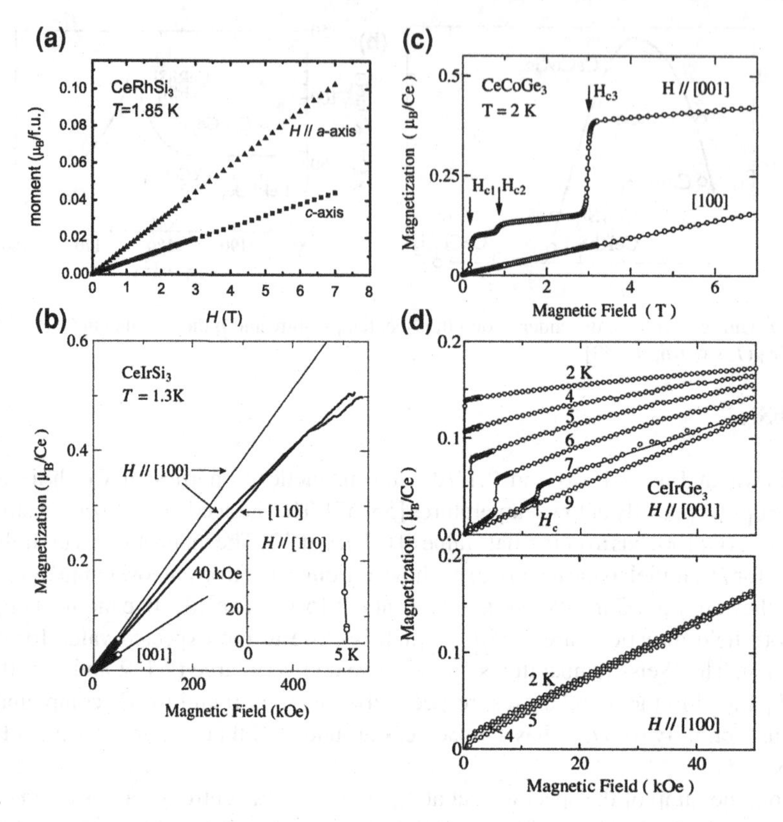

Fig. 2.10 Magnetization curves of $CeRhSi_3$, $CeIrSi_3$, $CeCoGe_3$ and $CeIrGe_3$ [44, 55, 68]

Considering that multiple magnetic transitions are observed in other magnetic CeTX_3 compounds, the magnetic structure realized in $CeRhSi_3$ at ambient pressure may change under pressure where superconductivity appears.

The electrical resistivity below T_N can be fitted by an antiferro-magnon model [72]. At sufficiently low temperatures ($T < 0.6$ K), the resistivity follows the Fermi-liquid description of $\rho(T) = \rho_0 + AT^2$, where ρ_0 is the residual resistivity. $A = 0.19\,\mu\Omega\cdot\text{cm/K}^2$ for the current J along the a axis and $A = 0.24\,\mu\Omega\cdot\text{cm/K}^2$ for $J \parallel c$. The ratios $A/\gamma_n^2 = 1.6\times10^{-5}$ $(J \parallel a)$, $2.0\times10^{-5}(J \parallel c)$ $\mu\Omega\cdot\text{cm}\cdot\text{K}^2\cdot\text{mol}\cdot\text{mJ}^{-2}$ are close to $1\times10^{-5}\mu\Omega\cdot\text{cm}\cdot\text{K}^2\cdot\text{mol}\cdot\text{mJ}^{-2}$ as given by the Kadowaki-Woods relation [73].

$CeIrSi_3$

Although the magnetic structure of $CeIrSi_3$ is unknown at present, the specific heat, magnetic susceptibility and electrical resistivity are similar to those of $CeRhSi_3$. The magnetization curve is anisotropic and a is the easy axis (Fig. 2.10(b)) [68]. The mag-

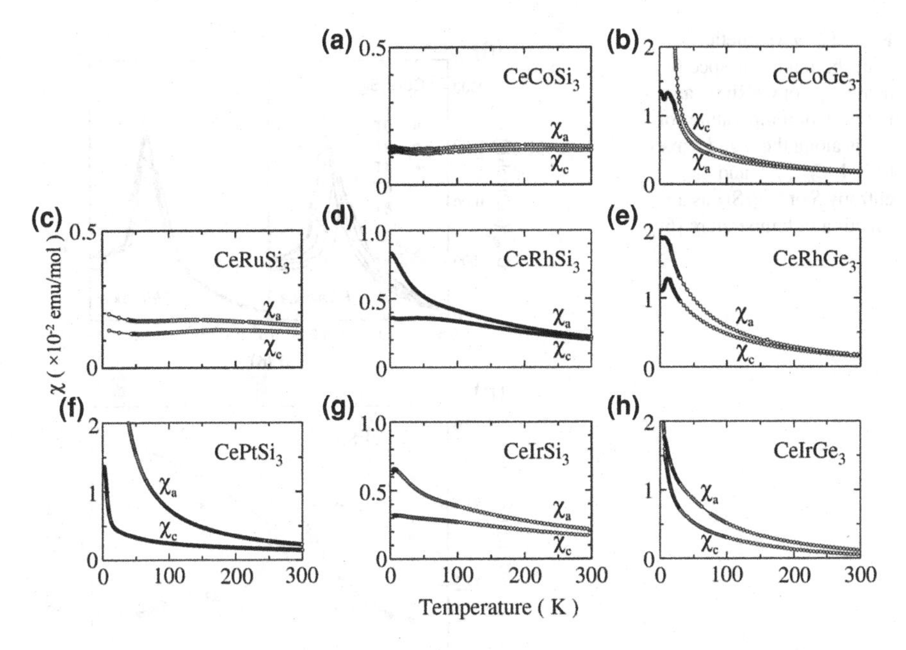

Fig. 2.11 Magnetic susceptibility of $CeTX_3$ compounds as a function of temperature [43]

netization increases almost linearly with magnetic field. Neither a saturation behavior nor metamagnetic transitions are observed for fields in the basal plane up to 50 T. The induced magnetization is very small and comparable to that of $CeRhSi_3$; $0.1\mu_B$ at 10 T for the easy axis. The temperature dependence of the magnetic susceptibility is anisotropic at low temperatures (Fig. 2.11(g)) [40, 68].

The entropy gain associated with the AFM transition is small, $0.2R\ln 2$, as in $CeRhSi_3$ (Fig. 2.12(b)) [68]. The AFM state is robust against a magnetic field, as shown in the inset of Fig. 2.10(b). The electrical resistivity below T_N can be fitted by an antiferro-magnon model as done in $CeRhSi_3$. The coefficient of the T^2 term $A = 0.04\,\mu\Omega\cdot\text{cm/K}^2$ for $J \parallel c$ and $J \perp c$ [68] is about one fifth of the coefficients of $CeRhSi_3$. The ratio $A/\gamma_n^2 = 3.6\times10^{-6}\mu\Omega\cdot\text{cm}\cdot\text{K}^2\cdot\text{mol}\cdot\text{mJ}^{-2}$ is much smaller than the ratios in $CeRhSi_3$ but still in the range of the Kadowaki-Woods relation.

$CeCoGe_3$

Unlike $CeRhSi_3$ and $CeIrSi_3$, $CeCoGe_3$ exhibits three successive AFM transitions [44]. Correspondingly, the magnetization for $H \parallel c$ shows three-step metamagnetic transitions. It reaches $M_s/4$, $M_s/3$ and M_s, where $M_s = 0.43\mu_B/\text{Ce}$, at each transition. The anisotropic magnetization curve indicates an Ising-like magnetism

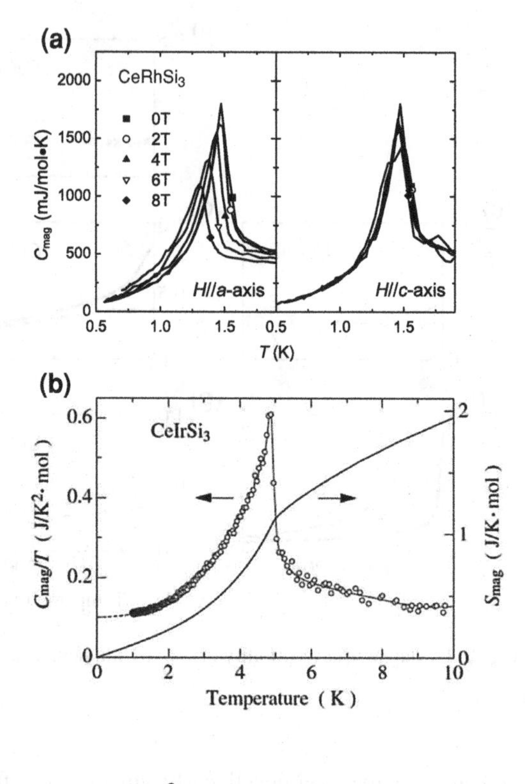

Fig. 2.12 **a** Magnetic contribution to the specific heat C_{mag} of $CeRhSi_3$ as a function of temperature for fields along the a and c axes [67]. **b** C_{mag}/T and the entropy S of $CeIrSi_3$ as a function of temperature [68]

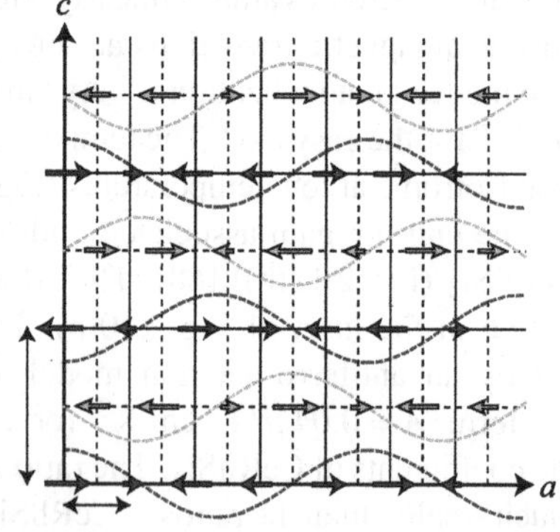

Fig. 2.13 Magnetic structure of $CeRhSi_3$. Only the Ce sites at (0, 0, 0) and (0.5, 0.5, 0.5) positions are projected on the ac plane. The arrows are depicted to show the size and the direction of the magnetic moment at the Ce sites [70]

with the easy axis along the c axis, which is different from what is found in $CeRhSi_3$ and $CeIrSi_3$ as seen in Fig. 2.11.

The neutron-diffraction experiments revealed that the magnetic structure of the ground state at ambient pressure consists of two components with dominant $q_1 = (0, 0, 1/2)$ and subordinate $q_2 = (0, 0, 3/4)$ [74]. Figure 2.14 shows a possible magnetic structure of the q_1 sublattice. The magnetic moments are parallel to the c axis and alternate in the up-up-down-down sequence. The magnitude of the mag-

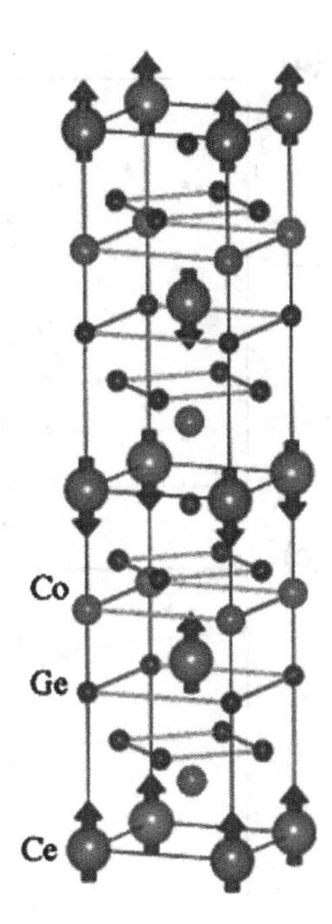

Fig. 2.14 Possible magnetic structure for $q_1 = (0, 0, 0.5)$ sublattice in the ground state of $CeCoGe_3$ [74]

netic moment in the sublattice μ_1 is estimated to be $0.5(1)\mu_B$. A more complete magnetic structure that includes the q_2 sublattice is not clear at present.

The magnetic susceptibility of $CeCoGe_3$ does not show a peak or a saturated behavior at low temperatures as observed in $CeRhSi_3$ and $CeIrSi_3$. Specific heat measurements revealed that the entropy gain reaches 68% of $R\ln 2$ at $T_{N1} = 21$ K and that 32% of entropy loss is recovered at 38 K. The magnetism of $CeCoGe_3$ is basically understood in terms of localized 4*f* electron.

$CeIrGe_3$

The magnetic structure of $CeIrGe_3$ seems to be more complicated. There are two successive magnetic transitions at $T_{N1} = 8.7$ K and $T_{N2} = 4.8$ K. The former is antiferromagnetic and the latter is unknown although magnetization measurements indicate weak ferromagnetism with a small moment below T_{N2} (see Fig. 2.10(d)). A parasitic ferromagnetism due to the Dzyaloshinsky-Moriya interaction caused by the broken space inversion symmetry is discussed in Ref. [43]. T_{N2} merges into T_{N1} with the application of pressure (see Fig. 2.15(d)).

2.2.2.2 Temperature-Pressure Phase Diagram

All known $CeTX_3$ superconductors need pressure to become superconducting. The temperature-pressure (*T*-*P*) phase diagrams of $CeRhSi_3$ [72], $CeIrSi_3$ [75], $CeCoGe_3$

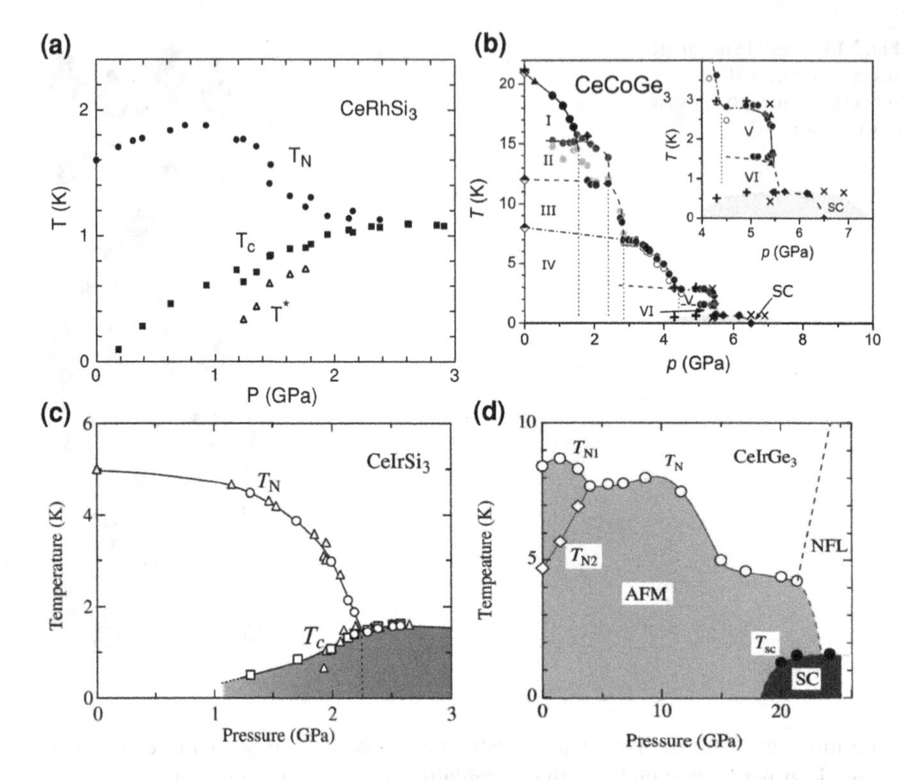

Fig. 2.15 Temperature-pressure phase diagrams of $CeRhSi_3$ [72], $CeIrSi_3$ [75], $CeCoGe_3$ [77] and $CeIrGe_3$ [42]. T^* in panel **a** denotes an anomaly observed in the resistivity and magnetic-susceptibility measurements [39]

[76] and $CeIrGe_3$ [42] are shown in Fig. 2.15. In $CeRhSi_3$, the Néel temperature T_N initially increases and subsequently decreases with applying pressure, while those of $CeIrSi_3$ decreases monotonically with pressure. The $T_N(P)$ of $CeCoGe_3$ and $CeIrGe_3$ exhibits step-like decreases probably relevant to successive magnetic phase transitions observed at ambient pressure. The superconducting transitions of these four compounds are observed at pressures at which the AFM order still exists. We define here three characteristic pressures: P_1^* where $T_N = T_c$, P_2^* where $T_N \rightarrow 0$ and P_3^* where the superconducting transition temperature T_c reaches a maximum. The values of these pressures are summarized in Table 2.3. P_2^* of $CeRhSi_3$ is unclear because T_N does not decrease steeply near P_1^*.

The resistivity drop at the superconducting transition of $CeRhSi_3$, $CeIrSi_3$ and probably $CeCoSi_3$ has its sharpest form at P_3^*. The resistivity drop in the AFM phase, especially far below P_3^*, is very broad. In this pressure region, the drop width depends on the applied current [39, 75]. These observations imply that superconductivity is inhomogeneous or fluctuating in the antiferromagnetic state and is optimum at P_3^*. Such a phenomenon is found in centrosymmetric HF superconductors as well

Table 2.3 Normal and superconducting parameters of CeTX_3 compounds

	$CeRhSi_3$	$CeIrSi_3$	$CeCoGe_3$	$CeIrGe_3$
Crystal structure			Tetragonal	
Space group			$I4mm$	
a [Å]	4.237	4.252	4.320	4.401
c [Å]	9.785	9.715	9.835	10.024
γ_n [mJ/mol· K^2]	110	120	32	80
$m^*[m_0]$	4 – 19	N/A	N/A	N/A
l [Å]	2400 – 3400	N/A	N/A	N/A
T_N [K]	1.6	5.0	21/12/8	8.7/4.8
$\mathbf{q}$	($\pm$0.215, 0, 0.5)	N/A	(0,0,1/2) (0,0,3/4)	N/A
$\mathbf{m_Q}$ orientation	[001]	[001]	[100]	[100]?
$\mu_s[\mu_B/\mathrm{Ce}]$	0.13	N/A	0.5	N/A
P_1^* (GPa)	2.4–2.5	2.25	$5.5 - 5.6^d$	> 21
P_2^* (GPa)	?	2.50	$5.5 - 5.7^d$	~ 24
P_3^* (GPa)	2.65	2.63	5.7^d or 6.5^e	≥ 24
T_c @ P_3^* [K]	1.09	1.56 – 1.59	0.66^d or 0.69^e	≥ 1.6
$\Delta C/\gamma_n T_c$	N/A	5.7	N/A	N/A
$H_{c2}(0)$ [T] ($H \parallel c$)	30 ± 2^a	45 ± 10^c	$22 \pm 8^{f,g}$	27 ± 10
$H_{c2}(0)$ [T] ($H \perp c$)	7.5^b	9.5^c	N/A	N/A
$\xi(0)$ [Å] ($H \parallel c$)	~ 33	~ 27	~ 39	~ 35
H'_{c2} [T/K] ($H \parallel c$)	23^b	17^c	20^d	16^h
H'_{c2} [T/K] ($H \perp c$)	27^b	14.5^c	N/A	N/A

$\xi(0)$ is estimated from Eq. (2.5). Here, we regard $H_{c2}(0)$ for $H \parallel c$ as H_{orb}; that is, the paramagnetic pair-breaking effect is absent. P_1^*, P_2^* and P_3^* are the pressures at which $T_N = T_c$, $T_N \to 0$ and T_c reaches a maximum, respectively. $\mathbf{q}$, $\mathbf{m_Q}$ and μ_s denote magnetic propagation vector, magnetic moment and value of ordered moment, respectively, and $H'_{c2} \equiv -dH_{c2}/dT|_{T=T_c}$

^{a} at 2.85 GPa

^{b} at 2.6 GPa

^{c} at 2.65 GPa

^{d} determined from heat-capacity measurements[77].

^{e} determined from electrical-resistivity measurements [43].

^{f} at 6.5 GPa

^{g} estimated from Fig. 2.25.

^{h} at 24 GPa

and, thus, seems to be realized irrespective of the presence or absence of inversion symmetry.

A similar evolution of superconductivity in the AFM state is seen by the heat capacity C (Fig. 2.16). The heat capacity jump (ΔC) at the superconducting transition below P_1^* is small and broad, while it becomes sharper with increasing pressure and is strongly enhanced above P_1^* [75]. In $CeIrSi_3$, as shown in Fig. 2.16, $\Delta C/C_n$, where C_n is the normal state value just above T_c, reaches 5.7 ± 0.1 at 2.58 GPa, which is much larger than the 1.43 value expected from the weak-coupling

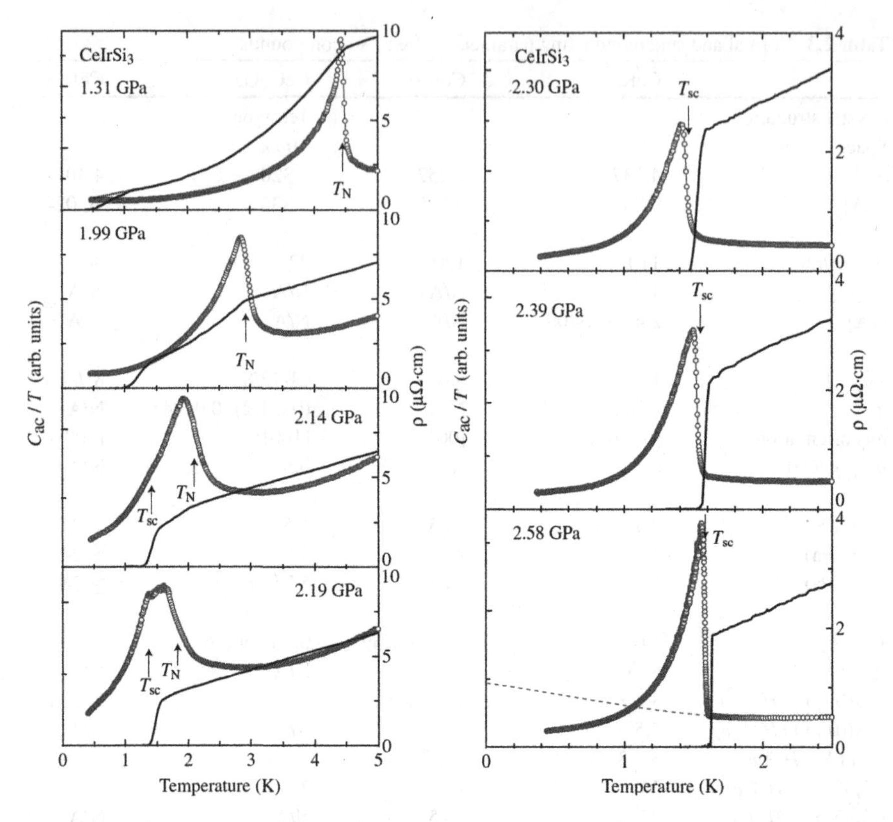

Fig. 2.16 Temperature dependence of the ac heat capacity C_{ac} (*circles, left side*) and the electrical resistivity ρ (*lines, right side*) at several pressures in $CeIrSi_3$. The *dotted line* in the panel at 2.58 GPa indicates the entropy balance below T_c [75]

BCS model and is probably the highest value among all known superconductors. On the other hand, the jump of the heat capacity associated with antiferromagnetism diminishes when approaching P_1^*, and is no longer observed above P_1^*. Apparently, the entropy gain of the AFM transition is transferred to the superconducting one.

2.2.2.3 Quantum Criticality and Non-Fermi Liquid

Superconductivity in *f*-electron materials often appears in the vicinity of a quantum critical point (QCP) at which the magnetic ordering temperature is reduced to zero by a nonthermal control parameter such as pressure, magnetic field, or chemical substitution. Often, a QCP accompanies the emergence of a non-Fermi liquid in which the temperature dependence of some physical properties obeys different behavior from that expected in the Fermi-liquid theory. As shown in Fig. 2.17, the electrical resistivity changes from the Fermi-liquid prediction $\rho(T) = \rho_0 + AT^2$ below P_1^* to $\rho(T) = \rho_0 + A'T$ above P_1^* in $CeRhSi_3$ and $CeIrSi_3$

[40, 72]. A similar crossover occurs in $CeCoGe_3$ at 6.9 GPa [43]. The T-linear dependence of the resistivity agrees with the prediction by the 2D spin fluctuation theory as indicated in Table 2.4. On the other hand, in $CeIrSi_3$ $1/T_1$-NMR measurements in the normal state reveal a $\sqrt{T}$ dependence at 2.7 GPa for $H \perp c$. This supports a prediction for 3D AFM fluctuations in this system: $1/T_1 \propto T\sqrt{\chi_Q(T)} \propto T/\sqrt{(T+\theta)}$ [78]. Here, $\chi_Q(T)$, the staggered susceptibility with the AFM propagation vector **Q**, follows the Curie-Weiss law and θ measures the deviation from the QCP. The specific heat divided by temperature C/T of $CeIrSi_3$ is estimated to be enhanced almost linearly with decreasing temperature from consideration of the entropy balance [75]. The enhancement of C/T is consistent with the spin fluctuation theory, even though from theory we cannot determine the proper dimensionality. The contradiction in the temperature responses of ρ and $1/T_1$ remains unsettled.

In general, it is unclear whether or not a QCP truly exists in $CeTX_3$, since the AFM transitions above P_1^* are prevented or masked by superconductivity. For example, the specific heat jump associated with the AFM transition seems to disappear suddenly just above P_1^* in $CeIrSi_3$ as mentioned above. This suggests that the QCP exists neither at P_2^* nor at higher pressures. However, this does not necessarily indicate that magnetic fluctuations vanish at these pressures. The fact that the maximum T_c and the sharpest resistivity drop take place at P_3^* implies that magnetic fluctuations survive even above P_1^* and rather develop toward P_3^*. Assuming that magnetic fluctuations stabilize the superconducting phase and become stronger at the QCP, P_3^* should be regarded as a *virtual* QCP.

In the vicinity of the QCP some pressure-induced HF superconductors display a strong enhancement of the effective mass of the conduction electrons; namely, via the γ_n value and the coefficient A of the T^2 term of the electrical resistivity [79–81]. In the cases of $CeRhSi_3$ and $CeIrSi_3$, however, such an enhancement is less obvious. The coefficient A in $CeRhSi_3$ is almost constant up to P_1^* [72] and the γ_n value in $CeIrSi_3$ is suggested to be unchanged up to P_3^* [75]. On the other hand, the coefficient A in $CeCoGe_3$ is strongly enhanced from 0.011 $\mu\Omega \cdot$ cm/K^2 at ambient pressure to 0.357 $\mu\Omega \cdot$ cm/K^2 at 5.4 GPa. In $CeIrGe_3$ $\rho(T)$ exhibits a complicated pressure dependence (as shown in Fig. 2.18) that makes unclear how A varies with pressure. The drastic changes at certain pressures between 8.6 GPa and 17 GPa may be associated with the step-like decrease of $T_N(P)$ around 13 GPa as seen in Fig. 2.15(d). The residual resistivity becomes maximum at a pressure near 17 GPa. Since such an enhancement of residual resistivity can be a signature of critical valence fluctuations [82], a valence transition or crossover may take place around 13 GPa in $CeIrGe_3$.

2.2.3 Superconducting State

2.2.3.1 Anisotropic Upper Critical Field: Two Limiting Fields

Thus far, the most interesting phenomenon in the $CeTX_3$ superconductors is the extremely high anisotropy in the upper critical magnetic field H_{c2}, with stunningly

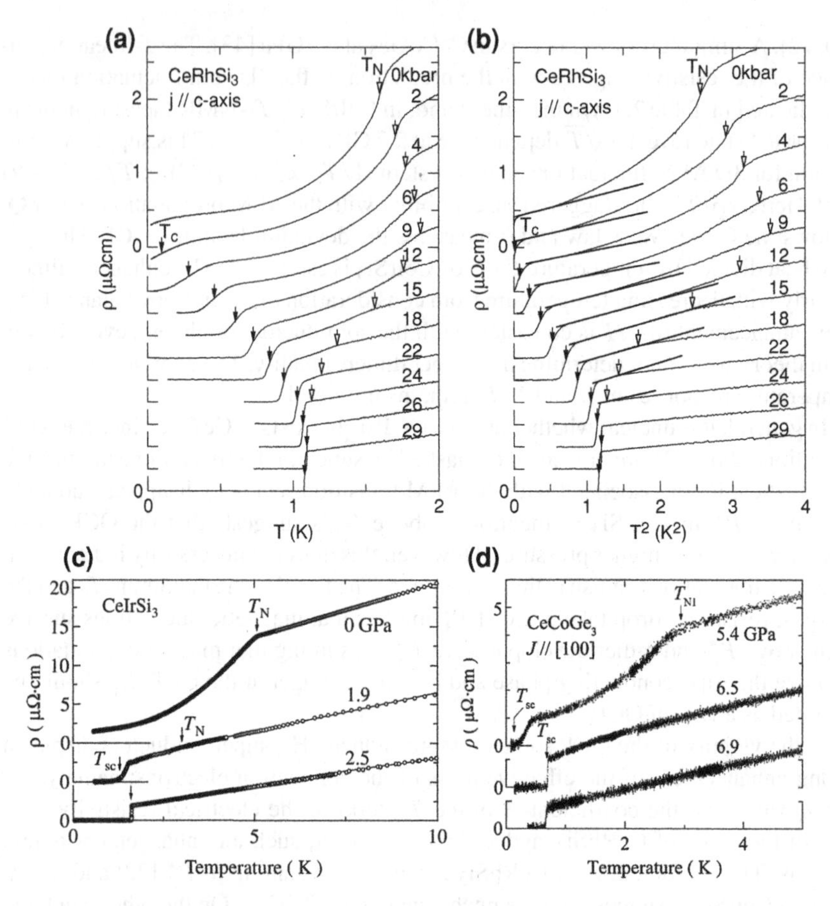

Fig. 2.17 Temperature dependence of the resistivity of **a** $CeRhSi_3$ [72], **c** $CeIrSi_3$ [40] and **d** $CeCoGe_3$ [43]. **b** Resistivity against T^2 in $CeRhSi_3$

Table 2.4 Theoretically predicted quantum critical behavior for 3D and 2D antiferromagnetic fluctuations and corresponding temperature dependence on CeTX_3 compounds

	ρ	$1/T_1$	C/T
3D AFM	$T^{3/2}$	$T/\sqrt{T+\theta}$	Const. $- T^{1/2}$
2D AFM	T	$T/(T+\theta)$	$-\ln T$
$CeRhSi_3$	T [39]		
$CeIrSi_3$	T [40]	$T/\sqrt{T+\theta}$ [78]	T ? [75]
$CeCoGe_3$	T [43]		
$CeIrGe_3$	T [42]		

high values for fields along the c axis. In $CeRhSi_3$ and $CeIrSi_3$, for example, H_{c2} exceeds 30 T along the c axis, whereas falls below 10 T along the plane. Considering that in these materials T_c is of the order of 1 K, this leads to very high H_{c2}/T_c ratios not known previously for any *centrosymmetric* superconductors (except for field-

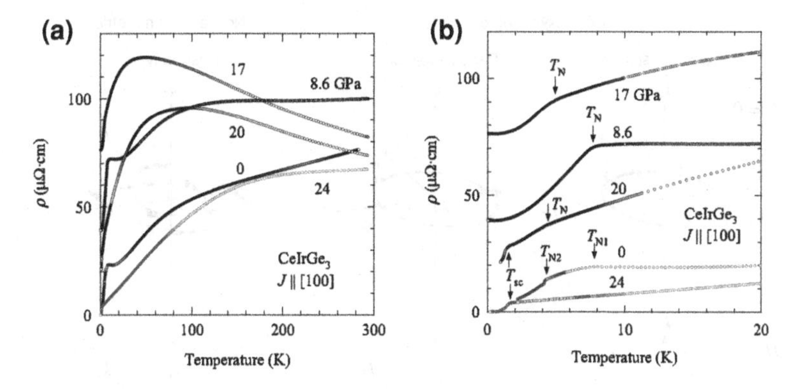

Fig. 2.18 Temperature dependence of the resistivity of $CeIrGe_3$ at several pressures up to 24 GPa below room temperature (*left*) and below 20 K (*right*) [42]

induced superconductors like URhGe [83] and organic superconductors [84]). The strong field anisotropy and the extremely high H_{c2}s are thought to be understood in terms of the anisotropy of the paramagnetic pair-breaking effect characteristic of non-centrosymmetric (Rashba-type) superconductors.

There are two pair-breaking mechanisms for Cooper pairs under magnetic fields: the paramagnetic and the orbital. The former mechanism is attributed to the spin polarization due to the Zeeman effect, which competes with the antiparallel-spin formation of the Cooper pair in spin-singlet superconductors. The influence of the paramagnetic effect depends on the symmetry of the Cooper pairs, as discussed later. On the other hand, the orbital effect is ascribed to the orbital motion in a magnetic field. The influence of this effect is thought to be independent of the pairing symmetry. The magnitude of $H_{c2}(0)$ is consequently restricted by both the paramagnetic (Pauli-Clogston-Chandrasekhar) limiting field H_P and the orbital limiting field H_{orb} [85]. Hereafter, we call H_P the Pauli limiting field for simplicity.

Pauli-Clogston-Chandrasekhar Limit

The paramagnetic effect in spin-singlet and spin-triplet pairing symmetries is illustrated schematically in Fig. 2.19. In centrosymmetric superconductors, the up-spin and down-spin bands of the conduction electrons are degenerate in zero field. The corresponding Fermi surfaces are also degenerate. The presence of a magnetic field inflates and deflates the down-spin and up-spin Fermi surfaces, respectively, due to the Zeeman effect. When the Cooper pair consists of antiparallel spins, e.g. the conventional singlet pair, the paramagnetic pair breaking occurs on the whole Fermi surface. When the Cooper pair comprises parallel spins, namely the triplet pair, the paramagnetic pair breaking does not occur, therefore the Pauli limit is absent. Since the spin direction of the Cooper pair is always aligned to the magnetic field, both cases are independent of the field direction unless the coupling between the orbital and spin parts of the pairing function is present.

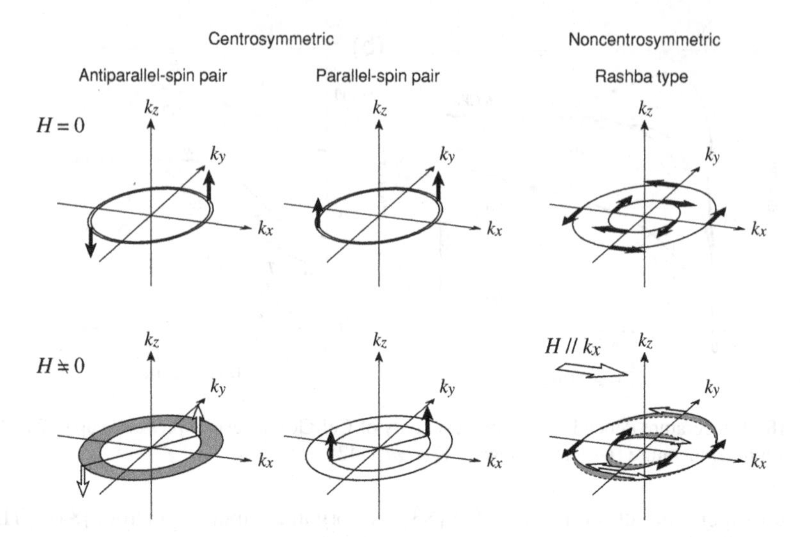

Fig. 2.19 Two-dimensional Fermi surfaces for centrosymmetric and non-centrosymmetric superconductors at zero and finite fields. The centrosymmetric superconductors can be classified into parallel- and antiparallel-spin pairs. In the non-centrosymmetric superconductor, only the Rashba type is displayed. The hollow arrows indicate a pairing no longer allowed. In the antiparallel-spin pair including the conventional singlet pair, the pair breaking occurs on the whole Fermi surface under magnetic fields. On the other hand, in the parallel-spin pair, all the pairing persist. In the Rashba-type superconductor, pair breaking occurs only for the pairs parallel to the applied magnetic field. When the magnetic field is applied along k_z, pair breaking does not occur

In non-centrosymmetric superconductors, the up-spin and down-spin Fermi surfaces are not degenerate even in zero field. In the Rashba-type (tetragonal) superconductors, the spins are perpendicularly aligned with the momenta in the k_z plane (Fig. 2.19) by the spin-orbit coupling, yielding a momentum-dependent effective magnetic field. The presence of an applied magnetic field along the k_x direction inflates and deflates the Fermi surfaces only along the k_y direction, since the spin component perpendicular to the k_x direction is not affected by this field. The paramagnetic pair-breaking effect is thus partial for field along the k_z plane. On the other hand, when the magnetic field is applied along the k_z axis, the paramagnetic pair-breaking effect is absent because all the spins aligned with the k_z plane are perpendicular to the field direction. The strongly anisotropic $H_{c2}(T)$, with high $H_{c2}(0)$ for $H \parallel c$, realized in CeTX_3 compounds is attributed to the anisotropic spin susceptibility expected to appear in non-centrosymmetric superconductors as discussed above in Spin State, Sect. 2.1.3.1.

Figure 2.20 shows $H_{c2}(0)$ versus T_c for CeTX_3 compounds and some well-known HF superconductors. The dashed line indicates H_P^{BCS} (see Eq. 2.4). The $H_{c2}(0)$ of the U-based superconductors UGe_2, URhGe and UPt_3 exceeds H_P^{BCS}. These superconductors are thought to form a parallel-spin pairing free from the paramagnetic pair-breaking effect. In some spin-singlet superconductors $H_{c2}(0)$ is located above the H_P^{BCS} line, which seems a contradiction. Two possibilities have been proposed:

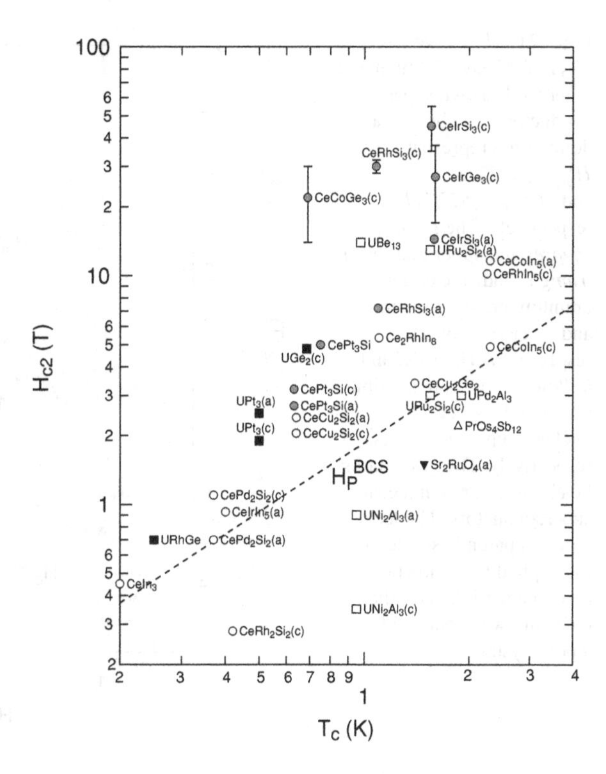

Fig. 2.20 $H_{c2}(0)$ versus T_c for heavy-fermion and some other well-known superconductors. The broken line represents $H_P^{BCS} = 1.86T_c$. The *circles*, *squares* and *upper* and *lower triangles* indicate cerium, uranium, praseodymium and other compounds, respectively. The *solid* and *hollow* symbols tag possible triplet and singlet (or unclear) superconductors, respectively. The gray marks label non-centrosymmetric superconductors. The letter *a* or *c* in parentheses denotes the applied field direction and no letter indicates that the result was obtained for a polycrystal

one is a reduction of the g-factor and the other is an enhancement of Δ_0 due to a strong-coupling effect. $H_{c2}(0)$s of CeTX_3 compounds do not only exceed the H_P^{BCS} line, but also surpass those of all other materials. To understand the high H_{c2} of the CeTX_3 compounds, it is necessary to consider the orbital pair-breaking effect that we discuss next.

Orbital Limit

As we discussed above, the orbital limiting field can be estimated from Eq. (2.6). The value $h(0)$ depends on both the ratio $\xi(0)/l$ and the strong-coupling parameter λ (l: mean free path). In the weak-coupling limit ($\lambda = 0$), namely the BCS model, $h(0) = 0.727$ for $(\xi(0)/l) \to 0$ (clean limit) and $h(0) = 0.693$ for $(\xi(0)/l) \to \infty$ (dirty limit). Most HF superconductors satisfy the clean limit. In the strong-coupling limit ($\lambda \longrightarrow \infty$), $h(0)$ approaches 1.57 for clean superconductors [86] and increases with λ for dirty superconductors. It is noted that the λ dependence of h is usually derived on the basis of the conventional electron-phonon model, but that in HF systems a corresponding electron-magnon approach should be employed since it is generally believed that the attractive interaction leading to Cooper pairs arises from coupling to spin excitations.

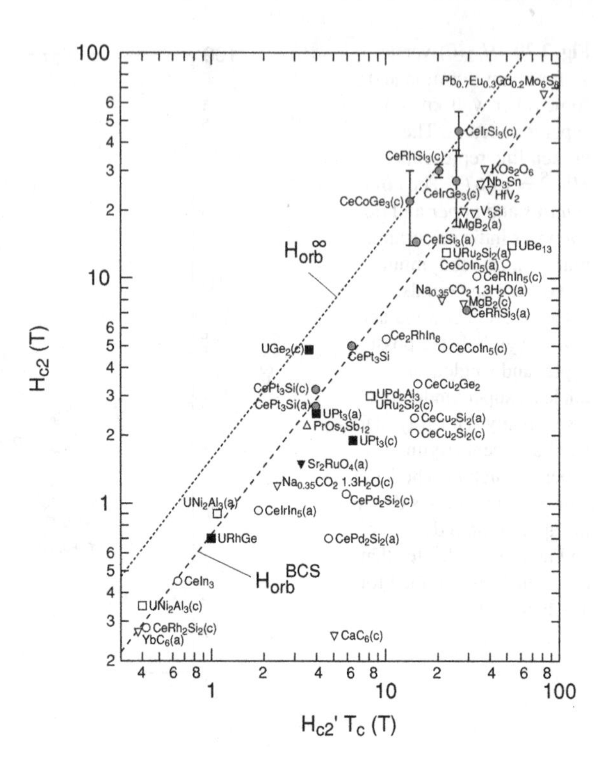

Fig. 2.21 $H_{c2}(0)$ versus $H'_{c2}T_c$ for heavy-fermion and other well-known superconductors. The broken and dotted lines represent $H_{orb}^{BCS} = 0.727 H'_{c2}T_c$ and $H_{orb}^{\infty} = 1.57 H'_{c2}T_c$, respectively. The *circles*, *squares* and *upper* and *lower triangles* indicate cerium, uranium, praseodymium and other compounds, respectively. The *solid* and *hollow* symbols tag possible triplet and singlet (or unclear) superconductors, respectively. The gray marks label non-centrosymmetric superconductors. The letter *a* or *c* in parentheses denotes the applied field direction and no letter indicates that the result was obtained for a polycrystal

The $H_{c2}(0)$s in Fig. 2.20 are plotted against $H'_{c2}T_c$ in Fig. 2.21. The dashed and dotted lines indicate the orbital limiting fields at $\lambda = 0$ (weak-coupling limit) and $\lambda \to \infty$ (strong-coupling limit) for clean superconductors; that is, $H_{orb}^{BCS} = 0.727 H'_{c2}T_c$ and $H_{orb}^{\infty} = 1.57 H'_{c2}T_c$, respectively. Some parallel-spin, namely triplet, superconductors like UPt_3 and URhGe are located just below H_{orb}^{BCS}. The compounds located far below the H_P^{BCS} line in Fig. 2.20, e.g. UNi_2Al_3 and $CeRh_2Si_2$, are seen in the vicinity of the H_{orb}^{BCS} line in Fig. 2.21. The H_{c2}s of these compounds may be mainly restricted by the orbital limit rather than by the Pauli limit. $CePt_3Si$ is also located near the H_{orb}^{BCS} line. Considering that H_{c2} is almost isotropic in $CePt_3Si$ [10], it may be also mainly constrained by the orbital limit rather than by the paramagnetic one.

The H_{c2}s of the CeTX_3 compounds for fields along the c axis well exceed the H_{orb}^{BCS} line. They seem to be close to the strong-coupling limit H_{orb}^{∞}. Although the high $H_{c2}/(H'_{c2}T_c)$ is a common feature in the CeTX_3 HF superconductors, it is not obvious that such a result can be associated with their non-centrosymmetric crystal structures. Note that UGe_2 also seems to be above the H_{orb}^{BCS} line, but this is due to the jump of the H_{c2} curve attributed to the metamagnetic transition [87, 88]. Therefore, this plot does not necessarily indicate that the intrinsic $H_{c2}(0)$ of UGe_2 exceeds the H_{orb}^{BCS} line.

2.2.3.2 Upper Critical Field for *c* Axis

In addition to the high values of $H_{c2}(0)$, the upward shape of the temperature dependence of $H_{c2}(T)$ for certain pressures seems to be also a characteristic of the CeTX_3 superconductors [43, 54, 72]. They mostly keep a positive curvature ($d^2H_{c2}/dT^2>0$) down to relatively low temperatures; for example, in $CeIrSi_3$ down to $T \approx 0.25T_c$ at 2.6 GPa (Fig. 2.23). Interestingly, the curve shapes of $CeRhSi_3$ and $CeIrSi_3$ vary with pressure in a complex manner. As for $CeIrSi_3$, a positive curvature is seen up to 2 GPa that gradually changes to a negative curvature at 2.4 GPa to a quasi-linear shape at 2.3 GPa. At 2.6 GPa and 2.65 GPa, a positive curvature is recovered that turns to a negative one at higher pressures. Similar behavior is seen in $CeRhSi_3$. Below and above 2.6 GPa, the curvatures of $H_{c2}(T)$ are positive. At 2.6 GPa, a quasi-linear change of $H_{c2}(T)$ is observed. Because of the lack of sufficient pressure data, it is not clear whether such a phenomenon is realized in $CeCoGe_3$ and $CeIrGe_3$.

In order to understand the superconducting phase diagram of the CeTX_3 compounds the pressure dependence of $H_{c2}(0)$ will be very helpful. Fig. 2.22(c) shows $H_{c2}(0)$ versus pressure in $CeIrSi_3$ [54]. $H_{c2}(0)$ increases with pressure and tends to diverge close to 2.65 GPa ($\approx P_3^*$). Above 2.65 GPa it falls steeply to half or less of the value of the maximum $H_{c2}(0)$. It is pointed out that such an acute enhancement of $H_{c2}(0)$ can be interpreted as an electronic instability arising at P_3^* [54]. This instability can cause a mass enhancement of the conduction electrons, giving rise to a reduction in the superconducting coherence length $\xi(0)$. At a first glance, a mass enhancement at P_3^* is consistent with the pressure dependence of the initial slope of the superconducting H-T phase diagram, $H'_{c2} = -dH_{c2}/dT|_{T=T_c}$. From Eqs. (2.5) and (2.6), we can derive

$$H'_{c2}T_c \sim H_{orb} \sim \xi^{-2}(0) \sim (\Delta_0 m^*)^2. \tag{2.10}$$

Here, we use $v_F = \hbar k_F/m^*$. As shown in Figs. 2.23–2.25, H'_{c2}s of $CeIrSi_3$ and $CeRhSi_3$ increase with increasing pressure and become maximum at about P_3^*. However, strong pressure dependence of the cyclotron effective mass is not observed in de Haas-van Alphen experiments in $CeRhSi_3$ [89]. This is consistent with the result of the less obvious pressure dependence of the resistivity coefficient A. A small enhancement of the effective mass is also suggested by heat-capacity measurements in $CeIrSi_3$ [75].

In order to explain the positive curvature of $H_{c2}(T)$ and the strong pressure dependence of $H_{c2}(0)$, Tada *et al.* considered the temperature and pressure dependencies of the correlation length of the spin fluctuations, ξ_{sf}, [90]. Since ξ_{sf} is expressed as $\xi_{sf}(T) = \frac{\tilde{\xi}_{sf}}{\sqrt{T+\theta}}$, in which $\theta \to 0$ toward the QCP and the effective pairing interaction is quadratically proportional to ξ_{sf}, the superconducting coherence length $\xi(0)$ is strongly reduced and H_{orb} enhanced at low temperatures. This model can explain the enhancement of the initial slope and is compatible with the weaker enhancement of the effective mass.

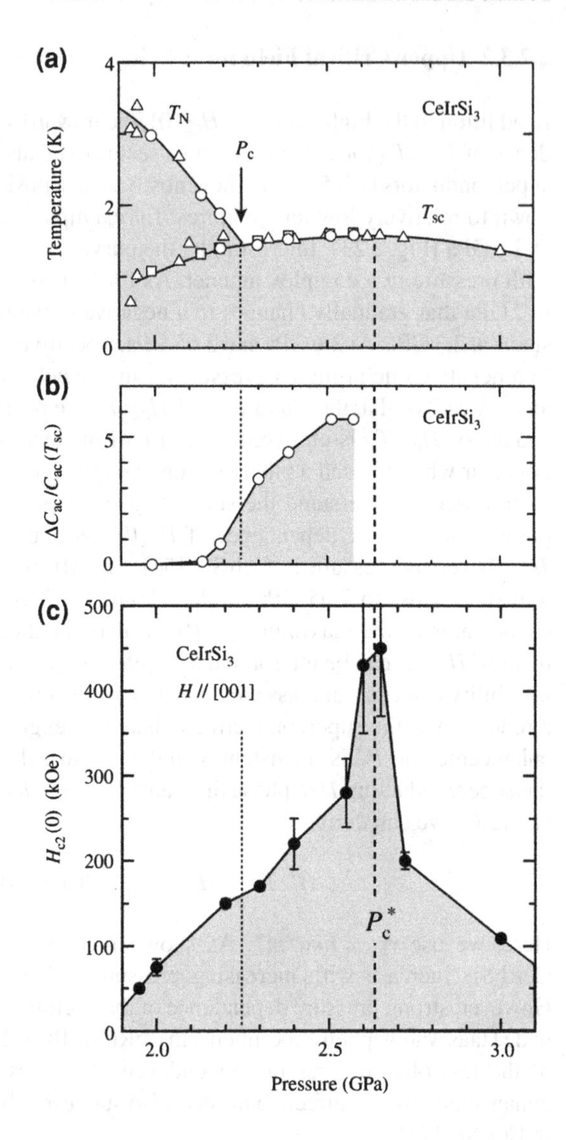

Fig. 2.22 Pressure dependence of **a** T_N and T_c, **b** specific heat jump $\Delta C/C(T_c)$ and **c** $H_{c2}(0)$ for $H \parallel c$ in $CeIrSi_3$ [54]

2.2.3.3 Superconducting Phase Diagram for Field in the Basal Plane

In contrast to the high $H_{c2}(T)$ for the c axis, $H_{c2}(T)$ in the basal plane is not too high. However, it still significantly exceeds H_P^{BCS} and is situated in the upper part of Fig. 2.20. As discussed in the section above, the high $H_{c2}(0)$ is attributed to the reduced paramagnetic pair-breaking effect which originates mainly from a non-vanishing spin susceptibility. In addition to this, Agterberg et al. pointed out that another characteristic mechanism, the helical vortex state, can also evade the

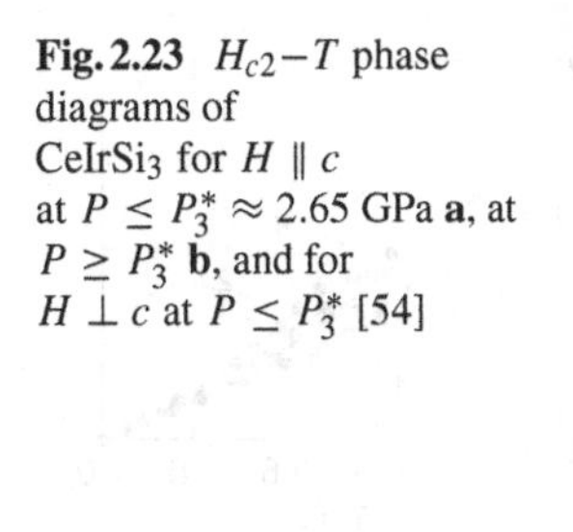

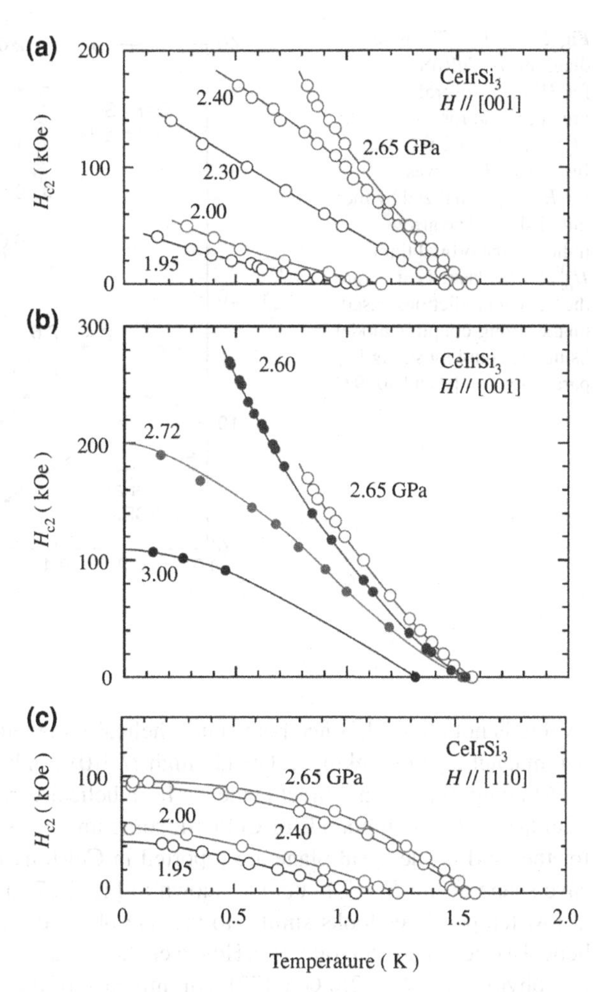

Fig. 2.23 $H_{c2}-T$ phase diagrams of $CeIrSi_3$ for $H \parallel c$ at $P \leq P_3^* \approx 2.65$ GPa **a**, at $P \geq P_3^*$ **b**, and for $H \perp c$ at $P \leq P_3^*$ [54]

paramagnetic pair-breaking effect when the magnetic field is applied in the basal plane [91, 92].

As discussed above, the paramagnetic pair-breaking effect operates partially, with the center-of-mass momentum of the Cooper pair remaining zero at any k as shown in Fig. 2.26(b). On the other hand, in the helical vortex state, the Fermi surfaces shift toward opposite directions perpendicular to the field direction. An application of a magnetic field does not break the Cooper pairs for all k. In this case, the center-of-mass momenta of the Cooper pairs belonging to each Fermi surface acquire a finite value $\pm q$ (Fig. 2.26(c)). The sign of q depends on the Fermi surface. In the Fulde-Ferrel-Larkin-Ovchinnikov (FFLO) state, a Cooper pair with a finite center-of-mass momentum is also realized. However, the pairing takes place in a limited region on the Fermi surface. In other regions, indicated by the shades in Fig. 2.26(c),

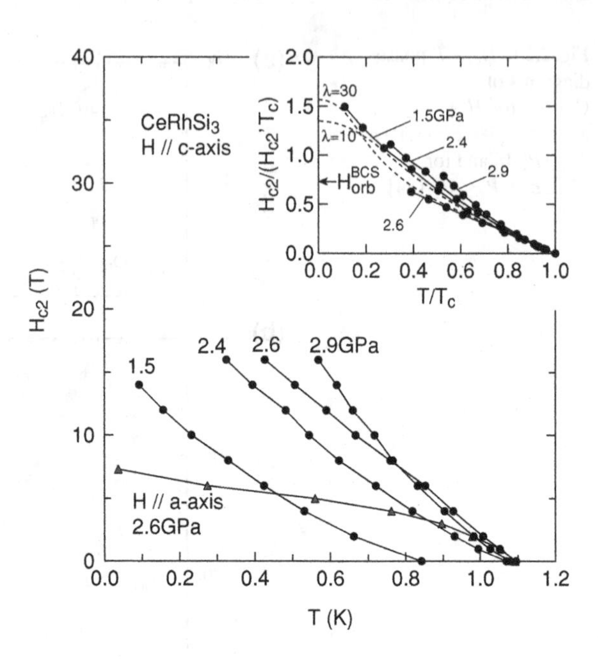

Fig. 2.24 $H_{c2}-T$ phase diagrams of $CeRhSi_3$ for $H \parallel c$ at several pressures and for $H \parallel a$ at $P_3^* \approx 2.6$ GPa. Inset: $H_{c2}(T)$ curves for $H \parallel c$ normalized by the initial slope. The arrow indicates the orbital limit H_{orb}^{BCS}. The dashed curves are theoretical predictions based on the strong-coupling model using the coupling strength parameter $\lambda = 10$ and 30 [93]

pairing is not allowed. Since both states, helical vortex and FFLO phase, evade the paramagnetic pair breaking, relatively high $H_{c2}(0)$ can be realized.

Although thus far no direct evidence for a helical vortex state has been detected in either $CeTX_3$ compounds or $CePt_3Si$, some unusual superconducting properties for the field in the basal plane are reported in $CeRhSi_3$ [39]. First, there is a concave shape of the $H_{c2}(T)$ curve as shown in Fig. 2.27(a). The rapid change of H_{c2} at low temperatures looks similar to the one observed in a theoretically predicted helical-vortex phase diagram [92]. However, this feature observed at $P < P_1^*$ becomes less obvious at $P_3^* \approx 2.6$ GPa [72]. An influence of the antiferromagnetic order to explain this unusual curve shape cannot be excluded. Second, the ac susceptibility in the superconducting state shows an unusual shape especially below P_1^*. The temperature at which a large drop occurs in the real part of the susceptibility χ' is far below the onset temperature of superconductivity. This might indicate that superconductivity develops gradually in the antiferromagnetic state. On the other hand, the imaginary part χ'', namely the energy dissipation associated with the dynamics of the superconducting flux, is large even above the temperature at which the large drop occurs in χ'. This contradicts the view of a gradual development of superconductivity. Although the influence of antiferromagnetism is unclear at present, this rare behavior of the flux may be a key feature to verify the helical vortex state.

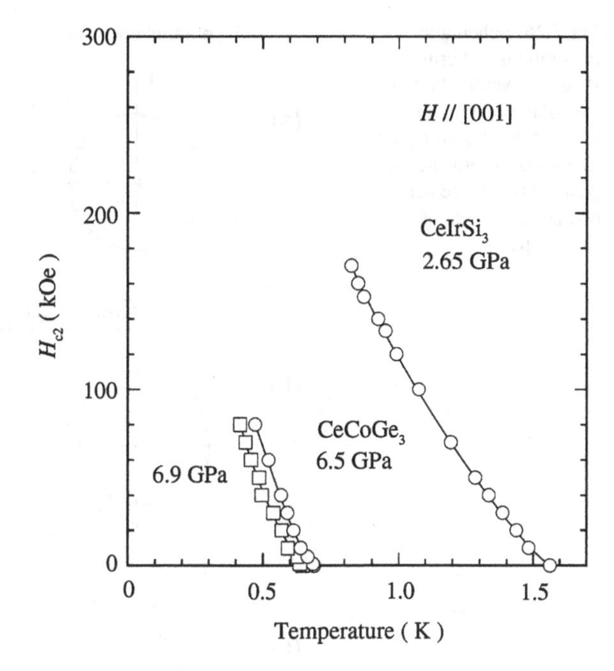

Fig. 2.25 $H_{c2}-T$ phase diagram of $CeCoGe_3$ for $H \parallel c$ at 6.5 GPa and 6.9 GPa. The phase diagram of $CeIrSi_3$ is also displayed [43]

2.2.3.4 Energy Gap Structure and Pairing Symmetry

In $CeIrSi_3$ the nuclear spin-lattice relaxation rate $1/T_1$ shows a T^3-like dependence below T_c without a coherence peak, as shown in Fig. 2.28 [78]. The data are well fitted by the line-node gap model $\Delta = \Delta_0 \cos 2\theta$. The fit yields $2\Delta_0/k_B T_c \approx 6$, which is much larger than the BCS weak-coupling value 3.53 and thus suggests strong-coupling superconductivity. Using an extended $s + p$ pairing state within a recent theoretical model a behavior indicative of a line-node gap can be predicted [94]. This $1/T_1$-NMR measurement is the only one carried out to determine the energy gap structure in $CeTX_3$ compounds. To identify the pairing symmetry in these materials, other measurements, like the Knight shift, are highly desirable.

2.2.4 Outlook

The high anisotropy and strong enhancement of H_{c2} seem to be unique to $CeTX_3$ superconductors. Interestingly, other HF and non-HF superconductors without inversion symmetry do not show these properties. The high orbital limiting field inherent to the non-centrosymmetric HF $CeTX_3$ superconductors discloses the absence of the paramagnetic pair-breaking effect. Conversely, absence of the effect unveils the unconventional nature of the upper critical fields probably associated with the quantum criticality of magnetism. $CeTX_3$ compounds have the potential to provide a

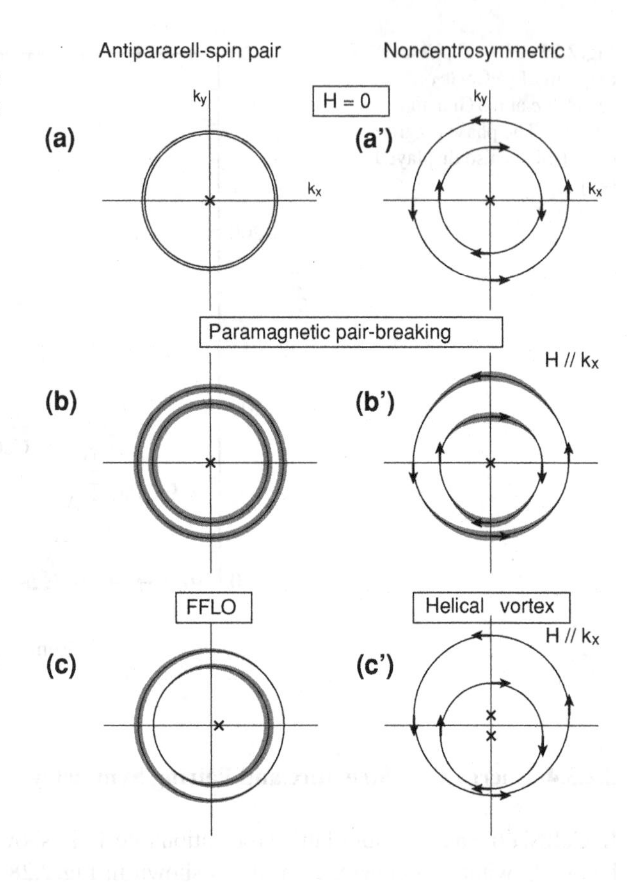

Fig. 2.26 Schematic illustrations of Fermi surface sliced perpendicular to the k_z axis. Pair breaking takes place at the shaded region. The x marks indicate the positions of the center-of-mass momenta of the Cooper pairs

vital clue to the underlying mechanism of superconductivity mediated by magnetic fluctuations.

To consider the relation between magnetism and superconductivity in CeTX_3, we need to keep in mind that the magnetic ground states of $CeCoGe_3$ and $CeIrGe_3$ are different from that of $CeRhSi_3$ and probably of $CeIrSi_3$. $CeCoGe_3$ seems to display localized *f*-electron magnetism, while $CeRhSi_3$ probably exhibits itinerant-electron magnetism. It is very interesting that in these compounds the H_{c2} behaviors are similar in spite of their different magnetism. The mass enhancement at P_3^* is suggested only in $CeCoGe_3$. The comparison of the superconducting properties of $CeCoGe_3$ and $CeIrGe_3$ with those of $CeRhSi_3$ and $CeIrSi_3$ will be important to elucidate the nature of HF superconductors. Identification of the gap structure in each compound is also a challenging issue, which could provide possible evidence for the parity mixing of the superconducting wavefunction. Moreover, some theoretically predicted phenomena, like a helical vortex phase and a novel magnetoelectric effect, remain to be verified in the future.

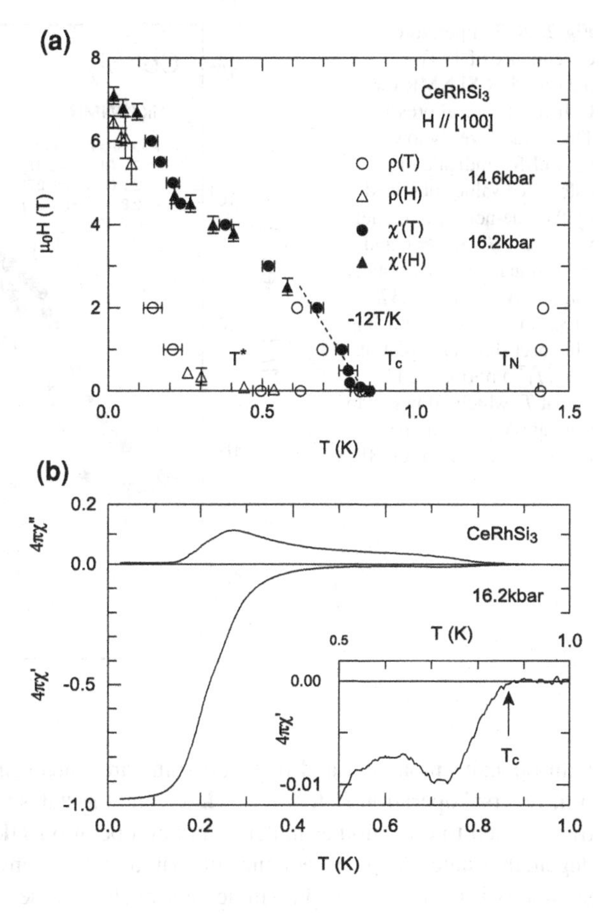

Fig. 2.27 (**a**) Magnetic field-temperature phase diagram of $CeRhSi_3$ for $H \parallel a$ below P_1^*. T^* denotes the temperature at which an anomaly is observed in the resistivity and magnetic ac susceptibility measurements. (**b**) ac susceptibility as a function of temperature at zero field below P_1^*. An enlarged view of the vicinity of T_c is shown in the inset [39]

2.3 UIr

Whereas $CeRhSi_3$, $CeIrSi_3$, $CeCoGe_3$, $CeIrGe_3$ and $CePt_3Si$ are 4*f*-electron antiferromagnets, UIr is a 5*f*-itinerant-electron ferromagnet with a Curie temperature $T_{c1} = 46$K at ambient pressure [95]. The superconducting state in UIr appears to develop within a higher pressure ferromagnetic phase at a critical temperature $T_c = 0.14$ K in a narrow pressure region around 2.6 GPa [96]. UIr is a moderate heavy-fermion compound with a cyclotron mass $m^* \sim 10 - 30\, m_0$ [97]. The coexistence of superconductivity and ferromagnetism imposes several theoretical challenges, such as the mechanism and the state of pairing. The pairing state in superconducting ferromagnets needs to be spin triplet, otherwise the internal exchange field would break the Cooper pairs. On the other hand, a ferromagnetic state has a broken time reversal symmetry. The superconducting BCS ground state is formed by Cooper pairs with zero total angular momentum. The electronic states are four-fold degenerate: $|\mathbf{k}\uparrow\rangle$, $|-\mathbf{k}\uparrow\rangle$, $|\mathbf{k}\downarrow\rangle$ and $|-\mathbf{k}\downarrow\rangle$ have the same energy $\varepsilon(\mathbf{k})$. The states

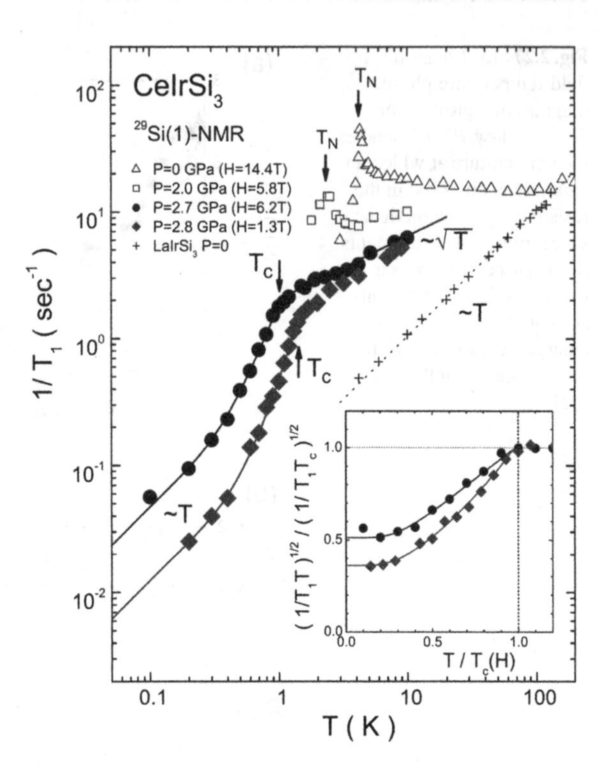

Fig. 2.28 Temperature dependence of $1/T_1$ measured by Si NMR for $CeIrSi_3$ at several pressures. The solid curves below T_c for $CeIrSi_3$ indicate the calculated values obtained by the line-node gap model with $2\Delta_0/(k_B T_c) \approx 6$ and the residual density-of-states fraction $N_{res}/N_0 \approx 0.37$ (0.52) in $H = 1.3$ T (6.2 T). The inset shows the plot of $\sqrt{1/(T_1 T)}$ normalized by that at T_c which allows us to evaluate N_{res}/N_0 in the low-temperature limit [78]

with opposite momenta and opposite spins are transformed to one another under time reversal operation $\hat{K}|\mathbf{k}\uparrow\rangle = |-\mathbf{k}\downarrow\rangle$, and the states with opposite momenta are transformed to one another under inversion operation $\hat{I}|\mathbf{k}\downarrow\rangle = |-\mathbf{k}\downarrow\rangle$. The four degenerate states are a consequence of spatial and time inversion symmetries. Parity symmetry is irrelevant for spin-singlet pairing, but is essential for spin-triplet pairing. Time reversal symmetry is required for spin-singlet configuration, but is unimportant for spin-triplet state [98, 99]. In UIr the lack of spatial and time inversion symmetries lifts the degeneracies and, therefore, superconductivity is not expected to occur. Thus, UIr differs from the other two known ferromagnetic superconductors UGe_2 [100] and URhGe [101], in which the spatial inversion symmetry allows degeneracy in the spin-triplet states. Theoretically and experimentally UIr is a very special and challenging superconductor.

2.3.1 Crystal Structure and Characteristic Parameters

UIr crystallizes in a monoclinic PbBi-type structure (space group $P2_1$) without inversion symmetry [95]. The lattice parameters are given in Table 2.5. The unit cell has eight formula units with four inequivalent U and Ir sites. The absence of inversion symmetry comes from the missing mirror plane (0, $\frac{1}{2}$, 0) perpendicular to the b axis (see Fig. 2.29). Magnetism is of the Ising type with the ordered magnetic moment oriented along the spin easy axis [1 0 $\bar{1}$] (Fig. 2.29).

Table 2.5 Normal and superconducting parameters of UIr

Crystal structure	Monoclinic
Space group	$P2_1$
Lattice parameters	$a = 5.62$ Å
	$b = 10.59$ Å
	$c = 5.60$ Å
	$\beta = 98.9°$
Sommerfeld value of specific heat	$\gamma_n = 40–49$ mJ/molK^2
Effective electron mass (Fermi sheet α)	$m^* \sim 10–30\, m_0$
Mean free path (Fermi sheet α)	$l = 1270$ Å
Ferromagnetic transition temperature (ambient pressure)	$T_{c1} = 46$ K
Magnetic propagation vector	$\mathbf{q} = (1, 0, -1)$
Magnetic moment $\mathbf{m}_Q$ along	$[10\bar{1}]$
Saturated moment per U atom	$\mu_s = 0.5\, \mu_B$
Superconducting transition temperature	$T_c = 0.14$ K
Upper critical field	$H_{c2}(0) = 26$ mT
Thermodynamic critical field	$H_c(0) = 8$ mT
Ginzburg-Landau coherence length	$\xi(0) = 1100$ Å
Ginzburg-Landau parameter	$\kappa \sim 2$

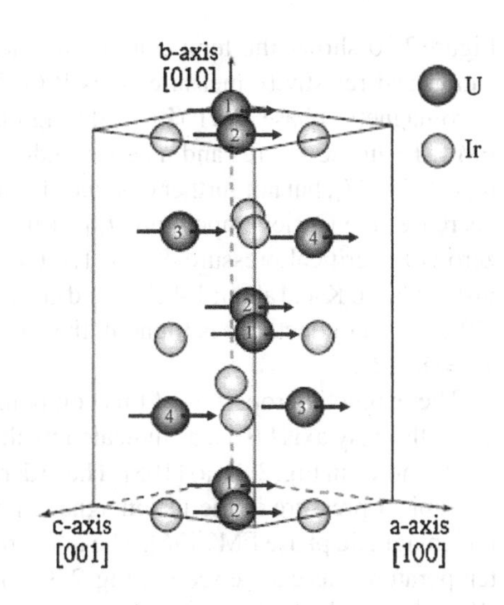

Fig. 2.29 Crystal and magnetic structures of UIr

Single crystals of UIr have been grown by the Czochralski method in a tetra-arc furnace [102–106]. After annealing, using the solid-state electrotransport technique under high vacuum of the order of 10^{-10} Torr, crystals become of very high quality with residual resistivity $\rho_0 \sim 0.5\, \mu\Omega$ and residual resistivity ratio (RRR) $\rho_{300K}/\rho_0 \approx 200$ at ambient pressure. Interestingly, single crystals of UIr seem to be of the highest quality amongst those of non-centrosymmetric heavy-fermion superconductors.

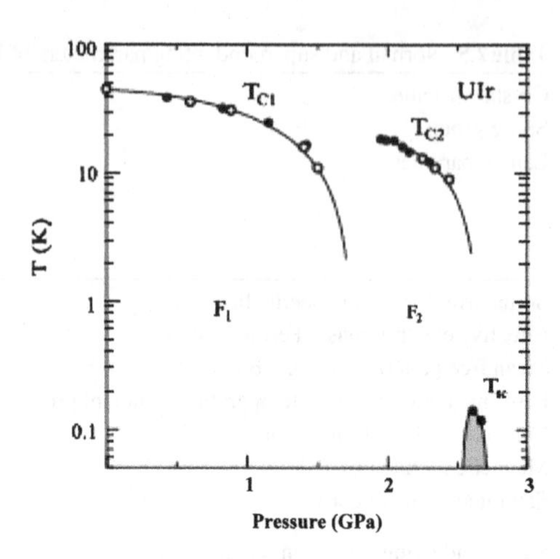

Fig. 2.30 Temperature–pressure phase diagram of UIr [96]. F_1 and F_2 are ferromagnetic phases

2.3.2 *Normal State*

2.3.2.1 Phase Diagram and Magnetic Properties

Figure 2.30 shows the temperature-pressure phase diagram as drawn by magnetization and resistivity measurements [96]. The diagram consists of a low-pressure ferromagnetic phase FM1 (F_1 in the figure), a high-pressure ferromagnetic phase FM2 (F_2 in the figure) and a superconducting phase. A third magnetic phase was reported [107], but not further evidence for it has been found. Application of pressure decreases the Curie temperature T_{c1} of the ferromagnetic phase FM1 eventually to zero at the critical pressure $P_{c1} \sim 1.7$ GPa. The FM2-paramagnetic curve appears just below 30 K and about 1.4 GPa, and goes away at a critical pressure $P_{c3} \sim 2.7-2.8$ GPa. Superconductivity is found in the narrow pressure range 2.55–2.75 GPa below $T_c = 0.14$ K.

The magnetic properties of this compound are governed by a saturation moment along the easy axis $[10\bar{1}]$, as indicated by the magnetization curve of a single crystal at 2 K shown in Fig. 2.31(a) [108]. The ordered magnetic moment goes from $0.5\mu_B$/U at ambient pressure in the ferromagnetic phase FM1 to $0.07\mu_B$/U at 2.4 GPa in the ferromagnetic phase FM2 [96]. The anisotropy of the magnetization remains at high temperatures, as can be seen in Fig. 2.31(b). The susceptibility data follow a Curie-Weiss law in the high-temperature paramagnetic region, with an effective magnetic moment around 3.57 μ_B/U that is pretty close to the $5f^2$ free-ion value 3.58 μ_B/U. The small value of the ordered moment $0.5\mu_B$/U has been taken as evidence for the itinerant character of the 5*f* electrons in the ferromagnetic phase, though such a low value could also be due to crystal-field effects [106, 109].

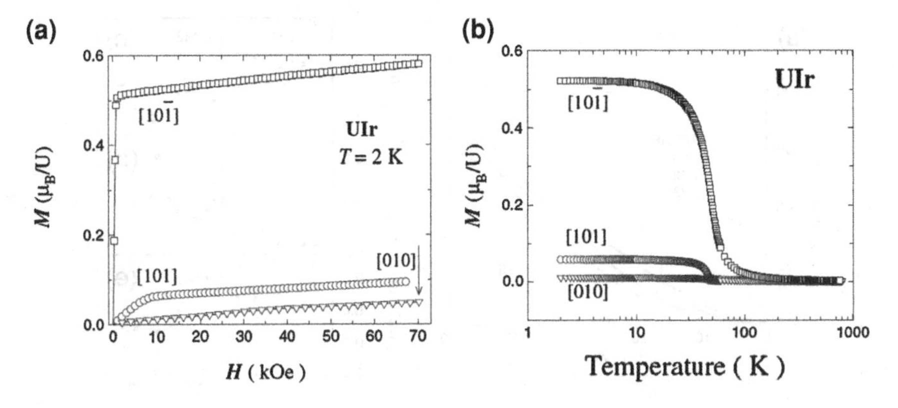

Fig. 2.31 Magnetization as a function of **a** field strength at 2 K and **b** temperature in 10 kOe in a single crystal of UIr [108]

2.3.2.2 Electronic States

Quantum-oscillation and resistivity measurements provide evidence that the low-temperature metallic state of UIr is a Fermi liquid at ambient pressure. The dHvA measurements suggest that the Fermi surface of UIr is two-dimensional and consists mainly of nearly cylindrical sheets [97]. It has an effective mass $m^* \sim 10-30\, m_0$ and a mean free path $l \sim 1270$ Å. Such a value of the effective mass of the 5*f* electrons leads to the classification of UIr as a moderate heavy-fermion compound. Since the linear coefficient of the heat capacity $\gamma_n \propto m^*$, summing for all the branches yields the electronic specific-heat coefficient of 40–49 mJ/K^2 mol [97, 110]. There are no band-structure calculations for this compound.

Figure 2.32(a) shows the variation of the electrical resistivity of UIr as a function of pressure [105, 110]. At low pressure, in the ferromagnetic phase FM1, the resistivity follows $\rho = \rho_0 + AT^n$, with $n \sim 2$ suggesting Fermi-liquid behavior. However, as pressure increases the behavior becomes non-Fermi-liquid like and eventually superconductivity appears in this regime. Figures 2.32(c–e) show the variation in n, A and ρ_0 as pressure increases. The non-Fermi-liquid behavior above $\sim$1GPa may be related to critical fluctuations near the different magnetic transitions.

2.3.3 Superconducting State

Because of its extremely low critical temperature $T_c = 0.14$ K (in most figures in this section this critical temperature is called T_{sc}), there is little information on the superconducting state of UIr. Superconductivity seems to occur inside and near the quantum critical point of the FM2 phase, in the very narrow pressure range of 2.6–2.75 GPa [96]. In the ferromagnet UGe_2 with inversion symmetry the superconducting phase exists inside the ferromagnetic phase as well. Figure 2.33(a) shows the temperature dependence of the resistivity below 10 K and at 2.61 GPa, where

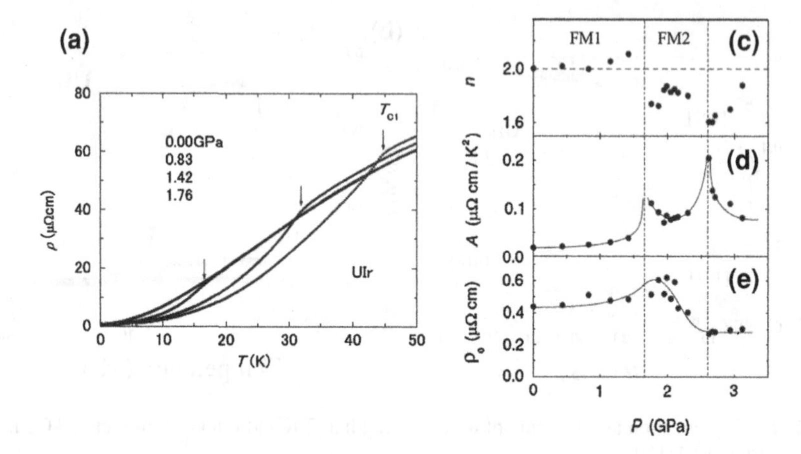

Fig. 2.32 **a** Resistivity as a function of temperature of UIr at different pressures, and **c**–**e** variations in the parameters n, A and ρ_0 of $\rho = \rho_0 + AT^n$ as pressure increases [108]

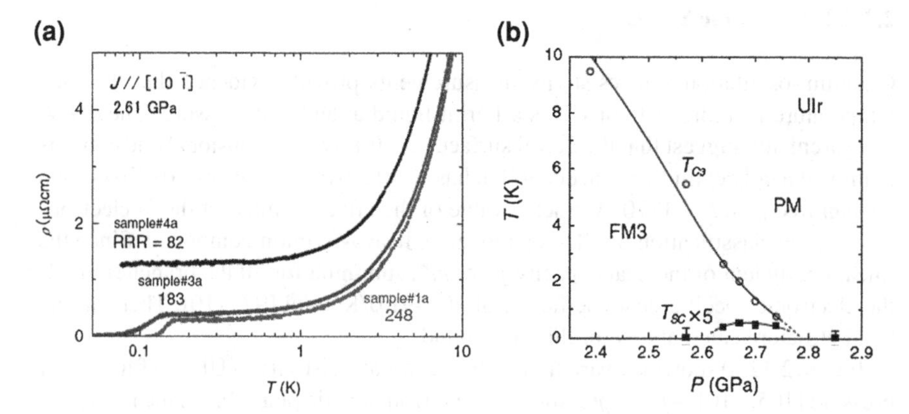

Fig. 2.33 **a** Resistivity below 10 K and at 2.61 GPa of UIr showing a non-Fermi-liquid behavior. The three samples possess different RRR-values: 82, 183 and 248. **b** Low-temperature region of the phase diagram of UIr [107]

it follows a non-Fermi-liquid form $T^{1.6}$ [105]. Figure 2.33(b) is a close-up of the low-temperature region of the phase diagram where superconductivity appears (in this figure, FM3 denotes the ferromagnetic phase FM2).

Up to now experimental indications of the existence of a superconducting phase in UIr come from measurements of resistivity in samples of different qualities. No definite diamagnetic signal has yet been observed in UIr (Fig. 2.34) [107]. The temperature dependence of the resistivity for three different samples is shown in Fig. 2.33(a). The data indicate that superconductivity becomes weaker as the residual resistivity ratio (RRR) of the samples drops. Such a strong suppression of T_c with increasing impurities/defects is typical of unconventional parity-conserving super-

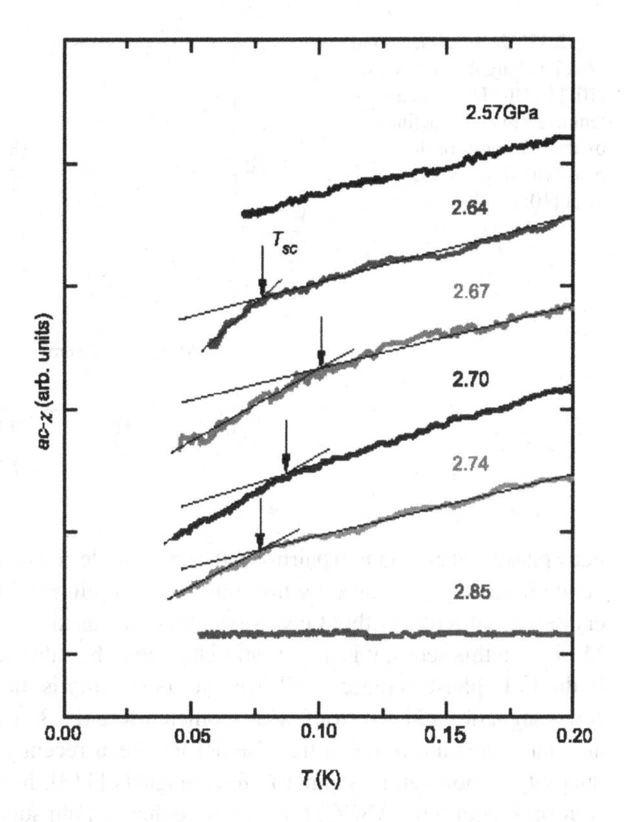

Fig. 2.34 Temperature dependence of the ac susceptibility of UIr below 0.2 K at different pressure. Arrows indicate the onset temperatures [107]

conductors [27]. Recent theoretical works [111, 112] considered impurity effects on the critical temperature of superconductors without inversion symmetry. It was found that impurity scattering leads to a functional form of T_c that, up to a prefactor, is the same as the one for unconventional superconductor with inversion symmetry: $\ln(T_c/T_{c0}) = \alpha\left[\psi(\frac{1}{2}) - \psi\left(\frac{1}{2} - \frac{\Gamma}{2\pi T_c}\right)\right]$. The suppression of T_c by impurities in non-centrosymmetric UIr agrees with this prediction [111, 112].

The upper critical field $H_{c2}(T)$ in the direction of the easy axis $[1\,0\,\bar{1}]$ in the high-temperature region was determined by resistivity measurements in a sample with a very high RRR (Fig. 2.35) [105]. T_c was defined as the midpoint of the resistivity drop. By assuming the standard empirical expression $H_{c2}(T) = H_{c2}(0)\left[1 - (T/T_c)^2\right]$, the zero-temperature upper critical field $H_{c2}(0)$ was estimated as 26 mT corresponding to a coherence length of $\xi(0) = 1100$ Å. Since this value of $H_{c2}(0)$ is smaller than the paramagnetic limiting field $H_p = 280$ mT orbital depairing is the likely mechanism for the upper critical field in UIr.

It is believed that near a ferromagnetic quantum critical point spin fluctuations lead to Cooper pairing in a spin-triplet channel. There are some possible scenarios for the unexpected realization of superconductivity in this compound. It needs to be confirmed that FM2 is indeed a ferromagnetic phase, and not a canted antiferromag-

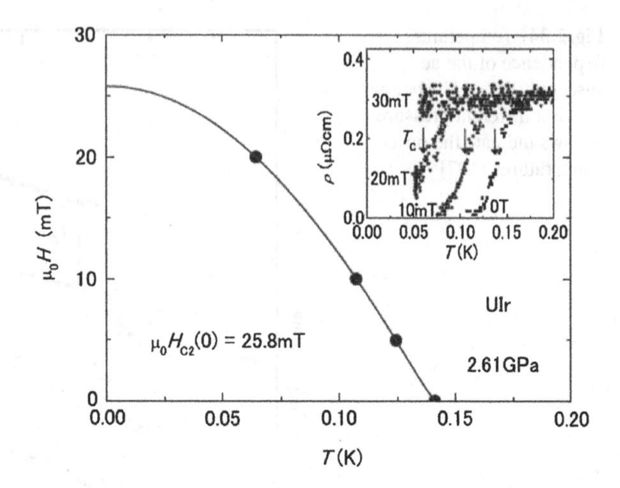

Fig. 2.35 Upper critical field $H_{c2}(T)$ along the easy axis $[10\bar{1}]$ in UIr. The critical temperatures were defined by the *midpoints* of the resistivity drops in the inset [105]

netic phase that could yield pairing in the spin-single channel. A canted antiferromagnetic phase may be caused by the spin-orbit coupling and the low symmetry of the crystalline structure without inversion symmetry, as discussed by Dzyaloshinky and Moriya. In this sense, it is important to note that the saturated moment of 0.07 μ_B/U in the FM2 phase is quite small. Another possibility is the FFLO state, in which at zero magnetic field electrons with momenta $\mathbf{k}$ and $-\mathbf{k}+\mathbf{q}$ can pair with nonzero angular momenta. A mean-field model has been recently proposed for superconductivity in non-centrosymmetric ferromagnets [113], in which the antisymmetric spin-orbit coupling (ASOC) turns out to enhance both superconductivity and ferromagnetism in all spin channels. Future experiments will be absolutely important for the understanding of this unique superconductor.

2.4 Comparison of the Superconducting States of $CePt_3Si$, $CeRhSi_3$, $CeIrSi_3$, $CeCoGe_3$, $CeIrGe_3$ and UIr

The superconducting properties of the non-centrosymmetric HF compounds have not been easy to determine. On the one hand, the CeTX_3 and UIr compounds only superconduct under pressure, making technically difficult to study them. On the other hand, $CePt_3Si$ does become superconducting at ambient pressure, but has drawbacks in the crystal quality available. In spite of this, it has been possible to establish some of the characteristics of these materials.

The CeTX_3 materials, with tetragonal space group $I4mm$, and $CePt_3Si$, with tetragonal space group $P4mm$, have the same generating point group C_{4v} which lacks a mirror plane and a two-fold axis normal to the c axis. Thus, a Rashba-type interaction appears in all these compounds due to the missing of inversion symmetry. In contrast, UIr has a monoclinic lattice.

The Sommerfeld coefficient is much larger in $CePt_3Si$ than in $CeTX_3$ and UIr. Electron correlations should hence be stronger in $CePt_3Si$. On the other hand, all compounds turn from Fermi-liquid to non-Fermi-liquid states as pressure is increased and become superconducting in the Fermi-liquid state, with the exception of UIr.

The behavior of the upper critical field H_{c2} in the $CeTX_3$ systems is very different from that in $CePt_3Si$. In $CeTX_3$, $H_{c2\|c} > 22$ T and $(-dH_{c2\|c}(T)/dT)_{T_c} > 17$ T/K are larger than $H_{c2\|c} \approx 3$ T and $(-dH_{c2\|c}(T)/dT)_{T_c} \sim 6.3$ T/K in $CePt_3Si$. Moreover, in $CeTX_3$ $H_{c2}(T)$ has a positive curvature, unlike in $CePt_3Si$. In $CeIrSi_3$, for example, superconductivity is anisotropic $(H_{c2\|c}/H_{c2\|ab}) > 3$, whereas in $CePt_3Si$ it is almost isotropic. There are clear signatures for unconventional superconductivity in $CeIrSi_3$, $CeRhSi_3$ and $CePt_3Si$, including evidence for line nodes in some cases. The fact that these compounds also support strong magnetic features and order suggests strongly that unconventional pairing mechanisms could be at work here. In this context it is particularly intriguing to analyze the role the antisymmetric spin-orbit coupling.

Acknowledgements We are grateful to M. Sigrist, S. Fujimoto, Y. Tada, D. F. Agterberg, K. Samokhin, F. Honda, Y. Matsuda, T. Sugawara, T. Terashima, H. Aoki, F. Lévy and I. Sheikin for helpful discussions. The work of N.K. was supported by KAKENHI (grant numbers 18684018 and 20102002) and partially by MEXT of Japan through Tohoku University GCOE program "Weaving Science Web beyond Particle-matter Hierarchy". I.B. acknowledges initial support by the Venezuelan FONACIT (grant number S1-2001000693).

References

1. Bauer, E., Hilscher, G., Michor, H., Paul, C., Scheidt, E.W., Gribanov, A., Seropegin, Y., Noel, H., Sigrist, M., Rogl, P.: Phys. Rev. Lett. **92**, 027003 (2004)
2. Amato, A., Bauer, E., Baines, C.: Phys. Rev. B **71**, 092501 (2005)
3. Metoki, N., Kaneko, K., Matsuda, T.D., Galatanu, A., Takeuchi, T., Hashimoto, S., Ueda, T., Settai, R., Ōnuki, Y., Bernhoeft, N.: J. Phys.: Conf. Ser. **16**, L207 (2004)
4. Bauer, E., Bonalde, I., Sigrist, M.: Low Temp. Phys. **31**, 748 (2005)
5. Bauer, E., Kaldarar, H., Prokofiev, A., Royanian, E., Amato, A., Sereni, J., Brämer-Escamilla, W., Bonalde, I.: J. Phys. Soc. Jpn. **76**, 051009 (2007)
6. Gribanov, A., Seropegin, Y.D., Tursina, A.I., Bodak, O.I., Rogl, P., Noel, H.: J. Alloys Compd. **383**, 286 (2004)
7. Kim, J.S., Mixson, D.J., Burnette, D.J., Stewart, G.R.: J. Low. Temp. Phys. **147**, 135 (2007)
8. Tateiwa, N., Haga, Y., Matsuda, T.D., Ikeda, S., Yasuda, T., Takeuchi, T., Settai, R., Ōnuki, Y.: J. Phys. Soc. Jpn. **74**, 1903 (2005)
9. Hashimoto, S., Yasuda, T., Kubo, T., Shishido, H., Ueda, T., Settai, R., Matsuda, T.D., Haga, Y., Harima, H., Ōnuki, Y.: J. Phys.: Condens. Matter **16**, L287 (2004)
10. Yasuda, T., Shishido, H., Ueda, T., Hashimoto, S., Settai, R., Takeuchi, T., Matsuda, T.D., Haga, Y., Ōnuki, Y.: J. Phys. Soc. Jpn. **73**, 1657 (2004)
11. Motoyama, G., Maeda, K., Oda, Y.: J. Phys. Soc. Jpn. **77**, 044710 (2008)
12. Mukuda, H., Nishide, S., Harada, A., Iwasaki, K., Yogi, M., Yashima, M., Kitaoka, Y., Tsujino, M., Takeuchi, T., Settai, R., Ōnuki, Y., Bauer, E., Itoh, K.M., Haller, E.E.: J. Phys. Soc. Jpn. **78**, 014705 (2009)

13. Bonalde, I., Ribeiro, R.L., Brämer-Escamilla, W., Rojas, C., Bauer, E., Prokofiev, A., Haga, Y., Yasuda, T., Ōnuki, Y.: New J. Phys. **11**, 055054 (2009)
14. Ribeiro, R.L., Bonalde, I., Haga, Y., Settai, R., Ōnuki, Y.: J. Phys. Soc. Jpn. **78**, 115002 (2009)
15. Samokhin, K.V., Zijlstra, E.S., Bose, S.K.: Phys. Rev. B **69**, 094514 (2004)
16. Sergienko, I.A., Curnoe, S.H.: Phys. Rev. B **70**, 214510 (2004)
17. Frigeri, P.A., Agterberg, D.F., Koga, A., Sigrist, M.: Phys. Rev. Lett. **92**, 097001 (2004)
18. Fujimoto, S.: J. Phys. Soc. Jpn. **76**, 051008 (2007)
19. Yanase, Y., Sigrist, M.: J. Phys. Soc. Jpn. **76**, 043712 (2007)
20. Yogi, M., Mukuda, H., Kitaoka, Y., Hashimoto, S., Yasuda, T., Settai, R., Matsuda, T.D., Haga, Y., Ōnuki, Y., Rogl, P., Bauer, E.: J. Phys. Soc. Jpn. **75**, 013709 (2006)
21. Frigeri, P.A., Agterberg, D.F., Sigrist, M.: New J. Phys. **6**, 115 (2004)
22. Samokhin, K.V.: Phys. Rev. Lett. **94**, 027004 (2005)
23. Kaur, R.P., Agterberg, D.F., Sigrist, M.: Phys. Rev. Lett. **94**, 137002 (2005)
24. Samokhin, K.V.: Phys. Rev. B **78**, 224520 (2008)
25. Helfand, E., Werthamer, N.R.: Phys. Rev. **147**, 288 (1966)
26. Bonalde, I., Brämer-Escamilla, W., Bauer, E.: Phys. Rev. Lett. **94**, 207002 (2005)
27. Mineev, V.P., Samokhin, K.V.: *Introduction to unconventional superconductivity.* Gordon and Breach Science Publishers, Amsterdam (1999)
28. Bonalde, I., Brämer-Escamilla, W., Haga, Y., Bauer, E., Yasuda, Y., Ōnuki, Y.: Physica C **460**, 659 (2007)
29. Izawa, K., Kasahara, Y., Matsuda, Y., Behnia, K., Yasuda, T., Settai, R., Ōnuki, Y.: Phys. Rev. Lett. **94**, 197002 (2005)
30. Takeuchi, T., Yasuda, T., Tsujino, M., Shishido, H., Settai, R., Harima, H., Ōnuki, Y.: J. Phys. Soc. Jpn. **76**, 014702 (2007)
31. Kübert, C., Hirschfeld, P.J.: Phys. Rev. Lett. **80**, 4963 (1998)
32. Lejay, P., Higashi, I., Chevalier, B., Etourneau, J., Hagenmuller, P.: Mater. Res. Bull. **19**, 115 (1984)
33. Steglich, F., Aarts, J., Bredl, C.D., Lieke, W., Meschede, D., Franz, W., Schäfer, H.: Phys. Rev. Lett. **43**, 1892 (1979)
34. Jaccard, D., Behnia, K., Sierro, J.: Phys. Lett. A **163**, 475 (1992)
35. Mathur, N.D., Grosche, F.M., Julian, S.R., Walker, I.R., Freye, D.M., Haselwimmer, R.K.W., Lonzarich, G.G.: Nature **394**, 39 (1998)
36. Movshovich, R., Graf, T., Mandrus, D., Thompson, J.D., Smith, J.L., Fisk, Z.: Phys. Rev. B **53**, 8241 (1996)
37. Palstra, T.T.M., Menovsky, A.A., Vandenberg, J., Dirkmaat, A.J., Kes, P.H., Nieuwenhuys, G.J., Mydosh, J.A.: Phys. Rev. Lett. **55**, 2727 (1985)
38. A. Szytula, In: *Handbook of Magnetic Materials*, vol. 6 ed. by K.H.J. Buschow (Elsevier Science Publishers B. V., Amsterdam, Tokyo, North-Holland, 1991), p. 152
39. Kimura, N., Ito, K., Saitoh, K., Umeda, Y., Aoki, H., Terashima, T.: Phys. Rev. Lett. **95**, 247004 (2005)
40. Sugitani, I., Okuda, Y., Shishido, H., Yamada, T., Thamizhavel, A., Yamamoto, E., Matsuda, T.D., Haga, Y., Takeuchi, T., Settai, R., Ōnuki, Y.: J. Phys. Soc. Jpn. **75**, 043703 (2006)
41. Settai, R., Sugitani, I., Okuda, Y., Thamizhavel, A., Nakashima, M., Ōnuki, Y., Harima, H.: J. Magn. Magn. Mater. **310**, 844 (2007)
42. Honda, F., Bonalde, I., Shimizu, K., Yoshiuchi, S., Hirose, Y., Nakamura, T., Settai, R., Ōnuki, Y.: Phys. Rev. B **81**, 140507 (2010)
43. Kawai, T., Muranaka, H., Measson, M.A., Shimoda, T., Doi, Y., Matsuda, T.D., Haga, Y., Knebel, G., Lapertot, G., Aoki, D., Flouquet, J., Takeuchi, T., Settai, R., Ōnuki, Y.: J. Phys. Soc. Jpn. **77**, 064716 (2008)
44. Thamizhavel, A., Takeuchi, T., Matsuda, T.D., Haga, Y., Sugiyama, K., Settai, R., Ōnuki, Y.: J. Phys. Soc. Jpn. **74**, 1858 (2005)
45. Mock, S., Pfleiderer, C., von Löhneysen, H.: J. Low. Temp. Phys. **115**, 1 (1999)
46. Paschen, S., Felder, E., Ott, H.R.: Europ. Phys. J. B **2**, 169 (1998)

47. Mentink, S.A.M., Bos, N.M., Vanrossum, B.J., Nieuwenhuys, G.J., Mydosh, J.A., Buschow, K.H.J.: J. Appl. Phys. **73**, 6625 (1993)
48. Kontani, M., Ido, H., Ando, H., Nishioka, T., Yamaguchi, Y.: J. Phys. Soc. Jpn. **63**, 1652 (1994)
49. Sugawara, H., Saha, S.R., Matsuda, T.D., Aoki, Y., Sato, H., Gavilano, J.L., Ott, H.R.: Physica B **261**, 16 (1999)
50. Nakashima, M., Tabata, K., Thamizhavel, A., Kobayashi, T.C., Hedo, M., Uwatoko, Y., Shimizu, K., Settai, R., Ōnuki, Y.: J. Phys.: Condens. Matter **16**, L255 (2004)
51. Morozkin, A.V., Seropegin, Y.D.: J. Alloys Compd. **237**, 124 (1996)
52. Ghosh, K., Ramakrishnan, S., Dhar, S.K., Malik, S.K., Chandra, G., Pecharsky, V.K., Gschneidner, K.A., Hu, Z., Yelon, W.B.: Phys. Rev. B **52**, 7267 (1995)
53. Takabatake, T., Maeda, Y., Fujii, H., Ikeda, S., Nishigori, S., Fujita, T., Minami, A., Oguro, I., Sugiyama, K., Oda, K., Date, M.: Physica B **188**, 734 (1993)
54. Settai, R., Miyauchi, Y., Takeuchi, T., Lévy, F., Sheikin, I., Ōnuki, Y.: J. Phys. Soc. Jpn. **77**, 073705 (2008)
55. Y. Muro, Ph.D. thesis, The University of Tokyo (2000)
56. Pecharsky, V.K., Hyun, O.B., Gschneidner, K.A.: Phys. Rev. B **47**, 11839 (1993)
57. Das, A., Kremer, R.K., Pöttgen, R., Ouladdiaf, B.: Physica B **378**(80), 837 (2006)
58. Rupp, B., Rogl, P., Hulliger, F.: J. Less-Common Met. **135**, 113 (1987)
59. Eom, D., Ishikawa, M., Kitagawa, J., Takeda, N.: J. Phys. Soc. Jpn. **67**, 2495 (1998)
60. Muro, Y., Eom, D., Takeda, N., Ishikawa, M.: J. Phys. Soc. Jpn. **67**, 3601 (1998)
61. Kawai, T., Nakashima, M., Ōnuki, Y., Shishido, H., Shimoda, T., Matsuda, T.D., Haga, Y., Takeuchi, T., Hedo, M., Uwatoko, Y., Settai, R., Ōnuki, Y.: J. Phys. Soc. Jpn. Suppl. A **76**, 166 (2007)
62. Kitagawa, J., Muro, Y., Takeda, N., Ishikawa, M.: J. Phys. Soc. Jpn. **66**, 2163 (1997)
63. Haen, P., Lejay, P., Chevalier, B., Lloret, B. J. Etourneau, Sera, M.: J. Less-Common Met. **110**, 321 (1985)
64. Kawai, T., Okuda, Y., Shishido, H., Thamizhavel, A., Matsuda, T.D., Haga, Y., Nakashima, M., Takeuchi, T., Hedo, M., Uwatoko, Y., Settai, R., Ōnuki, Y.: J. Phys. Soc. Jpn. **76**, 014710 (2007)
65. Yamamoto, H., Ishikawa, M., Hasegawa, K., Sakurai, J.: Phys. Rev. B **52**, 10136 (1995)
66. T. Kawai, Split fermi surface properties and superconductivity in the non-centrosymmetric crystal structure. Ph.D. thesis, Osaka University (2008)
67. Muro, Y., Ishikawa, M., Hirota, K., Hiroi, Z., TakedaN., , Kimura, N., Aoki, H.: J. Phys. Soc. Jpn. **76**, 033706 (2007)
68. Okuda, Y., Miyauchi, Y., Ida, Y., Takeda, Y., Tonohiro, C., Oduchi, Y., Yamada, T., Dung, N.D., Matsuda, T.D., Haga, Y., Takeuchi, T., Hagiwara, M., Kindo, K., Harima, H., Sugiyama, K., Settai, R., Ōnuki, Y.: J. Phys. Soc. Jpn. **76**, 044708 (2007)
69. Aso, N., Miyano, H., Yoshizawa, H., Kimura, N., Komatsubara, T., Aoki, H.: J. Magn. Magn. Mater. **310**, 602 (2007)
70. H. Miyano, Spin correlation in pressure-induced superconductors. Master's thesis, the University of Tokyo (2007)
71. Tomioka, F., Umehara, I., Ono, T., Hedo, M., Uwatoko, Y., Kimura, N.: Jpn. J. Appl. Phys. **46**, 3090 (2007)
72. Kimura, N., Muro, Y., Aoki, H.: J. Phys. Soc. Jpn. **76**, 051010 (2007)
73. Kadowaki, K., Woods, S.B.: Sol. State Commun. **58**, 507 (1986)
74. Kaneko, K., Metoki, N., Takeuchi, T., Matsuda, T.D., Haga, Y., Thamizhavel, A., Settai, R., Ōnuki, Y.: J. Phys.: Conf. Ser. **150**, 042082 (2009)
75. Tateiwa, N., Haga, Y., Matsuda, T.D., Ikeda, S., Yamamoto, E., Okuda, Y., Miyauchi, Y., Settai, R., Ōnuki, Y.: J. Phys. Soc. Jpn. **76**, 083706 (2007)
76. Settai, R., Okuda, Y., Sugitani, I., Ōnuki, Y., Matsuda, T.D., Haga, Y., Harima, H.: Int. J. Mod. Phys. B **21**, 3238 (2007)

77. Knebel, G., Aoki, D., Lapertot, G., Salce, B., Flouquet, J., Kawai, T., Muranaka, H., Settai, R., Ōnuki, Y.: J. Phys. Soc. Jpn. **78**, 074714 (2009)
78. Mukuda, H., Fujii, T., Ohara, T., Harada, A., Yashima, M., Kitaoka, Y., Okuda, Y., Settai, R., Ōnuki, Y.: Phys. Rev. Lett. **100**, 107003 (2008)
79. Fisher, R.A., Bouquet, F., Phillips, N.E., Hundley, M.F., Pagliuso, P.G., Sarrao, J.L., Fisk, Z., Thompson, J.D.: Phys. Rev. B **65**, 224509 (2002)
80. Araki, S., Nakashima, M., Settai, R., Kobayashi, T.C., Ōnuki, Y.: J. Phys.: Condens. Matter **14**, L377 (2002)
81. Knebel, G., Braithwaite, D., Canfield, P.C., Lapertot, G., Flouquet, J.: Phys. Rev. B **65**, 024425 (2002)
82. Miyake, K., Maebashi, H.: J. Phys. Soc. Jpn. **71**, 1007 (2002)
83. Lévy, F., Sheikin, I., Grenier, B., Huxley, A.D.: Science **309**, 1343 (2005)
84. Uji, S., Shinagawa, H., Terashima, T., Yakabe, T., Terai, Y., Tokumoto, M., Kobayashi, A., Tanaka, H., Kobayashi, H.: Nature **410**, 908 (2001)
85. Hake, R.R.: Appl. Phys. Lett. **10**, 189 (1967)
86. Bulaevskii, L.N., Dolgov, O.V., Ptitsyn, M.O.: Phys. Rev. B **38**, 11290 (1988)
87. Huxley, A., Sheikin, I., Ressouche, E., Kernavanois, N., Braithwaite, D., Calemczuk, R.: J. Flouquet, Phys. Rev. B **63**, 144519 (2001)
88. Sheikin, I., Huxley, A., Braithwaite, D., Brison, J.P., Watanabe, S., Miyake, K., Flouquet, J.: Phys. Rev. B **64**, 220503 (2001)
89. Terashima, T., Takahide, Y., Matsumoto, T., Uji, S., Kimura, N., Aoki, H., Harima, H.: Phys. Rev. B **76**, 054506 (2007)
90. Tada, Y., Kawakami, N., Fujimoto, S.: Phys. Rev. Lett. **101**, 267006 (2008)
91. Agterberg, D.F., Frigeri, P.A., Kaur, R.P., Koga, A., Sigrist, M.: Physica B **378**(80), 351 (2006)
92. Agterberg, D.F., Kaur, R.P.: Phys. Rev. B **75**, 064511 (2007)
93. Kimura, N., Ito, K., Aoki, H., Uji, S., Terashima, T.: Phys. Rev. Lett. **98**, 197001 (2007)
94. Tada, Y., Kawakami, N., Fujimoto, S.: J. Phys. Soc. Jpn. **77**, 054707 (2008)
95. Dommann, A., Hullinger, F., Siegrist, T., Fischer, P.: J. Magn. Magn. Mater. **67**, 323 (1987)
96. Akazawa, T., Hidaka, H., Fujiwara, T., Kobayashi, T.C., Yamamoto, E., Haga, Y., Settai, R., Ōnuki, Y.: J. Phys.: Condens. Matter **16**, L29 (2004)
97. Yamamoto, E., Haga, Y., Shishido, H., Nakawaki, H., Inada, Y., Settai, R., Ōnuki, Y.: Physica B **312**, 302 (2002)
98. Anderson, P.W.: J. Phys. Chem. Solids **11**, 26 (1959)
99. Anderson, P.W.: Phys. Rev. B **30**, 4000 (1984)
100. Saxena, S.S., Agarwal, P., Ahilan, K., Grosche, F.M., Haselwimmer, R.K.W., Steiner, M.J., Pugh, E., Walker, I.R., Julian, S.R., Monthoux, P., Lonzarich, G.G., Huxley, A., Sheikin, I., Braithwaite, D., Flouquet, J.: Nature **406**, 587 (2000)
101. Aoki, D., Huxley, A., Ressouche, E., Braithwaite, D., Flouquet, J., Brison, J.P., Lhotel, E., Paulsen, C.: Nature **413**, 613 (2001)
102. Ōnuki, Y., Settai, R., Sugiyama, K., Inada, Y., Takeuchi, T., Haga, Y., Yamamoto, E., Harima, H., Yamagami, H.: J. Phys.: Condens. Matter **19**, 125203 (2007)
103. Kobayashi, T.C., Hidaka, H., Fujiwara, T., Tanaka, M., Takeda, K., Akazawa, T., Shimizu, K., Kirita, S., Asai, R., Nakawaki, H., Nakashima, M., Settai, R., Yamamoto, E., Haga, Y., Ōnuki, Y.: J. Phys.: Condens. Matter **19**, 125205 (2007)
104. Yamamoto, E., Haga, Y., Ikeda, S., Matsuda, T.D., Akazawa, T., Kotegawa, H., Kobayashi, T.C., Ōnuki, Y.: J. Magn. Magn. Mater. **310**, e123 (2007)
105. Akazawa, T., Hidaka, H., Kotegawa, H., Kobayashi, T.C., Fujiwara, T., Yamamoto, E., Haga, Y., Settai, R., Ōnuki, Y.: J. Phys. Soc. Jpn. **73**, 3129 (2004)
106. Sakarya, S., van Dijk, N.H., de Visser, A., Bruck, E., Huang, Y., Perenboom, J., Rakoto, H., Broto, J.M.: J. Magn. Magn. Mater. **310**, 1564 (2007)
107. Kobayashi, T.C., Hori, A., Fukushima, S., Hidaka, H., Kotegawa, H., Akazawa, T., Takeda, K., Ohishi, Y., Yamamoto, E.: J. Phys. Soc. Jpn. **76**, 051007 (2007)

108. Galatanu, A., Haga, Y., Yamamoto, E., Matsuda, T.D., Ikeda, S., Ōnuki, Y.: J. Phys. Soc. Jpn. **73**, 766 (2004)
109. Galatanu, A., Haga, Y., Matsuda, T.D., Ikeda, S., Yamamoto, E., Aoki, D., Takeuchi, T., Ōnuki, Y.: J. Phys. Soc. Jpn. **74**, 1582 (2005)
110. Bauer, E.D., Freeman, E.J., Sirvent, C., Maple, M.B.: J. Phys.: Condens. Matter **13**, 5675 (2001)
111. Frigeri, P.A., Agterberg, D.F., Milat, I., Sigrist, M.: Europ. Phys. J. B **54**, 435 (2006)
112. Mineev, V.P., Samokhin, K.V.: Phys. Rev. B **75**, 184529 (2007)
113. Linder, J., Nevidomskyy, A.H., Sudbo, A.: Phys. Rev. B **78**, 172502 (2008)

Chapter 3
Electronic States and Superconducting Properties of Non-centrosymmetric Rare Earth Compounds

Yoshichika Ōnuki and Rikio Settai

Abstract The property of Fermi surface splitting in the non-centrosymmetric tetragonal compounds $R\mathrm{Pt_3Si}$ (R: La, Ce) and RTX_3 (T: Co, Rh, Ir; X: Si, Ge) are studied by de Haas-van Alphen experiments and compared with energy band calculations. Moreover superconducting properties are investigated in these compounds. In particular, in $\mathrm{CeIrSi_3}$ the unusual behavior of the upper critical field H_{c2} at pressures around 2.6 GPa is analyzed in detail. At 2.6 GPa, a huge value of $H_{c2}(0) \simeq 450$ kOe is found for magnetic fields along the [0 0 1] direction of the non-centrosymmetric tetragonal crystal structure, in contrast to the smaller $H_{c2}(0) \simeq 95$ kOe for $H \parallel [1\ 1\ 0]$.

3.1 Introduction

Since the discovery of the first heavy fermion superconductor $\mathrm{CeCu_2Si_2}$ in 1979, heavy fermion superconductivity has been observed in several cerium, praseodymium, uranium, and nowadays even neptunium and plutonium compounds. Heavy fermion superconductivity is found to coexist with antiferromagnetism as well as ferromagnetism. Furthermore, it is widely recognized that pressure P is a useful tuning parameter to find superconductivity in magnetically ordered f-electron compounds [1, 2]. With increasing pressure, the magnetic ordering temperature T_{mag} becomes zero at a critical pressure P_{c} in some compounds: $T_{\mathrm{mag}} \to 0$ for $P \to P_{\mathrm{c}}$. For example, an antiferromagnetic cerium compound is changed into a paramagnet at pressures higher than P_{c}. The heavy fermion state is formed around P_{c} as a result of the competition between the Ruderman-Kittel-Kasuya-Yosida (RKKY)

Y. Ōnuki (✉) · R. Settai
Graduate School of Science, Osaka University,
Toyonaka, Osaka 560-0043, Japan
email: onuki@phys.sci.osaka-u.ac.jp

R. Settai
email: settai@phys.sci.osaka-u.ac.jp

E. Bauer and M. Sigrist (eds.), *Non-centrosymmetric Superconductors*,
Lecture Notes in Physics 847, DOI: 10.1007/978-3-642-24624-1_3,

interaction and the Kondo effect. Heavy fermion superconductivity is often observed in this pressure region.

Phonon-mediated Cooper pairing is most likely not effective in heavy fermion superconductors due to the strongly repulsive interaction among the quasiparticles derived from the itinerant and strongly correlated *f*-electrons. Nevertheless, superconductivity is realized in these systems. To minimize the repulsive interactions, electrons preferentially choose an anisotropic channel, such as a *p*-wave spin-triplet state or a *d*-wave spin-singlet state, to form Cooper pairs [3]. Neutron-scattering experiments clearly indicate a close relationship between superconductivity and magnetic excitations in UPd_2Al_3 [4, 5, 6]. The magnetic excitation gap of UPd_2Al_3, which appears in inelastic neutron scattering below the superconducting transition temperature T_{sc}, results from the presence of the superconducting order parameter.

Figure 3.1 shows a schematic view of the superconducting order parameters with *s*-, *d*-, and *p*-wave pairings. The order parameter $\Psi(\boldsymbol{r})$ with even parity (*s*- and *d*-waves) is symmetric with respect to $\boldsymbol{r}$ where one electron with the up-spin state of the Cooper pair is simply considered to be located at the center of $\Psi(\boldsymbol{r})$, namely at $\boldsymbol{r} = 0$, and the other electron with the down-spin state is located at $\boldsymbol{r}$, revealing the spin-singlet state of $(|\uparrow\downarrow\rangle - |\downarrow\uparrow\rangle)/\sqrt{2}$ (total spin $S = 0$). The width of $\Psi(\boldsymbol{r})$ with respect to $\boldsymbol{r}$ is called the coherence length ξ, as shown in Fig 3.1a. Nuclear magnetic resonance (NMR) and nuclear quadrupole resonance (NQR) have proved to be useful tools for determining the symmetry of the superconducting condensate. For example, UPd_2Al_3 is considered to be a *d*-wave superconductor from the NMR experiment [7], which corresponds to the case in Fig. 3.1b. The origin of pairing has also been clarified by neutron-scattering experiments on UPd_2Al_3 [4, 5, 6], as mentioned above. On the other hand, $\Psi(\boldsymbol{r})$ with odd parity (*p*- or *f*-wave) is not symmetric with respect to $\boldsymbol{r}$, where the parallel spin state is shown in Fig. 3.1c. Namely, the spin state is of spin-triplet in nature: $|\uparrow\uparrow\rangle$ $(S_z = 1)$, $(|\uparrow\downarrow\rangle + |\downarrow\uparrow\rangle)/\sqrt{2}$ $(S_z = 0)$, and $|\downarrow\downarrow\rangle$ $(S_z = -1)$. From NMR, magnetization and thermal conductivity experiments, UPt_3 is considered to possess odd parity symmetry [8, 9, 10].

Recently superconductivity in several non-centrosymmetric heavy fermion compounds has been reported, in $CePt_3Si$ [11, 12] with the tetragonal structure ($P4mm$), UIr [13, 14] with the monoclinic structure ($P2_1$), $CeRhSi_3$ [15–17], $CeIrSi_3$ [1, 18, 19] and $CeCoGe_3$ [20] with the tetragonal $BaNiSn_3$-type structure ($I4mm$). We show in Fig. 3.2 the crystal structure of $CePt_3Si$ and $CeIrSi_3$, which lack inversion symmetry along the tetragonal [0 0 1] direction (*c*-axis). In $CeIrSi_3$, the Ce atoms occupy the four corners and the body center of the tetragonal structure, similar to the well-known tetragonal $CeCu_2Si_2$($ThCr_2Si_2$-type), but the Ir and Si atoms lack inversion symmetry along the [0 0 1] direction.

Inversion is an essential symmetry for the formation of Cooper pairs. In non-centrosymmetric metals a splitting of Fermi surfaces with different spin directions occurs, restricting the possible Cooper pair states which can be formed keeping the total momentum zero. In the case of $CePt_3Si$, $CeRhSi_3$, $CeIrSi_3$ and $CeCoGe_3$ split Fermi surfaces very similar to each other in topology but different in volume are formed due to the presence of a Rashba-type antisymmetric spin-orbit interaction:

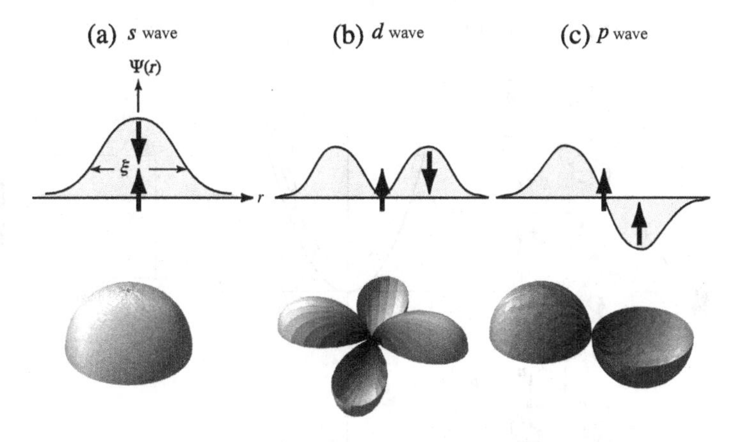

Fig. 3.1 Schematic view of the superconducting order parameters with *s*-, *d*- and *p*-wave pairings

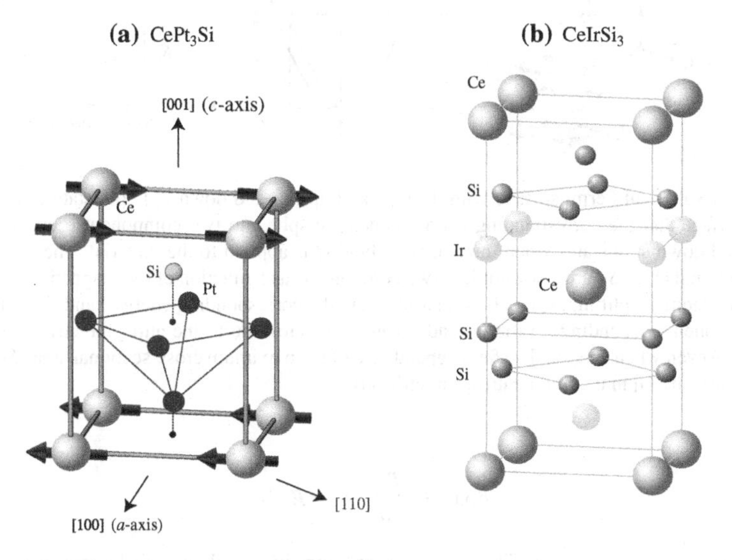

Fig. 3.2 **a** Crystal and magnetic structure of $CePt_3Si$, and the crystal structure of $CeIrSi_3$, which also lack inversion symmetry along the [0 0 1] direction

$$\begin{aligned}\mathcal{H}_{\rm so} &= -\frac{\hbar}{4m^{*2}c^2}(\nabla V(\boldsymbol{r}) \times \boldsymbol{p}) \cdot \boldsymbol{\sigma} \\ &= \alpha(\boldsymbol{n} \times \boldsymbol{p}) \cdot \boldsymbol{\sigma} \\ &= \alpha \boldsymbol{p}_{\perp} \cdot \boldsymbol{\sigma},\end{aligned} \tag{3.1}$$

where α denotes the strength of the spin-orbit coupling, $\boldsymbol{n}$ is a unit vector derived from $\nabla V(\boldsymbol{r})$, which lies along the [0 0 1] direction (*c*-axis) for these compounds, and $\boldsymbol{\sigma}$ is the Pauli matrix [21, 22]. This additional term in the electron Hamiltonian separates the spin degenerate bands into two given by [22–24],

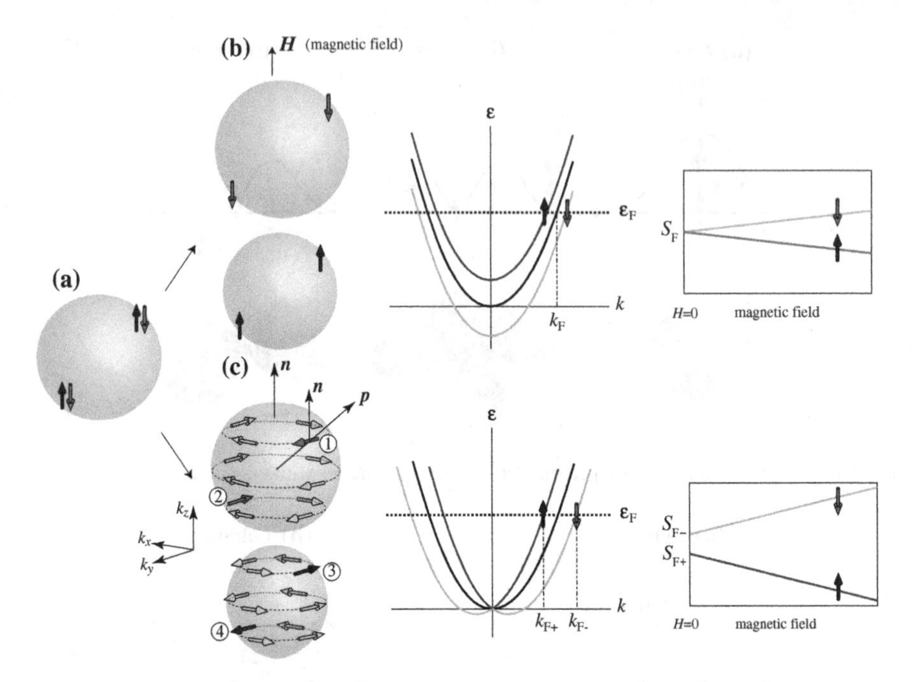

Fig. 3.3 **a** Spherical Fermi surface with degenerated up (↑) and down (↓) spin states, and **b** the Fermi surface and the corresponding energy bands are split into two components depending on the up and down spin states when the magnetic field H is applied to the material. The maximum cross-sectional areas S_F are also split into two components as a function of the magnetic field, well known as Zeeman splitting. **c** The Fermi surface and the corresponding energy band are split into two components depending on the up and down spin states due to the antisymmetric spin-orbit interaction even when $H = 0$. The field dependence of the maximum cross-sectional areas S_{F-} and S_{F+} are also shown in the non-centrosymmetric structure

$$\varepsilon_{p\pm} = \frac{\boldsymbol{p}^2}{2m^*} \mp |\alpha \boldsymbol{p}_\perp|, \tag{3.2}$$

where $\boldsymbol{p}_\perp = (p_y, -p_x, 0)$. This splitting appears in the absence of a magnetic field and introduces a characteristic momentum-dependent spin structure to the electronic states, as shown in Fig. 3.3(c). Note that the spins of the conduction electrons are rotated for the direction of the effective magnetic field, $\boldsymbol{n} \times \boldsymbol{p}$, clockwise or anticlockwise, depending on the up and down spin states. For comparison, in Fig. 3.3(b), we show the well-known Zeeman splitting, where the degenerated Fermi surface is split into two Fermi surfaces corresponding to a majority and minority spin, respectively, for a given quantization axis parallel to an applied magnetic field.

In a non-centrosymmetric metal most p-wave pairing states are prohibited because electrons would have to form zero-momentum Cooper pairs, which are separated by an energy of $2|\alpha \boldsymbol{p}_\perp| \sim 10 - 1000K$, much larger than the superconducting energy gap of a few Kelvin in heavy fermion superconductors. Frigeri *et al.* studied theoretically the possible existence of a spin-triplet pairing state compatible with

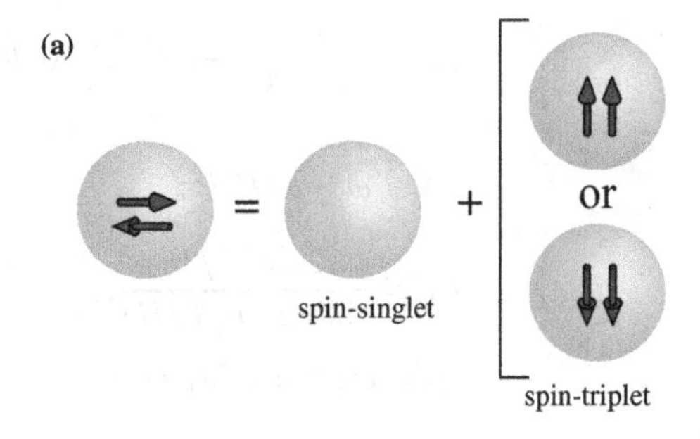

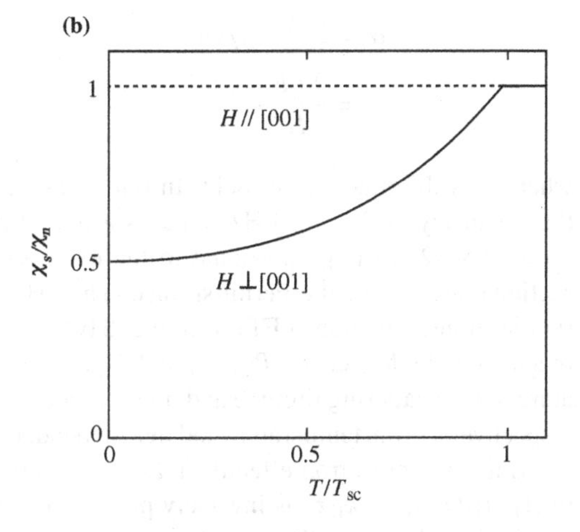

Fig. 3.4 **a** Schematic picture of parity-mixed Cooper pairs. The *left-hand side* represents the Cooper pair with the spin quantization axis parallel to $\boldsymbol{n} \times \boldsymbol{p}$. The *right-hand side* represents the sum of a spin-singlet state and a spin-triplet state with $S_z = \pm 1$ for the spin quantization axis parallel to the z-axis, or the [0 0 1] direction, Ref. [26]. **b** Temperature dependence of the spin susceptibility below T_{sc} for $H \parallel [0\,0\,1]$ and $H \perp [0\,0\,1]$, Ref. [25].

antisymmetric spin-orbit coupling [21, 25]. The $\boldsymbol{d}$ vector characterizing the corresponding spin-triplet state, is parallel to $\boldsymbol{p}_{\perp}$: $\boldsymbol{d}(\boldsymbol{k}) = \Delta(k_y \boldsymbol{x} - k_x \boldsymbol{y})$. In a complete description the order parameter is a mixture of spin-singlet and spin-triplet components, as shown schematically in Fig. 3.4(a) [26]. The corresponding spin susceptibility becomes non-zero at 0 K, as shown in Fig. 3.4(b): for the magnetic field H along the [0 0 1] direction, $\chi(H \parallel [0\,0\,1])$ is unchanged below the superconducting transition temperature T_{sc}, and $\chi(H \perp [0\,0\,1])$ becomes $\chi(H \parallel [0\,0\,1])/2$ at 0 K. The fact that the spin susceptibility remains constant below T_{sc} for $H \parallel [0\,0\,1]$ is a result of a van Vleck-type contribution to χ due to the spin-orbit coupling and the Fermi surface splitting.

The theoretical result in Fig. 3.4 suggests that there exists no paramagnetic limiting effect on the upper critical field H_{c2} for magnetic fields along the [0 0 1] direction (c-axis), while paramagnetic suppression is expected for $H \perp [0\,0\,1]$. On rather general grounds it has been shown theoretically that the mixing of even and odd parity states can give rise to nodes in the quasiparticle excitation gap.

We studied Fermi-surface properties by de Haas-van Alphen (dHvA) measurements and compared the results with energy band-structure calculations. The dHvA signal V_{osc}, in the usual 2ω-type field modulation method, is simply given as follows [27, 28]:

$$V_{\text{osc}} = A \left| \frac{\partial^2 S_{\text{F}}(k_z)}{\partial k_z^2} \right|^{-1/2} R_{\text{T}} R_{\text{D}} R_{\text{S}} \sin\left(\frac{2\pi F}{H} + \phi \right), \tag{3.3}$$

$$A \propto \omega J_2(x) H^{1/2}, \tag{3.4}$$

$$R_{\text{T}} = \frac{2\alpha m_{\text{c}}^* T/H}{\sinh(2\alpha m_{\text{c}}^* T/H)}, \tag{3.5}$$

$$R_{\text{D}} = \exp(-\alpha m_{\text{c}}^* T_{\text{D}}/H), \tag{3.6}$$

$$R_{\text{S}} = \cos(\pi m_{\text{c}}^* g/2m_0), \tag{3.7}$$

$$\alpha = 2\pi^2 c k_{\text{B}}/e\hbar, \tag{3.8}$$

$$x = \frac{2\pi F h}{H^2}, \tag{3.9}$$

where h is the modulation field. In our measurements we take $h = 100$ Oe and the frequency $\omega/2\pi = 11$ Hz, $J_2(x)$ is a Bessel function and the dHvA frequency $F(= c\hbar S_{\text{F}}/2\pi e)$ is proportional to the extremal (maximum or minimum) cross-sectional area S_{F} of the Fermi surface. The dHvA frequencies F, obtained by fast Fourier transformation (FFT) from the dHvA oscillations, are expressed in units of magnetic field. Moreover, R_{T}, R_{D} and R_{S} are reduction factors due to finite temperature, finite scattering lifetime and interference between up and down spin electrons, respectively. From temperature and field dependences of the dHvA amplitude, we can determine the cyclotron effective mass m_{c}^* and the Dingle temperature T_{D}, respectively. $T_{\text{D}}(= \hbar/2\pi k_{\text{B}}\tau)$ is inversely proportional to the scattering lifetime τ of the conduction electrons. We can also estimate the mean free path ℓ from the simple relations: $S_{\text{F}} = \pi k_{\text{F}}^2$, $\hbar k_{\text{F}} = m_{\text{c}}^* v_{\text{F}}$, and $\ell = v_{\text{F}}\tau$, where k_{F} is half of the diameter of a circular S_{F} and v_{F} is the Fermi velocity.

Here we comment on the relation between the Fermi surface and the spin states of conduction electrons. For an inversion-symmetric crystal structure the up and down spin states of conduction electrons are degenerate and have the same Fermi surface, as shown in Fig. 3.3(a). In a magnetic field, the degenerate Fermi surface is split into two sheets, as shown in Fig. 3.3(b), the well known Zeeman splitting. The split Fermi surfaces, however, yield the same dHvA frequency F corresponding to the frequency extrapolated to zero field, as shown in Fig. 3.3(b).

The relations of Eqs. (3.3)–(3.9) can also be extended to the dHvA oscillations for non-centrosymmetric metals. Due to the Fermi surface splitting the dHvA frequency F is split into two dHvA frequencies, F_+ and F_-, as shown in Fig. 3.3(c). Using the relations of $\varepsilon_{\text{F}} = \hbar^2 k_{\text{F}}^2/2m_{\text{c}}^*$, $S_{\text{F}} = \pi k_{\text{F}}^2$ and $S_{\text{F}} = (2\pi e/c\hbar)F$, we obtain from Eq. (3.2):

$$|F_+ - F_-| = \frac{2c}{\hbar e} |\alpha \boldsymbol{p}_\perp| m_c^*. \tag{3.10}$$

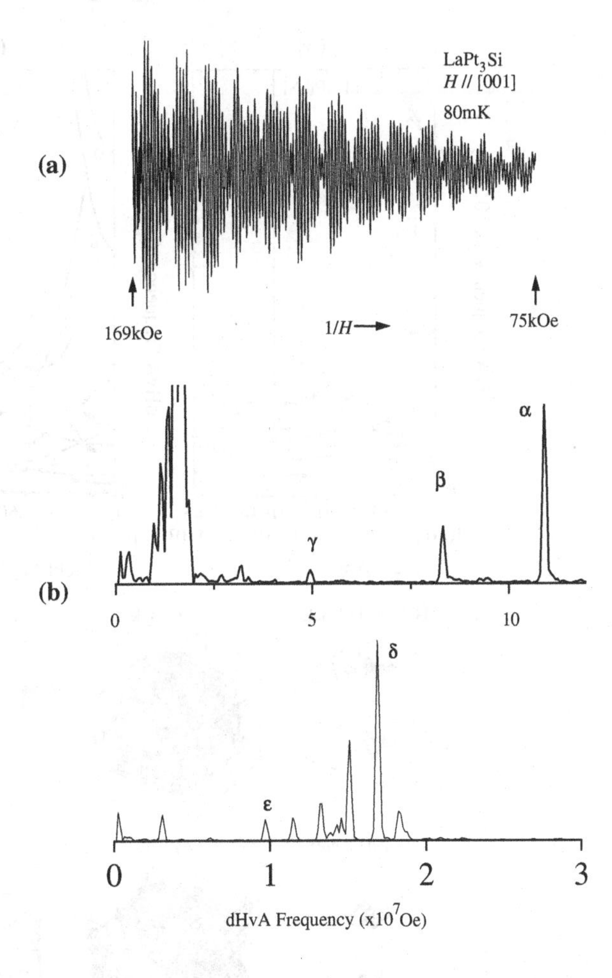

Fig. 3.5 **a** Typical dHvA oscillation for $H \parallel [0\,0\,1]$ and **b** its FFT spectrum in $LaPt_3Si$, Ref. [29]

We can thus determine the magnitude of the antisymmetric spin-orbit interaction $2|\alpha \boldsymbol{p}_{\perp}|$ via the dHvA experiment.

In this chapter, we show the split Fermi-surface properties of $R\mathrm{Pt_3Si}(R\!: \mathrm{La, Ce})$ and $\mathrm{R}TX_3$ (*T*: Co, Rh, Ir, *X*: Si, Ge), and also their corresponding characteristic superconducting properties.

3.2 Electronic States and Superconducting Properties of $LaPt_3Si$ and $CePt_3Si$

First we show in Fig. 3.5(a) typical dHvA oscillations for $LaPt_3Si$ in a magnetic field *H* along the [0 0 1] direction (*c*-axis) and its FFT spectrum [29]. The detected dHvA branches are named α, β, γ, δ and ε, as shown in Fig. 3.5(b).

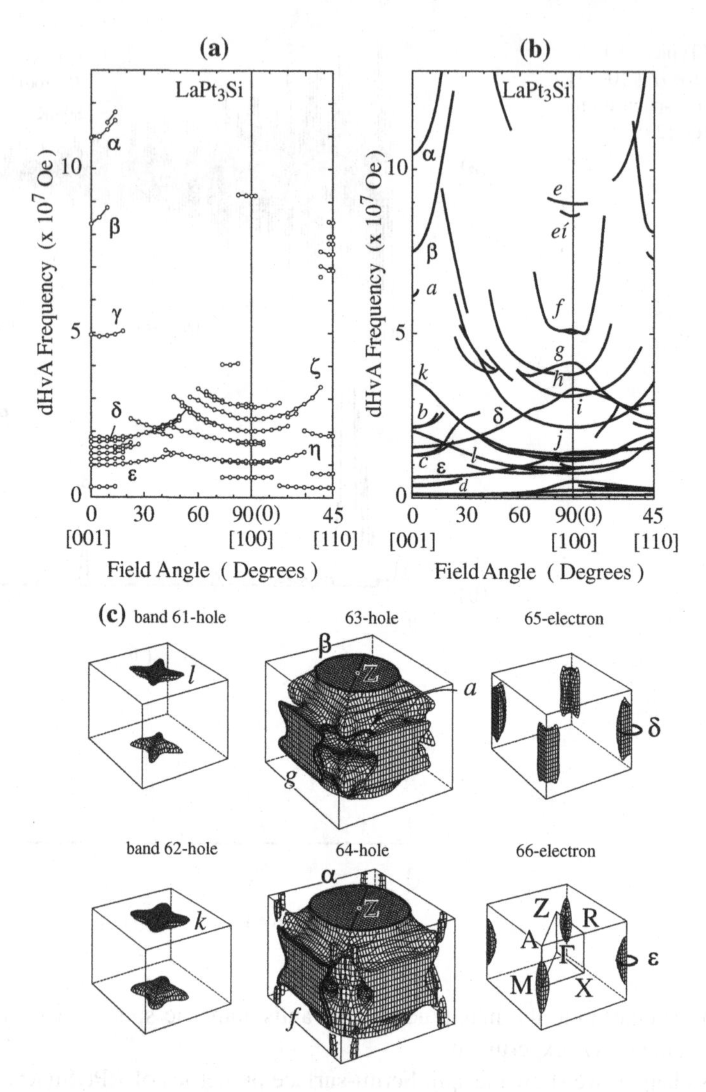

Fig. 3.6 **a** Experimental angular dependence of the dHvA frequency, **b** the theoretical one and **c** the corresponding theoretical Fermi surfaces in $LaPt_3Si$, Ref. [29]

Figures 3.6(a), 3.6(b) and 3.6(c) show the experimental angular dependence of the dHvA frequency, the theoretical one based on the full potential linearized augmented plane wave (FLAPW) method and the corresponding Fermi surfaces in $LaPt_3Si$, respectively. The branches α, β, δ and ε correspond to bands 64, 63 hole-Fermi surfaces, 65 and 66 electron-Fermi surfaces, respectively, as shown in Fig. 3.6(c). The two kinds of Fermi surfaces can be attributed to the splitting of the electron spectrum by antisymmetric spin-orbit coupling, as mentioned above. Therefore, the Fermi surfaces of the bands 61 and 62, bands 63 and 64, and bands 65 and 66 form

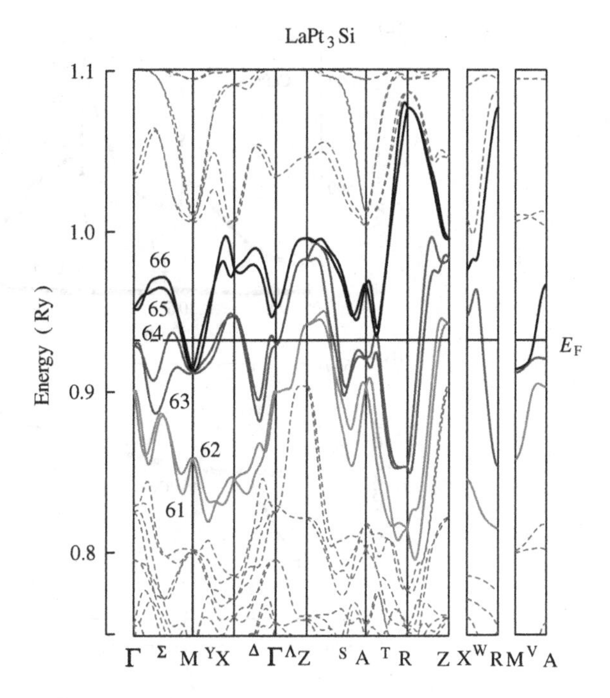

Fig. 3.7 Energy band structure in $LaPt_3Si$, Ref. [30]

such split pairs, respectively. The structure off spin-orbit coupling implies that the Fermi surface degeneracy exists only for k along the c-axis (within the Brillouin zone along the Γ–Z, X–R and M–A directions) in $LaPt_3Si$ ($CePt_3Si$), while splitting is realized along all other directions as shown in Figs. 3.6(c) and 3.7.

Using the observed dHvA frequencies and the averaged value of two cyclotron effective masses, we obtain the antisymmetric spin-orbit interaction $2|\alpha p_\perp|$ as 2400 K for branches α and β, and 800 K for branches δ and ε. The corresponding theoretical values are 4200 K for branches α and β, and 2000 K for branches δ and ε, where we used the theoretical dHvA frequencies and averaged band masses in Table 3.1. We summarize in Table 3.1 the experimental values of the dHvA frequency F, the cyclotron effective mass m_c^* and the antisymmetric spin-orbit interaction $2\,|\alpha p_\perp|$, together with the corresponding theoretical values of the dHvA frequency F_b and the band mass m_b and the estimated value $2\,|\alpha p_\perp|$ for branches α, β, δ and ε.

$CePt_3Si$ orders antiferromagnetically below $T_N = 2.3$ K [11, 32, 32], with a small ordered moment of 0.16 μ_B/Ce and a propagation vector $\boldsymbol{q} = (0\ 0\ \frac{1}{2})$, where the ordered moments are ferromagnetically directed along the [1 0 0] (a-axis) in each plane and antiferromagnetically stacked along the [0 0 1] direction, [33] as shown in Fig. 3.2(a). Within the antiferromagnetic state, $CePt_3Si$ becomes superconducting below the transition temperature $T_{sc} = 0.46$ K, as shown in Fig. 3.8(a). The electronic specific heat (C_e/T) is shown in Fig. 3.8(c) as a function of temperature for $H = 0$, after subtracting antiferromagnetic and phonon contributions from the raw C/T

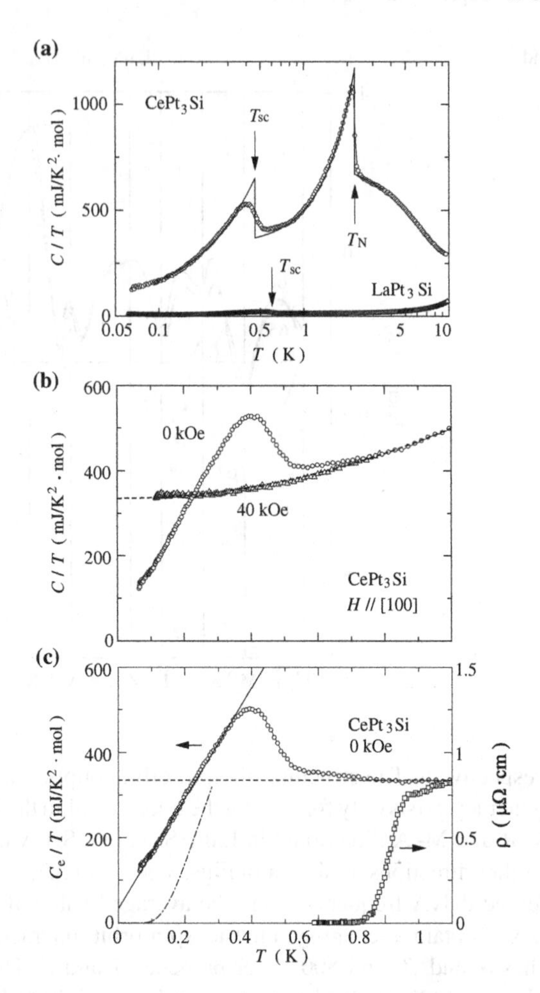

Fig. 3.8 **a** Temperature dependence of the specific heat divided by temperature C/T of $CePt_3Si$, together with the data of a non-4*f* reference compound, $LaPt_3Si$. Solid lines show idealized jumps at T_{sc} and T_N, Ref. [30]. **b** Specific heat divided by temperature C/T around T_{sc} of $CePt_3Si$. Circles (○) and triangles (△) show the data under magnetic fields $H = 0$ and 40 kOe, respectively. A nuclear Schottky contribution of 0.03[mJ· K/mol]/T^3 was subtracted from the raw C/T data measured at 40 kOe. The broken line indicates an extrapolation below T_N which is expressed as $C/T = 335$ [mJ/K^2· mol]+165[mJ/K^4· mol]T^2. **c** Electronic specific heat divided by temperature C_e/T for $H = 0$ after subtracting 165[mJ/K^4· mol]T^2 from the raw C/T data. The solid line shows the linear temperature dependence of C_s/T = 34.1[mJ/K^2· mol]+1290[mJ/K^3· mol]T. The broken line indicates the normal-state electronic specific-heat coefficient γ_n = 335 mJ/K^2· mol. The dashed-dotted line shows a prediction of BCS weak-coupling theory with $2\Delta/k_B T_{sc} = 3.53$. The electrical resistivity data of the same sample are also shown in the right-hand scale

Table 3.1 Experimental dHvA frequency F, the cyclotron effective mass m_c^* and the spin-orbit interaction $2\,|\alpha p_\perp|$, and the theoretical dHvA frequency F_b, the band mass m_b and the theoretical spin-orbit interaction $2\,|\alpha p_\perp|$ for $H \parallel [0\,0\,1]$ in $LaPt_3Si$, Ref. [30]

	Experiment			Theory		
Branch	$F(\times 10^7$ Oe)	$m_c^*(m_0)$	$2\,\lvert\alpha p_\perp\rvert$ (K)	$F_b(\times 10^7$ Oe)	$m_b(m_0)$	$2\,\lvert\alpha p_\perp\rvert$ (K)
α	11.0	1.4	2400 (α, β)	10.5	0.91	4200 (α, β)
β	8.41	1.5		7.50	0.98	
δ	1.68	1.2	800 (δ, ε)	1.54	0.72	2000 (δ, ε)
ε	0.97	1.0		0.63	0.51	

data in Fig. 3.8(b): $C_e/T = \gamma_s + \beta_s T$ with $\gamma_s = 34.1$ mJ/K$^2\cdot$mol and $\beta_s =$ 1290 mJ/K$^3\cdot$mol, as shown in Fig. 3.8(c), revealing a low-temperature power law consistent with the existence of line nodes in the superconducting energy gap. The normal-state electronic specific heat is obtained as 335 mJ/K$^2\cdot$mol. $LaPt_3Si$ also shows superconductivity below $T_{sc} = 0.6$ K. The specific heat below 0.5 K displays an exponential dependence as a function of temperature, following the standard BCS relation. $LaPt_3Si$ might be even a type-I superconductor, because superconductivity can be easily suppressed by a small magnetic field of 100 Oe. The electronic specific-heat coefficient is 11 mJ/K$^2\cdot$ mol in the normal state.

We show in Fig. 3.9 the magnetic and superconducting phase diagram. Open circles and squares in the phase diagram were determined from specific heat and electrical-resistivity measurements, respectively. Open triangles are due to the data obtained from thermal expansion and magnetostriction measurements. With regard to the superconducting phase, we observe a small anisotropy in the upper critical field $H_{c2}(0)$ between $H \parallel [1\;0\;0]$ and $[0\;0\;1]$ from the resistivity measurements, where the superconducting transition temperature T_{sc} is defined through the vanishing of electrical resistivity. As shown in Figs. 3.8(c) and 3.9, the resistivity becomes zero below 0.7 K, whereas bulk superconductivity inferred from the specific heat is realized only below $T_{sc} = 0.46$ K.

We show in Fig. 3.10 the pressure phase diagram of $CePt_3Si$ [31, 34]. In this figure, data shown by open and closed circles were obtained from ac-specific-heat measurements, data marked by squares from zero-resistivity measurements and data shown by triangles from ac-susceptibility measurements. By applying pressure, the Néel temperature $T_N = 2.3$ K decreases and becomes zero around $P_{AF} = 0.6$ GPa. Also the superconducting transition temperature T_{sc} decreases with increasing pressure, and becomes almost constant around 0.6–0.8 GPa, showing a shoulder-like feature in the pressure dependence of T_{sc}. Upon further increasing pressure, T_{sc} decreases again, and superconductivity disappears around $P_{sc} = 1.5$ GPa.

We performed the dHvA experiment under pressures crossing P_{AF} and P_{sc}, in order to study the change of the electronic states. Figures 3.11(a) and 3.11(b) show typical dHvA oscillations at 1.28 GPa for $H \parallel [0\,0\,1]$ and the FFT spectra at 0, 1.28 and 2.4 GPa, respectively, at about 80 mK. The only detected dHvA branch is branch δ at 0 and 1.28 GPa and its second harmonic. The spectrum at $P = 0$ GPa, obtained at

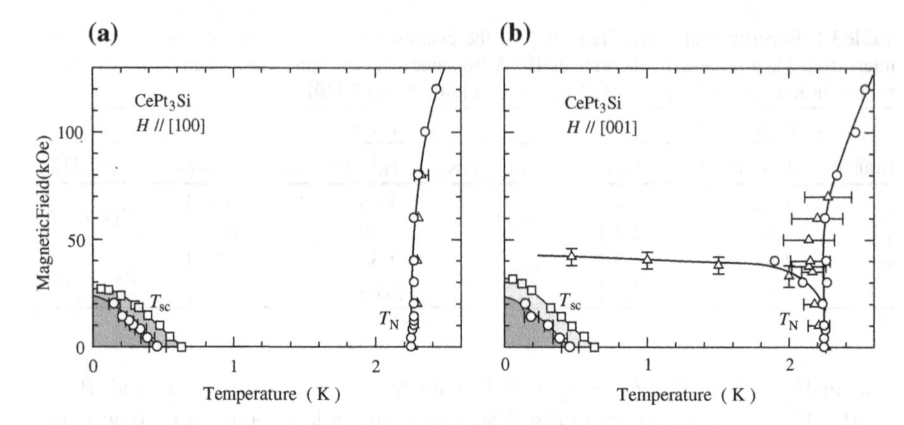

Fig. 3.9 Magnetic and superconducting phase diagrams for **a** $H \parallel [1\,0\,0]$ and **b** $[0\,0\,1]$ in $CePt_3Si$, Ref. [30]. Open circles and squares are obtained from specific heat and the electrical-resistivity measurements, respectively, and open triangles show data obtained from thermal expansion and magnetostriction measurements. *Solid lines* connecting the data are a guide to the eye

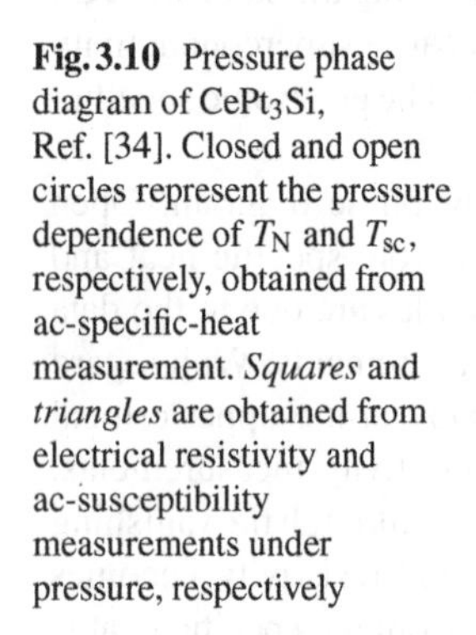

Fig. 3.10 Pressure phase diagram of $CePt_3Si$, Ref. [34]. Closed and open circles represent the pressure dependence of T_N and T_{sc}, respectively, obtained from ac-specific-heat measurement. *Squares* and *triangles* are obtained from electrical resistivity and ac-susceptibility measurements under pressure, respectively

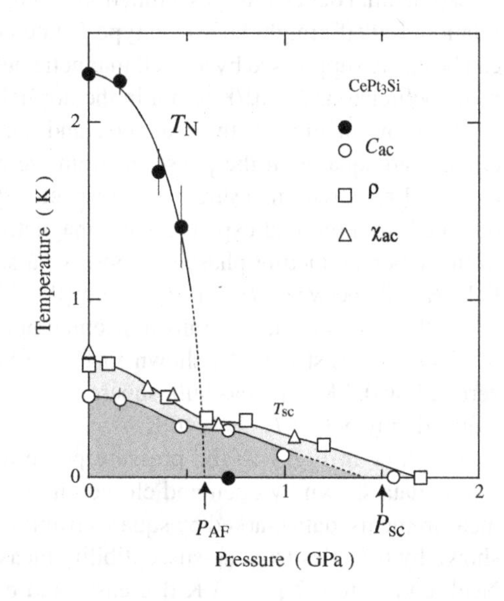

34 mK, contains one more branch with the dHvA frequency $F = 1.37 \times 10^7$ Oe, [29] although the corresponding dHvA peak could not be seen in the present pressure-cell experiment due to the very weak intensity of this branch. On the other hand, two branches are clearly observed at 2.4 GPa: $F = 1.64 \times 10^7$ Oe and $F = 2.21 \times 10^7$ Oe. The former branch might correspond to δ, but the latter branch is a new one.

We show in Figs. 3.12(a) and 3.12(b) the pressure dependence of the dHvA frequency and the cyclotron effective mass, respectively. The branch δ (open circles) might be observed from 0 GPa to 2.7 GPa, indicating a gradual increase of the dHvA

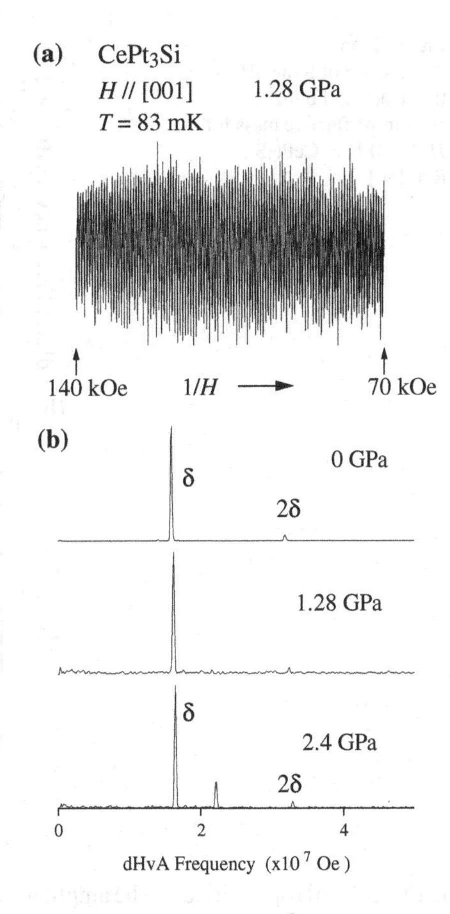

Fig. 3.11 **a** dHvA oscillation at 1.28 GPa and **b** the corresponding FFT spectra at 0, 1.28 and 2.4 GPa for $H \parallel [0\,0\,1]$ in $CePt_3Si$, Ref. [30]

frequency with increasing pressure with a rate of $(2.5 \pm 0.1) \times 10^2$ kOe/GPa, as shown in Fig. 3.12(a). Above 1.5 GPa, a new branch (open squares) is observed, and the dHvA frequency (2.21×10^7 Oe) of this new branch also increases with growing pressure by $(3.2 \pm 0.1) \times 10^2$ kOe/GPa. On the other hand, the cyclotron effective mass of branch δ decreases slightly as a function of pressure and shows a discontinuous change around 1.5 GPa. The cyclotron effective mass of the new branch is extremely large, $19m_0$ at 2.0 GPa, and decreases steeply with increasing pressure, $11m_0$ at 2.7 GPa. We could not determine the cyclotron effective mass of the new branch at 1.5 and 1.77 GPa because of a small dHvA signal and most likely a larger cyclotron mass.

We also carried out the dHvA experiment for $H \parallel [1\,0\,0]$. At ambient pressure and at temperatures of about 30 mK, three branches with dHvA frequencies $F = 1.49 \times 10^7$, 1.64×10^7, 1.83×10^7 Oe are observed, [29] while only one branch is observed in the pressure-cell experiment at $T = 80$ mK. The pressure dependences of the dHvA frequency and the effective mass are shown in Figs. 3.13(a) and 3.13(b), respectively. The dHvA frequency of $F = 1.83 \times 10^7$ Oe at ambient pressure decreases with increasing pressure up to 2 GPa with the rate $-(3.5 \pm 0.7) \times 10^2$ kOe/GPa, as shown

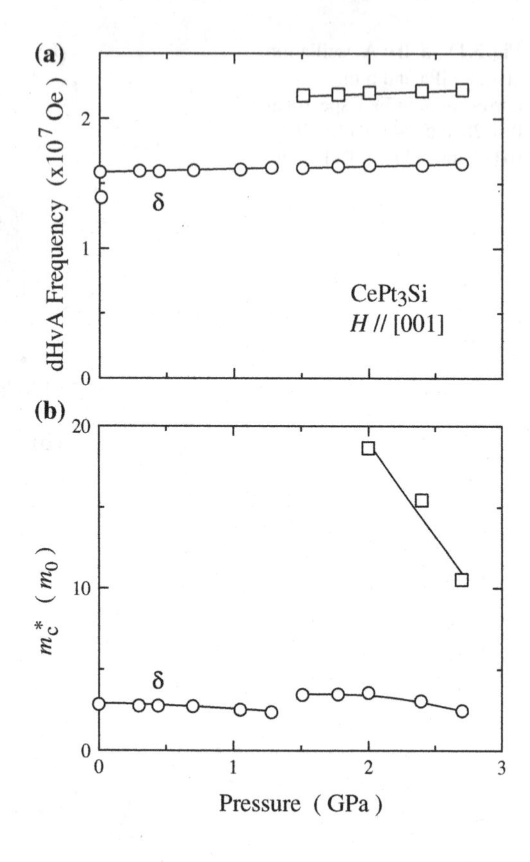

Fig. 3.12 Pressure dependence of **a** the dHvA frequency and **b** the cyclotron effective mass for $H \parallel [0\,0\,1]$ in $CePt_3Si$, Ref. [30]

in Fig. 3.13(a) (open circles). The amplitude of the dHvA signal at ambient pressure is relatively small compared with the one for $H \parallel [0\,0\,1]$. It rapidly decreases above 1.5 GPa, and was not observed anymore above 2 GPa. Although the obtained cyclotron effective mass has a relatively large ambiguity due to the small dHvA amplitude, it is clear that it increases as a function of pressure up to 1 GPa and becomes almost constant at higher pressures, as shown in Fig. 3.13(b) (open circles).

We discuss the pressure dependence of the Fermi-surface properties in $CePt_3Si$. At ambient pressure, we assume that the topology of the Fermi surface in $CePt_3Si$ is similar to that in $LaPt_3Si$. This might be the case up to the critical pressure $P_{AF} = 0.6$ GPa, where T_N becomes zero. Actually similar dHvA experiments under pressure for antiferromagnets such as $CeRh_2Si_2$, [35, 36] $CeRhIn_5$ [37] and $CeIn_3$ [38] indicate that their dHvA branches at ambient pressure persist up to P_{AF} and new dHvA branches appear only above P_{AF}. These new branches are well explained by an itinerant 4*f*-band model. We expect that in $CePt_3Si$ the 4*f*-Fermi surfaces shown in Fig. 3.14(c) are observed above P_{AF} as in $CeRh_2Si_2$, $CeRhIn_5$ and $CeIn_3$. However, no distinctive change in the dHvA frequency and cyclotron effective mass of $CePt_3Si$ is visible at P_{AF}, although a new branch with a dHvA frequency of 2.21×10^7 Oe and a large cyclotron effective mass of $19m_0$ appears above 2 GPa.

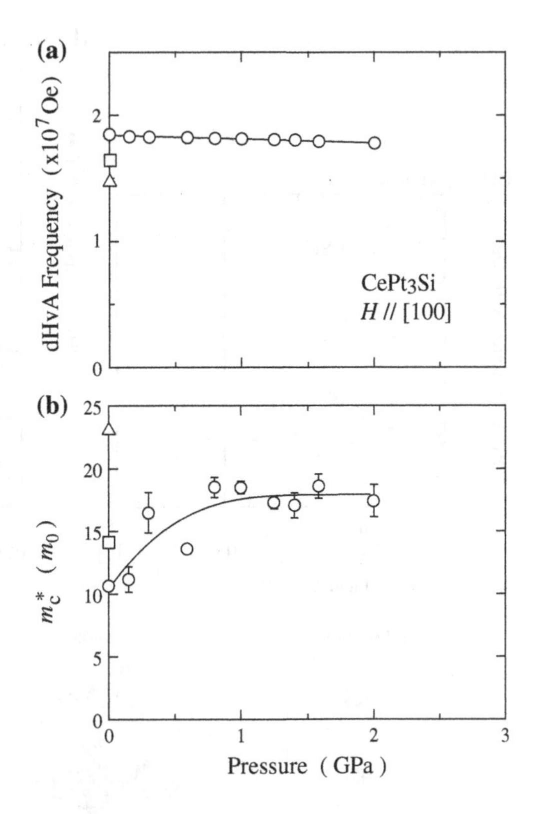

Fig. 3.13 Pressure dependence of **a** the dHvA frequency and **b** the cyclotron effective mass for $H \parallel [1\ 0\ 0]$ in $CePt_3Si$, Ref. [30]

We show the experimental angular dependence of the dHvA frequency at ambient pressure in Fig. 3.14(a). This should be compared with the theoretical angular dependence of the dHvA frequency of $CePt_3Si$ under the assumption that the 4*f* electrons are itinerant, because $CePt_3Si$ displays a paramagnetic state above P_{AF} (Fig. 3.14(b)). In the paramagnetic state, $CePt_3Si$ becomes a compensated metal with equal volumes of electron and hole Fermi surfaces. The theoretically determined main dHvA branches are $\alpha(F_b = 5.76 \times 10^7$ Oe and $m_b = 2.96m_0)$, $\delta(F_b = 5.67 \times 10^7$ Oe and $m_b = 6.25m_0)$, $\varepsilon(F_b = 3.52 \times 10^7$ Oe and $m_b = 2.52m_0)$, $\beta(F_b = 3.03 \times 10^7$ Oe and $m_b = 2.07m_0)$ and $\delta'(F_b = 2.01 \times 10^7$ Oe and $m_b = 1.50m_0)$. These branches were not observed in the experiment, although the new branch of 2.21×10^7 Oe is close to branch δ'. In the dHvA experiment under pressure for $CePt_3Si$, we found a change of the electronic states at about 1.5 GPa, but it cannot be concluded that itinerant 4*f*-electronic states are realized above 1.5 GPa. It is necessary to perform the dHvA experiment at ambient pressure as well as under pressure for a sample of much higher quality.

The electronic state is often reflected in the upper critical field in superconductivity. The superconducting transition temperature T_{sc} is 0.7 K, from electrical-resistivity measurements, as mentioned above. Using resistivity measurements the zero-temperature upper critical field is $H_{c2}(0) = 32$ kOe for $H \parallel [0\ 0\ 1]$ and

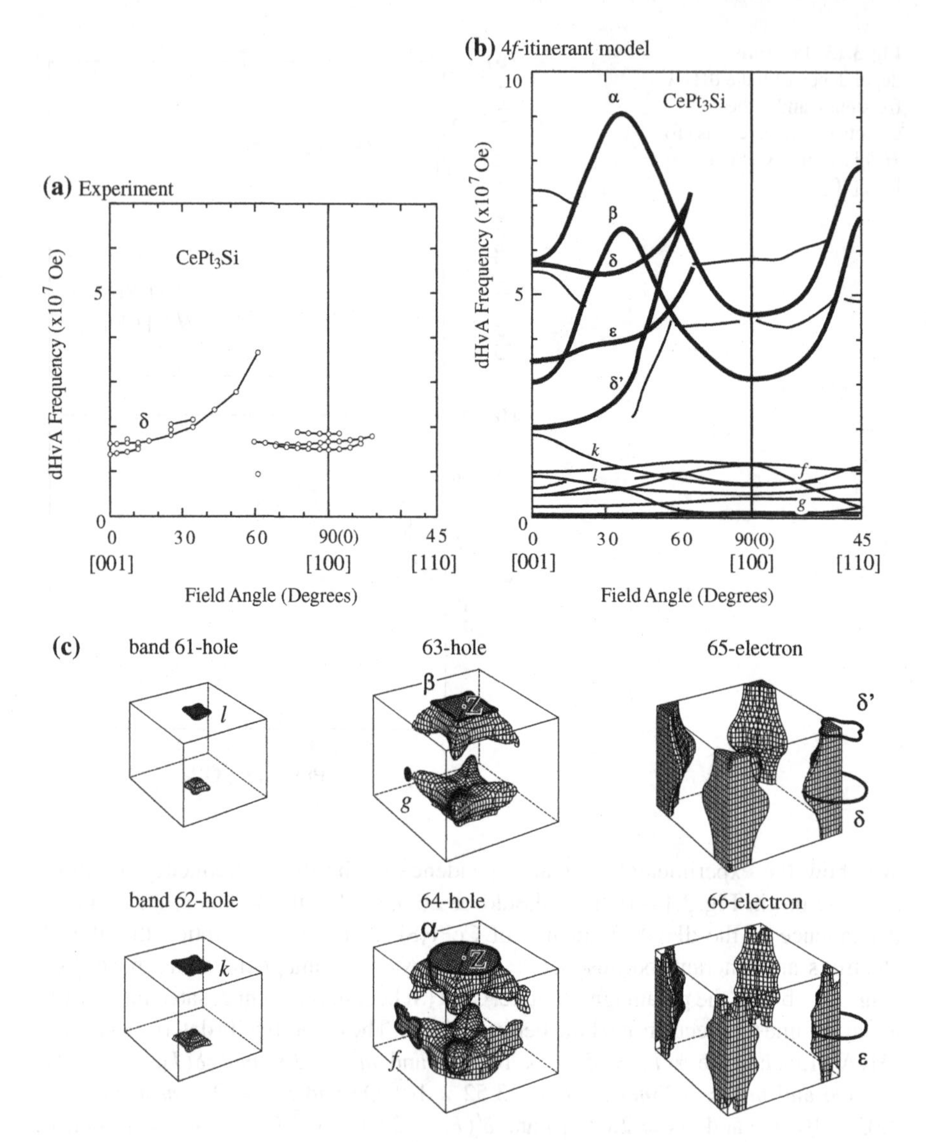

Fig. 3.14 **a** Experimental angular dependence of the dHvA frequency at ambient pressure, **b** the theoretical dependence and **c** the corresponding theoretical Fermi surfaces based on the 4f-itinerant band model in $CePt_3Si$, Refs. [29] and [30]

$H_{c2}(0) = 27$ kOe for $H \parallel [1\,0\,0]$ as shown in Fig. 3.9 (squares). Specific-heat measurements probe the onset of bulk superconductivity and point to a lower $T_{sc} = 0.46$ K and an upper critical field of 23 kOe, as shown by circles in Fig. 3.9.

We studied the upper critical field under pressure from resistivity measurement. Figure 3.15 shows the temperature dependence of the electrical resistivity at 0.1, 0.3

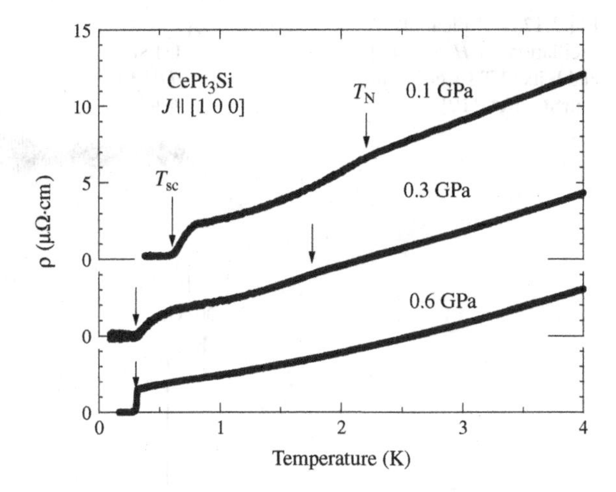

Fig. 3.15 Low-temperature electrical resistivity of $CePt_3Si$ at 0.1, 0.3 and 0.6 GPa, Ref. [39]

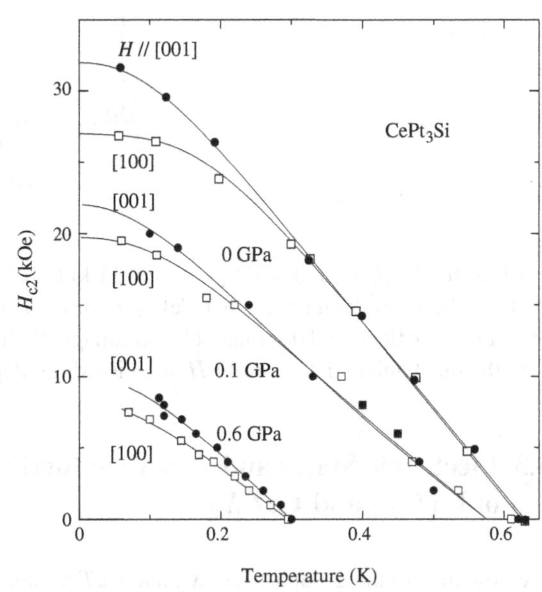

Fig. 3.16 Temperature dependence of the upper critical field in $CePt_3Si$ at 0.1 and 0.6 GPa, Ref. [39]

and 0.6 GPa. The electrical resistivity at 0.1 GPa decreases slightly below $T_N = 2.2$ K, and steeply below $T_N = 0.8$ K, becoming zero at $T_{sc} = 0.6$ K, as shown by arrows. At 0.3 GPa, a similar resistivity behavior is obtained, but the resistivity drop at $T_{sc} = 0.3$ K becomes sharp at 0.6 GPa, where the Néel temperature approaches zero.

We measured the electrical resistivity under magnetic fields to analyze the behavior of the upper critical field. Figure 3.16 gives the upper critical field for $H \parallel$ [0 0 1] and [1 0 0], shown by circles and squares, respectively, at 0, 0.1 and 0.6 GPa. The anisotropy of H_{c2} between $H \parallel$ [0 0 1] and [1 0 0] is found to be small even at 0.6 GPa: $-\mathrm{d}H_{c2}/\mathrm{d}T = 43$ kOe/K at $T_{sc} = 0.3$ K and $H_{c2}(0) \simeq$

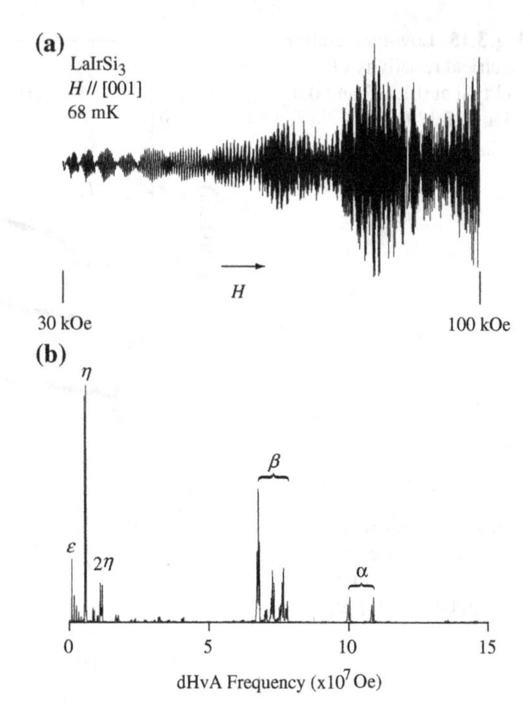

Fig. 3.17 **a** Typical dHvA oscillation for $H \parallel [0\,0\,1]$ and **b** its FFT spectrum in $LaIrSi_3$, Ref. [19]

9.5 kOe for $H \parallel [001]$ and $-\, dH_{c2}/dT = 33$ kOe/K and $H_{c2}(0) \simeq 7.9$ kOe for $H \parallel [1\,0\,0]$. These results indicate that $CePt_3Si$ does not show the characteristic features of an upper critical field dominated by paramagnetic limiting effects, suggesting that here the mechanism determining $H_{c2}(T)$ is orbital depairing.

3.3 Electronic States and Superconducting Properties of LaTX_3 and CeTX_3

We now turn to the group of LaTX_3 and CeTX_3 compounds (T = Co, Rh, Ir and X = Si, Ge). In Fig. 3.17(a) we show the typical dHvA oscillations of $LaIrSi_3$ for the magnetic field H along the [0 0 1] direction (c-axis) and its FFT spectrum. The observed dHvA branches, α, β, ε, and η, are depicted in Fig. 3.17(b). The signal α with the largest dHvA frequency is clearly split into two branches, each of which separates once more into two branches. The former splitting is due to antisymmetric spin-orbit coupling, while the latter one is mainly due to the slight corrugation of each Fermi surface, possessing two extremal (maximum and minimum) cross-sections.

Figures 3.18(a) and 3.18(b) show the angular dependence of the dHvA frequency in $LaIrSi_3$, together with the theoretical one. The detected dHvA branches are well explained by the result of the FLAPW energy band calculation. The corresponding theoretical Fermi surfaces are shown in Fig. 3.19.

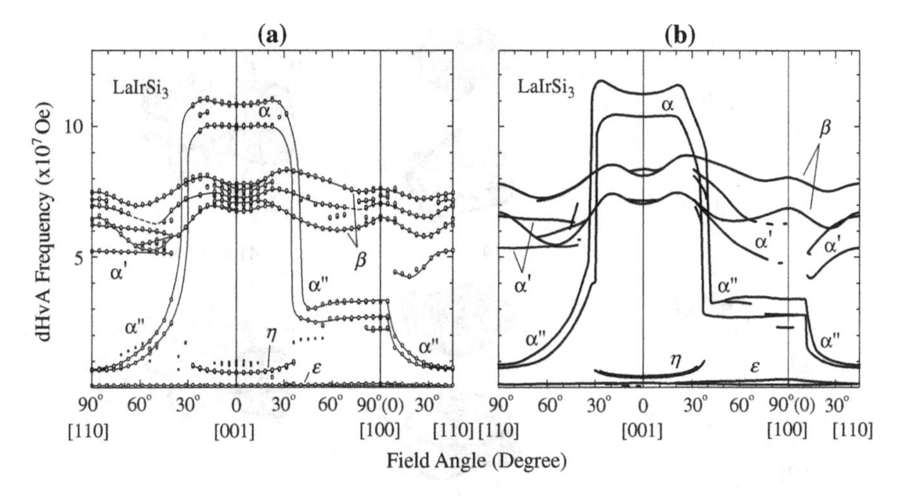

Fig. 3.18 **a** Angular dependence of the dHvA frequency and **b** theoretical angular dependence in $LaIrSi_3$, Ref. [19]

The dHvA branches are identified as follows:

1. branches α and η correspond to outer and inner orbits of the electron Fermi surfaces belonging to band 41 and 42, respectively. These Fermi surfaces have a void around the center of the Brillouin zone (Γ point). Also the branches α' and α'' are due to these Fermi surfaces.
2. branch β is due to the hole Fermi surfaces corresponding to band 39 and 40.
3. branch ε originates from the hole Fermi surface of band 38.

We determine the cyclotron effective mass m_c^* from the temperature dependence of the dHvA amplitude and find, for example, 1.03 m_0 for branch α in the magnetic field along the [0 0 1] direction. The Fermi-surface properties are summarized in Table 3.2.

The dHvA results of $LaIrSi_3$ are also compared with those of $LaIrGe_3$, as shown in Fig. 3.20. The valence-electron configurations are $3s^23p^2$ in Si and $4s^24p^2$ in Ge. The dHvA frequency of $LaIrGe_3$ is slightly smaller than that of $LaIrSi_3$, because the lattice constants of $a = 4.4343$ Å and $c = 10.0638$ Å in $LaIrGe_3$ are larger than $a = 4.2820$ Å and $c = 9.8391$ Å in $LaIrSi_3$ and the corresponding Brillouin zone of $LaIrGe_3$ has a smaller volume than that of $LaIrSi_3$. However, the $2|\alpha p_\perp|$ value is almost the same for the two compounds as listed in Table 3.2.

We now discuss the Fermi-surface properties and the magnitude of the antisymmetric spin-orbit coupling in $LaTGe_3$ (T: Co, Rh, Ir). Figure 3.21 shows the angular dependence of the dHvA frequency. The topologies of the Fermi surfaces are essentially the same in $LaTGe_3$ (T = Co, Rh, Ir). This is plausible in view of the fact that the valence electron configurations do not differ in $LaTGe_3$: $3d^74s^2$ in Co, $4d^85s^1$ in Rh and $5d^9$ in Ir. Note, however, that the dHvA frequency of branch α in $LaCoGe_3$ is smaller than those in $LaRhGe_3$ and $LaIrGe_3$, and the width of

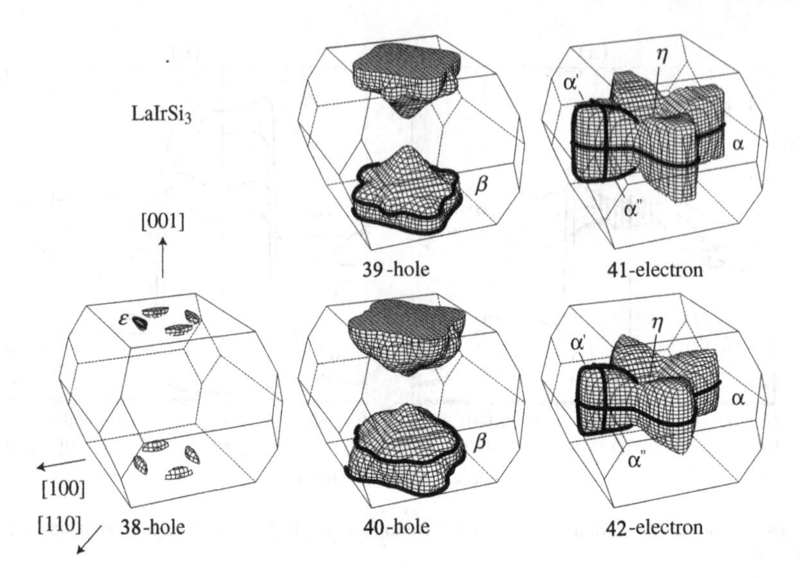

Fig. 3.19 Theoretical Fermi surfaces in $LaIrSi_3$, Ref. [19]

Table 3.2 Experimental dHvA frequency F, the cyclotron mass m_c^*, and the antisymmetric spin-orbit interaction $2|\alpha p_\perp|$ for $H \parallel [0\ 0\ 1]$ in La TX_3 ($T =$ Co, Rh, Ir and $X =$ Si, Ge) and $PrCoGe_3$, Ref. [40]

	Branch α			Branch β		
	$F(\times 10^7$ Oe)	$m_c^*(m_0)$	$2\|\alpha p_\perp\|$ (K)	$F(\times 10^7$Oe)	$m_c^*(m_0)$	$\|\alpha p_\perp\|$ (K)
$LaCoGe_3$	9.15	1.19	461	7.09	1.28	416
	8.74	1.20		6.72	1.11	
$LaRhGe_3$	10.4	1.04	511	6.98	0.83	505
	10.0	1.04		6.67	0.85	
$LaIrGe_3$	10.4	1.13	1090	7.25	1.32	1066
	9.29	1.51		6.22	1.29	
$LaIrSi_3$	10.9	0.97	1100	7.64	0.97	1250
	10.0	1.03		6.76	0.92	
$PrCoGe_3$	9.04	1.80	284	7.13	2.04	302
	8.64	1.97		6.71	1.70	

the split dHvA frequencies, $|F_+ - F_-|$, for branch α in $LaIrGe_3$ is larger than those in $LaCoGe_3$ and $LaRhGe_3$. Hence the Fermi surface in $LaCoGe_3$ is slightly smaller in volume than those in $LaRhGe_3$ and $LaIrGe_3$. Moreover, we may conclude that the antisymmetric spin-orbit coupling $2|\alpha p_\perp|$ in $LaIrGe_3$ is larger than in $LaCoGe_3$ and $LaRhGe_3$: $2|\alpha p_\perp| = 460$ K in $LaCoGe_3$, 510 K in $LaRhGe_3$ and

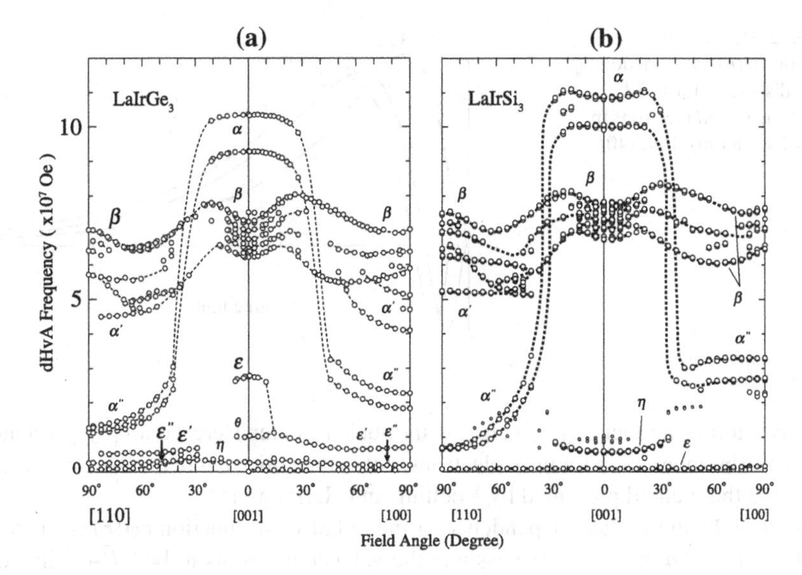

Fig. 3.20 Angular dependence of the dHvA frequency in **a** $LaIrGe_3$ and **b** $LaIrSi_3$, Ref. [40]

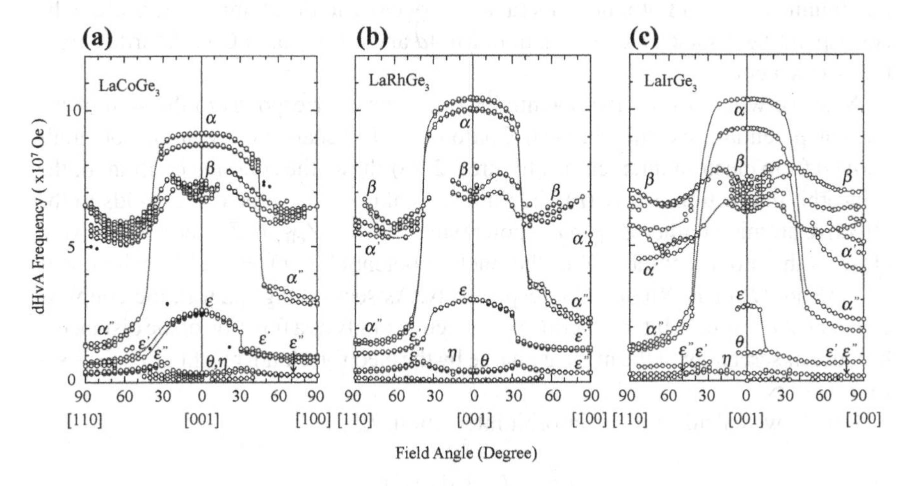

Fig. 3.21 Angular dependence of the dHvA frequency in **a** $LaCoGe_3$, **b** $LaRhGe_3$ and **c** $LaIrGe_3$, Ref. [40]

1090 K in $LaIrGe_3$ for branch α. Precise values for branches α and β are given in Table 3.2.

In this series of dHvA experiments, the potentials can be varied by changing the transition metal ions T = Co, Rh and Ir in La$T$$Ge_3$. This may explain the reason why the antisymmetric spin-orbit coupling in $LaIrGe_3$ is relatively large compared with those in $LaCoGe_3$ and $LaRhGe_3$. The difference is connected with both the characteristic radial wave function $\phi(r)$ of Ir-5d electrons and the relatively large

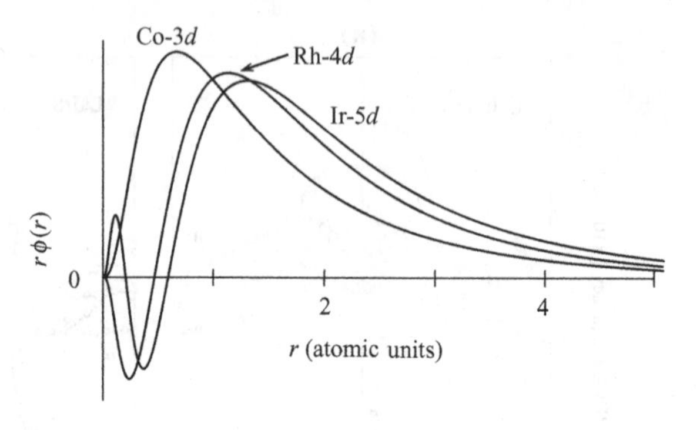

Fig. 3.22 Radial wave function $r\phi(r)$ as a function of the distance r for Ir-5d, Rh-4d and Co-3d electrons in the isolated atoms, Ref. [40]

effective atomic number Z_{eff} in Ir close to the nuclear center. Here we simply calculate the spin-orbit interaction for the d electrons, not in the lattice but in the isolated atom, following the method presented by Koelling and Harmon [41].

Figure 3.22 shows the r-dependence of the radial wave function $r\phi(r)$ for Ir-5d, Rh-4d and Co-3d electrons. We assume the valence electrons to be $3d^7 4s^2$ in Co, $4d^7 5s^2$ in Rh and $5d^7 6s^2$ in Ir. The $r\phi(r)$ function of Ir-5d electrons possesses a maximum at $r = 0.11$ atomic units (a. u.), very close to the atomic center, while the corresponding distance r is 0.37 a. u. in Rh-4d and 0.66 a. u. in Co-3d, farther from the atomic center.

Next we will consider the potential $V(r)$, which corresponds to the sum of the nuclear potential, and the classical Coulomb and exchange-correlation potentials derived from electron interactions. Figure 3.23(a) shows the coupling constant of the spin-orbit interaction, $r^2\,\mathrm{d}V(r)/\mathrm{d}r$. Simply thinking, this value corresponds to the effective atomic number Z_{eff} in the potential $V(r) = -Z_{\mathrm{eff}}/r$. Z_{eff} at $r = 0$ is very close to the atomic number Z in the nuclear potential $V(r) = -Z/r$, where Z is 77, 45 and 27 for Ir, Rh and Co, respectively. As seen in Fig. 3.23(a), the coupling constant of the spin-orbit interaction is reduced strongly as a function of the distance r, because of screening of the nuclear charge by the electron cloud, reaching $Z_{\mathrm{eff}} \longrightarrow 1$ for $r \longrightarrow \infty$.

Finally we calculate the spin-orbit interaction, I_{so}:

$$I_{\mathrm{so}}(r) = \frac{\hbar^2}{2m^2c^2}\int_0^r \frac{1}{r'}\frac{\mathrm{d}V(r')}{\mathrm{d}r'}|r'\phi(r')|^2\,\mathrm{d}r', \tag{3.11}$$

which is shown in Fig. 3.23(b) as a function of the distance r. I_{so} becomes constant at about 1.0 a. u., but approximately reaches this constant value at $r = 0.11$ a. u. for Ir-5d, 0.37 a. u. for Rh-4d and 0.66 a. u. for Co-3d, where the corresponding radial wave functions possess the extremal values, as mentioned above. The spin-orbit coupling is, thus, obtained as 38.0 mRy (6000 K) for Ir-5d, 12.8 mRy (2020 K) for Rh-4d and 5.72 mRy (900 K) for Co-3d. These calculations suggest that the radial wave function of Ir-5d electrons has a larger weight at distances close to the center, compared with those of Rh-4d and Co-3d, and develops a relatively strong spin-orbit

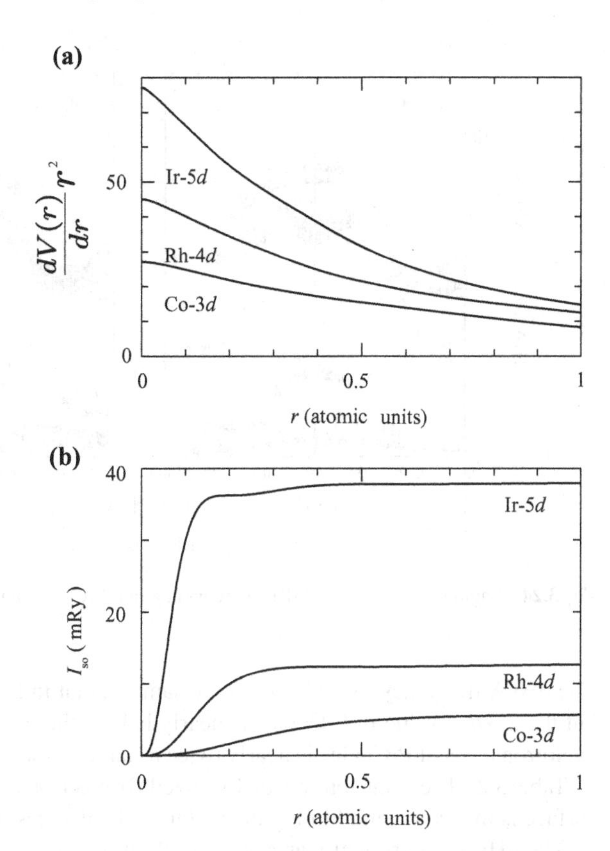

Fig. 3.23 **a** Coupling constant of the spin-orbit interaction $(\mathrm{d}^2 Vr)/\mathrm{d}r^2)r^2$ and **b** the spin-orbit interaction I_{so} as a function of the distance r for Ir-5d, Rh-4d and Co-3d electrons in the isolated atoms, Ref. [40]

coupling in Ir, closely connected with the relatively large effective atomic number Z_{eff} for Ir at the distance r close to the atomic center.

This result of the spin-orbit interaction for an isolated atom is applied to the non-centrosymmetric crystal lattice. In this case, the degenerate Fermi surface is split into two Fermi surfaces of which the magnitude of the antisymmetric spin-orbit interaction is approximately proportional to the spin-orbit coupling of Eq. (3.11), because the same potential is in principle used in the band structure calculation. The d electrons in the transition element as well as the 5d electrons in the La atom and the other electrons contribute to the conduction electrons in LaTGe$_3$. This is the main reason why the antisymmetric spin-orbit interaction $2|\alpha p_{\perp}|$ in LaIrGe$_3$ and LaIrSi$_3$ is larger than in LaCoGe$_3$ and LaRhGe$_3$. These results may also be applied to Li$_2$Pt$_3$B and Li$_2$Pd$_3$B, [42] where the spin-orbit coupling in Li$_2$Pt$_3$B, which originates mainly form Pt-5d electrons, is expected to be larger than the contribution from Pd-4d orbitals in Li$_2$Pd$_3$B. Note also that $2\,|\alpha p_{\perp}| = 2400$ K is large for the main Fermi surface in LaPt$_3$Si.

Next we will discuss spin-orbit coupling in PrCoGe$_3$, which shows no magnetic ordering and has a singlet ground state within the 4f crystalline electric-field scheme.

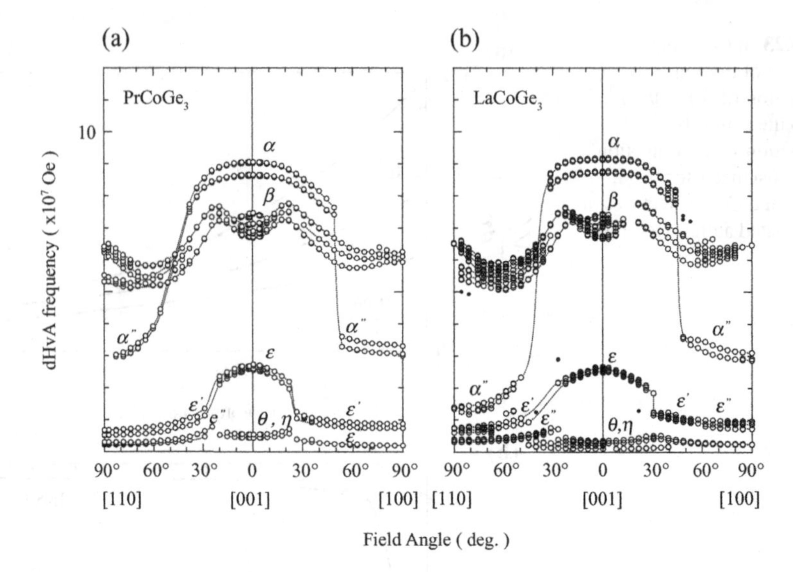

Fig. 3.24 Angular dependence of dHvA frequency in **a** $PrCoGe_3$ and **b** $LaCoGe_3$, Ref. [40]

The dHvA frequency in $PrCoGe_3$ is the same as that in $LaCoGe_3$ (see in Fig. 3.24), but the $2|\alpha p_\perp|$ value in $PrCoGe_3$ is nearly half of the one in $LaCoGe_3$ because the cyclotron mass of $PrCoGe_3$ is nearly twice as large as the one of $LaCoGe_3$, as shown in Table 3.2. The contribution of localized $4f$ electrons to the topology of Fermi surface is thus very small in $PrCoGe_3$, but still enhances the cyclotron mass.

The dHvA experiment was also carried out for the antiferromagnetic $CeCoGe_3$ [43]. Figure 3.25 shows the angular dependence of the dHvA frequency in the field-induced ferromagnetic state. The magnetization indicates a metamagnetic transition, and the antiferromagnetic state is changed into a field-induced ferromagnetic state with a magnetic moment of 0.42 μ_B/Ce. The width of the split dHvA frequencies, $|F_+ - F_-|$, is large due to the ferromagnetic exchange interaction and also to the antisymmetric spin-orbit coupling. The cyclotron mass for $H \parallel [0\ 0\ 1]$ is relatively large, 12 m_0 for branch β, for example, reflected in the large value $\gamma = 32$ mJ/K^2 · mol for the specific-heat coefficient. Note that the corresponding cyclotron mass in $LaCoGe_3$ is about 1 m_0, and yields a small γ value of 4.4 m_0 mJ/K^2 · mol.

Now we consider the Fermi-surface properties of $CeIrSi_3$, which is an antiferromagnet with $T_N = 5.0$ K. The magnetic structure has not been clarified in detail so far. In low-temperature specific-heat experiments a large γ value was found, $\gamma = 120$ mJ/K^2 · mol, while the corresponding magnetic entropy $S_{mag} = 0.2R \ln 2$ at T_N is small, indicating a heavy fermion antiferromagnet based on the Kondo screening effect [44].

Figure 3.26(a) shows the angular dependence of the dHvA frequency in the antiferromagnetic state of $CeIrSi_3$ at ambient pressure. No dHvA signal could be observed

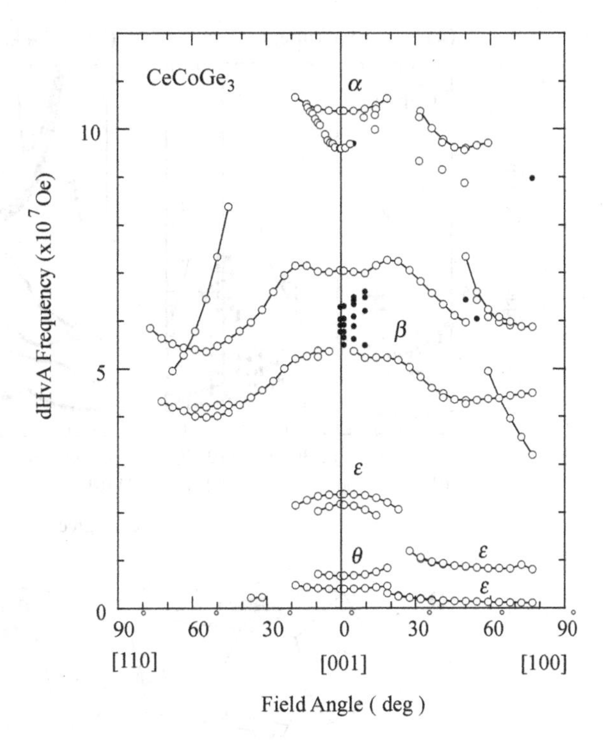

Fig. 3.25 Angular dependence of the dHvA frequency in the field-induced ferromagnetic state of an antiferromagnet $CeCoGe_3$, Ref. [43]

around the [0 0 1] direction, which is most likely due to the presence of the antiferromagnetic zone boundaries, indicating a complicated antiferromagnetic structure. The cyclotron mass is large: 40 m_0 ($F = 3.8 \times 10^7$ Oe) for a magnetic field along [1 1 0]. Figure 3.26(b) shows the angular dependence of the theoretical dHvA frequency based on the 4*f*-itinerant band model. The experimental result is different for localized 4*f*-electrons corresponding to the result of $LaIrSi_3$ shown in Fig. 3.18 and the heavy-fermion-like itinerant 4*f*-electrons in Fig. 3.26(b).

The magnitude of the antisymmetric spin-orbit coupling of $CeIrSi_3$ cannot be determined from the result of the present dHvA experiments, but can be roughly estimated from the γ value: we estimate $2|\alpha p_\perp| \simeq 40$ K from $\gamma = 120\,\text{mJ/K}^2\cdot$ mol in $CeIrSi_3$, assuming that $|\alpha p_\perp|m_c^*$ in Eq. (3.10) or $|\alpha p_\perp|\gamma$ are unchanged between $CeIrSi_3$ and $LaIrSi_3$, and $2|\alpha p_\perp|$ and γ values are $2|\alpha p_\perp| \simeq 1000$ K and $\gamma =$ 4.5 mJ/K$^2\cdot$ mol in $LaIrSi_3$ [19]. This value is much larger than the superconducting transition temperature $T_{sc} = 1.6$ K at 2.65 GPa in $CeIrSi_3$, as shown below.

Figure 3.27 shows the FFT spectra under pressure for the magnetic field along [0 0 1]. While there was no dHvA signal at ambient pressure, three signals appear above 1.62 GPa, named a, b and c. Here, the Néel temperature decreases monotonically and reaches zero at $P_c = 2.25$ GPa. Superconductivity appears above 1.9 GPa. The dHvA signals above 2.20 GPa were obtained in the superconducting mixed state. As shown in Fig. 3.27, the FFT spectra in the superconducting mixed state are extremely small in magnitude. Note that the scale of the FFT spectrum at 2.50 GPa

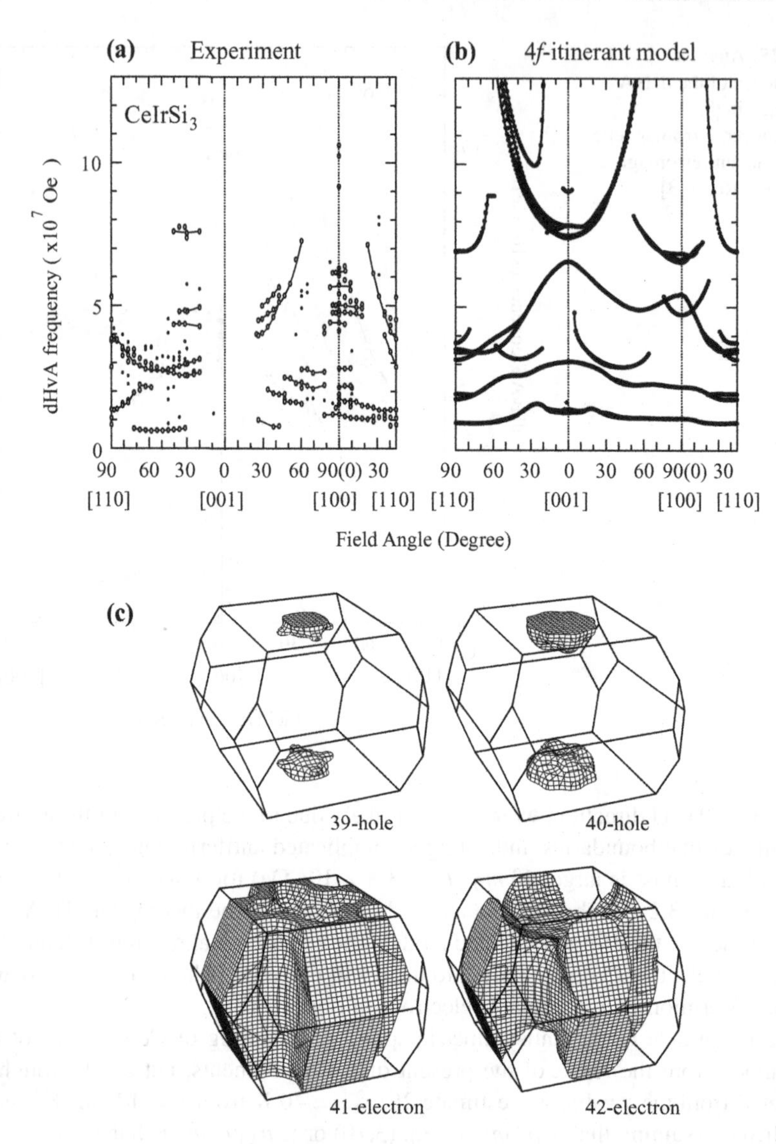

Fig. 3.26 **a** Experimental angular dependence of the dHvA frequency at ambient pressure, **b** the theoretical one, and **c** the corresponding theoretical Fermi surfaces based on the 4*f*-itinerant band model in $CeIrSi_3$, Ref. [45]

is enlarged twice in magnitude compared with the spectra below 2.35 GPa, and the spectrum at 2.60 GPa is enlarged by ten times. The dHvA frequencies are, however, unchanged as a function of pressure, as shown in Fig. 3.28(a), although branch "a" with $F = 6.85 \times 10^7$ Oe disappears above 2.25 GPa. The corresponding cyclotron masses slightly decrease as a function of pressure, as shown in Fig. 3.28(b).

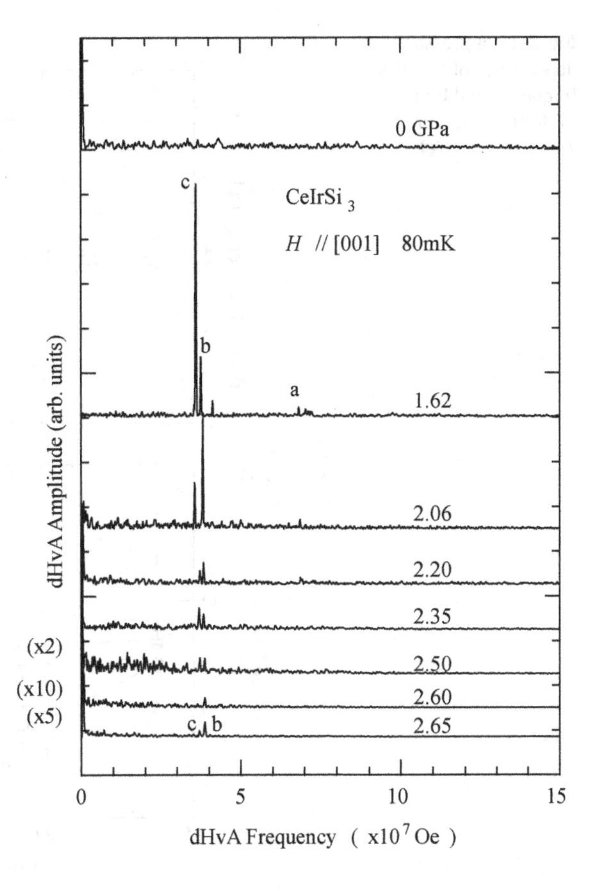

Fig. 3.27 FFT spectra for $H \parallel [0\,0\,1]$ under several pressures in $CeIrSi_3$, where the electronic state above 2.20 GPa is in the superconducting mixed state. The dHvA experiment was carried out in the magnetic field range from 90 kOe to 169 kOe

We investigated the electronic states of $CeTSi_3$ (T: Co, Rh, Ir) and $CeTGe_3$ to look for superconductivity under large enough pressure. The localization of $4f$-electrons is enhanced in $CeTGe_3$ (T: Co, Rh, Ir) compared with that in $CeTSi_3$. The corresponding Néel temperature in $CeTGe_3$ is larger than that in $CeTSi_3$. This is because the lattice constants of $CeTGe_3$ are larger than those of $CeTSi_3$: $a = 4.398$ Å and $c = 10.032$ Å in $CeRhGe_3$ and $a = 4.237$ Å and $c = 9.785$ Å in $CeRhSi_3$. Actually, Ge in $CeTGe_3$ increases the molar volume and enhances the antiferromagnetic ordering, as discussed also for CeT_2X_2 (T: transition metal, X: Ge, Si) [47].

We plot the Néel temperature and the electronic specific-heat coefficient γ as a function of the average lattice constant $\sqrt[3]{a^2c}$, in Figs. 3.29(a) and 3.29(b), respectively. Note that the lattice constant decreases from left to right (data from Refs. [17, 19, 44] and [46]). The observed relation of T_N vs $\sqrt[3]{a^2c}$ in Fig. 3.29(a) roughly corresponds to the Doniach phase diagram, [48] which indicates a competition between the RKKY interaction and the Kondo screening effect. The magnetic ordering temperature is shown as a function of $|J_{cf}|\,D(\varepsilon_F)$ in the Doniach phase diagram, where J_{cf} is the magnetic exchange interaction and $D(\varepsilon_F)$ the electronic density of states

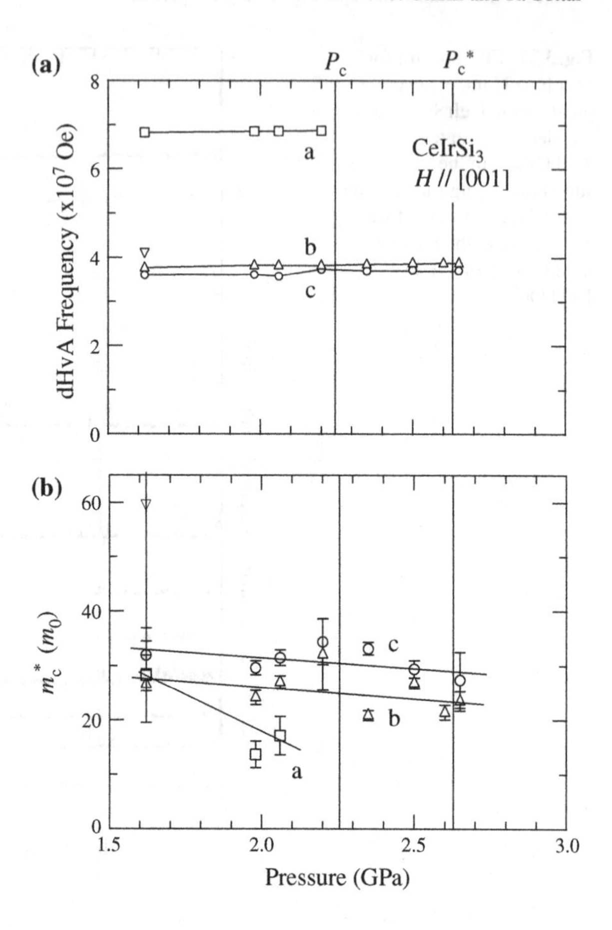

Fig. 3.28 **a** Pressure dependence of the dHvA frequency and **b** the cyclotron mass for $H \parallel [0\,0\,1]$ in CeIrSi$_3$

at Fermi energy ε_F. Experimentally, $|J_{cf}|\,D(\varepsilon_F)$ corresponds to pressure. In fact, the Néel temperatures in CeRhSi$_3$ and CeIrSi$_3$ become zero at a relatively small value of pressure, $P_c \simeq 2$ GPa, which corresponds to their magnetic quantum critical points.

The effect of pressure on the electronic states in CeTGe$_3$ compounds is very different from that in CeRhSi$_3$ and CeIrSi$_3$. The Néel temperature does not change appreciably upon application of pressure in CeRhGe$_3$ and CeIrGe$_3$ [50–51]. Figure 3.30 shows the pressure dependence of T_N in CeIrGe$_3$. High pressures are needed to reach the quantum critical region. The antiferromagnetic ordering temperature $T_{N1} = 8.7$ K possesses a maximum and then decreases with increasing pressure, while the magnetic transition temperature $T_{N2} = 4.7$ K increases with increasing pressure. Both temperatures merge at 4 GPa. The antiferromagnetic ordering temperature T_N remains unchanged up to pressures of 8 GPa, decreases steeply below 10 GPa, remains again unchanged up to 22 GPa, and becomes zero around 24 GPa, reaching a quantum critical region. Superconductivity with $T_{sc} = 1.3$–1.5 K is observed at pressures larger than 20 GPa.

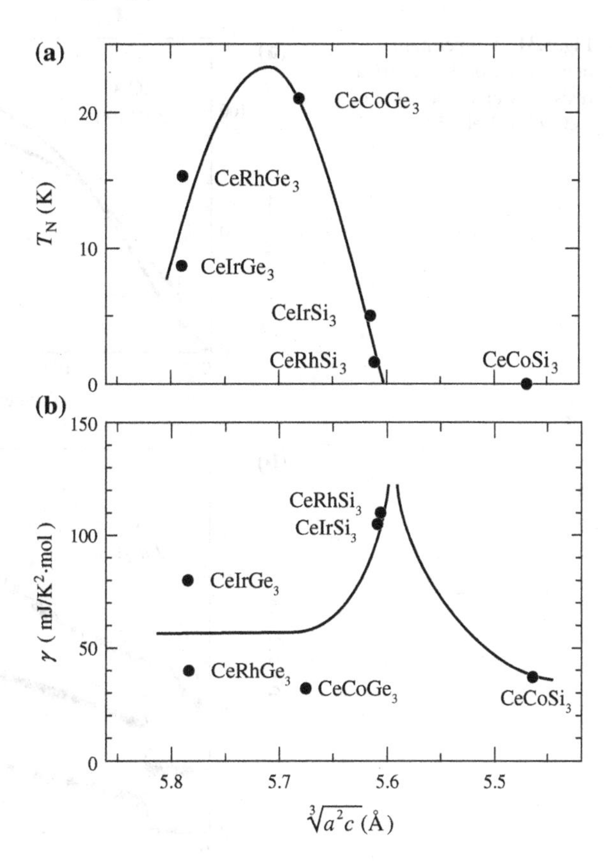

Fig. 3.29 $\sqrt[3]{a^2c}$-dependence of **a** the Néel temperature and **b** the γ value in Ce*T* Si$_3$ and Ce*T*Ge$_3$ (T: Co, Rh, Ir), Refs. [17, 19, 44] and [46]

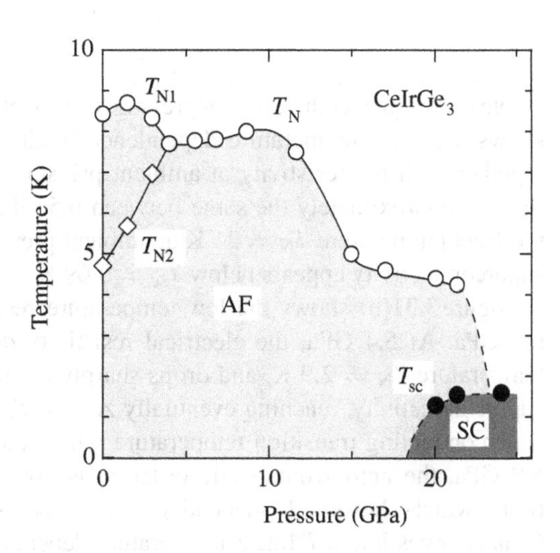

Fig. 3.30 Pressure-temperature phase diagram in CeIrGe$_3$, Ref. [51]

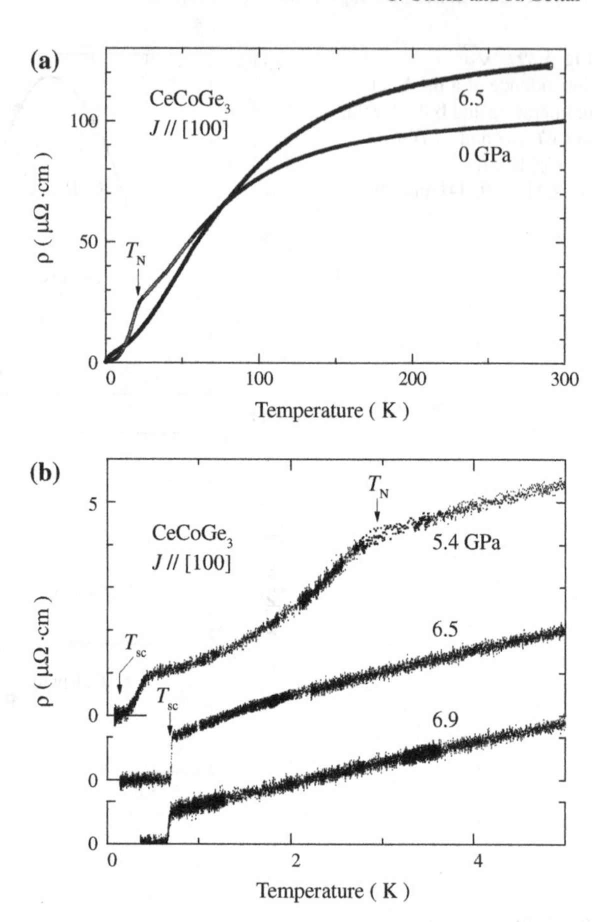

Fig. 3.31 Temperature dependence of the electrical resistivity under pressure in $CeCoGe_3$, Ref. [52]

Next we consider the effect of pressure on $CeCoGe_3$ [20, 52, 53, 54]. Figure 3.31(a) shows a typical temperature dependence of the electrical resistivity at 6.5 GPa, together with the resistivity at ambient pressure. The overall feature of the resistivity is approximately the same between 6.5 GPa and ambient pressure, although the Néel temperature $T_N = 21$ K at ambient pressure becomes zero at 6.5 GPa and superconductivity appears below $T_{sc} = 0.69$ K.

Figure 3.31(b) shows the low-temperature resistivity at pressures 5.4, 6.5 and 6.9 GPa. At 5.4 GPa, the electrical resistivity decreases steeply below the Néel temperature $T_N = 2.9$ K, and drops sharply below 0.43 K, indicating the onset of superconductivity, reaching eventually zero at $T_{sc} = 0.13$ K. We define T_{sc} as the superconducting transition temperature where the resistivity vanishes. At 6.5 and 6.9 GPa, the antiferromagnetic ordering is not seen clearly. The electrical resistivity, which shows a T^2-dependence, $\rho = \rho_0 + AT^2$, below about 2.5 K at 5.4 GPa, changes into a T-linear temperature dependence below about 4 K at 6.9 GPa, indicating non-Fermi-liquid character. The A value of $A = 0.357\,\mu\Omega \cdot \text{cm/K}^2$ at

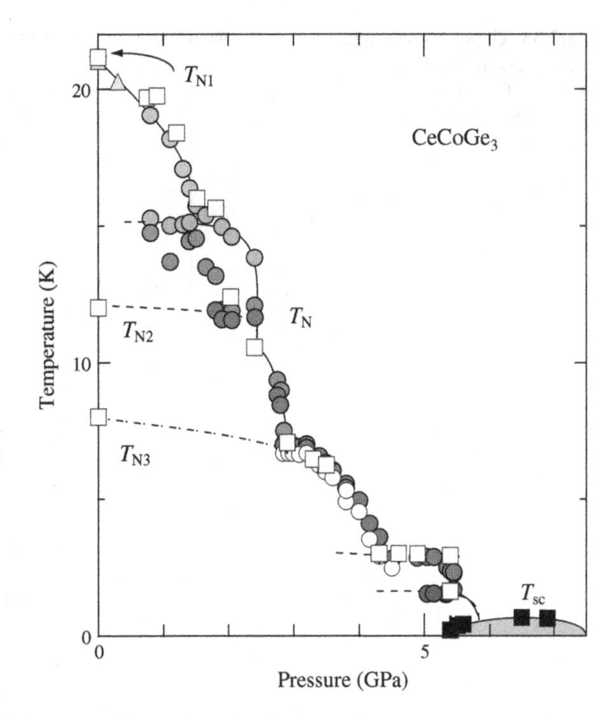

Fig. 3.32 Pressure-temperature phase diagram in CeCoGe$_3$, Refs. [20, 52, 53] and [54]

5.4 GPa corresponds to the electronic specific-heat coefficient $\gamma = 190\,\mathrm{mJ/K^2}\cdot$ mol, following the Kadowaki-Woods relation, [55] which is compared with $A = 0.011\,\mu\Omega\cdot\mathrm{cm/K^2}$ and $\gamma = 34\,\mathrm{mJ/K^2}\cdot$ mol at ambient pressure. Superconductivity in CeCoGe$_3$ is apparently realized in a moderately heavy fermion state. Note that the γ value of about 120 mJ/K$^2\cdot$ mol at ambient pressure in CeIrSi$_3$ is unchanged as a function of pressure, even at about 2.6 GPa where the superconducting state is realized, as described below [56].

Superconductivity in CeCoGe$_3$ is observed at $T_{sc} = 0.69$ K at 6.5 GPa and $T_{sc} = 0.65$ K at 6.9 GPa. The maximum superconducting transition temperature might be realized at about 6.5 GPa. We show in Fig. 3.32 the pressure-temperature phase diagram. The Néel temperature decreases with increasing pressure and superconductivity appears in the pressure region from 5.4 GPa to about 7.5 GPa. The critical pressure is estimated to be $P_c \simeq 6.5$ GPa.

In CeIrGe$_3$ and CeCoGe$_3$, the Néel temperature does not decrease monotonically, but with a few steps as a function of pressure, as shown in Figs. 3.30 and 3.32, respectively. The present step-like feature in the phase diagram might be a signature corresponding to a change of the magnetic spin structure. This is because the magnetic structure, where the ordered moment is oriented along the [0 0 1] direction, with $\boldsymbol{q} = (001/2)$ in CeCoGe$_3$, [57] is not favorable in superconductivity, especially in the tetragonal crystal structure. The magnetic structure might be changed

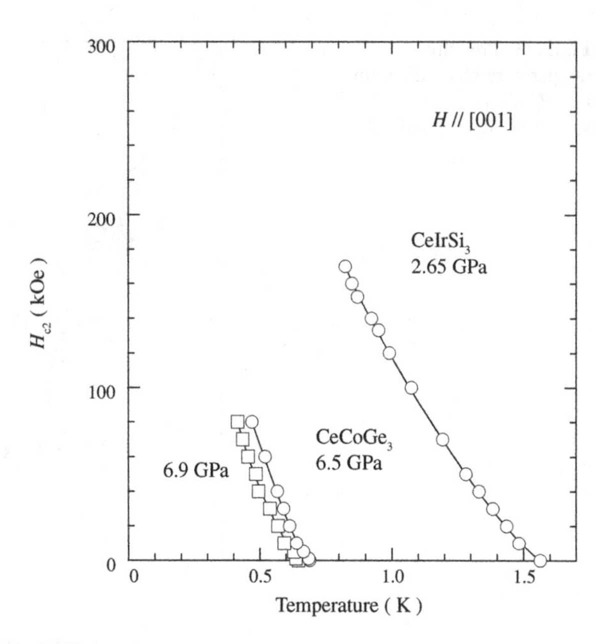

Fig. 3.33 Upper critical field under pressure for $H \parallel [0\,0\,1]$ in $CeCoGe_3$ and $CeIrSi_3$, Refs. [19] and [52]

as a function of pressure, and superconductivity is most likely realized in favorable magnetism.

Now we discuss the superconducting phase of $CeCoGe_3$ in a magnetic field. In this compound, superconductivity is very robust against magnetic fields. Actually, the slope of H_{c2} as a function of temperature is extremely large: $-\mathrm{d}H_{c2}/\mathrm{d}T = 200$ kOe/K at $T_{sc} = 0.69$ K under 6.5 GPa, larger than $-\mathrm{d}H_{c2}/\mathrm{d}T = 154$ kOe/K at $T_{sc} = 1.56$ K under 2.65 GPa in $CeIrSi_3$ [19, 20], shown below. The upper critical field has an upward curvature with decreasing temperature, as shown in Fig. 3.33, and superconductivity appears with an increasing slope of the upper critical field. In Fig. 3.33, the upper critical field at 6.9 GPa is also shown: $-\mathrm{d}H_{c2}/\mathrm{d}T = 190$ kOe/K at $T_{sc} = 0.65$ K.

Finally, we investigate the pressure-induced superconducting state of $CeIrSi_3$ [18, 19]. The effect of pressure on the electronic state was studied through resistivity measurements. Figure 3.34 shows the low-temperature resistivity at $P = 0$, 1.95, and 2.65 GPa. With increasing pressure, the Néel temperature, shown by arrows, decreases monotonically, although it is not clearly defined at pressures higher than 2 GPa, where pressure-induced superconductivity appears, as shown in Fig. 3.34 for 1.95 GPa. The antiferromagnetic ordering disappears completely at $P = 2.65$ GPa. The superconducting transition temperature T_{sc}, as shown by arrows, increases as a function of pressure and finally attains a value of $T_{sc} = 1.6$ K at 2.65 GPa. Note that the resistivity at 2.65 GPa does not show a T^2-dependence, but indicates a T-linear dependence, which persists up to 18 K.

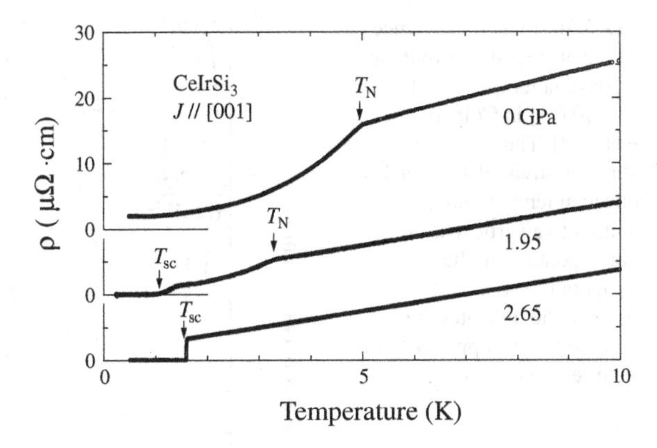

Fig. 3.34 Temperature dependence of the electrical resistivity in $CeIrSi_3$ at 0, 1.95 and 2.65 GPa

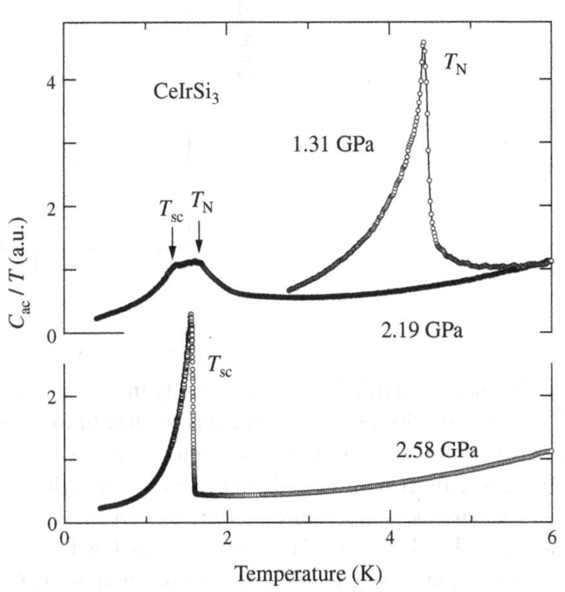

Fig. 3.35 Temperature dependence of the ac-specific heat in the form of C_{ac}/T at 1.31, 2.19 and 2.58 GPa in $CeIrSi_3$, Ref. [56]

Figure 3.35 shows the temperature dependence of the ac-specific heat at pressures 1.31, 2.19 and 2.58 GPa. At 1.31 GPa, the antiferromagnetic ordering is clearly observed at $T_N = 4.5$ K, but at 2.19 GPa antiferromagnetism with $T_N = 1.7$ K coexists with superconductivity with $T_{sc} = 1.4$ K. An exclusively superconducting phase is observed only above the critical pressure $P_c = 2.25$ GPa [56]. The specific heat has a huge jump at the superconducting transition above P_c, $\Delta C_{ac}/C_{ac}(T_{sc})$ at 2.58 GPa is 5.75 at $T_{sc} = 1.6$ K, which is extremely large compared with the BCS value of $\Delta C_{ac}/\gamma T_{sc} = 1.43$. This value is the largest in all the discussed superconductors. The antiferromagnet $CeIrSi_3$ is thus changed by pressure into a strong-coupling superconductor. The γ value at 2.58 GPa is roughly estimated as $\gamma = 100 \pm 20\,\mathrm{mJ/K^2 \cdot mol}$, which is approximately the same as $\gamma = 120\,\mathrm{mJ/K^2 \cdot mol}$ at ambient pressure.

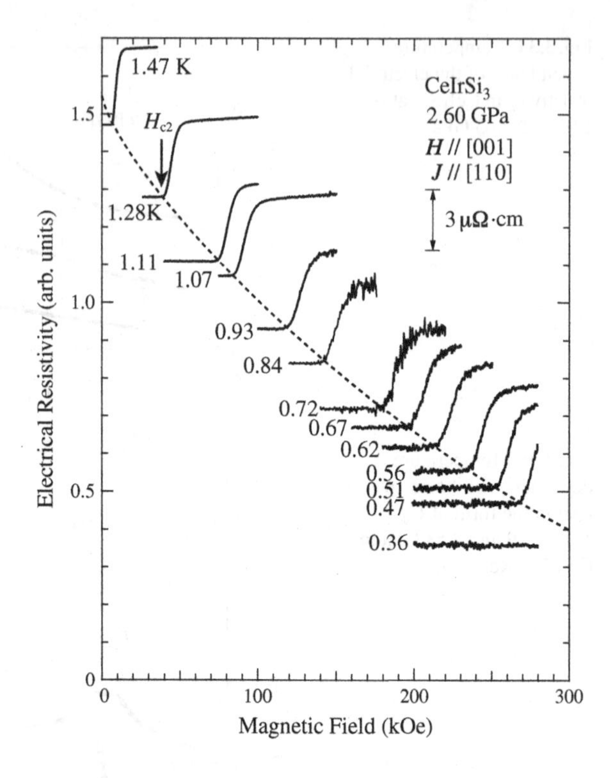

Fig. 3.36 Field dependence of the electrical resistivity at a constant temperature for $H \parallel [0\,0\,1]$ in $CeIrSi_3$, Ref. [58]. The zero-resistivity data at each constant temperature are shifted to a vertical scale corresponding to the temperature so that the dashed line indicates the temperature vs upper critical field relation

We measured the electrical resistivity in a magnetic field as a function of pressure [58]. Figure 3.36 shows the field dependence of the electrical resistivity at 2.60 GPa at a constant temperature, with the magnetic field applied along the [0 0 1] direction. These resistivity data are plotted in a way as to reflect the phase diagram of magnetic field versus the temperature at which the resistivity reaches zero. The dashed line indicates the phase boundary which extrapolated to zero temperature exceeds clearly 300 kOe, and is roughly estimated to be about 450 kOe (=450 ± 100 kOe).

The upper critical field H_{c2} is determined in a wide pressure range from 1.95 to 3.00 GPa, as shown in Fig. 3.37. $H_{c2}(0) \simeq 50$ kOe at 1.95 GPa is almost the same for both $H \parallel [0\,0\,1]$ and [1 1 0]. The temperature dependence of H_{c2}, however, differs strongly between $H \parallel [0\,0\,1]$ and [1 1 0]. The upper critical field for $H \parallel [0\,0\,1]$ displays an upward feature with decreasing temperature, while for $H \parallel [1\,1\,0]$ saturation is observed at low temperatures.

With further increasing pressure, the upper critical field deviates substantially between $H \parallel [0\,0\,1]$ and [1 1 0]. The superconducting properties become highly anisotropic: $-dH_{c2}/dT = 170$ kOe/K at $T_{sc} = 1.56$ K, and $H_{c2}(0) \simeq 450$ kOe for $H \parallel [0\,0\,1]$, and $-dH_{c2}/dT = 145$ kOe/K at $T_{sc} = 1.59$ K, and $H_{c2}(0) = 95$ kOe for $H \parallel [1\,1\,0]$ at 2.65 GPa, as shown in Fig. 3.38. The upper critical field H_{c2} for $H \parallel [1\,1\,0]$ shows strong signs of Pauli paramagnetic suppression with decreasing temperature because the orbital limiting field $H_{orb}(= -0.73(dH_{c2}/dT)T_{sc})$ is estimated to be

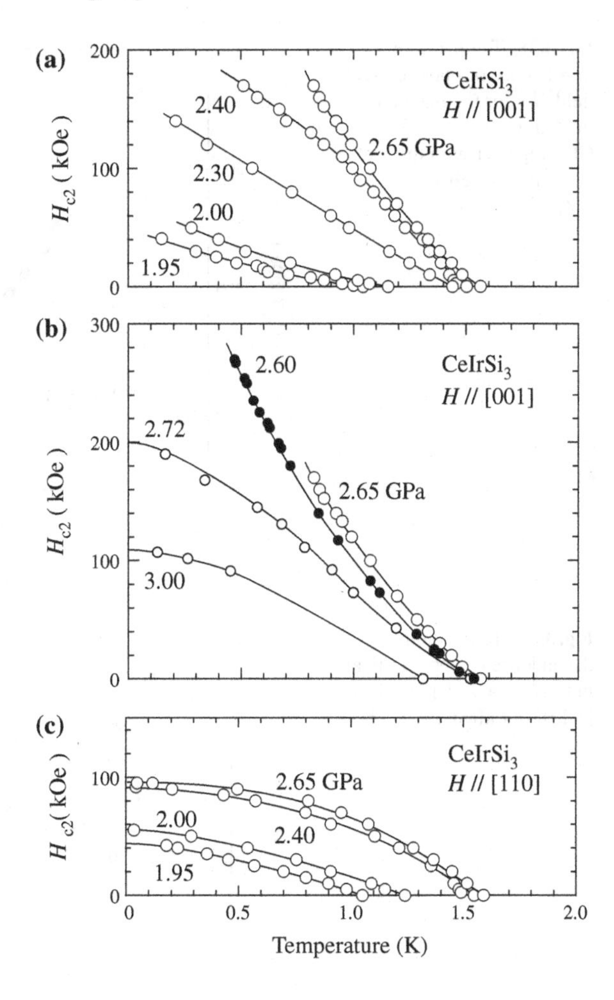

Fig. 3.37 Upper critical field H_{c2} for **a** $H \parallel [0\ 0\ 1]$ in pressure range from 1.95 GPa to 2.65 GPa, **b** 2.65 GPa to 3.0 GPa, and **c** $H \parallel [1\ 1\ 0]$ in pressure range from 1.95 GPa to 2.65 GPa in $CeIrSi_3$, Ref. [58]

170 kOe, [59] which is larger than $H_{c2}(0) = 95$ kOe for $H \parallel [1\ 1\ 0]$. On the other hand, the upper critical field $H \parallel [0\ 0\ 1]$ is not destroyed by spin polarization based on Zeeman coupling but possesses an upward curvature below 1 K. A similar result is also obtained for $CeRhSi_3$ [15, 16].

It is interesting to compare the results of H_{c2} with the upper critical field of the non-heavy-fermion reference compound $LaIrSi_3$. This is a conventional superconductor with $T_{sc} \simeq 0.9$ K, with an exponential dependence of specific heat as a function of temperature [19]. Figure 3.39 shows the temperature dependence of the upper critical field H_{c2} in $LaIrSi_3$, which was obtained by resistivity measurements in a magnetic field [58]. The anisotropy of H_{c2} is small between $H \parallel [0\ 0\ 1]$ and $[1\ 1\ 0]$. Note that the upper critical field for $H \parallel [1\ 1\ 0]$ is slightly larger than that for $H \parallel [0\ 0\ 1]$: $-dH_{c2}/dT = 2.6$ kOe/K and $H_{c2}(0) \simeq 1.7$ kOe for $H \parallel [1\ 1\ 0]$,

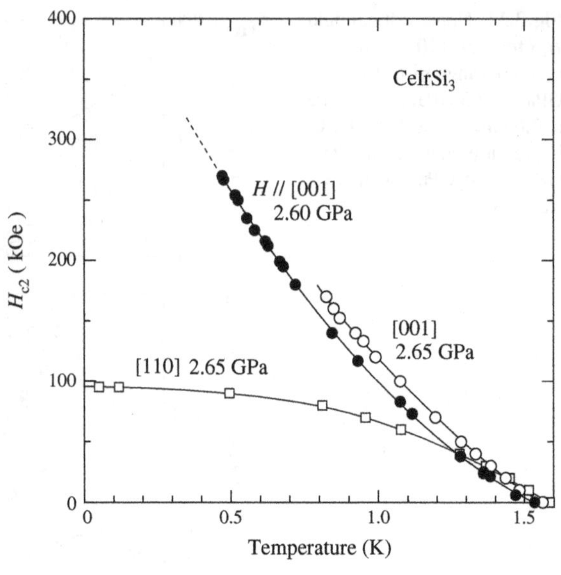

Fig. 3.38 Temperature dependence of upper critical field H_{c2} for the magnetic field along [0 0 1] at 2.60 GPa, together with those at 2.65 GPa in CeIrSi$_3$, Ref. [58]

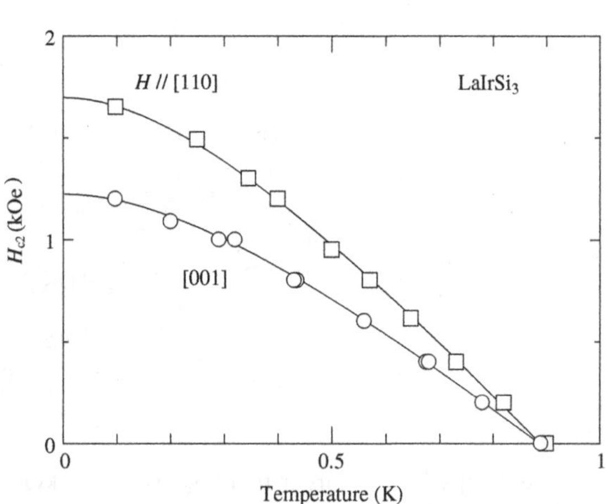

Fig. 3.39 Temperature dependence of upper critical field H_{c2} for $H \parallel [0\ 0\ 1]$ and [1 1 0] in LaIrSi$_3$, Ref. [58]

and $-dH_{c2}/dT = 1.9$ kOe/K and $H_{c2}(0) \simeq 1.25$ kOe for $H \parallel [0\ 0\ 1]$. The solid lines connecting the data are guidelines based on the WHH theory [60].

As shown above, the electronic structure as inferred from the Fermi surface of LaIrSi$_3$ is three-dimensional [19]. The small anisotropy of H_{c2} in LaIrSi$_3$ is most likely due to the corresponding anisotropy of effective mass. On the other hand, the extremely large $H_{c2}(0)$ for $H \parallel [0\ 0\ 1]$ in CeIrSi$_3$ cannot be explained by an effective-mass model, because the electronic states are also three-dimensional in CeIrSi$_3$. In fact, the electrical resistivity in CeIrSi$_3$ is approximately the same for $J \parallel [0\ 0\ 1]$ and [1 1 0] at ambient pressure as well as under pressure.

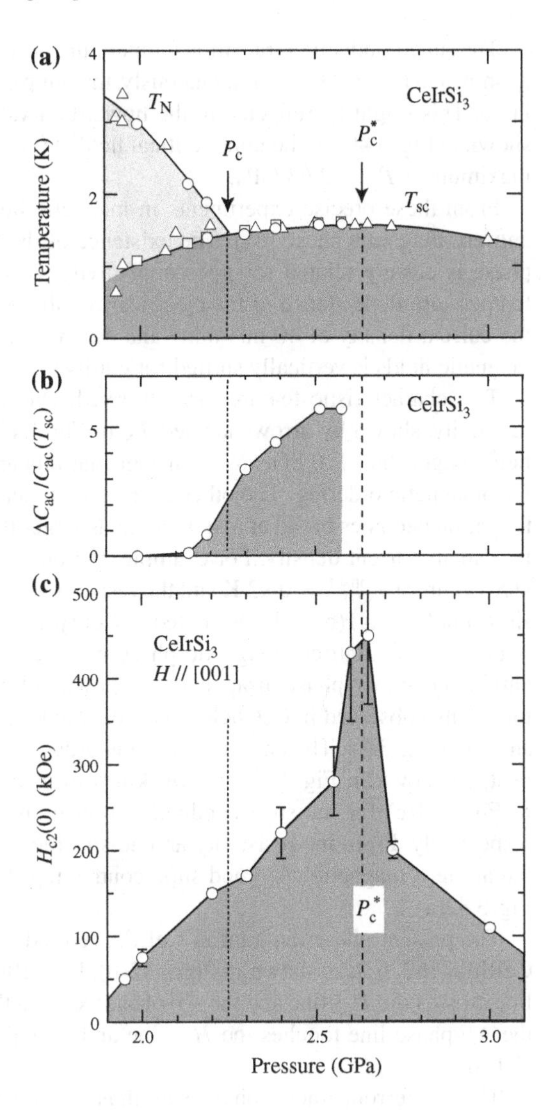

Fig. 3.40 Pressure dependences of **a** the Néel temperature T_N and superconducting transition temperature T_{sc}, **b** specific heat ju mp $\Delta C_{ac}/C_{ac}(T_{sc})$, and **c** the upper critical field $H_{c2}(0)$ for $H \parallel [0\,0\,1]$ in $CeIrSi_3$, Refs. [56] and [58]

We will discuss the reason why the $H_{c2}(0)$ for $H \parallel [0\,0\,1]$ in $CeIrSi_3$ becomes extremely large at 2.65 GPa. Figure 3.40 shows the pressure dependence of the Néel temperature T_N, the superconducting transition temperature T_{sc}, the specific-heat jump at T_{sc}, $\Delta C_{ac}/C_{ac}(T_{sc})$, and the upper critical field at 0 K $H_{c2}(0)$ for $H \parallel [0\,0\,1]$. The ac-specific-heat measurements indicate both the antiferromagnetic ordering at $T_N = 1.88$ K and the superconducting transition at $T_{sc} = 1.40$ K at 2.19 GPa [56]. The critical pressure, where the Néel temperature becomes zero, is estimated to be $P_c = 2.25$ GPa. Above $P_c = 2.25$ GPa, the antiferromagnetic ordering absent, and only the superconducting transition is observed in the ac-specific-heat measurement.

The superconducting transition temperature becomes maximum at about 2.6 GPa, as shown in Fig. 3.40(a). Simultaneously, the jump of the specific heat at T_{sc} becomes large. This might be reflected in the upward curvature of H_{c2}, as noted above. As shown in Fig. 3.40(c), the upper critical field at 0 K $H_{c2}(0)$ for $H \parallel [0\,0\,1]$ becomes maximum at $P_c^* \simeq 2.63$ GPa.

From these precise experiments in magnetic fields, we noticed a field-induced antiferromagnetic phase [61]. The existence of the field-induced antiferromagnetic phase is closely related to superconductivity. Figures 3.41(a) an3.41(b) show the temperature dependence of the electrical resistivity at 2.40 GPa for $H \parallel [0\,0\,1]$ in the current density of 40 mA/mm^2 and 4 mA/mm^2, respectively. Each data under magnetic fields is vertically shifted for clarity.

Two characteristic features are observed. One is a small kink in the electrical resistivity, shown by arrows labeled T_N in Fig. 3.41(a), which appears in magnetic fields larger than 120 kOe. We consider that this anomaly corresponds to the antiferromagnetic ordering. The other characteristic feature is that the superconducting transition becomes broad at magnetic fields larger than 100 kOe, which occurs even in a small current density. For example, the onset of superconductivity under 140 kOe occurs at $T_{sc}^{onset} = 0.93$ K and the resistivity-zero is attained at $T_{sc} = 0.62$ K, as shown in Fig. 3.41(b). It is also noted in $CePt_3Si$ that the superconducting transition is broad in the antiferromagnetic phase but becomes sharp at 0.6 GPa where the antiferromagnetic phase disappears, as shown in Figs. 3.10 and 3.15. A similar feature is also observed in $CeCoGe_3$, as shown in Fig. 3.31(b) and again in $CeIrSi_3$, as shown in Fig. 3.34. The antiferromagnetic ordering is more visible in the ac-specific heat, as shown in Fig. 3.41(c). Two kinds of arrows, up and down, are indicated in Fig. 3.41(c) for the superconducting transition and antiferromagnetic ordering, respectively. From the resistivity and ac-specific-heat experiments, we constructed the antiferromagnetic (AF) and superconducting (SC) phase diagram, as shown in Fig. 3.42(a).

The present phase diagram in $CeIrSi_3$ reminds us the similar phase diagram in $CeRhIn_5$ [62, 63], as shown in Fig. 3.42(b). In $CeRhIn_5$, the field-induced AF phase line crosses the H_{c2} line and the AF phase exists in the SC phase. On the other hand, the AF phase line touches the H_{c2} line and disappears at lower magnetic fields in $CeIrSi_3$.

The antiferromagnetic phase was thus investigated under magnetic fields and pressures in $CeIrSi_3$. We show in Fig. 3.43 the AF and SC phase diagram under various pressures. The antiferromagnetic phase is robust in magnetic fields. In other words, superconductivity is realized in the antiferromagnetic state. It is concluded that the H_{c2} value for $H \parallel [0\,0\,1]$ becomes maximum when the antiferromagnetic phase disappears completely in magnetic fields. This pressure corresponds to $P_c^* \simeq$ 2.63 GPa.

Also note that result of the ^{29}Si-NMR experiment at 2.7 GPa for $CeIrSi_3$ supports the present result [64]. In the normal state, the nuclear spin-lattice relaxation rate $1/T_1$ shows a $\sqrt{T}$ dependence, as shown in Fig. 3.44. When the system is close to an antiferromagnetic quantum critical point, the isotropic antiferromagnetic spin-fluctuation model predicts the relation of $1/T_1 \propto T/\sqrt{T+\theta}$, [65] where θ is a

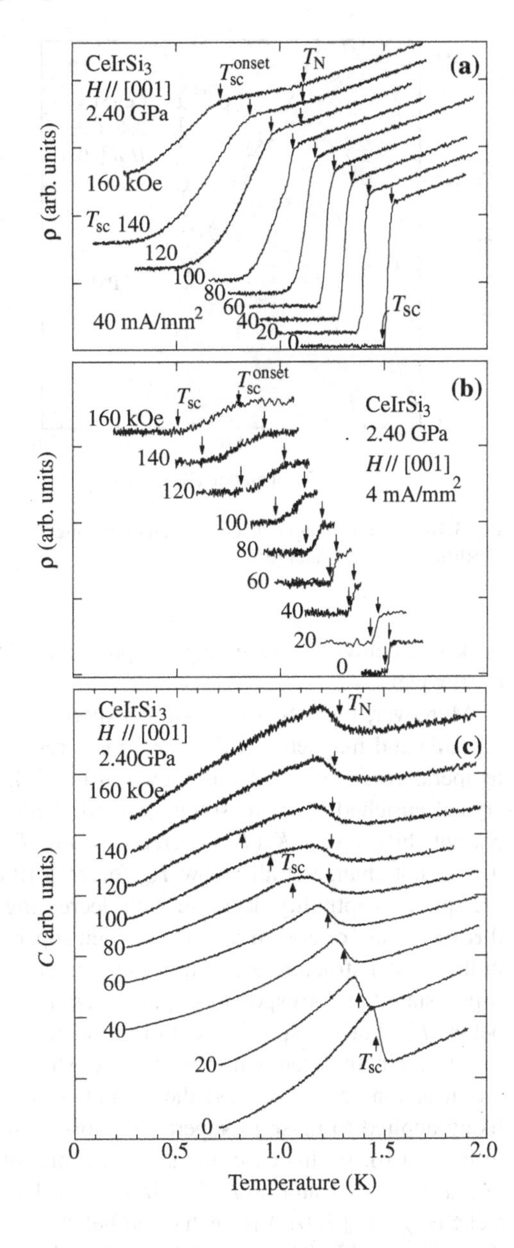

Fig. 3.41 Temperature of the electrical resistivity in the current density of **a** 40 mA/mm^2, **b** 4 mA/mm^2 at 2.40 GPa and ac specific heat for $H \parallel [0\,0\,1]$ at 2.40 GPa, Ref. [61]

measure of the proximity of the system to the quantum critical point. If $\theta = 0$, $1/T_1$ shows a $1/\sqrt{T}$ dependence. In this context, the NMR experiment shows that the electronic state at 2.7 GPa is very close to the quantum critical point. Moreover, the temperature dependence of $1/T_1$ below T_{sc} is a T^3 dependence without a coherence

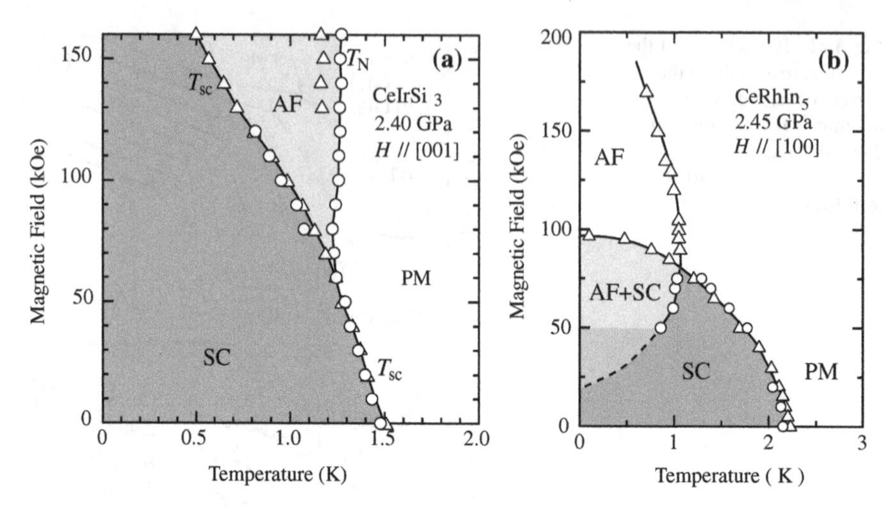

Fig. 3.42 Antiferromagnetic and superconducting phase diagram in $CeIrSi_3$, Ref. [61] and **b** $CeRhIn_5$, Refs. [62] and [63]

peak just below T_{sc}, revealing the presence of line nodes in the superconducting energy gap.

Moreover, the ^{29}Si Knight-shift experiment was carried out under pressure of 2.8 GPa and magnetic field of 13.26 Oe, revealing the superconducting transition temperature $T_{sc} = 1.2$ K, as shown in Fig. 3.45 [66]. In this experiment, a single crystal enriched by ^{29}Si isotope was used for $CeIrSi_3$. It is remarkable that the Knight-shift $K_s(T)/K_s(T_{sc})$ decreases below $T_{sc} = 1.2$ K for $H \parallel [1\ 1\ 0]$, whereas it does not change at all below T_{sc} for $H \parallel [0\ 0\ 1]$. Note that the Knight-shift or the spin susceptibility decrease with decreasing temperature and becomes zero at 0 K regardless of crystal directions when superconductors with an inversion center in the crystal structure are in the spin-singlet state. On the other hand, in the spin-triplet state the corresponding spin susceptibility for $\boldsymbol{H} \perp \boldsymbol{d}$ vector is unchanged below T_{sc}, but the spin susceptibility decreases to zero for $\boldsymbol{H} \parallel \boldsymbol{d}$. The present results are inconsistent with the characteristic features in both superconductors with the inversion center, because the Knight shift does not become zero, but are most likely applied to those in superconductors without inversion symmetry, as shown in Fig. 3.4(b). In this case, the spin susceptibility $\chi(H \parallel [110])/\chi(T_{sc})$ becomes 1/2 at $T \longrightarrow 0$ and $\chi(H \parallel [001])$ is unchanged. In the present experiment of $CeIrSi_3$, $\chi(H \parallel [001])$ is unchanged below T_{sc} but $\chi(H \parallel [110])/\chi(T_{sc})$ is 0.9 at T_{sc} at $T \longrightarrow 0$ K. The present discrepancy between experiment and theoretical prediction is unclear, but the present Knight-shift experiment reveals that paramagnetic suppression of the upper critical field H_{c2} would be absent only for $H \parallel [0\ 0\ 1]$.

In conclusion, we found a strongly increasing upper critical field H_{c2} at pressure $P_c^* \simeq 2.63$ GPa for $H \parallel [0\ 0\ 1]$, indicating a huge $H_{c2}(0) \simeq 450$ kOe, while the upper critical field H_{c2} for $H \parallel [1\ 1\ 0]$ indicates Pauli paramagnetic suppression. The electronic instability, non-centrosymmetry and strong-coupling superconductiv-

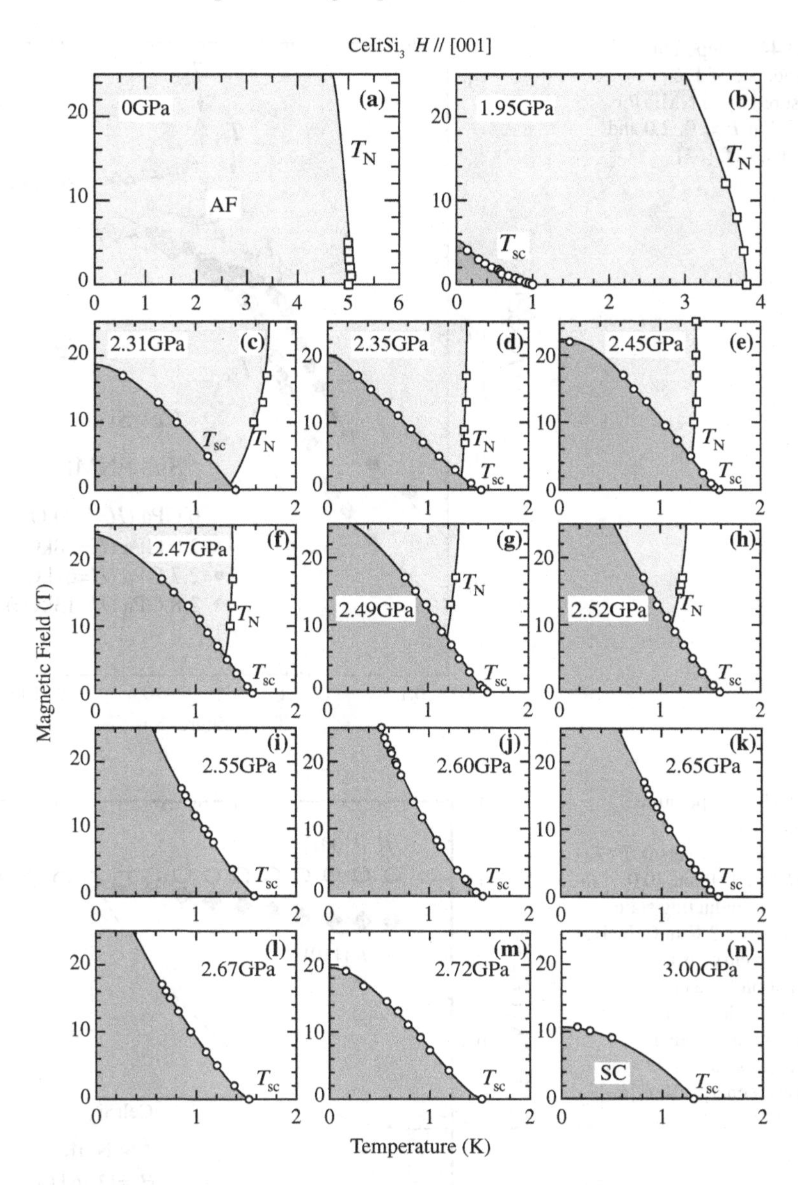

Fig. 3.43 Antiferromagnetic and superconducting phase diagram in CeIrSi$_3$ under various pressures in CeIrSi$_3$, Ref. [61]

ity are combined into the huge $H_{c2}(0)$ value for $H \parallel [0\,0\,1]$. The electronic instability produces a large slope of the upper critical field, $-dH_{c2}/dT$ at T_{sc}, where $P_c^* = 2.63$ GPa is a quantum critical point. It has been predicted on theoretical grounds, together with the results of the ^{29}Si Knight-shift experiment, that paramagnetic suppression of H_{c2} would be absent only for $H \parallel [0\,0\,1]$, based on the temperature dependence

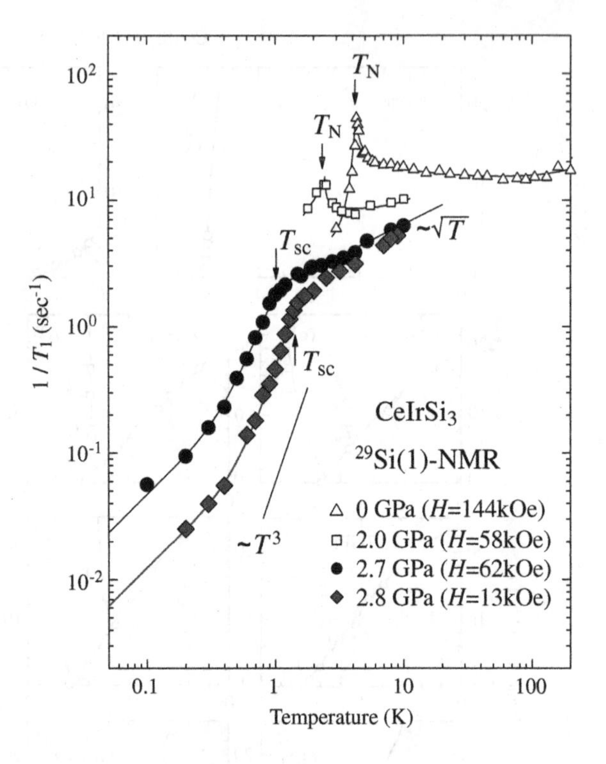

Fig. 3.44 Temperature dependence of $1/T_1$ measured by Si NMR for $CeIrSi_3$ at $P = 0$, 2.0 and 2.8 GPa, Ref. [64]

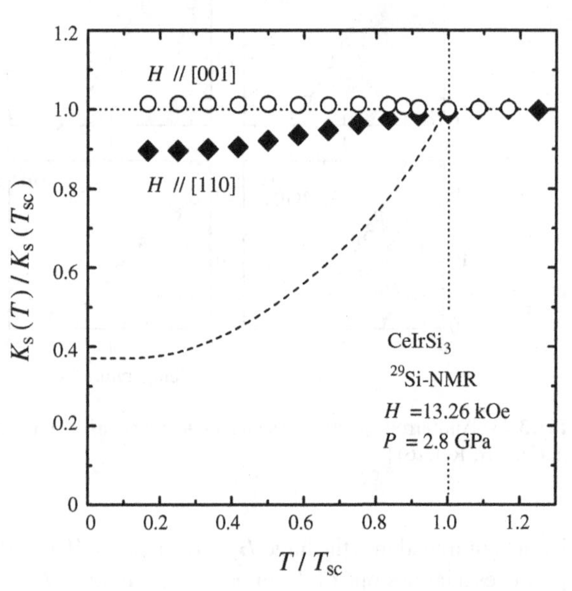

Fig. 3.45 Temperature dependence of $K_s(T)/K_s(T_{sc})$ versus T/T_{sc} for $H \parallel [1\,1\,0]$ and $[0\,0\,1]$ in the superconducting state with $T_{sc} = 1.2$ K in $CeIrSi_3$. A broken curve is a calculation for a case of $\chi \longrightarrow 0$ at low-T limit, assuming the residual density of states of 37% at the Fermi energy, Ref. [66]

of the spin susceptibility below T_{sc} for this type of non-centrosymmetric superconductor. It turns out that $CeIrSi_3$ is a strong-coupling superconductor, looking at the results of specific-heat measurements. These result in the huge H_{c2} values for $H \parallel [0\,0\,1]$ and a large anisotropy of $H_{c2}(0)$ between $H \parallel [1\,1\,0]$ and $[0\,0\,1]$. The present experimental results are discussed in the recent theoretical work [67].

3.4 Summary

We studied the splitting of Fermi surfaces by dHvA measurements and obtained estimate for the magnitude of the antisymmetric spin-orbit coupling $2|\alpha p_{\perp}|$ in RPt_3Si (R: La, Ce) and RTX_3 (T: Co, Rh, Ir, X: Si, Ge) which all have a non-centrosymmetric tetragonal crystal structure. The $2|\alpha p_{\perp}|$ value varies with changing T from Co to Rh and Ir for $LaTGe_3$, but remains unaffected when replacing Si by Ge in $LaIrSi_3$. The value of $2|\alpha p_{\perp}|$ is largest for the compounds $LaIrSi_3$ and $LaIrGe_3$ and decreases for Rh and Co. This effect can be attributed to the large effective atomic number of Ir and a large weight of the radial wave function of the Ir-5d electrons close to the nuclear center, compared with those of Co and Rh. $LaPt_3Si$ with Pt-5d electrons also possesses a large value of $2\,|\alpha p_{\perp}| = 2400$ K for a main Fermi surface. The magnitude of the antisymmetric spin-orbit coupling of $CeIrSi_3$ and $CePt_3Si$ can be roughly estimated under consideration of the electronic specific-heat coefficient, which is one order smaller than that of the corresponding La compound.

We also studied the superconducting properties such as the upper critical field H_{c2} under pressure for $CeIrSi_3$ and $CePt_3Si$.A magnetic quantum phase transition occurs in $CeIrSi_3$ at about 2.6 GPa, coinciding with a huge H_{c2} value of $H_{c2}(0) \simeq 450$ kOe for $H \parallel [0\,0\,1]$, while $H_{c2}(0) \simeq 95$ kOe for $H \parallel [1\,1\,0]$. Quantum critical behavior and the specific form of spin-orbit coupling combined are likely responsible for the extraordinary enhancement of the out-of-plane upper critical field in this superconductor. On the other hand, no such drastic behavior is observed in $CePt_3Si$. The corresponding $H_{c2}(0)$ value is not so large and determined by orbital depairing. Therefore this superconductor does not show paramagnetic limiting effects for any field direction. The important task to estimate the relative magnitude of even and odd parity components in the Cooper pairing state is left for future studies. The huge $H_{c2}(0)$ values found in $CeIrSi_3$ is a unique phenomenon reflecting the drastic influence of the non-centrosymmetric crystal structure on electronic properties.

Acknowledgements We are very grateful to Y. Yanase, S. Fujimoto, and K. Miyake for helpful discussions, and to T. Kawai, Y. Okuda, T. Takeuchi, F. Honda, K. Sugiyama, A. Sumiyama, Y. Haga, T. D. Matsuda, E. Yamamoto, N. Tateiwa, K. Kaneko, I. Sheikin, M.-A. Measson, I. Bonalde, K. Shimizu, D. Aoki, G. Knebel, J. Flouquet, H. Harima, H. Mukuda, and Y. Kitaoka for collaborations.

References

1. Settai, R., Takeuchi, T., Ōnuki, Y.: J. Phys. Soc. Jpn. **76**, 051003 (2007)
2. Haga, Y., Sakai, H., Kambe, S.: J. Phys. Soc. Jpn., **76**, 051012 (2007)
3. Sigrist, M., Ueda, K.: Rev. Mod. Phys. **63**, 239 (1991)
4. Metoki, N., Haga, Y., Koike, Y., Ōnuki, Y.: Phys. Rev. Lett. **80**, 5417 (1998)
5. Bernhoeft, N., Sato, N., Roessli, B., Aso, N., Heiss, A., Lander, G.H., Endoh, Y., Komatsubara, T.: Phys. Rev. Lett. **81**, 4244 (1998)
6. Sato, N.K., Aso, N., Miyake, K., Siina, R., Thalmeier, P., Varelogiannis, G., Geibel, C., Steglich, F., Fulde, P., Komatsubara, T.: Nature **410**, 340 (2001)
7. Kyogaku, M., Kitaoka, Y., Asayama, K., Geibel, C., Schank, C., Steglich, F.: J. Phys. Soc. Jpn. **62**, 4016 (1993)
8. Tou, H., Kitaoka, Y., Ishida, K., Asayama, K., Kimura, N., Ōnuki, Y., Yamamoto, E., Haga, Y., Maezawa, K.: Phys. Rev. Lett. **80**, 3129 (1998)
9. Tenya, K., Ikeda, M., Sakakibara, T., Yamamoto, E., Maezawa, K., Kimura, N., Settai, R., Ōnuki, Y.: Phys. Rev. Lett. **77**, 3193 (1996)
10. Brison, J.- P., Suderow, H., Rodiére, P., Huxley, A., Kambe, S., Rullier-Albenque, F.: J Flouquet: Physica B **281–282**, 872 (2000)
11. Bauer, E., Hilscher, G., Michor, H., Paul, C., Scheidt, E.W., Gribanov, A., Seropegin, Y., Noel, H., Sigrist, M., Rogl, P.: Phys. Rev. Lett. **92**, 027003 (2004)
12. Bauer, E., Kaldarar, H., Prokofiev, A., Royanian, E., Amato, A., Sereni, J., Brämer-Escamilla, W., Bonalde, I.: J. Phys. Soc. Jpn. **76**, 051009 (2007)
13. Akazawa, T., Hidaka, H., Fujiwara, T., Kobayashi, T.C., Yamamoto, E., Haga, Y., Settai, R., Ōnuki, Y.: Condens. Matter **16**, L29 (2004)
14. Kobayashi, T.C., Hori, A., Fukushima, S., Hidaka, H., Kotegawa, H., Akazawa, T., Takeda, K., Ohishi, Y., Yamamoto, E.: J. Phys. Soc. Jpn. **76**, 051007 (2007)
15. Kimura, N., Ito, K., Saitoh, K., Umeda, Y., Aoki, H., Terashima, T.: Phys. Rev. Lett. **95**, 247004 (2005)
16. Kimura, N., Ito, K., Aoki, H., Uji, S., Terashima, T.: Phys. Rev. Lett. **98**, 197001 (2007)
17. Kimura, N., Muro, Y., Aoki, H.: J. Phys. Soc. Jpn., **76**, 051010 (2007)
18. Sugitani, I., Matsuda, T.D., Haga, Y., Takeuchi, T., Settai, R., Ōnuki, Y.: J. Phys. Soc. Jpn., **76**, 043703 (2006)
19. Ōkuda, Y., Miyauchi, Y., Ida, Y., Takeda, Y., Tonohiro, C., Oduchi, Y., Yamada, T., Dung, N.D., Matsuda, T.D., Haga, Y., Takeuchi, T., Hagiwara, M., Kindo, K., Harima, H., Sugiyama, K., Settai, R., Ōnuki, Y.: J. Phys. Soc. Jpn., **76**, 044708 (2007)
20. Settai, R., Okuda, Y., Sugitani, I., Ōnuki, Y., Matsuda, T.D., Haga, Y., Harima, H.: Int. J. of Mod. Phys. B **21**, 3238 (2007)
21. Frigeri, P.A., Agterberg, D.F., Koga, A., Sigrist, M.: Phys. Rev. Lett. **92**, 097001 (2004) Errata; **93** (2004) 099903
22. Rashba, E.I.: Sov. Phys. Solid State **2**, 1109 (1960)
23. Samokhin, K.V., Zijlstra, E.S., Bose, S.K.: Phys. Rev. B **69**, 094514 (2004)
24. Mineev, V.P., Samokhin, K.V.: Phys. Rev. B **72**, 212504 (2005)
25. Frigeri, P.A., Agterberg, D.F., Sigrist, M.: New J. Phys. **6**, 115 (2004)
26. Fujimoto, S.: J. Phys. Soc. Jpn. **76**, 051008 (2007)
27. Shoenberg, D.: Magnetic Oscillations in Metals. Cambridge University Press, Cambridge (1984)
28. Y. Ōnuki, T. Goto, and T. Kasuya: in Materials Science and Technology, ed. K. H. J. Buschow (VCH, Weinheim, 1991) Vol. 3A, Part I, Chap. 7, p. 545.
29. Hashimoto, S., Yasuda, T., Kubo, T., Shishido, H., Ueda, T., Settai, R., Matsuda, T.D., Haga, Y., Harima, H., Ōnuki, Y.: J. Phys. Condens. Matter **16**, L287 (2004)
30. Takeuchi, T., Yasuda, T., Tsujino, M., Shishido, H., Settai, R., Harima, H., Ōnuki, Y.: J. Phys. Soc. Jpn. **76**, 014702 (2007)

31. Yasuda, T., Shishido, H., Ueda, T., Hashimoto, S., Settai, R., Takeuchi, T., Matsuda, T.D., Haga, Y., Ōnuki, Y.: J. Phys. Soc. Jpn. **73**, 1657 (2004)
32. Takeuchi, T., Hashimoto, S., Yasuda, T., Shishido, H., Ueda, T., Yamada, M., Obiraki, Y., Shiimoto, M., Kohara, H., Yamamoto, T., Sugiyama, K., Kindo, K., Matsuda, T.D., Haga, Y., Aoki, Y., Sato, H., Settai, R., Ōnuki, Y.: . J. Phys. Condens Matter, **16**, L333 (2004)
33. Metoki, N., Kaneko, K., Matsuda, T.D., Galatanu, A., Takeuchi, T., Hashimoto, S., Ueda, T., Settai, R., Ōnuki, Y., Bernhoeft, N.: J. Phys.: Condens. Matter **16**, L207 (2004)
34. Tateiwa, N., Haga, Y., Matsuda, T.D., Ikeda, S., Yasuda, T., Takeuchi, T., Settai, R., Ōnuki, Y.: J. Phys. Soc. Jpn. **74**, 1903 (2005)
35. Araki, S., Settai, R., Kobayashi, T.C., Harima, H., Ōnuki, Y.: Phys. Rev. B **64**, 224417 (2001)
36. Ōnuki, Y., Settai, R., Sugiyama, K., Takeuchi, T., Kobayashi, T.C., Haga, Y., Yamamoto, E.: J. Phys. Soc. Jpn. **73**, 769 (2004)
37. Shishido, H., Settai, R., Harima, H., Ōnuki, Y.: J. Phys. Soc. Jpn. **74**, 1103 (2005)
38. Settai, R., Kubo, T., Shiromoto, T., Honda, D., Shishido, H., Sugiyama, K., Haga, Y., Matsuda, T.D., Betsuyaku, K., Harima, H., Kobayashi, T.C., Ōnuki, Y.: J. Phys. Soc. Jpn. **74**, 3016 (2005)
39. Ōnuki, Y., Miyauchi, Y., Tsujino, M., Ida, Y., Settai, R., Takeuchi, T., Tateiwa, N., Matsuda, T.D., Haga, Y., Harima, H.: J. Phys. Soc. Jpn. **77** (suppl. A), 37 (2008)
40. Kawai, T., Muranaka, H., Endo, T., Dung, N.D., Doi, Y., Ikeda, S., Matsuda, T.D., Haga, Y., Harima, H., Settai, R., Ōnuki, Y.: J. Phys. Soc. Jpn. **77**, 064717 (2008)
41. Koelling, D.D., Harmon, B.N.: J. Phys. C: Solid State Phys. **10**, 3107 (1977)
42. Nishiyama, M., Inada, Y., Zheng, G.-Q.: Phys. Rev. Lett. **98**, 047002 (2007)
43. Thamizhavel, A., Shishido, H., Okuda, Y., Harima, H., Matsuda, T.D., Haga, Y., Settai, R., Ōnuki, Y.: J. Phys. Soc. Jpn. **75**, 044711 (2006)
44. Muro, Y., Eom, D.H., Takeda, N., Ishikawa, M.: J. Phys. Soc. Jpn. **67**, 3601 (1998)
45. H. Harima: private communication.
46. Thamizhavel, A., Takeuchi, T., Matsuda, T.D., Haga, Y., Sugiyama, K., Settai, R., Onuki, Y.: J. Phys. Soc. Jpn. **74**, 1858 (2005)
47. Jaccard, D., Behnia, K., Sierro, J.: Phys. Lett. A **163**, 475 (1992)
48. Doniach, S.: Physica **91B**, 231 (1977)
49. Kawai, T., Nakashima, M., Okuda, Y., Shishido, H., Shimoda, T., Matsuda, T.D., Haga, Y., Takeuchi, T., Hedo, M., Uwatoko, Y., Settai, R., Ōnuki, Y.: J. Phys. Soc. Jpn. **76** (Suppl. A), 166 (2007)
50. Okuda, Y., Sugitani, I., Shishido, H., Yamada, T., Thamizhavel, A., Yamamoto, E., Matsuda, T.D., Haga, Y., Takeuchi, T., Settai, R., Ōnuki, Y.: J. Magn. Magn. Mat. **310**, 563 (2007)
51. Honda, F., Bonalde, I., Shimizu, K., Yoshiuchi, S., Hirose, Y., Nakamura, T., Settai, R., Ōnuki, Y.: Phys. Rev. B **81**, 140507(R) (2010)
52. Kawai, T., Muranaka, H., Measson, M.-A., Shimoda, T., Doi, Y., Matsuda, T.D., Haga, Y., Knebel, G., Lapertot, G., Aoki, D., Flouquet, J., Takeuchi, T., Settai, R., Ōnuki, Y.: J. Phys. Soc. Jpn. **77**, 064716 (2008)
53. Knebel, G., Aoki, D., Lapertot, G., Salce, B., Flouquet, J., Kawai, T., Muranaka, H., Settai, R., Ōnuki, Y.: J. Phys. Soc. Jpn. **77**, 074714 (2009)
54. Méasson, M.-A., Muranaka, H., Kawai, T., Ota, Y., Sugiyama, K., Hagiwara, M., Kindo, K., Takeuchi, T., Shimizu, K., Honda, F., Settai, R., Ōnuki, Y.: J. Phys. Soc. Jpn. **78**, 124713 (2009)
55. Kadowaki, K., Woods, S.B.: Solid State Commun. **58**, 507 (1986)
56. Tateiwa, N., Haga, Y., Matsuda, T.D., Ikeda, S., Yamamoto, E., Okuda, Y., Miyauchi, Y., Settai, R., Ōnuki, Y.: J. Phys. Soc. Jpn. **76**, 083706 (2007)
57. Kaneko, K., Metoki, N., Takeuchi, T., Matsuda, T.D., Haga, Y., Thamizhavel, A., Settai, R., Ōnuki, Y.: J. Phys.: Conf. Series **150**, 042082 (2009)
58. Settai, R., Miyauchi, Y., Takeuchi, T., Sheikin, I., Lévy, F., Ōnuki, Y.: J. Phys. Soc. Jpn. **77**, 073705 (2008)
59. Helfand, E., Werthamer, N.R.: Phys. Rev. **147**, 288 (1966)
60. Werthamer, N.R., Helfand, E., Hohenberg, P.C.: Phys. Rev. **147**, 295 (1966)

61. Settai, R., Katayama, K., Aoki, D., Sheikin, I., Knebel, G., Flouquet, J., Ōnuki, Y.: J. Phys. Soc. Jpn. **80**, 094703 (2011)
62. Knebel, G., Aoki, D., Braithwaite, D., Cherroret, N., Salce, B., Flouquet, J.: J. Phys. Soc. Jpn. **76** (Suppl. A), 124 (2007)
63. Ida, Y., Settai, R., Ota, Y., Honda, F., Ōnuki, Y.: J. Phys. Soc. Jpn., **77**, 084708 (2008)
64. Mukuda, H., Fujii, T., Ohara, T., Harada, A., Yashima, M., Kitaoka, Y., Okuda, Y., Settai, R., Ōnuki, Y.: Phys. Rev. Lett. **100**, 107003 (2008)
65. Moriya, T., Ueda, K.: Adv. Phys. **49**, 555 (2000)
66. Mukuda, H., Ohara, T., Yashima, M., Kitaoka, Y., Settai, R., Ōnuki, Y., Itoh, K.M., Haller, E.E.: Phys. Rev. Lett. **104**, 017002 (2010)
67. Tada, Y., Kawakami, N., Fujimoto, S.: Phys. Rev. Lett. **101**, 267006 (2008)

Part II
Theory of Non-centrosymmetric Superconductors

Chapter 4
Basic Theory of Superconductivity in Metals Without Inversion Center

V. P. Mineev and M. Sigrist

Abstract This Chapter gives a brief introduction to some basic aspects metals and superconductors in crystals without inversion symmetry. In the first part we analyze some normal state properties which arise through antisymmetric spin-orbit coupling existing in non-centrosymmetric materials and show its influence on the de Haas–van Alphen effect. For the superconducting phase we introduce a multi-band formulation which naturally arises due the spin splitting of the bands by spin-orbit coupling. It will then be shown how the states can be symmetry classified and their relation to the original classification in even-parity spin-singlet and odd-parity spin-triplet pairing states. The general Ginzburg–Landau functional will be derived and applied to the nucleation of superconductivity in a magnetic field. It will be shown that magneto-electric effects can modify the standard paramagnetic limiting behavior drastically.

4.1 Introduction

Motivated by the discovery of the non-centrosymmetric heavy fermion superconductor $CePt_3Si$ [1], the physics of unconventional superconductivity in materials without inversion symmetry has recently become a subject of growing interest. The lack of inversion symmetry, a key symmetry for Cooper pairing, combined with unconventional pairing symmetry is responsible for a number of intriguing novel properties. In a short time the list of new superconductors in this class has been enlarged by compounds such as UIr [2], $CeRhSi_3$ [3], $CeIrSi_3$ [4], Y_2C_3 [5] and $Li_2(Pd_{1-x}Pt_x)_3B$ [6–8]. In all listed heavy fermion compounds superconductivity appears in combination with a magnetic quantum phase transition suggesting the presence of

V. P. Mineev (✉)
Commissariat à l'Energie Atomique, INAC/SPSMS, 38054 Grenoble, France
email: vladimir.mineev@cea.fr

M. Sigrist
Theoretische Physik, ETH-Zurich, 8093 Zurich, Switzerland
email: sigrist@itp.phys.ethz.ch

E. Bauer and M. Sigrist (eds.), *Non-centrosymmetric Superconductors*, Lecture Notes in Physics 847, DOI: 10.1007/978-3-642-24624-1_4,

strong electron correlation effects. Thus, it is widely believed that magnetic fluctuations are likely responsible for inducing here unconventional Cooper pairing. For other materials correlation effects seem to be less relevant. Nevertheless, some of them show unexpectedly features of unconventional pairing. Li_2Pt_3B belongs to this class. While in some cases experimental results give rather clear suggestions on the gap symmetry, the definite identification of pairing states is far from concluded.

The microscopic theory of superconductivity in metals without inversion has a long history predating these recent experimental developments [9–12]. Specific aspects such as the possibilities of inhomogeneous superconducting states [13–15] and the magneto-electric effect [16, 17] in this type of materials have been discussed already in the nineties of the last century. Moreover, general symmetry aspects of non-centrosymmetric superconductors have been addressed rather early in Refs. [18–20]. A wide variety of physical phenomena connected with non-centrosymmetricity have since been studied by many groups:

- paramagnetic limitations of superconductivity and the helical vortex state [21–32];
- paramagnetic susceptibility [33–35, 66] and the magnetic field induced superconducting gap structure [36];
- Josephson and quasiparticle tunneling [37, 38], surface bound states [39, 40], and vortex bound states [41];
- London penetration depth [42] and the magnetic field distribution [43];
- effects of impurities [44–46];
- upper critical field [47, 48];
- nuclear magnetic relaxation rate [49, 50];
- general forms of pairing interaction [51];
- inhomogeneous superconducting states in the absence of external field [52].

In this Chapter we give a brief introduction to several topics in this context, leaving most of the special aspects of non-centrosymmetric superconductivity to other Chapters.

4.2 Normal State

The absence of inversion symmetry is imprinted into the electronic structure through spin-orbit coupling effects. Already the normal state of non-centrosymmetric metals bears intriguing features which result from the specific form of spin-orbit coupling. In this section we discuss the electronic spectrum.

4.2.1 Electronic States in Non-centrosymmetric Metals

Our starting point is the following Hamiltonian of non-interacting electrons in a crystal without inversion center:

$$H_0 = \sum_{\mathbf{k}} \sum_{\alpha\beta=\uparrow,\downarrow} [\xi(\mathbf{k})\delta_{\alpha\beta} + \gamma(\mathbf{k}) \cdot \sigma_{\alpha\beta}] a^{\dagger}_{\mathbf{k}\alpha} a_{\mathbf{k}\beta} \tag{4.1}$$

where $a^{\dagger}_{\mathbf{k}\alpha}$ ($a_{\mathbf{k}\alpha}$) creates (annihilates) an electronic state $|\mathbf{k}\alpha\rangle$. Furthermore, $\xi(\mathbf{k}) = \varepsilon(\mathbf{k}) - \mu$ denotes the spin-independent part of the spectrum measured relative to the chemical potential μ, $\alpha, \beta = \uparrow, \downarrow$ are spin indices and σ are the Pauli matrices. The sum over $\mathbf{k}$ is restricted to the first Brillouin zone. The second term in Eq. (4.1) describes the *antisymmetric* spin-orbit (SO) coupling whose form depends on the specific non-centrosymmetric crystal structure [53–56]. The pseudovector $\gamma(\mathbf{k})$ satisfies $\gamma(-\mathbf{k}) = -\gamma(\mathbf{k})$ and $g\gamma(g\mathbf{k}) = \gamma(\mathbf{k})$, where g is any symmetry operation in the generating point group $\mathscr{G}$ of the crystal (see below). The usual symmetric spin-orbit coupling which is present also in centrosymmetric crystals yields a new spinor basis (pseudospinor) α, β in Eq. (4.1), which retains the ordinary spin-1/2 structure with complete SU(2) symmetry. This is different for the antisymmetric spin-orbit coupling. The effect of the antisymmetric spin-orbit coupling is a spin splitting of the band energy with $\mathbf{k}$-dependent spin quantization axis which removes the SU(2) symmetry.

Depending on the purpose it is more convenient to express the Hamiltonian (4.1) in the initial 2x2 matrix form (*spinor representation*) or in its diagonal form (*band representation*). The energy bands are given by

$$\xi_{\pm}(\mathbf{k}) = \xi(\mathbf{k}) \pm |\gamma(\mathbf{k})| \tag{4.2}$$

with the Hamiltonian

$$H_0 = \sum_{\mathbf{k}} \sum_{\lambda=\pm} \xi_{\lambda}(\mathbf{k}) c^{\dagger}_{\mathbf{k}\lambda} c_{\mathbf{k}\lambda}, \tag{4.3}$$

where the two sets of electronic operators are connected by a unitary transformation,

$$a_{\mathbf{k}\alpha} = \sum_{\lambda} u_{\alpha\lambda}(\mathbf{k}) c_{\mathbf{k}\lambda}, \tag{4.4}$$

with

$$(u_{\uparrow\lambda}(\mathbf{k}), u_{\downarrow\lambda}(\mathbf{k})) = \frac{(|\gamma| + \lambda\gamma_z,\ \lambda(\gamma_x + i\gamma_y))}{\sqrt{2|\gamma|(|\gamma| + \lambda\gamma_z)}}. \tag{4.5}$$

The normal-state electron Green's functions in the spinor representation can be written as

$$\hat{G}(\mathbf{k}, \omega_n) = \sum_{\lambda=\pm} \hat{\Pi}_{\lambda}(\mathbf{k}) G_{\lambda}(\mathbf{k}, \omega_n), \tag{4.6}$$

where

$$\hat{\Pi}_\lambda(\mathbf{k}) = \frac{1 + \lambda\hat{\gamma}(\mathbf{k})\sigma}{2} \tag{4.7}$$

are the band projection operators and $\hat{\gamma} = \gamma/|\gamma|$. The Green's functions in the band representation have then the simple form

$$G_\lambda(\mathbf{k}, \omega_n) = \frac{1}{i\omega_n - \xi_\lambda(\mathbf{k})}, \tag{4.8}$$

where $\omega_n = \pi T(2n+1)$ is the Matsubara frequency.

The Fermi surfaces defined by the equations $\xi_\pm(\mathbf{k}) = 0$ are split, except at specific points or lines where $\gamma(\mathbf{k}) = 0$ is satisfied. The band dispersion functions $\xi_\lambda(\mathbf{k})$ are invariant with respect to all operations of $\mathscr{G}$ and the time reversal operations $K = i\hat{\sigma}_2 K_0$ (K_0 is the complex conjugation). The states $|\mathbf{k}, \lambda\rangle$ and $K|\mathbf{k}, \lambda\rangle$ belonging to the band energies $\xi_\lambda(\mathbf{k})$ and $\xi_\lambda(-\mathbf{k})$, respectively, are degenerate, since the time reversal operation yields $K|\mathbf{k}, \lambda\rangle = t_\lambda(\mathbf{k})| - \mathbf{k}, \lambda\rangle$, where $t_\lambda(\mathbf{k}) = -t_\lambda(-\mathbf{k})$ is a nontrivial phase factor [12, 19]. For the eigenstates of H_0, defined by (4.5), this phase factor takes the form,

$$t_\lambda(\mathbf{k}) = -\lambda \frac{\gamma_x(\mathbf{k}) - i\gamma_y(\mathbf{k})}{\sqrt{\gamma_x^2(\mathbf{k}) + \gamma_y^2(\mathbf{k})}}. \tag{4.9}$$

Finally we turn to the basic form of the antisymmetric spin-orbit coupling as it results from the non-centrosymmetric crystal structures. Here we ignore the Brillouin zone structure and use only the expansion for small momenta $\mathbf{k}$ leading to basis functions satisfying the basic symmetry requirements of $\gamma(\mathbf{k})$. For the cubic group $\mathscr{G} = \mathbf{O}$, the point group of $Li_2(Pd_{1-x}, Pt_x)_3B$, the simplest form compatible with symmetry requirements is

$$\gamma(\mathbf{k}) = \gamma_0 \mathbf{k}, \tag{4.10}$$

where γ_0 is a constant. For point groups containing improper elements, i.e. reflections and rotation-reflections, expressions become more complicated. The full tetrahedral group $\mathscr{G} = \mathbf{T}_d$, which is relevant for Y_2C_3 and possibly KOs_2O_6,the expansion of $\gamma(\mathbf{k})$ starts with third order in the momentum,

$$\gamma(\mathbf{k}) = \gamma_0[k_x(k_y^2 - k_z^2)\hat{x} + k_y(k_z^2 - k_x^2)\hat{y} + k_z(k_x^2 - k_y^2)\hat{z}]. \tag{4.11}$$

This is sometimes called Dresselhaus spin-orbit coupling [53, 54], and was originally discussed for bulk semiconductors of zinc-blend structure.

The tetragonal point group $\mathscr{G} = \mathbf{C}_{4v}$, relevant for $CePt_3Si$, $CeRhSi_3$ and $CeIrSi_3$, yields the antisymmetric spin-orbit coupling

$$\gamma(\mathbf{k}) = \gamma_\perp(k_y\hat{x} - k_x\hat{y}) + \gamma_\parallel k_x k_y k_z(k_x^2 - k_y^2)\hat{z}. \tag{4.12}$$

In the purely two-dimensional case, setting $\gamma_{\|} = 0$ one recovers the Rashba interaction [55, 56] which is often used to describe the effects of the absence of mirror symmetry in semiconductor quantum wells.

4.2.2 de Haas–van Alphen Effect

An experimental way of observing the spin-splitting of the Fermi surface is the de Haas–van Alphen (dHvA) effect which can help to estimate the magnitude of the antisymmetric spin-orbit coupling [57]. The single-electron Hamiltonian (4.1) can be extended to include the magnetic field as follows:

$$H_0 = \sum_{\mathbf{k}} \sum_{\alpha\beta=\uparrow,\downarrow} [\xi(\mathbf{k})\delta_{\alpha\beta} + \boldsymbol{\gamma}(\mathbf{k})\boldsymbol{\sigma}_{\alpha\beta} - \mu_B \mathbf{H}\boldsymbol{\sigma}_{\alpha\beta}] a^{\dagger}_{\mathbf{k}\alpha} a_{\mathbf{k}\beta}. \tag{4.13}$$

The last term describes the Zeeman interaction for an external magnetic field $\mathbf{H}$, with μ_B being the Bohr magneton. The orbital effect of the field can be included by replacing $\mathbf{k} \to \mathbf{k} + (e/\hbar c)\mathbf{A}(\hat{\mathbf{r}})$, [58] where $\hat{\mathbf{r}} = i\nabla_{\mathbf{k}}$ is the position operator in the $\mathbf{k}$-representation.

The eigenvalues of the Hamiltonian (4.13) are

$$\xi_\lambda(\mathbf{k}, \mathbf{H}) = \xi(\mathbf{k}) + \lambda|\boldsymbol{\gamma}(\mathbf{k}) - \mu_B \mathbf{H}|. \tag{4.14}$$

There are two Fermi surfaces determined by the equations

$$\xi_\lambda(\mathbf{k}, \mathbf{H}) = 0. \tag{4.15}$$

For certain directions and magnitudes of $\mathbf{H}$ there may be accidental degeneracies of the Fermi surfaces, determined by the equation $\boldsymbol{\gamma}(\mathbf{k}) = \mu_B \mathbf{H}$. However, there are no symmetry reasons for such intersections.

An important property of the Fermi surfaces (4.15) is the fact that their shapes depend on the magnetic field in a characteristic way, which can be directly probed by dHvA experiments. Note that while at $H = 0$ time reversal symmetry guarantees $\xi_\lambda(-\mathbf{k}) = \xi_\lambda(\mathbf{k})$, the loss of time reversal symmetry for $H \neq 0$ yields, in general, $\xi_\lambda(-\mathbf{k}, \mathbf{H}) \neq \xi_\lambda(\mathbf{k}, \mathbf{H})$, i.e. the Fermi surfaces do not have inversion symmetry.

Including now the coupling of the magnetic field to the orbital motion of the electrons we derive in quasi-classical approximation the Lifshitz–Onsager quantization rules [58]:

$$S_\lambda(\varepsilon, k_H) = \frac{2\pi e H}{\hbar c}\,[n + \alpha_\lambda(\Gamma)]\,. \tag{4.16}$$

Here S_λ is the area of the quasi-classical orbit Γ, in the $\mathbf{k}$-space defined by the intersection of the constant-energy surface $\varepsilon_\lambda(\mathbf{k}) = \varepsilon$ with the plane $\mathbf{k} \cdot \hat{\mathbf{h}} = k_H$ ($\hat{\mathbf{h}} = \mathbf{H}/H$). Moreover, n is an integer number ($n \gg 1$), and $0 \le \alpha_\lambda(\Gamma) < 1$ is connected with the Berry phase of the electron as it moves along Γ [59, 60]. The value of $\alpha_\lambda(\Gamma)$ does not affect the dHvA frequency discussed below.

The dHvA signal contains contributions from both bands and can be approximately decomposed into the form,

$$M_{osc} = \sum_\lambda A_\lambda \cos\left(\frac{2\pi F_\lambda}{H} + \phi_\lambda\right), \tag{4.17}$$

where A_λ and ϕ_λ are the amplitudes and the phases of the oscillations. The amplitudes are given by the standard Lifshitz–Kosevich formula and the dHvA frequencies F_λ are related to the extremal (with respect to k_H) cross-section areas of the two Fermi surfaces,

$$F_\lambda = \frac{\hbar c}{2\pi e} S_\lambda^{ext}. \tag{4.18}$$

In addition to the fundamental harmonics (4.17), the observed dHvA signal also contains higher harmonics with frequencies given by multiple integers of F_λ.

It is interesting to consider the field dependence of the band energies (4.14), which yield

$$S_\lambda^{ext}(\mathbf{H}) = S_\lambda^{ext}(0) + A_\lambda(\hat{\mathbf{h}})H + B_\lambda(\hat{\mathbf{h}})H^2 + \cdots . \tag{4.19}$$

Inserting this in Eq. (4.17) the term linear in H contributes to the phase shift, similar to the paramagnetic splitting of Fermi surfaces in centrosymmetric metals. The quadratic term is responsible for the magnetic field dependence of the dHvA frequencies. This is a specific feature of non-centrosymmetric metals which could be observable, if the Zeeman energy is at most of comparable magnitude as the spin-orbit coupling.

For illustration, let us look at the example of a three-dimensional elliptic Fermi surface with $\xi(\mathbf{k}) = \frac{\hbar^2 k_\perp^2}{2m_\perp} + \frac{\hbar^2 k_z^2}{2m_z} - \varepsilon_F$, where $m_\perp$ and m_z are the effective masses. The extremal (maximum) cross-sections of the Fermi surfaces (4.15) correspond to $k_z = 0$. Introducing the Fermi wave vector k_F via $\varepsilon_F = \hbar^2 k_F^2/2m_\perp$, we obtain

$$S_\lambda^{ext}(\mathbf{H}) = \pi k_F^2 \left[1 - \lambda \frac{|\gamma_\perp| k_F}{\varepsilon_F}\left(1 + \frac{\mu_B^2 H^2}{2\gamma_\perp^2 k_F^2}\right)\right]. \tag{4.20}$$

In this approximation we assumed that the Zeeman energy is small compared to the spin-orbit band splitting, which in turn is much smaller than the Fermi energy: $\mu_B H \ll |\gamma_\perp| k_F \ll \varepsilon_F$. Based on this result it is also possible to obtain an estimate of the strength of the spin-orbit coupling.

We use the expressions (4.18) and (4.20) to calculate the difference of the dHvA frequencies for the split bands:

$$F_- - F_+ = \frac{2c}{\hbar e} |\gamma_\perp| k_F m_\perp \left(1 + \frac{\mu_B^2 H^2}{2\gamma_\perp^2 k_F^2}\right). \tag{4.21}$$

For example, from the frequencies of the "α" and "β" dHvA frequency branches in $LaPt_3Si$ [61], $F_\alpha = 1.10 \times 10^8$Oe and $F_\beta = 8.41 \times 10^7$Oe, and $m_\perp \simeq 1.5m$, we obtain for the spin-orbit splitting of the Fermi surfaces: $|\gamma_\perp| k_F \simeq 10^3$K, which is in reasonable agreement with the results of band structure calculations [18, 61]. According to Eq. (4.21), the magnetic field effect on $F_- - F_+$ in the range of fields used in Ref. [61] (up to 17T) should be of the order of a few percent. In this way, the interplay of the Zeeman splitting and the spin-orbit coupling which results in a deformation of the Fermi surface, is responsible for a field dependence of the dHvA frequencies, an effect absent in centrosymmetric metals.

4.3 Superconducting State

In this section we turn to the discussion of some novel aspects of the superconducting state in non-centrosymmetric materials. Here we can consider only a few examples, while a wider range of other phenomena will be discussed in other Chapters of this book.

4.3.1 Basic Equations

After our introductory discussion of single-electron properties we now include electron–electron interactions to examine the implications of non-centrosymmetricity on Cooper pairing. Therefore we retain among all interactions only those terms corresponding to the Cooper channel and formulate it in the band representation. The general form is given by

$$H_{int} = \frac{1}{2\mathscr{V}} \sum_{\mathbf{k},\mathbf{k}'} \sum_{\lambda_1\lambda_2\lambda_3\lambda_4} V_{\lambda_1\lambda_2\lambda_3\lambda_4}(\mathbf{k},\mathbf{k}') c^\dagger_{\mathbf{k},\lambda_1} c^\dagger_{-\mathbf{k},\lambda_2} c_{-\mathbf{k}',\lambda_3} c_{\mathbf{k}',\lambda_4}. \tag{4.22}$$

It is reasonable to assume $\lambda_1 = \lambda_2$ and $\lambda_3 = \lambda_4$, such that only intra-band pairing is considered. Inter-band pairing is usually suppressed, since the spin-orbit coupling induced band splitting would require that electrons far from the Fermi surfaces would have to pair, which is unlikely, if the energy scale of the band splitting strongly exceeds the superconducting energy scale.[1] Introducing the notation $\lambda_1 = \lambda_2 = \lambda$ and $\lambda_3 = \lambda_4 = \lambda'$, we obtain:

---

[1] The interband pairing leads to the interesting possibility of existence of nonuniform superconducting states even in the absence of external magnetic fields [52]. These states, however, could be realized in the noncentrosymmetric compounds with the SO band splitting smaller than the superconducting critical temperature. To the best of our knowledge, in all noncentrosymmetric compounds discovered to date the relation between the two energy scales is exactly the opposite: the SO band splitting exceeds all superconducting energy scales by order of magnitude, completely suppressing the interband pairing, both uniform and nonuniform.

$$H_{int} = \frac{1}{2\mathscr{V}} \sum_{\mathbf{k}\mathbf{k}'\mathbf{q}} \sum_{\lambda\lambda'} V_{\lambda\lambda'}(\mathbf{k}, \mathbf{k}') c^{\dagger}_{\mathbf{k},\lambda} c^{\dagger}_{-\mathbf{k},\lambda} c_{-\mathbf{k}',\lambda'} c_{\mathbf{k}',\lambda'}, \tag{4.23}$$

where

$$V_{\lambda\lambda'}(\mathbf{k}, \mathbf{k}') = t_{\lambda}(\mathbf{k}) t^{*}_{\lambda'}(\mathbf{k}') \tilde{V}_{\lambda\lambda'}(\mathbf{k}, \mathbf{k}'). \tag{4.24}$$

Since under time reversal the creation and annihilation operators behave as

$$K a^{\dagger}_{\mathbf{k},\lambda} = t_{\lambda}(\mathbf{k}) a^{\dagger}_{-\mathbf{k},\lambda}, \quad K a_{\mathbf{k},\lambda} = t^{*}_{\lambda}(\mathbf{k}) a_{-\mathbf{k},\lambda}, \tag{4.25}$$

$\tilde{V}_{\lambda\lambda'}(\mathbf{k}, \mathbf{k}')$ represents the pairing interaction between time-reversed states. The amplitude $\tilde{V}_{\lambda\lambda'}(\mathbf{k}, \mathbf{k}')$ is even in both $\mathbf{k}$ and $\mathbf{k}'$ due to the anticommutation of fermionic operators and is invariant under the point group operations: $\tilde{V}_{\lambda\lambda'}(g\mathbf{k}, g\mathbf{k}') = \tilde{V}_{\lambda\lambda'}(\mathbf{k}, \mathbf{k}')$. The gap functions of the superconducting state can be expressed as, $\Delta_{\lambda}(\mathbf{k}) = t_{\lambda}(\mathbf{k}) \tilde{\Delta}_{\lambda}(\mathbf{k})$, in each band, where $\tilde{\Delta}_{\lambda}$ transforms according to one of the irreducible representations of the crystal point group [63, 64].

The *Gor'kov equations* in each band read

$$(i\omega_n - \xi_{\lambda}(\mathbf{k}))\, G_{\lambda}(\mathbf{k}, \omega_n) + \tilde{\Delta}_{\mathbf{k}\lambda} F^{\dagger}_{\lambda}(\mathbf{k}, \omega_n) = 1 \tag{4.26}$$

$$(i\omega_n + \xi_{\lambda}(-\mathbf{k}))\, F^{\dagger}_{\lambda}(\mathbf{k}, \omega_n) + \tilde{\Delta}^{\dagger}_{\mathbf{k}\lambda} G_{\lambda}(\mathbf{k}, \omega_n) = 0. \tag{4.27}$$

The gap functions obey the self-consistency equations

$$\tilde{\Delta}_{\mathbf{k}\lambda} = -T \sum_{n} \sum_{\mathbf{k}'} \sum_{\lambda'} \tilde{V}_{\lambda\lambda'}\left(\mathbf{k}, \mathbf{k}'\right) F_{\lambda'}(\mathbf{k}', \omega_n). \tag{4.28}$$

The resulting Green's functions are then,

$$G_{\lambda}\left(\mathbf{k}, \omega_n\right) = \frac{i\omega_n + \xi_{\lambda}(-\mathbf{k})}{(i\omega_n - \xi_{\lambda}(\mathbf{k}))(i\omega_n + \xi_{\lambda}(-\mathbf{k})) - \tilde{\Delta}_{\mathbf{k}\lambda} \tilde{\Delta}^{\dagger}_{\mathbf{k}\lambda}} \tag{4.29}$$

$$F_{\lambda}\left(\mathbf{k}, \omega_n\right) = \frac{-\tilde{\Delta}_{\mathbf{k}\lambda}}{(i\omega_n - \xi_{\lambda}(\mathbf{k}))(i\omega_n + \xi_{\lambda}(-\mathbf{k})) - \tilde{\Delta}_{\mathbf{k}\lambda} \tilde{\Delta}^{\dagger}_{\mathbf{k}\lambda}}. \tag{4.30}$$

and the quasiparticle excitation energies for each band have the form

$$E_{\mathbf{k}\lambda} = \frac{\xi_{\lambda}(\mathbf{k}) - \xi_{\lambda}(-\mathbf{k})}{2} + \sqrt{\left(\frac{\xi_{\lambda}(\mathbf{k}) + \xi_{\lambda}(-\mathbf{k})}{2}\right)^2 + \tilde{\Delta}_{\mathbf{k}\lambda} \tilde{\Delta}^{\dagger}_{\mathbf{k}\lambda}} \tag{4.31}$$

which becomes in case of time reversal symmetry,

$$E_{\mathbf{k}\lambda} = \sqrt{\xi_{\lambda}(\mathbf{k})^2 + \tilde{\Delta}_{\mathbf{k}\lambda} \tilde{\Delta}^{\dagger}_{\mathbf{k}\lambda}}. \tag{4.32}$$

These Green's function are analogous to those of a multi-band superconductor [62], apart from the fact that in the non-centrosymmetric case the two bands do not possess spin degeneracy. They rather correspond to a type of spinless fermions, since their spinors on each band are subject to a momentum dependent projection. This distinction becomes more apparent, if we write the Gor'kov equations in the initial spinor basis (spin up and down),

$$(i\omega_n - \xi(\mathbf{k}) - \gamma_{\mathbf{k}}\sigma)\,\hat{G}(\mathbf{k},\omega_n) + \hat{\Delta}_{\mathbf{k}}\hat{F}^{\dagger}(\mathbf{k},\omega_n) = \hat{1} \tag{4.33}$$

$$\left(i\omega_n + \xi(-\mathbf{k}) + \gamma_{-\mathbf{k}}\sigma^t\right)\hat{F}^{\dagger}(\mathbf{k},\omega_n) + \hat{\Delta}^{\dagger}_{\mathbf{k}}\hat{G}(\mathbf{k},\omega_n) = 0, \tag{4.34}$$

where $\xi(\mathbf{k}) = \varepsilon_{\mathbf{k}} - \mu$,

$$\hat{G}(\mathbf{k},\omega_n) = \hat{\Pi}_+ G_+(\mathbf{k},\omega_n) + \hat{\Pi}_- G_-(\mathbf{k},\omega_n), \tag{4.35}$$

$$\hat{F}^{\dagger}(\mathbf{k},\omega_n) = \hat{g}^t\left\{\hat{\Pi}_+ F^{\dagger}_+(\mathbf{k},\omega_n) + \hat{\Pi}_- F^{\dagger}_-(\mathbf{k},\omega_n)\right\}, \tag{4.36}$$

$$\hat{\Delta}_{\mathbf{k}} = \left\{\hat{\Pi}_+\tilde{\Delta}_{\mathbf{k},+} + \Pi_-\tilde{\Delta}_{\mathbf{k},-}\right\}\hat{g}, \tag{4.37}$$

with $\hat{g} = i\hat{\sigma}_y$. Examining the form of the gap function reveals that in the non-centrosymmetric superconductor both even-parity spin-singlet and odd-parity spin-triplet pairing are mixed, since no symmetry is available to distinguish between the two. Therefore, we may write

$$\hat{\Delta}_{\mathbf{k}} = \frac{\tilde{\Delta}_{\mathbf{k},+} + \tilde{\Delta}_{\mathbf{k},-}}{2}\hat{g} + \frac{\tilde{\Delta}_{\mathbf{k},+} - \tilde{\Delta}_{\mathbf{k},-}}{2}\hat{\gamma}_{\mathbf{k}}\sigma\hat{g}. \tag{4.38}$$

The odd-parity component is represented by a vector which is oriented along the vector $\gamma_{\mathbf{k}}$ of the spin-orbit coupling. In this discussion the only approximation entering so far is the absence of inter-band pairing.

4.3.2 Critical Temperature

For the discussion of the instability condition at the critical temperature and the topology of the quasiparticle gap we can use the symmetry properties of the pairing interaction matrix $\tilde{V}_{\lambda\lambda'}(\mathbf{k},\mathbf{k}')$ mentioned earlier. The momentum dependence of the matrix elements can be represented in a spectral form decomposed in products of the basis functions of irreducible representations of $\mathcal{G}$. It is generally sufficient to consider only the part of the pairing potential based on one irreducible representation Γ corresponding to the superconducting state with maximal critical temperature [63]. In a simplified formulation this could be represented by even basis functions on the two bands, $\phi_{+,i}(\mathbf{k})$ and $\phi_{-,i}(\mathbf{k})$,

$$\tilde{V}_{\lambda\lambda'}(\mathbf{k},\mathbf{k}') = -V_{\lambda\lambda'}\sum_{i=1}^{d_\Gamma}\phi_{\lambda,i}(\mathbf{k})\phi^*_{\lambda',i}(\mathbf{k}'), \tag{4.39}$$

While $\phi_{+,i}(\mathbf{k})$ and $\phi_{-,i}(\mathbf{k})$ both belong to the same symmetry representation, their momentum dependence does not have to be exactly the same. The basis functions are assumed to satisfy the following orthogonality conditions: $\langle\phi^*_{\lambda,i}(\mathbf{k})\phi_{\lambda,j}(\mathbf{k})\rangle_\lambda = \delta_{ij}$, where the angular brackets denote the averaging over the λth Fermi surface. The coupling constants $V_{\lambda\lambda'}$ form a Hermitian matrix, which becomes real symmetric, if the basis functions are real. The gap functions take the form

$$\tilde{\Delta}_\lambda(\mathbf{k}) = \sum_{i=1}^{d_\Gamma}\eta_{\lambda,i}\phi_{\lambda,i}(\mathbf{k}), \tag{4.40}$$

and $\eta_{\lambda,i}$ are the superconducting order parameter components in the λth band.

As an example, consider a superconducting state with the order parameter transforming according to a one-dimensional representation $\tilde{\Delta}_\lambda(\mathbf{k}) = \eta_\lambda\phi_\lambda(\mathbf{k})$. The linearized gap equations Eq. (4.28) acquire simple algebraic form

$$\begin{aligned}\eta_+ &= (g_{++}\eta_+ + g_{+-}\eta_-)S_1(T),\\ \eta_- &= (g_{-+}\eta_+ + g_{--}\eta_-)S_1(T),\end{aligned} \tag{4.41}$$

where

$$g_{\lambda\lambda'} = V_{\lambda\lambda'}N_{0\lambda'}, \tag{4.42}$$

and $N_{0\lambda} = \langle|\phi_\lambda(\mathbf{k})|^2 N_{0\lambda}(\hat{\mathbf{k}})\rangle_\lambda$ is the weighted average angular dependent density of states over the λth Fermi surface. Note that for multidimensional representations (dimensional d_Γ), due to the crystal point symmetry, the values of $N_{0\lambda} = \langle|\phi_{\lambda,i}(\mathbf{k})|^2 N_{0\lambda}(\hat{\mathbf{k}})\rangle_\lambda$ are equal for all components $i = 1,\dots,d_\Gamma$ and all components $\eta_{\lambda,1},\dots,\eta_{\lambda,d_\Gamma}$ separately satisfy the same system of equations (4.41).

The function $S_1(T)$ is

$$S_1(T) = 2\pi T\sum_{n\geq 0}\frac{1}{\omega_n} = \ln\frac{2\gamma\varepsilon_c}{\pi T}, \tag{4.43}$$

where $\ln\gamma = 0,577\dots$ is the Euler constant. Moreover, ε_c is an energy cutoff for the pairing interaction, which we assume to be the same for both bands. From Eq. (4.41) we obtain then the following expression for the critical temperature:

$$T_{c0} = \frac{2\gamma\varepsilon_c}{\pi}\exp\left(-\frac{1}{g}\right), \tag{4.44}$$

where

$$g = \frac{g_{++}+g_{--}}{2} + \sqrt{\left(\frac{g_{++}-g_{--}}{2}\right)^2 + g_{+-}g_{-+}} \tag{4.45}$$

is the effective coupling constant. For multidimensional representations the critical temperature is the same for all d_Γ components of $\eta_{\lambda,i}$ of the order parameter. The particular combination of amplitudes $\eta_{\lambda,i}$ in the superconducting state below T_c is determined by the nonlinear terms in the free energy or self-consistent equation, which depend on the symmetry of the dominant pairing channel.

The solution of Eq. (4.41) (η_+, η_-) corresponding to the eigenvalue $S_1(T_c)$ determines two unequal order parameter components $\tilde{\Delta}_\lambda(\mathbf{k}) = \eta_\lambda \phi_\lambda(\mathbf{k})$. In the spinor representation (4.38) both singlet and triplet parts of the order parameter are present. Pure singlet or pure triplet pairing occurs only under rather restrictive conditions. First, the momentum dependence of the gap function in both bands is the same $\tilde{\Delta}_\lambda(\mathbf{k}) = \eta_\lambda \phi(\mathbf{k})$. Second, $g_{++} = g_{--}$ and $g_{+-} = g_{-+}$ is realized. Then we obtain two solutions of equations Eq. (4.41) with

$$\eta_+ = \eta_-, \tag{4.46}$$

$$\eta_+ = -\eta_-. \tag{4.47}$$

The critical temperature of the state (4.46) corresponding to the singlet part is $T^s_{c0} = (2\varepsilon_c/\pi)e^{-1/g_s}$, where $g_s = g_{++} + g_{+-}$. The critical temperature of the spin triplet state (4.47) is $T^t_{c0} = (2\varepsilon_c/\pi)e^{-1/g_t}$, where $g_t = g_{++} - g_{+-}$. If $g_{+-} > 0$ then $T^s_{c0} > T^t_{c0}$ and the phase transition occurs to the state (4.46). While at $g_{+-} < 0$ we see that $T^t_{c0} > T^s_{c0}$ and the phase transition occurs to the state (4.47).

4.3.3 Zeros in the Quasiparticle Gap

On the one hand, the zeros in the gap for elementary excitations are dictated by the symmetry of the superconducting state or its *superconducting class* which is a subgroup $\mathscr{H}$ of the group of symmetry of the normal state $\mathscr{G} \times \mathscr{K} \times U(1)$. Here $\mathscr{G}$ is the point group, $\mathscr{K}$ is the group of time reversal, $U(1)$ is the gauge group. The procedure to find symmetry dictated nodes is described in Ref. [63]. Let us consider the possible superconducting states (4.40) and their nodes for $CePt_3Si$ with point group symmetry C_{4v}. This group has four one-dimensional irreducible representations, A_1, A_2, B_1, B_2, and one two-dimensional representation, E. Examples of even basis functions of these irreducible representations are

Γ	$\phi_\Gamma(\mathbf{k})$	nodes
A_1	$k_x^2 + k_y^2 + ck_z^2$	–
A_2	$k_x k_y (k_x^2 - k_y^2)$	$k_x = 0, k_y = 0, k_x = \pm k_y$
B_1	$k_x^2 - k_y^2$	$k_x = \pm k_y$
B_2	$k_x k_y$	$k_x = 0, k_y = 0$

For the E state the basis functions are $\phi_{E1}(\mathbf{k}) = k_x k_z$ and $\phi_{E2}(\mathbf{k}) = k_y k_z$ leading to the order parameter for the Fermi surfaces λ,

$$\tilde{\Delta}_\lambda(\mathbf{k}) = \eta_{\lambda,1}\phi_{\lambda,E1}(\mathbf{k}) + \eta_{\lambda,2}\phi_{\lambda,E2}(\mathbf{k}). \tag{4.48}$$

The symmetry of superconducting state and the corresponding node positions depend on the particular choice of amplitudes $\eta_{\lambda,i}$. Under weak-coupling conditions the combination generating least nodes is most stable, corresponding here to $(\eta_{\lambda,1}, \eta_{\lambda,2}) = \eta_\lambda(1, \pm i)$. This is a time reversal symmetry violating phase with a line node on the plane $k_z = 0$ and point nodes at $k_x = k_y = 0$.

On the other hand, the fact that their gaps on the two Fermi surfaces are composed of an even- and an odd-parity part, can also lead to nodes which are not symmetry protected, as discussed in Ref. [45].

4.3.4 The Amplitude of Singlet and Triplet Pairing States

The coupling constants $V_{\lambda\lambda'}$ we have used in previous considerations can be expressed through the real physical interactions between the electrons naturally introduced in the initial spinor basis where BCS-type Hamiltonian has the following form [63]

$$\begin{aligned} H_{int} =& \frac{1}{4\mathcal{V}} \sum_{\mathbf{k}\mathbf{k}'\mathbf{q}} \sum_{\alpha\beta\gamma\delta} [V^g(\mathbf{k},\mathbf{k}')(i\sigma_2)_{\alpha\beta}(i\sigma_2)^\dagger_{\gamma\delta} \\ &+ V^u_{ij}(\mathbf{k},\mathbf{k}')(i\sigma_i\sigma_2)_{\alpha\beta}(i\sigma_j\sigma_2)^\dagger_{\gamma\delta}] c^\dagger_{\mathbf{k}+\mathbf{q},\alpha} c^\dagger_{-\mathbf{k},\beta} c_{-\mathbf{k}',\gamma} c_{\mathbf{k}'+\mathbf{q},\delta}, \end{aligned} \tag{4.49}$$

here the amplitudes $V^g(\mathbf{k},\mathbf{k}')$ and $V^u_{ij}(\mathbf{k},\mathbf{k}')$ are even and odd with respect to their arguments, correspondingly. The unitary transformation (4.4) transforms the pairing Hamiltonian (4.49) to the band representation (4.22). If we neglect inter-band pairing, it is reduced to (4.23) and (4.24) with the amplitudes given by the following expression

$$\tilde{V}_{\lambda\lambda'}(\mathbf{k},\mathbf{k}') = \frac{1}{2}V^g(\mathbf{k},\mathbf{k}')(\sigma_0 + \sigma_x)_{\lambda\lambda'} + \frac{1}{2}V^u_{ij}(\mathbf{k},\mathbf{k}')\hat{\gamma}_i(\mathbf{k})\hat{\gamma}_j(\mathbf{k}')(\sigma_0 - \sigma_x)_{\lambda\lambda'}. \tag{4.50}$$

The explicit derivation is given in [51], where a similar procedure was also made for more general interactions mediated by phonons or spin fluctuations. It can be shown that the pairing given by the amplitude $V^g(\mathbf{k},\mathbf{k}')$ in the initial spinor basis including the simple s-wave pairing $V^g(\mathbf{k},\mathbf{k}') = const$ does not induce any inter-band pairing channel.

To illustrate the origin of the singlet and triplet pairing channels, let us consider a superconductor with tetragonal symmetry C_{4v} and Rashba spin-orbital coupling $\gamma(\mathbf{k}) = \gamma_\perp(\hat{z} \times \mathbf{k})$, for a spherical Fermi surface. We describe the pairing by the following model which is compatible with all symmetry requirements:

$$\begin{aligned} V^g(\mathbf{k},\mathbf{k}') &= -V_g, \\ V^u_{ij}(\mathbf{k},\mathbf{k}') &= -V_u(\hat{\gamma}_i(\mathbf{k})\hat{\gamma}_j(\mathbf{k}')), \end{aligned} \tag{4.51}$$

where V_g and V_u are constants. This type of pairing interaction yields the superconducting state with full symmetry of the tetragonal group C_{4v} transforming according to unit representation A_1 both in singlet and in triplet channels.

With Eq. (4.50) we arrive at the band representation:

$$\tilde{V}_{\lambda\lambda'}(\mathbf{k},\mathbf{k}') = -\frac{1}{2}V_g(\sigma_0+\sigma_x)_{\lambda\lambda'} - \frac{1}{2}V_u(\sigma_0-\sigma_x)_{\lambda\lambda'}, \tag{4.52}$$

Thus, this pairing interaction is even simpler than that considered in the previous subsection (Eq. (4.39)). So, in our model the gap functions in the two bands (Eq. (4.40)) are: $\tilde{\Delta}_\lambda(\mathbf{k}) = \eta_\lambda\varphi_{A_1}(\mathbf{k})$ with $\mathbf{k}$ independent functions $\varphi_{A_1}(\mathbf{k}) = 1$. The amplitudes η_λ satisfy the equations

$$\eta_\lambda = \sum_{\lambda'} g_{\lambda\lambda'}\pi T\sum_n \frac{\eta_{\lambda'}}{\sqrt{\omega_n^2+\eta_{\lambda'}^2}}, \tag{4.53}$$

where

$$g_{\pm\pm} = \frac{V_g+V_u}{2}N_{0\pm}, \qquad g_{\pm\mp} = \frac{V_g-V_u}{2}N_{0\mp}, \tag{4.54}$$

and the critical temperature is given by Eq. (4.44).

According to the Eq. (4.38) the singlet and triplet parts of the order parameter are determined by the order parameter amplitudes in different bands. For the ratio of triplet to singlet amplitudes in the vicinity of T_c we find:

$$r \equiv \frac{\eta_+-\eta_-}{\eta_++\eta_-} = \frac{2g_{+-}+g_{++}-g_{--}-\sqrt{\mathcal{D}}}{2g_{+-}+g_{--}-g_{++}+\sqrt{\mathcal{D}}}, \tag{4.55}$$

where $\mathcal{D} = (g_{++}-g_{--})^2 + 4g_{+-}g_{-+}$. It is easy to see that for $V_u = 0$ the triplet component of the order parameter vanishes identically and $r = 0$. On the other hand, for $V_g = 0$ the singlet component of the order parameter disappears and $r^{-1} = 0$. Generally the relative weight of singlet and triplet component in the order parameter depends on the ratio of pairing interactions decomposed into even- and odd-parity channels.

A simple BCS-type of model with

$$V^g(\mathbf{k},\mathbf{k}') = -V_g \quad \text{and} \quad V^u_{ij}(\mathbf{k},\mathbf{k}') = 0 \tag{4.56}$$

yields in the band representation

$$\tilde{V}^{BCS}_{\lambda\lambda'}(\mathbf{k},\mathbf{k}') = -\frac{1}{2}V_g(\sigma_0+\sigma_x)_{\lambda\lambda'}. \tag{4.57}$$

and gives rise to purely spin-singlet pairing within our notion.

4.3.5 Ginzburg–Landau Formulation

The Ginzburg–Landau theory is a very efficient tool to discuss a wide variety of phenomena of the superconducting state, in particular, the instability conditions at the critical temperature. We will derive the Ginzburg–Landau functional from a microscopic starting point, with the aim to address in the following chapter the influence of the magneto-electric effect on the nucleation of superconductivity in a magnetic field, i.e. the modification of paramagnetic limiting in a non-centrosymmetric metal.

For this purpose we extend the self-consistent equation (4.28) to the case where magnetic fields are present and the superconducting order parameter has a weak spatial dependence,

$$\tilde{\Delta}_\lambda(\mathbf{k},\mathbf{q}) = T\sum_n\sum_{\mathbf{k}'}\sum_\nu \tilde{V}_{\lambda\nu}\left(\mathbf{k},\mathbf{k}'\right) G_\nu(\mathbf{k}',\omega_n)G_\nu(-\mathbf{k}'+\mathbf{q},-\omega_n)\tilde{\Delta}_\nu(\mathbf{k}',\mathbf{q}). \tag{4.58}$$

Near the critical temperature one can use the normal metal Green functions $G^0_\lambda(\mathbf{k},\omega_n)$, which yield then the linearized gap equation, to examine the instability condition. This equation can be derived from the free energy functional of the form

$$\begin{aligned} F = \frac{1}{2}\int \frac{d^3\mathbf{q}}{(2\pi)^3} & \left\{ \sum_{\lambda\nu} \eta^*_{\lambda,i}(\mathbf{q}) \tilde{V}^{-1}_{\lambda\nu} \eta_{\nu,i}(\mathbf{q}) \right. \\ & \left. - \sum_\nu T \sum_n \int \frac{d^3\mathbf{k}}{(2\pi)^3} \tilde{\Delta}^*_\nu(\mathbf{k},\mathbf{q}) G_\nu(\mathbf{k},\omega_n) G_\nu(-\mathbf{k}+\mathbf{q},-\omega_n)\tilde{\Delta}_\nu(\mathbf{k},\mathbf{q}) \right\}. \end{aligned} \tag{4.59}$$

The corresponding normal-metal electron Green function $G_\lambda(\mathbf{k},\omega_n) = (i\omega - \xi_\lambda(\mathbf{k},\mathbf{H}))^{-1}$ in a magnetic field is determined by the electron energies (4.14),

$$\xi_\lambda(\mathbf{k},\mathbf{H}) = \xi(\mathbf{k}) + \lambda|\gamma(\mathbf{k}) - \mu_B\mathbf{H}| \approx \xi_\lambda(\mathbf{k}) - \mathbf{m}_\lambda(\mathbf{k})\mathbf{H}, \tag{4.60}$$

where $\xi_\lambda(\mathbf{k}) = \xi(\mathbf{k}) + \lambda|\gamma(\mathbf{k})|$ and the second term on the right-hand side is the analog of the Zeeman interaction for non-degenerate bands [34] with the form:

$$\mathbf{m}_\lambda(\mathbf{k}) = \lambda\mu_B\hat{\gamma}(\mathbf{k}), \tag{4.61}$$

which is valid everywhere except for the vicinity of band crossing points, where the approximation of independent non-degenerate bands fails. In standard centrosymmetric metals the magnetic field splits the Fermi surfaces into majority and minority spin surfaces. Here the spin-splitting is imposed at the outset by the spin-orbit coupling. The effect of the magnetic field is a deformation of the band and shape of the Fermi surfaces. As we will see this will influence the superconducting condensate nucleated in a magnetic field.

The normal electron Green function is then approximated as

$$G_\lambda(\mathbf{k},\omega_n) = \frac{1}{i\omega_n - \xi_\lambda(\mathbf{k}) + \mathbf{m}_\lambda(\mathbf{k})\mathbf{H}}. \tag{4.62}$$

Since the gap function depends weakly on energy in the vicinity of the Fermi surface, one can integrate the products of two Green's functions with respect to $\xi_\lambda = \xi_\lambda(\mathbf{k})$:

$$N_{0\lambda} \int d\xi_\lambda G_\lambda(\mathbf{k}, \omega_n) G_\lambda(-\mathbf{k}+\mathbf{q}, -\omega_n) = \pi N_{0\lambda} L_\lambda(\mathbf{k}, \mathbf{q}, \omega_n), \tag{4.63}$$

where

$$L_\lambda(\mathbf{k}, \mathbf{q}, \omega_n) = \frac{1}{|\omega_n| + i\Omega_\lambda(\mathbf{k}, \mathbf{q}) \operatorname{sign} \omega_n} \tag{4.64}$$

depends only on $\hat{k}$, the direction of $\mathbf{k}$,

$$\Omega_\lambda(\mathbf{k}, \mathbf{q}) = \frac{\mathbf{v}_\lambda(\mathbf{k})\mathbf{q}}{2} - \mathbf{m}_\lambda(\mathbf{k})\mathbf{H}, \tag{4.65}$$

with $\mathbf{v}_\lambda(\mathbf{k}) = \partial\xi_\lambda(\mathbf{k})/\partial\mathbf{k}$ being the Fermi velocity in the λth band.

The Ginzburg–Landau free energy in usual coordinate representation (as well as the Ginzburg–Landau equations) can be obtained from the Taylor expansion of Eqs. (4.58) and (4.59) in powers of $\Omega_\lambda(\mathbf{k}, \mathbf{q})$, by the replacement

$$\mathbf{q} \rightarrow \mathbf{D} = -i\nabla_{\mathbf{r}} + 2e\mathbf{A}(\mathbf{r}) \tag{4.66}$$

in the final expressions. In the following we will put $\hbar = c = 1$. The special form of $\Omega_\lambda(\mathbf{k}, \mathbf{q})$ introduces a novel gradient term in the free energy of non-centrosymmetric superconductors. Instead of powers of $\mathbf{q}$ it contains powers of $\Omega_\lambda(\mathbf{k}, \mathbf{q})$. This can can lead to the formation of a nonuniform superconducting state known as *helical phases* and to a *magneto-electric effect* in a magnetic field.

4.4 Magneto-Electric Effect and the Upper Critical Field

The term "magneto-electric effect" in non-centrosymmetric superconductors encompasses several intriguing features. It has been discussed on a phenomenological level by introducing additional linear gradients terms to the Ginzburg–Landau free energy, so-called Lifshitz invariants, like

$$\eta^*(\mathbf{r})\tilde{K}_{ij} H_i D_j \eta(\mathbf{r}) \tag{4.67}$$

Here $\eta(\mathbf{r})$ denotes the superconducting order parameter, $\mathbf{H}$ is the magnetic field and $\mathbf{D} = -i\nabla + 2e\mathbf{A}$ is the gauge-invariant gradient. First predicted by Levitov, Nazarov and Eliashberg [10], the magneto-electric effect was studied microscopically by several authors [11, 16, 17, 21, 26]. In this context several observable effects have been predicted: (i) the existence of a helically twisted superconducting order parameter in a magnetic field in two and three dimensional cases and spontaneous supercurrents in a 2D geometry [17, 21, 23, 24, 26, 29, 32] and near a superconductor surface

[31] as well as along junctions of two superconductors with opposite directions of polarization [30], (ii) the enhancement of the upper critical field oriented perpendicular to the direction of the space parity breaking [26, 29], (iii) magnetic interference patterns of the Josephson critical current for a magnetic field applied perpendicular to the junction [29].

The presence of Lifshitz invariants (4.67), however, can mislead to invalid conclusions so that a careful analysis of different contributions is mandatory. Especially the question of the influence of the magneto-electric effect on paramagnetic limiting deserves special attention in this context. Moreover, the notion of helical phase has to be considered with caution as it may seduce to wrong pictures. In this section we would like to give insight into these subtleties by discussing the magnetic field dependence of the effective critical temperature in the Ginzburg–Landau framework.

4.4.1 One-Band Case

Before considering the intrinsic multi-band situation due to the spin splitting of the electron band, we restrict ourselves, for simplicity, to a one-band situation, i.e. we ignore one of the two bands. This band shall be characterized by an isotropic density of states at the Fermi energy, $N_{0+}(\hat{\mathbf{k}}) = N_+$.

The Ginzburg–Landau free energy for this one-band case with a one-component order parameter can be derived from Eq. (4.59)

$$F = \frac{1}{2} \int \frac{d^3q}{(2\pi)^3} \left\{ \frac{2}{V_{++}} - N_{0+} S_1(T) + N_{0+} S_3(T) \langle (\phi^2(\mathbf{k}) \Omega(\mathbf{k}, \mathbf{q}))^2 \rangle \right\} |\eta(\mathbf{q})|^2, \tag{4.68}$$

where we restrict to the second order terms [26, 46]. This is sufficient to analyze the instability conditions. Here, $\phi(\mathbf{k})$ describes the superconducting state and is an even function belonging to one of the one-dimensional representations of the point group of the crystal, Ω is given by (4.65), $\langle \ldots \rangle$ means the averaging over the Fermi surface, the function $S_1(T)$ is given by Eq. (4.43) and

$$S_3(T) = \pi T \sum_n \frac{1}{|\omega_n|^3} = \frac{7\zeta(3)}{4\pi^2 T^2}. \tag{4.69}$$

The Ginzburg–Landau free energy functional in real space can be obtained through a Fourier transformation and leads to

$$F = \int d^3r \left\{ \alpha(T - T_{c0})|\eta|^2 + \eta^* \left[K_1(D_x^2 + D_y^2) + K_2 D_z^2 + K_{ij} H_i D_j + Q_{ij} H_i H_j \right] \eta \right\}, \tag{4.70}$$

where

$$\alpha = N_{0+}/2T_{c0}, \quad T_{c0} = (2\gamma \varepsilon_0/\pi) \exp(-2/V_{++} N_{0+}), \tag{4.71}$$

$$K_1 = \frac{N_{0+}S_3}{8}\langle\phi^2(\mathbf{k})v_x^2(\mathbf{k})\rangle, \quad K_2 = \frac{N_{0+}S_3}{8}\langle\phi^2(\mathbf{k})v_z^2(\mathbf{k})\rangle, \tag{4.72}$$

$$K_{ij} = -\frac{\mu_B N_{0+}S_3}{2}\langle\phi^2(\mathbf{k})\hat{\gamma}_i(\mathbf{k})v_j(\mathbf{k})\rangle, \quad Q_{ij} = \frac{\mu_B^2 N_{0+}S_3}{2}\langle\phi^2(\mathbf{k})\hat{\gamma}_i(\mathbf{k})\hat{\gamma}_j(\mathbf{k})\rangle. \tag{4.73}$$

The term linear in $\mathbf{H}$ incorporates the magneto-electric effects, while the term quadratic in $\mathbf{H}$ describes the paramagnetic effect. This means also that $Q_{ij}|\eta|^2$ is connected with the change of the paramagnetic susceptibility in the superconducting phase compared with the normal state (Pauli) susceptibility. In particular, Q_{ij} vanishes when there is no change of the paramagnetic susceptibility. These coefficients have to be compared with those of a spin singlet state in a centrosymmetric superconductor, $Q_{ij}^{(0)} = \delta_{ij}\mu_B^2 N_0 S_3/2$. Assuming $N_{0+} = N_0$, the coefficients $Q_{ij}^{(0)}$ are larger than the above Q_{ij} due to the fact that $1 = \langle\phi^2(\mathbf{k})\rangle \geq \langle\phi^2(\mathbf{k})\hat{\gamma}_i(\mathbf{k})\hat{\gamma}_j(\mathbf{k})\rangle$.

We consider now two illustrative cases, the point groups C_{4v} and D_4 which are characterized by the pseudovectors

$$\begin{aligned} \gamma(\mathbf{k}) &= \gamma_\perp(\hat{x}k_y - \hat{y}k_x) + \gamma_\parallel k_x k_y k_z(k_x^2 - k_y^2) \quad \text{for} \quad C_{4v}, \\ \gamma(\mathbf{k}) &= \gamma_\perp(k_x\hat{x} + k_y\hat{y}) + \gamma_\parallel k_z\hat{z} \quad \text{for} \quad LD_4. \end{aligned} \tag{4.74}$$

For symmetry arguments and using above expressions we find the following relations for the coefficients,

$$\begin{aligned} &K_{xy} = -K_{yx} \neq 0 && \text{and} \quad K_{ij} = 0 \quad \text{otherwise} \\ &Q_{xx} = Q_{yy} \neq Q_{zz} > 0 && \text{and} \quad Q_{ij} = 0 \quad \text{otherwise} \end{aligned} \tag{4.75}$$

for C_{4v} where presumably $|K_{zz}| \ll |K_{xy}|$ and $Q_{zz} \ll Q_{xx}$ due to the large number of nodes in the $\mathbf{k}$-dependence of the $\gamma_\parallel$-part of $\gamma(\mathbf{k})$, and

$$\begin{aligned} K_{xx} = K_{yy} \neq 0, K_{zz} \neq 0 \quad &\text{and} \quad K_{ij} = 0 \quad \text{otherwise} \\ Q_{xx} = Q_{yy} \neq Q_{zz} > 0 \quad &\text{and} \quad Q_{ij} = 0 \quad \text{otherwise} \end{aligned} \tag{4.76}$$

for the point group D_4.

4.4.1.1 Symmetry C_{4v}, $\mathbf{H} \parallel \hat{z}$

In the case of C_{4v} for the field directed parallel to the z-axis $\mathbf{H} = H(0, 0, 1)$ the terms linear in gradients and $\mathbf{H}$ are absent. The standard solution $\eta = e^{iq_y y}f(x)$ of the GL equation

$$\left\{\alpha(T - T_{c0}) + K_1\left[-\frac{\partial^2}{\partial x^2} + \left(-i\frac{\partial}{\partial y} + 2eHx\right)^2\right] + Q_{zz}H^2\right\}\eta = 0, \tag{4.77}$$

is degenerate in respect to q_y. The magnetic field dependence of the critical temperature is

$$T_c = T_{c0} - \frac{2eK_1}{\alpha}H - \frac{Q_{zz}}{\alpha}H^2 \qquad (4.78)$$

Both the orbital (linear in H) and paramagnetic (quadratic in H) depairing effects are present. Compared to the ordinary spin-singlet case, however, the effect of the paramagnetic limiting is weaker here due to $Q_{zz} < Q_{zz}^{(0)}$. It is important to note that no magneto-electric effect comes into play here.

4.4.1.2 Symmetry D_4, $\mathbf{H} \parallel \hat{z}$

The situation is quite different for uniaxial crystals with point symmetry group D_4 (or D_6). The GL equation includes gradient terms in the field direction and acquires the form

$$\left\{\alpha(T - T_{c0}) + K_1\left[-\frac{\partial^2}{\partial x^2} + \left(-i\frac{\partial}{\partial y} + 2eHx\right)^2\right]\right.$$
$$\left. + iK_{zz}H\frac{\partial}{\partial z} - K_2\frac{\partial^2}{\partial z^2} + +Q_{zz}H^2\right\}\eta = 0. \qquad (4.79)$$

The solution can be written as

$$\eta = e^{iq_y y}e^{iq_z z}f(x), \qquad (4.80)$$

which remains degenerate with respect to the wavevector q_y, but not with respect to q_z which is used to maximize the critical temperature to

$$T_c = T_{c0} - \frac{2eK_1}{\alpha}H + \left(\frac{K_{zz}^2}{4K_2} - Q_{zz}\right)\frac{H^2}{\alpha}. \qquad (4.81)$$

This corresponds to the finite wavevector

$$q_z = \frac{K_{zz}H}{2K_2}. \qquad (4.82)$$

Note that this wave vector could also be absorbed into the vector potential without changing the physically relevant results: $\mathbf{A} \to \mathbf{A} + \nabla\chi$ with $\chi = -q_z z/2e$.

The simple paramagnetic depairing effect is weakened due to the magneto-electric response of the system. Adjusting the nucleation of the superconducting phase to the shifted Fermi surface, as incorporated in the wavevector q_z, recovers some of the strength of the nucleating condensate. This is a specific effect of the non-centrosymmetric superconductor and has its conceptional analogue in the Fulde–Ferrel–Larking–Ovchinnikov (FFLO) phase for centrosymmetric spin singlet superconductors, where the condensate also nucleates with finite-momentum Cooper pairs in order to optimize the pairing of degenerate quasiparticles on the split Fermi surface.

4.4.1.3 Symmetry C_{4v}, $\mathbf{H} \perp \hat{z}$

Now we turn the magnetic field into the basal plane, $\mathbf{H} = H(\cos\varphi, \sin\varphi, 0)$, and impose a gauge to have the vector potential $\mathbf{A} = Hz(\sin\varphi, -\cos\varphi, 0)$. The corresponding GL equation takes the form

$$\left\{\alpha(T - T_{c0}) + K_1(D_x^2 + D_y^2) - K_2\frac{\partial^2}{\partial z^2} + K_{xy}(H_x D_y - H_y D_x) + Q_{xx}H^2\right\}\eta = 0, \tag{4.83}$$

where

$$D_x = -i\frac{\partial}{\partial x} + 2eH_y z, \qquad D_y = -i\frac{\partial}{\partial y} - 2eH_x z. \tag{4.84}$$

Like in ordinary superconductors the solution of this equation have the Abrikosov form

$$\eta(\mathbf{r}) = \exp\left[i(\mathbf{p}\times\mathbf{r})_z\right]f(z), \tag{4.85}$$

where we write $\mathbf{p} = p\mathbf{H}/H$ as a vector parallel to the magnetic field $(\mathbf{p}\times\mathbf{r})_z$ denoting the z-component of the vector $(\mathbf{p}\times\mathbf{r})$, and $f(z)$ satisfies the resulting renormalized harmonic oscillator equation

$$\left\{\alpha(T - T_{c0}) + K_1(2eH)^2(z - z_0)^2 - K_2\frac{\partial^2}{\partial z^2} + \left(Q_{xx} - \frac{K_{xy}^2}{4K_1}\right)H^2\right\}f(z) = 0, \tag{4.86}$$

with the shifted equilibrium position

$$z_0 = (2eH)^{-1}\left(p + \frac{K_{xy}}{2K_1}H\right). \tag{4.87}$$

Thus, the vector $\mathbf{p}$ is absorbed into the shift z_0 and does not appear anywhere else in the equation. Then the corresponding eigenvalue determines the magnetic field dependence of the optimized critical temperature:

$$T_c = T_{c0} - \frac{2e\sqrt{K_1K_2}}{\alpha}H + \left(\frac{K_{xy}^2}{4K_1} - Q_{xx}\right)\frac{H^2}{\alpha}. \tag{4.88}$$

In the used gauge the eigenstates are degenerate with respect to p and acquire the same structure as the usual Landau degeneracy. Nevertheless, the characteristics of non-centrosymmetricity incorporated in the K_{ij} terms appear in the expression of T_c. Similar to the previous case of D_4 with $\mathbf{H} \parallel z$ the magneto-electric effect yields a reduction of the paramagnetic limiting term. This renormalization is surprisingly strong in general, as we can see when we return to the expressions we derived for the different coefficients. We obtain for the last term in Eq. (4.88),

$$\left[\frac{K_{xy}^2}{4K_1} - Q_{xx}\right]\frac{H^2}{\alpha} = \left[\frac{\langle\phi^2(\mathbf{k})\hat{\gamma}_x(\mathbf{k})v_y(\mathbf{k})\rangle^2}{\langle\phi^2(\mathbf{k})v_x^2(\mathbf{k})\rangle} - \langle\phi^2(\mathbf{k})\hat{\gamma}_x^2(\mathbf{k})\rangle\right]\frac{\mu_B^2 H^2 N_+ S_3}{2\alpha} \tag{4.89}$$

Considering the simplified picture of a parabolic band with $\mathbf{v}(\mathbf{k}) = \mathbf{k}/m^*$ and a Rashba spin-orbit coupling $\hat{\gamma} = \mathbf{k} \times \hat{z}$ (setting $\gamma_z(\mathbf{k}) = 0$) we find the amazing result that the two terms cancel exactly and the paramagnetic effect is completely suppressed. This result can be immediately obtained, if we perform the gauge transformation

$$\mathbf{q} \to \mathbf{q} + \frac{2\mu_B m^*(\hat{z} \times \mathbf{H})}{k_F}. \tag{4.90}$$

already in Eq. (4.65) and eliminate the paramagnetic term at the outset. However, it is important to notice that this exact cancellation is a consequence of the simplified forms of the band structure and the spin-orbit coupling term. Taking more realistic band structure effects into account it is obvious that this identity does not hold anymore in general. Nevertheless, our results suggest that the magneto-electric effect can, in principle, yield a substantial contribution to eliminate the paramagnetic limiting also for fields in the basal plane.

4.4.1.4 Symmetry D_4, $\mathbf{H} \perp \hat{z}$

It is easy to see that this case is analogue to the situation for the field along the z-axis and has only quantitative differences. Thus also here we encounter a reduction of the paramagnetic limit due to the magneto-electric effect yielding

$$T_c = T_{c0} - \frac{2e\sqrt{K_1 K_2}}{\alpha}H + \left(\frac{K_{xx}^2}{3K_1} - Q_{xx}\right)\frac{H^2}{\alpha}, \tag{4.91}$$

where also the same considerations concerning the gauge freedom apply as in the case of $\mathbf{H} \parallel \hat{z}$.

4.4.1.5 Two-Dimensional Case, Symmetry C_{4v}, $\mathbf{H} \perp \hat{z}$

The simplest way to pass from the 3D to the 2D situation is to introduce a $\delta(z)$-function potential well into the 3D GL equation (4.83). This is equivalent to the theory used by Tinkham [65] for the calculation of the upper critical field in a thin film with thickness $d << \xi$ for a field parallel to the film. Thus, we consider the instability equation

$$\left\{\alpha(T - T_{c0}) - K_1(D_x^2 + D_y^2) - K_2\frac{\partial^2}{\partial z^2} \right.$$
$$\left. + K_{xy}(H_x D_y - H_y D_x) + Q_{xx}H^2 - \frac{2K_2}{d}\delta(z)\right\}\eta = 0, \tag{4.92}$$

where d is a length of the order of the film thickness that in the pure 2D case it is an atomic scale length. This eigenvalue equation has the solution

$$\eta(\mathbf{r}) = A \exp\left[i(\mathbf{p} \times \mathbf{r})_z\right] \exp\left(-\frac{|z|}{d}\right), \tag{4.93}$$

where $\mathbf{p} = p\mathbf{H}/H$ is a vector with arbitrary length directed along the magnetic field. This then determines the critical temperature as a function of the applied magnetic field.

$$\alpha(T - \tilde{T}_{c0}) + K_1(2eH)^2\langle(z - z_0)^2\rangle + \left(Q_{xx} - \frac{K_{xy}^2}{4K_1}\right)H^2 = 0. \tag{4.94}$$

Here $\tilde{T}_{c0}$ is the critical temperature in the absence of a magnetic field, corresponding to $d^2 = K_2/\alpha(\tilde{T}_{c0} - T_{c0})$. Moreover, the brackets $\langle ... \rangle$ denote the expectation value using the wave function $\exp(-|z|/d)$ and z_0 is determined by the same expression as in the 3D case

$$z_0 = (2eH)^{-1}\left(p + \frac{K_{xy}}{2K_1}H\right). \tag{4.95}$$

Hence, we obtain for the critical temperature

$$T_c = \tilde{T}_{c0} + \left[\frac{K_{xy}^2}{4K_1} - Q_{xx}\right]\frac{H^2}{\alpha} - \frac{K_1}{\alpha}(z_0^2 + d^2/2)(2eH)^2. \tag{4.96}$$

The critical temperature reaches obviously a maximal value at $z_0 = 0$, i.e. for

$$p = -\frac{K_{xy}}{2K_1}H. \tag{4.97}$$

The upper critical field shows also here the square root temperature dependence usual for thin films in a parallel magnetic field [65]. Under special conditions (e.g. rotation symmetry around the normal vector of the film) the expression in the square parenthesis in Eq. (4.96) may vanish, as described above. Then, unlike in usual superconductors, non-centrosymmetric superconductors follow the standard Tinkham behavior unchanged by paramagnetic contributions.

In view of strong inequality $d \ll 1/\sqrt{2eH}$ the complete suppression of the 2D superconducting state ($T_c(H) = 0$) is reached in the field which exceeds the orbital critical field in the 3D case (4.88).

4.4.2 Two-Band Case

While the one-band picture discussed so far gives useful insights into the influence of the magneto-electric effect on the upper critical field, in particular, in the context of paramagnetic limiting, in reality there are at least two split bands whose Fermi surface allows for the nucleation of a condensate in a finite magnetic field. In the two-band picture the situation is somewhat more complex, so that we restrict ourselves here to a few aspects only which, we believe, are relevant in this context without attempting to give a complete overview. We base our analysis on the formalism introduced for the homogeneous superconducting phase in Sect. 4.2.5. We use also an order parameter belonging to a one-dimensional representation on the two Fermi surfaces, $\tilde{\Delta}_\lambda(\mathbf{k}, \mathbf{r}) = \eta_\lambda(\mathbf{r})\phi_\lambda(\mathbf{k})$ with $\lambda = \pm$. Moreover we restrict our discussion to the case of the point group C_{4v} with an in-plane magnetic field. Then the linearized Ginzburg–Landau equation is given by

$$\begin{aligned} \eta_+ &= g_{++}[S_1(T) - \hat{L}_+]\eta_+ + g_{+-}[S_1(T) - \hat{L}_-]\eta_-, \\ \eta_- &= g_{-+}[S_1(T) - \hat{L}_+]\eta_+ + g_{--}[S_1(T) - \hat{L}_-]\eta_-, \end{aligned} \tag{4.98}$$

with the operators $\hat{L}_\lambda$,

$$\hat{L}_\lambda = N_{0\lambda}^{-1}\left[K_{1\lambda}(D_x^2 + D_y^2) - K_{2\lambda}\frac{\partial^2}{\partial z^2} + \lambda K_{xy\lambda}(H_x D_y - H_y D_x) + Q_{xx\lambda}H^2\right], \tag{4.99}$$

where the coefficients are defined through the straightforward generalization of Eqs. (4.72) and (4.73) to the two-band case with the gap functions $\phi_\lambda(\mathbf{k})$, the Fermi velocity components $v_{\lambda,i}(\mathbf{k})$ and the densities of states N_λ taken in the corresponding band.

Similar to the one-band case the solutions of this equation system can be cast into the Abrikosov form

$$\begin{pmatrix} \eta_+(\mathbf{r}) \\ \eta_-(\mathbf{r}) \end{pmatrix} = \begin{pmatrix} f_+(z) \\ f_-(z) \end{pmatrix} \exp\left[i(\mathbf{p} \times \mathbf{r})_z\right] \tag{4.100}$$

where again $\mathbf{p} = p\mathbf{H}/H$ and the functions $f_+(z)$, $f_-(z)$ satisfy the system of equations

$$\begin{aligned} f_+ &= g_{++}[S_1(T) - \hat{M}_+]f_+ + g_{+-}[S_1(T) - \hat{M}_-]f_-, \\ f_- &= g_{-+}[S_1(T) - \hat{M}_+]f_+ + g_{--}[S_1(T) - \hat{M}_-]f_-. \end{aligned} \tag{4.101}$$

Using the same gauge as in the one-band example, the new operator $\hat{M}_\lambda$ is then

$$\hat{M}_\lambda = N_{0\lambda}^{-1}\left[K_{1\lambda}(2eH)^2(z - z_{\lambda 0})^2 - K_{2\lambda}\frac{\partial^2}{\partial z^2} + \left(Q_{xx\lambda} - \frac{K_{xy\lambda}^2}{4K_{1\lambda}}\right)H^2\right], \tag{4.102}$$

$$z_{0\lambda} = (2eH)^{-1}\left(p + \lambda\frac{K_{xy\lambda}}{2K_{1\lambda}}H\right). \tag{4.103}$$

As in the one-band case the eigenstates of this system possess the Landau degeneracy represented through the equilibrium positions of the coupled harmonic oscillators, z_{0+} and z_{0-}, which both depend on p. Through the substitution

$$z = Z + \frac{p}{2eH}$$

we can formulate the equation system so that p is eliminated and $z_{0\lambda} \to Z_{0\lambda}$,

$$Z_{0\lambda} = \lambda \frac{K_{xy\lambda}}{4eK_{1\lambda}}. \tag{4.104}$$

The general solution of Eq. (4.101) can be found only numerically. Here we limit ourselves to a variational solution of the form,

$$\begin{aligned} f_+(Z) &= C_+ \exp\left\{-\frac{eH\sqrt{K_{1+}}(Z - Z_{0+})^2}{\sqrt{K_{2+}}}\right\}, \\ f_-(Z) &= C_- \exp\left\{-\frac{eH\sqrt{K_{1-}}(Z - Z_{0-})^2}{\sqrt{K_{2-}}}\right\}. \end{aligned} \tag{4.105}$$

In the following calculations, taking into account that the band splitting is much less than the Fermi energy $|\gamma_\perp| k_F \ll \varepsilon_F$, we neglect the difference between the Fermi velocities $v_\lambda(\mathbf{k})$ and the densities of states $N_{0\lambda}$ of the two bands. In this case, the values of $Z_{0\lambda} = \lambda Z_0$ for different bands differ each other only by sign. Returning to the free energy functional and integrating over Z we obtain new variational equations for the coefficients C_+ and C_-:

$$\begin{aligned} C_+ &= g_{++}[S_1(T) - M(H)]C_+ + I g_{+-}[S_1(T) - M(H)]C_-, \\ C_- &= I g_{-+}[S_1(T) - M(H)]C_+ + g_{--}[S_1(T) - M(H)]C_- \end{aligned} \tag{4.106}$$

with

$$M(H) = N_0^{-1}\left[2eH\sqrt{K_1 K_2} + \left(Q_{xx} - \frac{K_{xy}^2}{4K_1}\right)H^2\right], \tag{4.107}$$

and

$$I = \exp(-2eHZ_0^2). \tag{4.108}$$

The Eqs. (4.106) have the same form as Eq. (4.41) in the absence of magnetic field. Hence, for the critical temperature we obtain

$$T_c = \tilde{T}_{c0}(1 - M) \tag{4.109}$$

where the temperature $\tilde{T}_{c0}$ is given by the same formula (4.44) as T_{c0}, taking in mind the substitutions $g_{+-} \to \tilde{g}_{+-} = I g_{+-}, g_{-+} \to \tilde{g}_{-+} = I g_{-+}$. The product

$2eHZ_0^2 \approx eHm^{*2}/(k_F m)^2)$ is much less than unity generally, except for heavy fermion or layered superconductors.

Thus, in the two-band situation the paramagnetic suppression of the superconducting state $\propto -Q_{xx}H^2$ is substantially weakened by the magneto-electric effect $\propto K_{xy}^2 H^2/4K_1$. The latter has a finite value as long as we work in the limit $\mu_B H \ll |\gamma_\perp| k_F$.

This conclusion is qualitatively valid in general, although our discussion was limited to a simple variational approach only. Moreover, assuming $g_{++} = g_{--}$ and $g_{+-} = g_{-+}$, as it was done in the absence of magnetic field (see Eqs. (4.46) and (4.47)), we come to the solution of (4.106) with either pure singlet or with pure triplet pairing. The effect of paramagnetic limiting is identical in both cases. This underlines directly that the weakening of the paramagnetic limiting in the non-centrosymmetric superconductors is not connected to the formation of a mixed singlet–triplet state. The important point is the spin-orbital splitting of the bands. The Pauli spin susceptibility of the quasiparticles in whole space between the two Fermi surfaces is not changed in the superconducting state in comparison of its normal state value. This leads, therefore, to the weakening of the paramagnetic suppression of the superconducting state.

Various approximations to the two-band model have been used in the literature, some of which can obscure the subtleties of non-centrosymmetric superconductors. In particular, the discussion of the spin susceptibility in the superconducting phase given in [33–35] has to be considered with caution in view of the magneto-electric effects which are neglected there. The adjustment of the superconducting state to the field-induced shifts of the Fermi surface yields a correction to the spin susceptibility which is not negligible as our discussion in the Ginzburg–Landau regime show. The subtle two-band effects, however, often make quantitative predictions difficult [66].

4.5 Conclusion

In this chapter we have given an overview on some theoretical aspects of non-centrosymmetric superconductors. Unlike symmetric spin-orbit coupling found in centrosymmetric metals, the antisymmetric spin-orbit coupling has a spectacular influence on the electronic bands through a specific spin splitting of the quasiparticle states. Superconductivity as a Fermi-surface instability is naturally influenced by such a modification of the electronic states. Under these circumstances it has always multi-band character. Moreover, parity does no longer provide a good quantum number to classify the superconducting phases.

One of the physically most remarkable aspects of non-centrosymmetric superconductivity is connected with magneto-electricity, the peculiar connection between supercurrents and spin polarization. We have considered one aspect in this context, namely its influence on paramagnetic limiting. This effect is of interest in strongly correlated electron systems where the coherence length is generally small due to the enhanced masses as in heavy fermion compounds. Here ordinary orbital depairing in a magnetic field is weak, such that the upper critical field reaches magnitudes where paramagnetic limiting through spin polarization becomes visible. Interestingly, already on the basis of symmetry considerations, it is possible to arrive at important predictions which are borne out in some of the non-centrosymmetric heavy fermion superconductors.

References

1. Bauer, E., Hilscher, G., Michor, H., Paul Ch., Scheidt, E.W., Gribanov, A., Seropegin Yu., Noël, H., Sigrist, M., Rogl, P.: Phys. Rev. Lett. **92**, 027003 (2004)
2. Akazawa, T., Hidaka, H., Fujiwara, T., Kobayashi, T.C., Yamamoto, E., Haga, Y., Settai, R., Onuki, Y.: J. Phys.: Condens. Matter **16**, L29 (2004)
3. Kimura, N., Ito, K., Saitoh, K., Umeda, Y., Aoki, H., Terashima, T.: Phys. Rev. Lett. **95**, 247004 (2005)
4. Sugitani, I., Okuda, Y., Shishido, H., Yamada, T., Thamizhavel, A., Yamamoto, E., Matsuda, T.D., Haga, Y., Takeuchi, T., Settai, R., Onuki, Y.: J. Phys. Soc. Jpn. **75**, 043703 (2006)
5. Amano, G., Akutagawa, S., Muranaka, T., Zenitani, Y., Akimitsu, J.: . J. Phys. Soc. Jpn. **73**, 530 (2004)
6. Togano, K., Badica, P., Nakamori, Y., Orimo, S., Takeya, H., Hirata, K.: Phys. Rev. Lett. **93**, 247004 (2004)
7. Badica, P., Kondo, T., Togano, K.: J. Phys. Soc. Jpn. **74**, 1014 (2005)
8. Schuck, G., Kazakov, S.M., Rogacki, K., Zhigadlo, N. D., Karpinski, J.: Phys. Rev. **73**, 144506 (2006)
9. Bulaevskii, L.N., Guseinov, A.A., Rusinov, A.I.: Sov. Phys. JETP **44**, 1243 (1976)
10. Levitov, L.S., Nazarov Yu.V., Eliashberg, G.M.: JETP Letters **41**, 445 (1985)
11. Edel'stein, V.M.: Sov. Phys. JETP **68**, 1244 (1989)
12. Gor'kov, L.P., Rashba, E.I.: Phys. Rev. Lett. **87**, 037004 (2001)
13. Mineev, V.P.: JETP Letters **57**, 680 (1993)
14. Mineev, V.P.: Physica B **199–200**, 215 (1994)
15. Mineev, V.P., Samokhin, K.V.: Sov. Phys. JETP **78**, 401 (1994)
16. Edel'stein, V.M.: Phys.Rev.Lett. **75**, 2004 (1995)
17. Edel'stein, V.M.: . J. Phys. Condens. Matter **8**, 339 (1996)
18. Samokhin, K.V., Zijlstra, E.S., Bose, S.K.: Phys. Rev. B **69**, 094514 (2004) See also Erratum **70**, 069902(E) (2004)
19. Sergienko, I.A., Curnoe, S.H.: Phys. Rev. B **70**, 214510 (2004)
20. Mineev, V.P.: Int. J. Mod. Phys. B **18**, 2963 (2004)
21. Yip, S.K.: Phys. Rev. B **65**, 144508 (2002)
22. Barzykin, V., Gor'kov, L.P.: Phys. Rev. Lett. **89**, 227002 (2002)
23. Dimitrova, O.V., Feigel'man, M.V.: Pis'ma v ZhETF **78**, 1132 (2003)
24. Dimitrova, O.V., Feigel'man, M.V.: Phys. Rev. B **76**, 014522 (2007)
25. Edel'stein, V.M.: Phys. Rev. B **67**, 020505 (2003)
26. Samokhin, K.V.: Phys. Rev. B **70**, 104521 (2004)
27. Frigeri, P.A., Agterberg, D.F., Koga, A., Sigrist, M.: Phys. Rev. Lett. **92**, 097001 (2004) See also Erratum 93: 099903(E) (2004)
28. Mineev, V.P.: Phys. Rev B **71**, 012509 (2005)
29. Kaur, R.P., Agterberg, D.F., Sigrist, M.: Phys. Rev. Lett. **94**, 137002 (2005)
30. Fujimoto, S.: Phys. Rev B **72**, 024515 (2005)
31. Oka, M., Ishioka, M., Machida, K.: Phys. Rev B **73**, 214509 (2006)
32. Agterberg, D.F., Kaur, R.P.: Phys. Rev B **75**, 064511 (2007)
33. Frigeri, P.A., Agterberg, D.F., Sigrist, M.: New J. Phys. **6**, 115 (2004)

34. Samokhin, K.V.: Phys. Rev. Lett. **94**, 027004 (2005)
35. Samokhin, K.V.: Phys. Rev. B **76**, 094516 (2007)
36. Fujimoto, S.: Phys. Rev. B **76**, 184504 (2007)
37. Yokoyama, T., Tanaka, Y., Inoue, J.: Phys. Rev. B **72**, 220504(R) (2005)
38. Borkje, K., Sudbo, A.: Phys. Rev. B **74**, 054506 (2006)
39. Iniotakis, C., Hayashi, N., Sawa, Y., Yokoyama, T., May, U., Tanaka, Y., Sigrist, M.: Phys. Rev. B **76**, 012501 (2007)
40. Vorontsov, A.V., Vekhter, i., Eshrig, M.: Phys. Rev. Lett. **101**, 127003 (2008)
41. Chi-Ken Lu, Sungkip Yip: Phys. Rev. B **78**, 132502 (2008)
42. Hayashi, N., Wakabayashi, K., Frigeri, P.A., Sigrist, M.: Phys. Rev. B **73**, 024504 (2006)
43. Chi-Ken Lu, Sungkip Yip: Phys. Rev. B **77**, 054515 (2008)
44. Edel'stein, V.M.: Phys. Rev. B **72**, 172501 (2005)
45. Frigeri, P.A., Agterberg, D.F., Milat, I., Sigrist, M.: Eur. Phys. J. B **54**, 435 (2006)
46. Mineev, V.P., Samokhin, K.V.: Phys. Rev. B **75**, 184529 (2007)
47. Samokhin, K.V.: Phys. Rev. B **78**, 224520 (2008)
48. Samokhin, K.V.: Phys. Rev. B **78**, 144511 (2008)
49. Samokhin, K.V.: Phys. Rev. B **72**, 054514 (2005)
50. Hayashi, N., Wakabayashi, K., Frigeri, P.A., Sigrist, M.: Phys. Rev. B **73**, 092508 (2006)
51. Samokhin, K.V., Mineev, V.P.: Phys. Rev. B **77**, 104520 (2008)
52. Mineev, V.P., Samokhin, K.V.: Phys. Rev. B **78**, 144503 (2008)
53. Dresselhaus, G.: Phys. Rev. **100**, 580 (1955)
54. Roth, M.: Phys. Rev. **173**, 755 (1968)
55. Rashba, E.I.: Fiz. Tverd. Tela (Leningrad) **2**, 1224 (1960)
56. Rashba, E.I.: Sov. Phys. Solid State **2**, 1109 (1960)
57. Mineev, V.P., Samokhin, K.V.: Phys. Rev. B **72**, 212504 (2005)
58. Lifshitz, E.M., Pitaevskii, L.P.: Statistical Physics Part 2. Butterworth-Heinemann, Oxford (1995)
59. Mikitik, G.P., Sharlai Yu.V.: Phys. Rev. Lett. **82**, 2147 (1999)
60. Haldane F.D.M., Phys. Rev. Lett. **93**, 206602, (2002)
61. Hashimoto, S., Yasuda, T., Kubo, T., Shishido, H., Ueda, T., Settai, R., Matsuda, T.D., Haga, Y., Harima, H., Onuki, Y.: J. Phys.: Condens. Matter **16**, L287 (2004)
62. Suhl, H., Matthias, B.T., Walker, L.R.: Phys. Rev. Lett. **3**, 552 (1959)
63. Mineev, V. P., Samokhin, K.V.: Introduction to Unconventional Superconductivity. Gordon and Breach, London (1999)
64. Sigrist, M., Ueda, K.: Rev. Mod. Phys. **63**, 239 (1991)
65. Tinkham, M.: Introduction to Superconductivity. 2nd edn. McGraw-Hill, New York (1996)
66. Yanase, Y., Sigrist, M.: J. Phys. Soc. Jpn. **76**, 124709 (2007)

Chapter 5
Magnetoelectric Effects, Helical Phases, and FFLO Phases

D. F. Agterberg

Abstract This chapter emphasizes new magnetic properties that arise when inversion symmetry is broken in a superconductor. There are two aspects that will be covered in detail. The first topic encompasses physics related to superconducting magnetoelectric effects that arise from broken inversion symmetry. Broken inversion symmetry allow for Lifshitz invariants in the free energy which can be viewed as a coupling between the magnetic induction and the supercurrent. There are similarities between these invariants and the better known Dzyaloshinskii-Moyira interaction in magnetic systems. These Lifshitz invariants give rise to anomalous magnetic properties as well as new phases in the presence of magnetic fields. Here, we will describe the consequences of these Lifshitz invariants, provide estimates for the relative magnitudes of the novel effects, and discuss the important role that crystal symmetry plays in understanding this physics. Finally, we provide a discussion of the fate of Fulde-Ferrell-Larkin-Ovchinnikov (FFLO) phases in broken inversion superconductors. In particular, we show how broken inversion symmetry can have a profound effect on the stability, existence, and properties of FFLO phases.

5.1 Introduction

One important way in which non-centrosymmetric superconductors differ from conventional superconductors is in the response to magnetic fields. In particular, the removal of inversion symmetry leads to new terms in the free energy that give rise to magneto-electric effects. These effects are closely related to the appearance of magnetic field generated helical phase in which the superconducting order develops a periodic spatial variation. Here we review this physics beginning with a

D. F. Agterberg (✉)
Department of Physics, University of Wisconsin-Milwaukee,
Milwaukee, WI 53211, USA
email: agterber@uwm.edu

E. Bauer and M. Sigrist (eds.), *Non-centrosymmetric Superconductors*,
Lecture Notes in Physics 847, DOI: 10.1007/978-3-642-24624-1_5,

detailed examination of the phenomenological theory followed by an overview of microscopic treatments of these problems which include an overview an of the interplay of the helical phase and Fulde-Ferrell-Larkin-Ovchinnikov (FFLO) phases [1, 2].

5.2 Phenomenology of Single Component Superconductors

This section reviews the phenomenology relating Lifshitz invariants in the the free energy to magnetoelectric effects, vortex structures, and the helical phase.

5.2.1 Ginzburg–Landau Free Energy

A key new feature of non-centrosymmetric superconductors is the existence of Lifshitz invariants in the Ginzburg–Landau (GL) free energy [3–8]. These give rise to magnetoelectric effects [5, 9–13], helical phases [6, 7, 14–16], and novel magnetic properties [7, 9, 12, 17–19] discussed in this chapter. To examine the consequences of these invariants we initially consider a GL theory for a single component order parameter (for example, an s-wave superconductor) and add the most general Lifshitz invariant allowed by broken inversion symmetry. Specific Lifshitz invariants are tabulated in Table 5.1 for different point group symmetries of the material in question. Since the primary goal is to reveal the new physics arising from these invariants, we ignore the role of any anisotropy that might appear in the usual GL free energy. Under these conditions the GL free energy under consideration is (we work in units such that $\hbar = c = 1$):

$$F = \int d^3r \left\{ \alpha|\eta|^2 + K\eta^* \mathbf{D}^2 \eta + K_{ij} B_i [\eta^*(D_j \eta) + \eta(D_j \eta)^*] + \frac{\beta}{2}|\eta|^4 + \frac{B^2}{8\pi} \right\}, \tag{5.1}$$

where $\alpha = \alpha_0(T - T_c)$, $D_i = -i\nabla_i - 2eA_i$ and $\mathbf{B} = \nabla \times \mathbf{A}$. From this free energy, the GL equations can be found by varying the above with respect to $\mathbf{A}$ and η. This results in the following:

$$\alpha\eta + \beta|\eta|^2\eta + K\mathbf{D}^2\eta + K_{ij}[2h_i(D_j\eta) + i\eta\nabla_j B_i] = 0 \tag{5.2}$$

and

$$\mathbf{J}_i = \frac{1}{4\pi}[\nabla \times (\mathbf{B} - 4\pi\mathbf{M})]_i = 2eK[\eta^*(D_i\eta) + \eta(D_i\eta)^*] + 4eK_{ji}B_j|\eta|^2 \tag{5.3}$$

where

$$\mathbf{M}_i = -K_{ij}[\eta^*(D_j\eta) + \eta(D_j\eta)^*]. \tag{5.4}$$

Table 5.1 Allowed Lifshitz invariants for different point groups

Point group	Lifshitz invariants
O	$K(B_x j_x + B_y j_y + B_z j_z)$
T	$K(B_x j_x + B_y j_y + B_z j_z)$
D_6	$K_1(B_x j_x + B_y j_y + B_z j_z) + K_2 B_z j_z$
C_{6v}	$K(B_x j_y - B_y j_x)$
C_6	$K_1(B_x j_x + B_y j_y + B_z j_z) + K_2 B_z j_z + K_3(B_x j_y - B_y j_x)$
D_4	$K_1(B_x j_x + B_y j_y + B_z j_z) + K_2 B_z j_z$
C_{4v}	$K(B_x j_y - B_y j_x)$
D_{2d}	$K(B_x j_y - B_y j_x)$
C_4	$K_1(B_x j_x + B_y j_y + B_z j_z) + K_2 B_z j_z + K_3(B_x j_y - B_y j_x)$
S_4	$K_1(B_x j_x - B_y j_y) + K_2(B_y j_x + B_x j_y)$
D_3	$K_1(B_x j_x + B_y j_y + B_z j_z) + K_2 B_z j_z$
C_{3v}	$K(B_x j_y - B_y j_x)$
C_3	$K_1(B_x j_x + B_y j_y + B_z j_z) + K_2 B_z j_z + K_3(B_x j_y - B_y j_x)$
D_2	$K_1 B_x j_x + K_2 B_y j_y + K_3 B_z j_z$
C_{2v}	$K_1 B_x j_y + K_2 B_y j_x$
C_2	$K_1 B_x j_x + K_2 B_y k_y + K_3 B_z j_z + K_4 B_y j_x + K_5 B_x j_y$
C_s	$K_1 B_z k_x + K_2 B_z j_j + K_3 B_x j_z + K_4 B_y j_z$
C_1	all components allowed

Here $j_i = \eta^*(D_i\eta) + \eta(D_i\eta)^*$

These equations are joined by the boundary conditions (which follow from the surface terms that arise from integration by parts in the variation of F):

$$[K\hat{n}_i(D_i\eta) + K_{ij}B_i\hat{n}_j\eta]_{boundary} = 0 \tag{5.5}$$

where $\hat{n}_j$ is the component of the surface normal along $\hat{j}$, and the usual Maxwell boundary conditions on the continuity of the normal component of **B** and the transverse components of $\mathbf{H} = \mathbf{B} - 4\pi\mathbf{M}$ (the appearance of **M** due to the Lifshitz invariants makes this boundary condition non-trivial). Note that adding the complex conjugate of Eq. 5.5 multiplied by η^* to Eq. 5.5 multiplied by η yields $\mathbf{J}\cdot\hat{n}|_{boundary} = 0$.

The appearance of **M** in Eq. 5.4 and the associated magnetization current leads to new physics in non-centrosymmetric superconductors. Also note, as is the case for centrosymmetric superconductors, the boundary conditions are valid on a length scale greater that ξ_0, the zero-temperature coherence length. In the following few subsections, we present the solution to some common problems to provide insight into the role of the Lifshitz invariants.

5.2.2 Solution with a Spatially Uniform Magnetic Field: Helical Phase

In situations when the magnetic field is spatially uniform, the GL equations describing the physics can be greatly simplified by introducing the following new order parameter:

$$\tilde{\eta} = \eta \exp\left(i\mathbf{q}\cdot\mathbf{x}\right) = \eta \exp\left(i\frac{iB_j K_{jk} x_k}{K}\right). \tag{5.6}$$

The GL free energy for $\tilde{\eta}$ no longer has any Lifshitz invariants and is

$$F = \int d^3r \left\{ \left[\alpha - B_l K_{lm} B_j K_{jm}\right] |\tilde{\eta}|^2 + K_1 \tilde{\eta}^* \mathbf{D}^2 \tilde{\eta} + \frac{\beta}{2} |\tilde{\eta}|^4 + \frac{B^2}{8\pi} \right\}. \tag{5.7}$$

The resulting new GL equations are now those of a single component superconductor with a magnetic field induced enhancement of T_c (this magnetic field enhancement is discussed in more detail in Chap. 1). These new GL equations follow from a minimization of Eq. 5.7 with respect to $\mathbf{A}$ and $\tilde{\eta}$. Note that the phase factor introduced above cancels the additional current contribution from the Lifshitz invariants in Eq. 5.3 and also cancels the related Lifshitz invariant contribution to the boundary condition. Furthermore, the magnetization that follows from Eq. 5.7 by taking the derivative with respect to B_i coincides with that due Eq. 5.4 found prior to the redefinition of the order parameter. This modified free energy of Eq. 5.7 immediately implies that some results from the usual GL theory apply. In particular:

(i) the vortex lattice solution near the upper critical field is the same as that of Abrikosov.
(ii) the surface critical field H_{c3} is the same as that of DeGennes. The order B^2 corrections to T_c do not change H_{c3} to leading order in $(T_c - T)/T_c$).
(iii) the critical current in this wires will show no unusual asymmetry (this conclusion differs from that of Ref. [4]).

5.2.2.1 Helical Phase

The main new feature that appears in a uniform magnetic field is the spatial modulation of the order parameter. Since η develops a helical spatial dependence in the complex plane, the resulting thermodynamic phase has been named the helical phase. Since helicity of the order parameter is related to its phase, an interference experiment based on the Josephson effect would provide the most reliable test to observe this. Indeed, such an experiment has been proposed [7]. In particular, consider the example of a 2D non-centrosymmetric superconductor (with a Rashba spin-orbit interaction) with a Zeeman field applied in the 2D plane. Then consider a Josephson junction between this and another thin film superconductor that is centrosymmetric. For a magnetic field applied in the plane of the film *perpendicular* to the junction and with the non-centrosymmetric superconductor oriented so that the helicity $\mathbf{q}$ is perpendicular to the field ; we find this gives rise to an interference effect analogous to the standard Fraunhofer pattern. For this experiment, the film must be sufficiently thin that the magnetic field and the magnitude of the order parameter are spatially uniform.

To illustrate this, consider the following free energy of the junction

$$H_J = -t \int dx[\Psi_1(\mathbf{x})\Psi_2^*(\mathbf{x}) + c.c.] \tag{5.8}$$

where the integral is along the junction. The resulting Josephson current is

$$I_J = Im\Big[t \int dx \Psi_1(\mathbf{x})\Psi_2^*(\mathbf{x})\Big] \tag{5.9}$$

Setting the junction length equal to $2L$, and integrating yields a maximum Josephson current of

$$I_J = 2t|\Psi_1^0||\Psi_2^0|\frac{|\sin(qL)|}{|qL|} \tag{5.10}$$

This demonstrates that the Josephson current will display an interference pattern for a field *perpendicular* to the junction. Note that in the usual case the Fraunhofer pattern would be observed for a magnetic field perpendicular to the thin film for which a finite flux passes through the junction.

5.2.2.2 Magnetoelectric Effect

Amongst the early theoretical studies of non-centrosymmetric superconductors, it was pointed out that a supercurrent must be accompanied by a spin polarization of the carriers [10]. Within the macroscopic theory given above, this spin polarization is described by the magnetization in Eq. 5.4. This magnetization appears when the supercurrent is non-vanishing due to a finite phase gradient. Subsequent to this proposal, it was suggested that the converse effect would also appear: a Zeeman field would induce a supercurrent [5]. This would follow from the expression for the current of Eq. 5.3 when the usual GL current $(2eK[\eta^*(D_i\eta) + \eta(D_i\eta)^*])$ vanishes. However, the latter proposal does not include the possibility discussed above that the order parameter develops a spatial modulation in the presence of a spatially homogeneous magnetic field (which leads to a nonvanishing $2eK[\eta^*(D_i\eta) + \eta(D_i\eta)^*]$). Indeed, this new equilibrium state ensures that the resultant supercurrent is vanishing. Nevertheless, as pointed out in Ref. [14], it is possible to create this current using a geometry similar to that used to observe Little-Parks oscillations. In particular, the supercurrent has two contributions, one is the current due to the Lifshitz invariants and the other is the usual GL current $2eK[\eta^*(D_i\eta) + \eta(D_i\eta)^*]$. In the helical phase, these two contributions exactly cancel. By wrapping the superconductor in a cylinder, the condition that the order parameter is single valued does not allow the helical phase to fully develop since arbitrary spatial oscillations are not allowed. Consequently, when a magnetic field is applied along the cylindrical axis, a non-zero current can flow. The resulting current will develop a periodic dependence on the applied magnetic field [14].

5.2.3 London Theory and Meissner State

We now turn to situations in which the magnetic field is not spatially uniform. The Lifshitz invariants lead to new physics for both the single vortex solution and for the usual penetration depth problem. To see this, we begin with the London limit and set $\eta=|\eta|e^{i\theta}$ and assume that the magnitude $|\eta|$ is fixed. The GL free energy is then minimized with respect to θ and $\mathbf{A}$. The minimization with respect to θ yields

$$K_1\nabla\cdot(\nabla\theta-2e\mathbf{A})+K_{ij}\nabla_i B_j=0 \tag{5.11}$$

which is equivalent to the continuity equation for the current ($\nabla\cdot\mathbf{J}=0$). The minimization with respect to $\mathbf{A}$ yields

$$\mathbf{J}_i=\frac{1}{4\pi}[\nabla\times(\mathbf{B}-4\pi\mathbf{M})]_i=-\frac{1}{4\pi\lambda^2}\left[\mathbf{A}_i-\frac{1}{2e}\nabla_i\theta-\sum_j\sigma_{ji}\mathbf{B}_j\right] \tag{5.12}$$

with

$$4\pi\mathbf{M}_i=\frac{1}{\lambda^2}\sum_j\sigma_{ij}\left(\mathbf{A}_j-\frac{1}{2e}\nabla_j\theta\right), \tag{5.13}$$

$1/\lambda^2=8\pi(2e)^2K|\eta|^2$ and $\sigma_{ij}=16\pi e\lambda^2K_{ij}$. We take the surface normal is along the $\hat{z}$ direction and that the applied field is oriented along the $\hat{y}$ direction. Note that by applying an appropriate rotation to the fields in the free energy, this geometry results in no loss of generality. We assume that there are spatial variations only along the direction of the surface normal (z). We therefore have from $\nabla\cdot\mathbf{B}=0$ that $B_z=0$. We further choose $\mathbf{A}=[A_x(z),A_y(z),0]$ so that $\mathbf{B}=(-\partial A_y/\partial z,\partial A_x/\partial z,0)$ and work in a gauge where $\nabla\theta=0$. The three components of Eq. 5.12 yields

$$\frac{\partial B_y}{\partial z}=\frac{1}{\lambda^2}\frac{\partial}{\partial z}[\sigma_{yy}A_y+\sigma_{zy}A_z]+\frac{1}{\lambda^2}A_x-\frac{1}{\lambda^2}\sigma_{xx}B_x \tag{5.14}$$

$$\frac{\partial B_x}{\partial z}=\frac{1}{\lambda^2}\frac{\partial}{\partial z}[\sigma_{xx}A_x+\sigma_{zx}A_z]-\frac{1}{\lambda^2}A_y-\frac{1}{\lambda^2}\sigma_{yy}B_x \tag{5.15}$$

$$4\pi J_z=0=A_z-\sigma_{zx}B_x-\sigma_{zy}B_y. \tag{5.16}$$

Note that contributions from σ_{xy} and σ_{yx} cancel in the above. Taking derivatives of Eqs. 5.14 and 5.15 with respect to z, using Eq. 5.16 to eliminate A_z, we find

$$\left(1-\frac{\sigma_{zy}^2}{\lambda^2}\right)\frac{\partial^2B_y}{\partial z^2}=\frac{1}{\lambda^2}B_y-\frac{\sigma_{xx}+\sigma_{yy}}{\lambda^2}\frac{\partial B_x}{\partial z}+\frac{\sigma_{zy}\sigma_{zx}}{\lambda^2}\frac{\partial^2B_x}{\partial z^2} \tag{5.17}$$

$$\left(1-\frac{\sigma_{zx}^2}{\lambda^2}\right)\frac{\partial^2B_x}{\partial z^2}=\frac{1}{\lambda^2}B_x+\frac{\sigma_{xx}+\sigma_{yy}}{\lambda^2}\frac{\partial B_y}{\partial z}+\frac{\sigma_{zy}\sigma_{zx}}{\lambda^2}\frac{\partial^2B_y}{\partial z^2}. \tag{5.18}$$

The above must be solved with the boundary conditions $B_i(z\rightarrow\infty)=0$ and

$$H_y = B_y(z=0) - 4\pi M_y(z=0) \tag{5.19}$$

$$0 = B_x(z=0) - 4\pi M_x(z=0) \tag{5.20}$$

where H_y is the applied field. M_x, M_y can be found using Eqs. 5.13, 5.14, 5.15, and 5.16 to eliminate A_x, A_y, and A_z in favor of B_x and B_y and their derivatives. By setting $B_i = B_{i0}\exp(-\delta z/\lambda)$, the solution can be found analytically. The general form of the solution is quite involved, so here we present the solution for point groups O and C_{4v}.

5.2.3.1 O Point Group

A representative material is Li_2Pt_3B [20–22]. This problem has been solved in Refs. [9, 12]. In this case there is only one Lifshitz invariant: $K_1\mathbf{B}\cdot\mathbf{j}$. Since this is a scalar under rotations the solution is the same for any orientation of the surface normal. The equations for $\mathbf{B}$ become:

$$\frac{\partial^2 B_y}{\partial z^2} = \frac{1}{\lambda^2}B_y + \frac{\delta}{\lambda^2}\frac{\partial B_x}{\partial z} \tag{5.21}$$

$$\frac{\partial^2 B_x}{\partial z^2} = \frac{1}{\lambda^2}B_x - \frac{\delta}{\lambda^2}\frac{\partial B_y}{\partial z}. \tag{5.22}$$

where $\delta = -2\sigma_{xx}$ (note $\sigma_{xx} = \sigma_{yy}$ in this case). This coupled set of equations can solved for $B_{\pm} = B_x \pm iB_y$ [9, 12] with the result that to first order in δ/λ:

$$B_y = H_y\big[\cos\frac{\delta z}{\lambda^2} + \frac{\delta}{\lambda}\sin\frac{\delta z}{\lambda^2}\big]e^{-z/\lambda} \tag{5.23}$$

$$B_x = H_y\big[\frac{\delta}{\lambda}\cos\frac{\delta z}{\lambda^2} - \sin\frac{\delta z}{\lambda^2}\big]e^{-z/\lambda}. \tag{5.24}$$

Physically, this implies that the the magnitude of the B_x is discontinuous as it crosses the surface (though not that of B_y) and that $\mathbf{B}$ also rotates inside the superconductor. Note that in a slab geometry, B_x is of opposite sign on the two sides of the slab. It may be possible to observe this through muon spin resonance experiments.

5.2.3.2 C_{4v} Point Group

A representative material is $CePt_3Si$ [23]. In this case, the single Lifshitz invariant is generated by a Rashba spin-orbit coupling and is given by $K_1\hat{z}\cdot\mathbf{B}\times\mathbf{j}$. This implies $\sigma = \sigma_{xy} = -\sigma_{yx} \neq 0$. The solution of the London problem now depends upon surface orientation and has been considered in Ref. [18]. We consider two situations here: the surface normal along and perpendicular to $\hat{z}$ (the four-fold symmetry axis). Consider first the normal along the $\hat{z}$ direction (in this case the applied field is H_y and we find that $B_x = 0$), then we have the usual London equation

$$\frac{\partial^2 B_y}{\partial z^2} = \frac{1}{\lambda^2} B_y \tag{5.25}$$

with the unusual boundary condition $H_y|_{z=0} = (B_y + \frac{\sigma}{\lambda} B_y)|_{z=0}$. This yields the solution

$$B_y(z) = \frac{H_y}{1 + \frac{\sigma}{\lambda}} e^{-z/\lambda} \tag{5.26}$$

These equations show that there is no rotation of **B** across the sample surface. However, the magnetic induction **B** is discontinuous as the surface is crossed. Again, in a slab geometry, the discontinuity in B_y is opposite for the two sides of the slab.

For the surface normal perpendicular to the $\hat{z}$ direction, the situation is different. To be concrete, consider the normal along the $\hat{x}$ direction and the applied field along the $\hat{y}$ direction (for the field along the $\hat{z}$ direction the usual London Equations result). In this case, it is again permissible to set $B_z = 0$ and solve for B_y to find

$$B_y = \frac{H_y}{1 - \frac{\sigma^2}{\lambda^2}} e^{-z/\tilde{\lambda}} \tag{5.27}$$

where $\sigma = \sigma_{xy}$ and $\tilde{\lambda} = \lambda \left(1 - \frac{\sigma^2}{\lambda^2}\right)$.

5.2.4 Spatial Structure of a Single Vortex

The London theory can also be used to examine the field distribution of a vortex in a strongly type II superconductor. Again, the lack of inversion symmetry introduce some new physics. Here we focus (as above) on two examples with point groups O and C_{4v} and provide the solutions of Refs. [12, 13, 18, 19]. The approach used in these publications is to consider the parameter σ_{ij}/λ to be small and then the Lifshitz invariants perturb the usual London solution. When there are no Lifshitz invariants, the solution to the London equations are $\theta = -\phi$ (ϕ is the polar angle) and the field is applies along the $\hat{n}$ direction

$$\mathbf{B} = \frac{1}{2e\lambda^2} K_0(r/\lambda)\hat{n} \tag{5.28}$$

where $K_0(x)$ is a modified Bessel function. The perturbative solutions depend upon the specific form of the Lifshitz invariants and we turn to a discussion of two case in turn.

5.2.4.1 *O* Point Group

The solution in this case was found in Ref. [12, 13]. The modified London equation is (the problem does not depend upon field direction)

$$\nabla \times \nabla \times \mathbf{A} + \frac{1}{\lambda^2}\mathbf{A} = \frac{\nabla \phi}{2e\lambda^2} + 2\frac{\delta}{\lambda^2}\nabla \times \mathbf{A} - \frac{\pi\delta}{e\lambda^2}\delta^2(\mathbf{r})\hat{z}. \tag{5.29}$$

The new term implies that, in addition to the field along $\hat{z}$, there is an additional component along $\hat{\phi}$. The authors of Ref. [12, 13] find that to first order in δ/λ the additional field is

$$B_\phi^{(1)}(x = r/\lambda) = \frac{\delta}{e\lambda^3}\left\{K_1(x)\int_0^x x'dx'I_1(x')K_1(x') + I_1(x)\int_x^\infty x'dx'[K_1(x')]^2\right\} - \frac{\delta}{2e\lambda^2}K_1(x) \tag{5.30}$$

where I_1 and K_1 are modified Bessel functions of the first kind.

5.2.4.2 C_{4v} Point Group

The solution in this case was found in Ref. [18, 19]. The fields that appear due to the Lifschitz invariants depend in this case upon the orientation of the field. For the field along the $\hat{y}$ direction, it is found that the solution for $\mathbf{B}$ is given by (correct to first order in σ/λ) [18, 19]

$$\mathbf{B} = \frac{1}{2e\lambda^2}K_0\left(|\mathbf{r} + \frac{\sigma}{\lambda}\hat{z}|/\lambda\right)\hat{y}. \tag{5.31}$$

Physically, this implies that the maximum value of B_y is shifted from the vortex center. This shift has also been seen in a full numerical solution of the Ginzburg Landau equations [17]. For the field along the $\hat{z}$ direction (the four-fold symmetry axis), the $\mathbf{B}$ field is unchanged and there is an induced magnetization along the radial direction [18] (this redial magnetization was also found in the vortex lattice solution near H_{c2} [7]).

5.2.5 Vortex Lattice Solutions

For fields near the upper critical field, there have been a variety of studies on the Abrikosov vortex lattice [7, 24–26]. Some of these studies predict multiple phase transitions in the vortex lattice state [24–26]. These studies are based on microscopic weak-coupling theories and involve an interplay of paramagnetism, orbital diamagnetism, gap symmetry, band structure, and spin-orbit coupling [24–26]. While this chapter will not address these vortex lattice transitions, we will address some of the microscopic issues in the next chapter. Here we focuss on the GL theory, for which the predictions are more straightforward. In particular, near the upper critical field, the magnetic field is approximately uniform and the considerations above imply that the vortex lattice is hexagonal (perhaps distorted by uniaxial anisotropy). Consequently (following the arguments of Section II B), the order parameter solution near the upper critical field is $\eta(\mathbf{r}) = cnst \exp(i\mathbf{q}\cdot\mathbf{r})\phi_0(x,y)$ where $\phi_0(x,y)$ is a lowest Landau level (LLL) solution. This solution, combining a phase factor and a (LLL) solution, has been called the helical vortex phase. The primary consequence of this solution is that the upper critical field is enhanced due to the presence of the Lifshitz invariants [7]. We note that due to the degeneracy of the LLL solution, there is ambiguity in the existence of the phase factor. In particular, the LLL solution $\tilde{\phi}_0(x,y) = e^{i\tau_y x/l_H^2}\phi_0(x, y-\tau_y)$ (l_H is the magnetic length) is degenerate with $\phi_0(x,y)$, consequently in some circumstances the wavevector $\mathbf{q}$ can be removed in favor of a shift of origin. This can be done whenever $\mathbf{q}$ is perpendicular to the applied magnetic field (this is the case for C_{4v} point group symmetry but not for O point group symmetry). We feel that is still meaningful to speak of the helical vortex phase for the point group C_{4v} because the same phase factor implies an increase of the in-plane critical field in two-dimensions for which this ambiguity does not exist. The name helical vortex phase reveals the link between the solutions in two and three dimensions.

In addition to studies near the upper critical field, there has been one numerical study of the time dependent GL equations in the vortex phase [17]. This study found the surprising result that the vortices flow spontaneously, in spite of the lack of an applied current. The claim is that the paramagnetic supercurrent (the magnetization current $\nabla\times\mathbf{M}$) is the origin of this spontaneous flux flow. We note that in this study the following boundary condition was used: $\mathbf{B}_{outside} = \mathbf{B}_{inside}$. This differs from the continuity of $\mathbf{H} = \mathbf{B} - 4\pi\mathbf{M}$ discussed above. In the problem that was studied, $\mathbf{M}$ is non-trivial and an examination of its neglect in the boundary condition can be seen to be equivalent to having a current flow. We argue that this current is cause the spontaneous flux flow. We note that the boundary conditions discussed here should be used in problems where the minimum length scale is ξ_0, the zero temperature coherence length. However, at lengths scale smaller than this, a microscopic theory is required and the single particle quantum mechanical wavefunctions will obey quite different boundary conditions.

5.2.6 Multi-Component Order Parameters

There have not been as many studies on Lifshitz invariants in non-centrosymmetric superconductors in cases when the order parameter contains more than one complex degree of freedom. There has been one noteworthy result, which is the appearance of the helical phase when no magnetic fields are applied [8, 22]. In particular, if the ground state of the multi-component order parameter breaks time-reversal symmetry [27, 28], then the lack of both parity and time-reversal symmetries allows the helical phase to appear. As an example, consider the three dimensional irreducible representation of the point group O, with an order parameter η where the components transform as the (x, y, z) component of a vector. The following Lifschitz invariant exists [8]

$$iK(\eta_1^* D_y \eta_3 + \eta_2^* D_z \eta_1 + \eta_3^* D_x \eta_2 - c.c.). \tag{5.32}$$

This Lifschitz invariant leads to a ground state order parameter $\eta = e^{iqz}(1, i, 0)$. The state $\eta = (1, i, 0)$ breaks time reversal symmetry and thus mimics the role of the magnetic field in the single component case.

5.3 Microscopic Theory

The phenomenological arguments of the previous section have also been the subject of many microscopic calculations. These calculations, while all related, focus and extend different aspects of the phenomenological theory above. In particular, four points of contact exist between the phenomenological theories and the microscopic theories. These are: direct calculations of the Lifshitz invariants in the free energy in Eq. 5.1; calculations of the magnetization in Eq. 5.4; calculations of the current in Eq. 5.3; and calculations of the helical wavevector $\mathbf{q}$ in Eq. 5.6. We briefly review the first three of these and then turn to a more complete overview of microscopic studies of the helical phase since this turns out to be closely linked to the FFLO phases.

5.3.1 Contact Between Microscopic and Macroscopic Theories: Lifshitz Invariants

The direct calculation of the Lifschitz invariants in Eq. 5.1 has been carried out by a few authors [4, 5, 7, 8] and can be found in Chap. 1 of this book. In particular, the non-interacting Hamiltonian is

$$H_0 = \sum_{\mathbf{k}} \sum_{\alpha\beta=\uparrow,\downarrow} [\xi(\mathbf{k})\delta_{\alpha\beta} + \gamma(\mathbf{k}) \cdot \sigma_{\alpha\beta}] a^\dagger_{\mathbf{k}\alpha} a_{\mathbf{k}\beta} \tag{5.33}$$

where $a^\dagger_{\mathbf{k}\alpha}$ $(a_{\mathbf{k}\alpha})$ creates (annihilates) an electronic state $|\mathbf{k}\alpha\rangle$, $\xi(\mathbf{k}) = \varepsilon(\mathbf{k}) - \mu$ denotes the spin-independent part of the spectrum measured relative to the chemical potential μ, $\alpha, \beta = \uparrow, \downarrow$ are spin indices, σ are the Pauli matrices, and the sum over $\mathbf{k}$ is restricted to the first Brillouin zone. In the helicity basis, this Hamiltonian is diagonalized with energy bands given by

$$\xi_\pm(\mathbf{k}) = \xi(\mathbf{k}) \pm |\gamma(\mathbf{k})| \tag{5.34}$$

with the Hamiltonian

$$H_0 = \sum_{\mathbf{k}} \sum_{\lambda=\pm} \xi_\lambda(\mathbf{k}) c^\dagger_{\mathbf{k}\lambda} c_{\mathbf{k}\lambda}, \tag{5.35}$$

where the two sets of electronic operators are connected by a unitary transformation,

$$a_{\mathbf{k}\alpha} = \sum_\lambda u_{\alpha\lambda}(\mathbf{k}) c_{\mathbf{k}\lambda}, \tag{5.36}$$

with

$$(u_{\uparrow\lambda}(\mathbf{k}), u_{\downarrow\lambda}(\mathbf{k})) = \frac{(|\gamma| + \lambda\gamma_z, \lambda(\gamma_x + i\gamma_y))}{\sqrt{2|\gamma|(|\gamma| + \lambda\gamma_z)}}. \tag{5.37}$$

In the limit that only one of the bands cross the the Fermi energy (this can be realized for superconductivity at the surface of a topological insulator [29]), the following weak-coupling result for the coefficients defining the Lifishitz invariants of Eq. 5.1 is found

$$K_{ij} = -\frac{\mu_B N_0 S_3}{2} \langle \phi^2(\mathbf{k}) \hat{\gamma}_i(\mathbf{k}) v_j(\mathbf{k}) \rangle \tag{5.38}$$

where N_0 is the density of states of the band at the chemical potential, $\phi(\mathbf{k})$ describes the superconducting state and is an even function belonging to one of one-dimensional representations of the point group of the crystal, $\langle \ldots \rangle$ means the averaging over the Fermi surface, μ_B is the Bohr magneton, and

$$S_3(T) = \pi T \sum_n \frac{1}{|\omega_n|^3} = \frac{7\zeta(3)}{4\pi^2 T^2}. \tag{5.39}$$

Equation 5.38 is valid when there is only a single band present. When two bands are present (as is often the case), and assuming that $\phi(\mathbf{k})$ is the same for both bands, then Eq. 5.38 must be multiplied by the factor

$$\delta N = (N_+ - N_-)/(N_+ + N_-). \tag{5.40}$$

where $N_\pm$ are the density of states of the two bands ($N_0 = N_+ + N_-$). Microscopic calculations of the Lifshitz invariants are limited to the regime near T_c where the GL theory is valid.

5.3.2 Contact Between Microscopic and Macroscopic Theories: Current and Magnetization

In the limit of small magnetic fields ($\mathbf{B}$) and small phase gradients ($\nabla\theta$) in the superconducting order parameter, it it possible to find microscopic extensions to

Eq. 5.3 and Eq. 5.4 that are valid for all temperatures. This has been carried out in Refs. [5, 10]. Here, we follow the notation of Ref. [5]. In the clean limit, for 2D cylindrical bands with a Rashba interaction ($\gamma(\mathbf{k}) = \alpha\hat{n} \times \mathbf{p}(\mathbf{k})$) Eq. 5.3 and Eq. 5.4 can be rewritten as

$$\begin{aligned} J_x &= \rho_s \frac{\hbar \nabla_x \theta}{2m} - \kappa B_y \\ M_y &= \frac{\kappa}{2} \hbar \nabla_x \theta \end{aligned} \tag{5.41}$$

where M_y is the magnetic moment, ρ_s is the superfluid density, and

$$\kappa(T) = \frac{\mu}{4\pi\hbar^2}[p_{F+}\{1 - Y(T, \Delta_+\} - p_{F-}\{1 - Y(T, \Delta_-)\}] \tag{5.42}$$

where $p_{F,\pm}$ are the Fermi momenta for the two bands, $\Delta_\pm$ are the gaps on the two bands, μ is the Fermi energy, and $Y(x)$ is the Yoshida function. Note that Eq. 5.42 is proportional to δN in the limit $\delta N << 1$.

The role of Fermi liquid corrections has also been examined [11] in this context. This study has found the that the only Fermi liquid corrections that alter the current contribution from the Lifshitz invariants are ferromagnetic correlations. If there are no ferromagentic correlations, then Eq. 5.42 is unchanged. This is important in heavy Fermion materials, where the effective mass enhancement suppresses the usual supercurrent but does not change Eq. 5.42 [11].

5.3.3 Microscopic Theory of the Helical and FFLO Phases

The helical phase has received a great deal of attention from the microscopic point of view [7, 14–16, 24, 26, 29–31]. One reason for this is that it is closely related to the FFLO phase [1, 2] in which the superconducting order parameter develops a periodic spatial structure. The interplay between these two phases is not trivial. It is perhaps not surprising that spatially oscillating superconductor solutions readily appear in non-centrosymmetric superconductors when magnetic fields are applied. In particular, a state with momentum $\mathbf{k}$ at the Fermi surface will generally not have a degenerate partner at $-\mathbf{k}$ with which to form a Cooper pair when both parity and time reversal symmetries are broken. The state $\mathbf{k}$ would rather pair with a degenerate state $-k + \mathbf{q}$ and in this way generate a spatially oscillating superconducting order parameter.

The microscopic origin of the spatially oscillating states can be understood by an examination of the single particle eigenstates when a Zeeman field $\mathbf{H}$ is included (for now we ignore the vector potential $\mathbf{A}$)

$$H_Z = -\sum_{\mathbf{k},\alpha,\beta} \mu_B \mathbf{H} \cdot \boldsymbol{\sigma}_{\alpha\beta} a^\dagger_{\mathbf{k}\alpha} a_{\mathbf{k}\beta}. \tag{5.43}$$

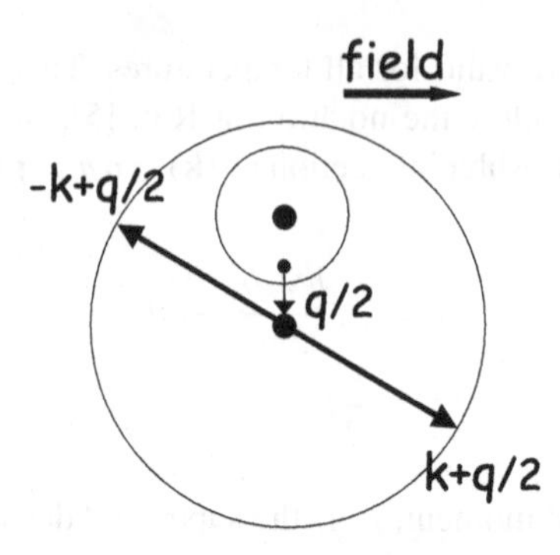

Fig. 5.1 A magnetic field directed as shown together with a Rashba spin-orbit interaction shifts the center of the large and small Fermi surfaces by $\pm\mathbf{q}/2$. The smaller dot represent the point (0,0) (center of Fermi surfaces without field) and the two larger dots represent the points $(0, -\mathbf{q}/2)$ and $(0, \mathbf{q}/2)$ (centers of the new Fermi surfaces). Pairing occurs between states of $\mathbf{k}+\mathbf{q}/2$ and $-\mathbf{k}+\mathbf{q}/2$, leading to a gap function that has a spatial variation $\Delta(\mathbf{x}) = \Delta_0 \exp(i\mathbf{q}\cdot\mathbf{x})$. From Ref. [16]

The single particle excitations now become

$$\xi_\pm(\mathbf{k}, \mathbf{H}) = \xi(\mathbf{k}) \pm \sqrt{\gamma^2(\mathbf{k}) - 2\mu_B\gamma(\mathbf{k})\cdot\mathbf{H} + \mu_B^2\mathbf{H}^2}. \tag{5.44}$$

In the limit $|\gamma| >> |H|$, this becomes (we ignore the small regions of phase space for which $\gamma = 0$)

$$\xi_\pm(\mathbf{k}, \mathbf{H}) \approx \xi(\mathbf{k}) \pm \mu_B\hat{\gamma}(\mathbf{k})\cdot\mathbf{H}. \tag{5.45}$$

The origin of pairing states with non-zero $\mathbf{q}$ (that is $\Delta(\mathbf{x}) \propto e^{i\mathbf{q}\cdot\mathbf{x}}$) follow from this expression. As an example, consider a Rashba interaction $\gamma = \gamma_\perp(k_y\hat{x} - k_x\hat{y})$ for a cylindrical Fermi surface and a magnetic field along $\hat{x}$. In this case, as shown in Fig. 5.1, the Fermi surfaces remain circular and the centers are shifted along the $\hat{y}$ direction. A finite center of mass momentum Cooper pair is stable because the same momentum vector $\mathbf{q}$ can be used to pair *every* state on one of the two Fermi surfaces. In the more general case, for a non-zero $\mathbf{q}$ state to be stable, the paired states should be degenerate: $\xi_\pm(\mathbf{k}+\mathbf{q}, \mathbf{H}) = \xi_\pm(-\mathbf{k}+\mathbf{q}, \mathbf{H})$, this gives the condition $\hbar\mathbf{q}\cdot\mathbf{v}_F = \mu_B\mathbf{H}\cdot\hat{\gamma}(\mathbf{k})$. This differs from the condition for the usual FFLO phase, for which $\hbar\mathbf{q}\cdot\mathbf{v}_F = \mu_B|\mathbf{H}|$. The optimal paring state corresponds to finding $\mathbf{q}$ that satisfies the pairing condition for the largest possible region on the Fermi surface.

The above paragraph also reveals the origin of the interplay between the helical and FFLO phases. In particular, the two Fermi surface sheets prefer pairing states with opposite sign of $\mathbf{q}$. Choosing a particular $\mathbf{q}$ allows pairing on one Fermi surface, but not on the other. This naturally leads to competition between single-$\mathbf{q}$ (helical) and multiple-$\mathbf{q}$ (FFLO-like) states. Which state appears depends upon the details of the system. Without going into further microscopic details, which can be found in Refs. [7, 14–16, 24, 26, 29–31], we summarize some of the main results here. One important result is that since there are two sources of the modulation $\mathbf{q}$ (FFLO-like

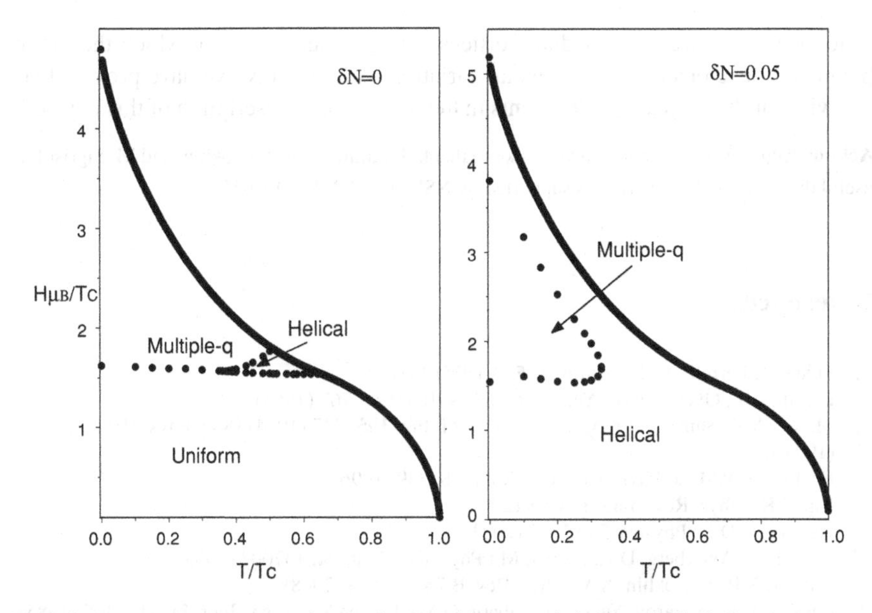

Fig. 5.2 Typical phase diagram showing both multiple-**q** and single-**q** (helical phase) phases as a function of Zeeman field in a clean non-centrosymmetric superconductor for two different values of δN. These calculations where carried out with a Rashba spin-orbit interaction and a 3D spherical Fermi surface (a 2D cylindrical Fermi surface gives similar results). For fields $H\mu_B/T_c < 1.5, q \approx \delta N H\mu_B/v_F$ (for the $\delta N = 0$, this leads to $\mathbf{q}=0$), while for higher fields $q \approx H\mu_B/v_F$. From Ref. [16]

physics and Lifshitz invariants), there are two typical values for the magnitude of q [15, 16, 26, 31] that both appear in different regions of the temperature/magnetic field phase diagram. In particular $q \approx H\mu_B/v_F$ stems from FFLO-like physics related to Fig. 5.1 and is the value of q in the high-field regime (in clean materials). While $q \approx \delta N H\mu_B/v_F$ stems from the Lifschitz invariants and is the typical magnitude of q in the low-field regime [15, 16, 31]. As shown in Fig. 5.2, in the clean limit, both single-**q** and multiple-**q** phases exist [15, 16]. However, the multiple-**q** phase become less stable as δN increases [16]. We note that in the case of superconductivity at the surface of a topological insulator, which is akin to $\delta N = 1$, only the single-**q** exists [29]. In the dirty limit the multiple-**q** phases no longer appear, while the single-**q** phase with $q \approx \delta N H\mu_B/v_F$ is robust [15, 31]. Finally we note that when the vector potential is also included then novel vortices and vortex phases may appear [15, 24–26, 32].

5.4 Conclusions

In this chapter we have examined the role of Lifshitz invariants that appear in the Ginzburg Landau free energy of non-centrosymmetric superconductors. These invariants lead to magnetoelectric effects, novel London physics in the Meissner state, new structure in individual vortices, and a helical phase in which the order parameter develops a periodic spatial variation. Additionally, we have provided an overview of theoretical developments in the microscopic description of this physics.

Acknowledgments The author would like to thank S. Fujimoto, K. Samokhin, and M. Sigrist for useful discussions. This work was supported by NSF grant DMR-0906655.

References

1. Fulde, P., Ferrell, A.: Phys. Rev. **135**, A550 (1964)
2. Larkin, A.I., Ovchinnikov, Yu.N.: Sov. Phys. JETP **20**, 762 (1965).
3. Mineev V.P., Samokhin K.V.: Zh. Eksp. Teor. Fiz. **105**, 747 (1994) [Sov. Phys. JETP **78**, 401 (1994)].
4. Edel'stein, V.M.: J. Phys.: Condens. Matter **8**, 339 (1996)
5. Yip, S.K.: Phys. Rev. B **65**, 144508 (2002)
6. Agterberg, D.F.: Physica C **387**, 13 (2003)
7. Kaur, R.P., Agterberg, D.F., Sigrist, M.: Phys. Rev. Lett. **94**, 137002 (2005)
8. Mineev, V.P., Samokhin, K.V.: Phys. Rev. B **78**, 144503 (2008)
9. Levitov, L.S., Nazarov, Yu.V., Eliashberg, G.M.: Pis'ma Zh. Eksp. Teor. Fiz. **41**, 365 (1985) [JETP Letters **41**, 445 (1985)].
10. Edel'stein, V.M.: Phys. Rev. Lett. **75**, 2004 (1995)
11. Fujimoto, S.: Phys. Rev B **72**, 024515 (2005)
12. Lu, C.K., Yip, S.: Phys. Rev. B **77**, 054515 (2008)
13. Lu, C.K., Yip, S.: J. Low Temp. Phys. **155**, 160 (2009)
14. Dimitrova, O.V., Feigel'man, M.V.: Pis'ma v ZhETF **78**, 1132 (2003).
15. Dimitrova, O., Feigel'man, M.V.: Phys. Rev. B **76**, 014522 (2007)
16. Agterberg, D.F., Kaur, R.P.: Phys. Rev. B **75**, 064511 (2007)
17. Oka, M., Ishioka, M., Machida, K.: Phys. Rev B **73**, 214509 (2006)
18. Yip, S.: J. Low. Temp. Phys **140**, 67 (2005).
19. Yip S. arXiv:0502477v2.
20. Togano, K., Badica, P., Nakamori, Y., Orimo, S., Takeya, H., Hirata, K.: Phys. Rev. Lett. **93**, 247004 (2004)
21. Badica, P., Kondo, T., Togano, K.: J. Phys. Soc. Jpn. **74**, 1014 (2005)
22. Yuan, H.Q., Agterberg, D.F., Hayashi, N., Badica, P., Vandervelde, D., Togano, K., Sigrist, M., Salamon, M.B.: Phys. Rev. Lett. **97**, 017006 (2006).
23. Bauer, E., Hilscher, G., Michor, H., Paul, Ch., Scheidt, E.W., Gribanov, A., Seropegin, Yu., Noël, H., Sigrist, M., Rogl, P.: Phys. Rev. Lett. **92**, 027003 (2004)
24. Matsunaga, Y., Hiasa, N., Ikeda, R.: Phys. Rev. B **78**, 220508 (2008)
25. Hiasa, N., Ikeda, R.: Phys. Rev. B **78**, 224514 (2008)
26. Hiasa, N., Saiki, T., Ikeda, R.: Phys. Rev. B **80**, 014501 (2009)
27. Sigrist, M., Ueda, K.: Rev. Mod. Phys. **63**, 239 (1991)
28. Mineev, V. P., Samokhin, K. V.: Introduction to unconventional superconductivity. Gordon and Breach, London (1999)
29. Santos, L., Neuport, T., Chamon, C., Murdy, C.: Phys. Rev. B **81**, 184502 (2010)
30. Barzykin, V., Gor'kov, L. P.: Phys. Rev. Lett. **89**, 227002 (2002)
31. Samokhin, K. V.: Phys. Rev. B **78**, 224520 (2008)
32. Agterberg, D.F., Zheng, Z., Mukherjee, S.: Phys. Rev. Lett. **100**, 017001 (2008)

Chapter 6
Microscopic Theory of Pairing Mechanisms

Y. Yanase and S. Fujimoto

Abstract This chapter deals with the microscopic theory of pairing mechanisms for noncentrosymmetric superconductors. One of curious questions is how to understand microscopically the parity mixing arising from the interplay between pairing interactions and the spin-orbit interactions that stem from broken inversion symmetry. Here, some examples of microscopic models which exhibit this phenomenon are presented. Our argument is mainly concentrated on the heavy fermion superconductors $CePt_3Si$ and $CeRh(Ir)Si_3$, for which ample experimental evidence confirms the realization of unconventional superconducticvity caused by non-phonon mechanisms. In these heavy fermion systems, superconductivity appears in or in the vicinity of an antiferromagnetic phase. Thus, it is important to clarify the role of magnetism in the pairing state. We examine the scenario that magnetic interaction is the origin of the pairing interaction for these systems. The influences of the coexisting magnetic order are also investigated.

6.1 Introduction

The BCS theory is the standard microscopic description of superconductivity. It is natural to expect that the central notion of the BCS theory is still applicable to noncentrosymmetric superconductors (NCS), of which the crystal structures lack inversion symmetry. Some distinct novel features associated with broken inversion symmetry, however, appear in their pairing states. For instance, because of parity violation, the admixture of spin-singlet pairing states and spin-triplet pairing states

Y. Yanase (✉)
Department of Physics, University of Tokyo,
email: yanase@hosi.phys.s.u-tokyo.ac.jp

S. Fujimoto
Department of Physics, Kyoto University,
email: fuji@scphys.kyoto-u.ac.jp

E. Bauer and M. Sigrist (eds.), *Non-centrosymmetric Superconductors*,
Lecture Notes in Physics 847, DOI: 10.1007/978-3-642-24624-1_6,

occurs [1–4]. We here illustrate this phenomenon briefly, before entering the main part of this chapter. For systems without inversion symmetry, there is an asymmetric potential gradient caused by nuclei located at asymmetric positions. The asymmetric potential gradient ∇V gives rise to a spin-orbit (SO) interaction $(\mathbf{k}\times\nabla V)\cdot\boldsymbol{\sigma}$ acting on electrons with momentum $\mathbf{k}$ and spin $\boldsymbol{\sigma}$. This interaction is odd in $\mathbf{k}$. In the following, we refer to this SO interaction as an antisymmetric SO interaction to distinguish it from usual SO interactions caused by spherical Coulomb potentials. To make our argument concrete, we consider the case that there is no mirror symmetry with respect to a (001)-plane, and that $\nabla V \propto \mathbf{n} = (001)$; i.e. the antisymmetric SO interaction is the Rashba SO interaction. This SO interaction splits the bands into two parts, each of which is an eigenstate of spin chirality, and hence the spin rotational symmetry is broken; i.e. in one band, an electron with momentum $\mathbf{k}$ $(-\mathbf{k})$ is in the spin up (down) state $|\uparrow\rangle$ $(|\downarrow\rangle)$, while in the other band the spin state is reversed. Here the spin quantization axis is taken to be parallel to $\mathbf{k}\times\mathbf{n}$. When the magnitude of the SO split E_{SO} is sufficiently larger than the superconducting gap Δ, as in the case of any NCS discovered so far for which E_{SO} is $10\sim100$ meV, interband Cooper pairing between electrons in the two SO split bands is strongly suppressed. In one of the two SO split bands, a Cooper pair formed between electrons with momenta $\mathbf{k}$ and $-\mathbf{k}$ is in the state $|\mathbf{k}\uparrow\rangle|-\mathbf{k}\downarrow\rangle$, while in the other band, it is $|\mathbf{k}\downarrow\rangle|-\mathbf{k}\uparrow\rangle$. If the density of states of the one band is different from the other, the superposition between the state $|\mathbf{k}\uparrow\rangle|-\mathbf{k}\downarrow\rangle$ and the state $|\mathbf{k}\downarrow\rangle|-\mathbf{k}\uparrow\rangle$ is suppressed. Then, the pairing state in one band $|\mathbf{k}\uparrow\rangle|-\mathbf{k}\downarrow\rangle = \frac{1}{2}(|\mathbf{k}\uparrow\rangle|-\mathbf{k}\downarrow\rangle - |\mathbf{k}\downarrow\rangle|-\mathbf{k}\uparrow\rangle) + \frac{1}{2}(|\mathbf{k}\uparrow\rangle|-\mathbf{k}\downarrow\rangle + |\mathbf{k}\downarrow\rangle|-\mathbf{k}\uparrow\rangle)$ is the admixture of the spin-singlet pairing and the spin-triplet pairing with the spin projection of the total spin of the Cooper pair $S^z = 0$ for the spin quantization axis parallel to $\mathbf{k}\times\mathbf{n}$. In addition to this effect, the antisymmetric SO interactions also give rise to pairing interactions which conserve neither spin nor parity; i.e. the spin-singlet channel and the spin-triplet channel are directly mixed by the spin-non-conserving pairing interactions. Because of these two effects, the parity-mixing of Cooper pairs occurs. In this argument, an important parameter which controls the parity-mixing is the ratio $E_{\mathrm{SO}}/E_{\mathrm{F}}$ [1–3]. For any NCS discovered so far, $E_{\mathrm{SO}}/E_{\mathrm{F}}$ is at most ~ 0.1. Despite such a small magnitude of $E_{\mathrm{SO}}/E_{\mathrm{F}}$, it is possible that, in some particular situations, the parity mixing becomes a substantial effect for pairing states. For instance, when the attractive pairing interactions in the spin-singlet channel and spin-triplet channel mixed by the antisymmetric SO interaction are comparable in magnitude to each other, the transition temperatures for the spin-singlet state and the spin-triplet state in the absence of the SO interaction are close to each other, and even for the small magnitude of the SO interaction, strong mixing of these two states occurs. Another possible scenario for strong parity mixing is that the spin-non-conserving pairing interactions which directly couple the odd parity channel with the even parity channel are enhanced by some specific mechanisms such as strong correlation effects. This results in the substantial admixture of spin-singlet pairs and spin-triplet pairs, even when the magnitude of $E_{\mathrm{SO}}/E_{\mathrm{F}}$ is small.

In the parity-mixed pairing state, when the interband pairings between the two SO split bands are completely suppressed, the superconducting gap function has the following form [1–3]

$$\Delta_{s_1 s_2}(\mathbf{k}) = \Delta_s(\mathbf{k}) i(\sigma_2)_{s_1 s_2} + \mathbf{d}(\mathbf{k}) \cdot i(\sigma\sigma_2)_{s_1 s_2}, \tag{6.1}$$

where $\Delta_s(\mathbf{k})$ is the gap for the spin-singlet pairing, and $\mathbf{d}(\mathbf{k})$ is the **d**-vector for the spin-triplet pairing. It is noted that the **d**-vector in this case is parallel to a vector $\mathbf{g}(\mathbf{k})$, which is defined by the average of $\mathbf{k} \times \nabla V$ with respect to Bloch wave functions. This simple form of the gap function (6.1), however, is applicable only when the dominant pairing interaction in the spin-triplet channel has the momentum dependence compatible with that of the antisymmetric SO interaction. Generally, it may occur that the spin-triplet pairing interaction which is mismatched with the antisymmetric SO interaction is substantially strong, giving rise to frustration between these two kinds of interactions. In this situation, the structure of the parity-mixed pairing state crucially depends on microscopic details of pairing interactions and the antisymmetric SO interactions. Thus, for understanding pairing states in NCS, it is vitally important to investigate the microscopic nature of pairing interactions.

In this chapter, we present some examples of microscopic studies on pairing mechanisms of NCS. In particular, we mainly consider two families of heavy fermion NCS, $CePt_3Si$ and $CeRh(Ir)Si_3$. The reason why we choose these systems is as follows. According to accumulated experimental evidence, it is well established that unconventional superconductivity is realized in these systems, and it is suggested that non-phonon mechanisms for providing the pairing glue play a crucial role. Actually, in these strongly correlated electron systems, superconductivity occurs in or in the vicinity of antiferromagnetically ordered phases. This observation strongly supports the idea that Cooper pairs are mediated by spin fluctuations. It is plausible to expect that strongly anisotropic pairing interactions mediated by spin fluctuations give rise to situations advantageous for parity mixing of Cooper pairs involving different angular momentum channels. In fact, as will be discussed in the next section, the strong parity mixing is realized for $CePt_3Si$ because of enhanced spin-non-conserving pairing interactions caused by a cooperative effect between spin fluctuations and the antisymmetric SO interaction. Furthermore, the interplay between broken inversion symmetry and magnetic correlations leads to various rich physics, as will be discussed in the following sections; e.g. a helical state with spatially modulated superconducting order parameters which is expected to be realized in $CePt_3Si$, and extremely large upper critical fields observed for $CeRh(Ir)Si_3$.

The organization of this chapter is as follows. In Sect. 6.2, we discuss the microscopic pairing mechanism of $CePt_3Si$. We also discuss the relation between the chiral magnetism and superconductivity for this system. In Sect. 6.3, the cases of $CeRh(Ir)Si_3$ are considered. The upper critical field of $CeRh(Ir)Si_3$ is also discussed on the basis of the microscopic theory. Sections 6.2 and 6.3 are independently readable. In fact, $CePt_3Si$ and $CeRh(Ir)Si_3$ are quite different from each other in their fundamental properties, as will be explained in the following sections. Thus, the approximation schemes used for the calculations of superconducting properties are different between these two classes of the heavy fermion systems. Conclusions are given in Sect. 6.4.

6.2 Case of $CePt_3Si$

Among non-centrosymmetric heavy fermion superconductors, $CePt_3Si$ has been investigated most extensively because its superconductivity occurs at ambient pressure [5]; others superconduct only under substantial pressure. In $CePt_3Si$, superconductivity with $T_c \sim 0.5$ K appears in the AF state with a Neél temperature $T_N = 2.2$ K [5–7]. The AF order microscopically coexists with superconductivity [8, 9]. Neutron scattering measurements characterize the AF order with an ordering wave vector $\mathbf{Q} = (0, 0, \pi)$ and magnetic moments in the *ab*-plane of a tetragonal crystal lattice [10]. The AF order is suppressed by pressure and vanishes at a critical pressure $P_c \sim 0.6$ GPa. Superconductivity is more robust against pressure and therefore a purely superconducting phase is present above the critical pressure $P > 0.6$ GPa [11, 12]. The coexistence of superconductivity and AF order gives rise to some intriguing phenomena, but complicates the situation. Therefore, we discuss the pairing state and the microscopic mechanism of superconductivity in the absence of AF order in Sect. 6.2.3. We study the NCS in the AF ordered state in Sect. 6.2.4. The Sect. 6.2.1 is devoted to the microscopic derivation of the Rashba SO interaction based on the tight-binding scheme. In Sect. 6.2.2, we describe the formulation to study the pairing states.

The nature of the superconducting phase has been clarified by several experiments. The low-temperature properties of thermal conductivity [13], superfluid density [14], specific heat [6], and NMR $1/T_1T$ [15] indicate line nodes in the gap. The upper critical field $H_{c2} \sim 3 - 4$ T exceeds the standard paramagnetic limit [5, 11], which seems to be consistent with the Knight shift data displaying no decrease in spin susceptibility below T_c for any field direction [16, 17]. The combination of these features is incompatible with the usual pairing states such as the *s*-wave, *p*-wave, or *d*-wave state. We show that these features are consistent with NCS which coexists with AF order (Sect. 6.2.4) [18–20].

6.2.1 Derivation of SO Interaction in Heavy Fermions

The antisymmetric SO interaction $\hat{H}_{ASO}$ stems from asymmetric potential gradients ∇V caused by broken inversion symmetry of crystal structures; i.e. $\hat{H}_{ASO} = \frac{e}{4m^2c^2}\langle\phi_A|\boldsymbol{\sigma}\cdot(\boldsymbol{k}\times\nabla V)|\phi_B\rangle$ with $|\phi_{A,B}\rangle$ a Bloch state. In heavy fermion systems which consist of both well-localized *f*-electrons and conduction electrons, the antisymmetric SO interaction involves matrix elements between different orbital states as well as those between the same orbital states. Here, we show how the Rashba SO interaction is microscopically derived by taking the multi-orbital nature of heavy fermion systems into account [20]. In the following, we omit the matrix element of the SO interaction between the same orbital states in order to focus on the specific feature of the antisymmetric SO interaction inherent in heavy fermion systems. The following recipe can be generally used for non-centrosymmetric heavy fermion systems. The generalization to the other systems, such as transition metals, is also straightforward.

We here construct a periodic Anderson model and Hubbard model on the basis of the tight-binding approximation. The localized $4f$ states in the Ce-based heavy fermion systems, such as $CePt_3Si$, $CeRhSi_3$, and $CeIrSi_3$ are described by the $J = 5/2$ manifold whose degeneracy is split by the crystal electric field. For instance, the following derivation is based on the level scheme proposed for $CePt_3Si$ [10]

$$|\Gamma_7\pm\rangle = \sqrt{\frac{5}{6}}|\pm\frac{5}{2}\rangle - \sqrt{\frac{1}{6}}|\mp\frac{3}{2}\rangle, \tag{6.2}$$

$$|\Gamma_6'\pm\rangle = |\pm\frac{1}{2}\rangle, \tag{6.3}$$

$$|\Gamma_7'\pm\rangle = \sqrt{\frac{1}{6}}|\pm\frac{5}{2}\rangle + \sqrt{\frac{5}{6}}|\mp\frac{3}{2}\rangle. \tag{6.4}$$

According to ref. [10], the ground state is $|\Gamma_7\pm\rangle$, and the others are excited states. Similar, but different crystal field levels have been proposed for $CePt_3Si$ [7, 21]. The Rashba SO interaction is derived for these levels in the same way, but the momentum dependence of g-vector $g(\mathbf{k})$ depends on the low-lying crystal field level, as we show below.

We here assume the low-lying $|\Gamma_7\pm\rangle$ doublet and take into account the hybridization with other electron states. Because the mirror symmetry is broken with respect to the ab-plane in $CePt_3Si$, the odd-parity $4f$ orbital is hybridized with the even-parity s and d orbitals in the same Ce site. Owing to the symmetry of the $|\Gamma_7\pm\rangle$ state, this $4f$ state is hybridized with the d_{xy}, d_{xz} and d_{yz} orbitals. Then, the wave function of the localized state is expressed as

$$|\mathrm{f}\pm\rangle = \kappa|\Gamma_7\pm\rangle + \mathrm{i}\epsilon|\mathrm{d_{xy}}\rangle\chi_\pm + \eta(|\mathrm{d_{xz}}\rangle \mp \mathrm{i}|\mathrm{d_{yz}}\rangle)\chi_\mp, \tag{6.5}$$

where ϵ, η and $\kappa = \sqrt{1-\epsilon^2-2\eta^2}$ are real number and $\chi_\pm$ describes the wave function of the spin. A periodic Anderson Hamiltonian is constructed for localized $|\mathrm{f}\pm\rangle$ states and conduction electrons. We here consider conduction electrons arising from the Ce $5s$ orbital for simplicity. Taking into account the inter-site hybridization between the s, d and f orbitals, we obtain the tight-binding Hamiltonian $H_0 = \sum_{\mathbf{k}} \hat{\psi}_{\mathbf{k}}^\dagger \hat{H}_0(\mathbf{k}) \hat{\psi}_{\mathbf{k}}$, where $\hat{\psi}_k^\dagger = (f_{\mathbf{k}+}^\dagger, f_{\mathbf{k}-}^\dagger, c_{\mathbf{k}\uparrow}^\dagger, c_{\mathbf{k}\downarrow}^\dagger)$, and

$$\hat{H}_0(\mathbf{k}) = \begin{pmatrix} \hat{\varepsilon}_{\mathrm{f}}(\mathbf{k}) & \hat{V}(\mathbf{k}) \\ \hat{V}(\mathbf{k})^\dagger & \hat{\varepsilon}_{\mathrm{c}}(\mathbf{k}) \end{pmatrix}. \tag{6.6}$$

The 2×2 matrices $\hat{\varepsilon}_{\mathrm{f}}(\mathbf{k})$, $\hat{\varepsilon}_{\mathrm{c}}(\mathbf{k})$, and $\hat{V}(\mathbf{k})$ are obtained as

$$\hat{\varepsilon}_{\mathrm{f}}(\mathbf{k}) = \begin{pmatrix} \varepsilon_{\mathrm{f}}(\mathbf{k}) & \alpha_1(\mathrm{i}\sin k_{\mathrm{x}} + \sin k_{\mathrm{y}}) \\ \alpha_1(-\mathrm{i}\sin k_{\mathrm{x}} + \sin k_{\mathrm{y}}) & \varepsilon_{\mathrm{f}}(\mathbf{k}) \end{pmatrix}, \tag{6.7}$$

$$\hat{\varepsilon}_{\mathrm{c}}(\mathbf{k}) = \begin{pmatrix} \varepsilon_{\mathrm{c}}(\mathbf{k}) & 0 \\ 0 & \varepsilon_{\mathrm{c}}(\mathbf{k}) \end{pmatrix}, \tag{6.8}$$

$$\hat{V}(\vec{\mathbf{k}}) = \begin{pmatrix} (8V_3 \sin k_z + 4\mathrm{i}\epsilon V_4) \sin k_x \sin k_y & (2\mathrm{i}V_5 - 4\eta V_6 \sin k_z)(\sin k_x + \mathrm{i} \sin k_y) \\ (2\mathrm{i}V_5 - 4\eta V_6 \sin k_z)(\sin k_x - \mathrm{i} \sin k_y) & (8V_3 \sin k_z + 4\mathrm{i}\epsilon V_4) \sin k_x \sin k_y \end{pmatrix}. \tag{6.9}$$

We ignored the off-diagonal terms in the second order with respect to small parameters ϵ and η. We obtain $\varepsilon_f(\mathbf{k}) = \kappa^2 \varepsilon_{\Gamma_7}(\mathbf{k}) + \epsilon^2 \varepsilon_{xy}(\mathbf{k}) + \eta^2(\varepsilon_{xz}(\mathbf{k}) + \varepsilon_{yz}(\mathbf{k}))$ where $\varepsilon_A(\mathbf{k})$ is the dispersion relation for the $|A\rangle$ state. In Eq. (6.8), we ignored the SO interaction in the conduction electrons to focus on f electron physics. It is straightforward to take into account the SO interaction in conduction electrons, and also to construct a model for conduction electrons in p and d orbitals.

Equation (6.7) has the Rashba SO interaction term and the coefficient is obtained as

$$\alpha_1 = -4\epsilon V_2 - 4\eta V_1. \tag{6.10}$$

The hybridization parameters in Eqs. (6.9) and (6.10) are obtained as

$$V_1 = \kappa \sqrt{\frac{5}{21}} V^{100}_{2-,yz}, \tag{6.11}$$

$$V_2 = \kappa \left(\sqrt{\frac{15}{42}} V^{100}_{y^3-3x^2y,yz} - \sqrt{\frac{1}{42}} V^{100}_{y(5z^2-r^2),yz} \right), \tag{6.12}$$

$$V_3 = \kappa \sqrt{\frac{5}{21}} V^{111}_{2-,s}, \tag{6.13}$$

$$V_4 = V^{110}_{xy,s}, \tag{6.14}$$

$$V_5 = \kappa \left(\sqrt{\frac{15}{42}} V^{100}_{x^3-3xy^2,s} - \sqrt{\frac{1}{42}} V^{100}_{x(5z^2-r^2),s} \right), \tag{6.15}$$

$$V_6 = V^{101}_{xz,s}, \tag{6.16}$$

where $V^{abc}_{A,B}$ is the hopping matrix element between the $|A\rangle$ and $|B\rangle$ states along the [abc]-axis.

Applying an appropriate unitary transformation to the conduction electron, $(c^\dagger_{\mathbf{k}+}, c^\dagger_{\mathbf{k}-}) = (c^\dagger_{\mathbf{k}\uparrow}, c^\dagger_{\mathbf{k}\downarrow})\hat{U}_c(\mathbf{k})$, the hybridization matrix is transformed as

$$\tilde{V}(\mathbf{k}) = \hat{V}(\mathbf{k})\hat{U}_c(\mathbf{k}) = \begin{pmatrix} V_{cf}(\mathbf{k}) & \alpha_2(\mathbf{k})(\mathrm{i}\sin^2 k_y \sin k_x + \sin^2 k_x \sin k_y) \\ \alpha_2(\mathbf{k})(-\mathrm{i}\sin^2 k_y \sin k_x + \sin^2 k_x \sin k_y) & V_{cf}(\mathbf{k}) \end{pmatrix}, \tag{6.17}$$

where

$$\alpha_2(\mathbf{k}) = 4(\epsilon V_4 V_5 - 4\eta V_3 V_6 \sin^2 k_z) \Big/ \sqrt{16V_3^2 \sin^2 k_x \sin^2 k_y \sin^2 k_z + V_5^2(\sin^2 k_x + \sin^2 k_y)}, \tag{6.18}$$

is a real and even function with respect to k_x, k_y, and k_z.

Taking into account the on-site repulsion in the $|\mathrm{f}\pm\rangle$ state, we obtain the periodic Anderson model with a Rashba SO interaction as

$$H = H_k + H_{SO} + H_I, \tag{6.19}$$

$$H_k = \sum_{k,s=\pm} \varepsilon_f(\mathbf{k}) f^{\dagger}_{\mathbf{k},s} f_{\mathbf{k},s} + \sum_{k,s=\pm} \varepsilon_c(\mathbf{k}) c^{\dagger}_{\mathbf{k},s} c_{\mathbf{k},s} + \sum_{k,s=\pm} [V_{cf}(\mathbf{k}) f^{\dagger}_{\mathbf{k},s} c_{\mathbf{k},s} + h.c.], \tag{6.20}$$

$$H_{SO} = \alpha_1 \sum_k \mathbf{g}_f(\mathbf{k}) \cdot \mathbf{S}_{ff}(\mathbf{k}) + \sum_k [\alpha_2(\mathbf{k}) \mathbf{g}_{cf}(\mathbf{k}) \cdot \mathbf{S}_{cf}(\mathbf{k}) + h.c.], \tag{6.21}$$

$$H_I = U \sum_i n^f_{i,+} n^f_{i,-}, \tag{6.22}$$

where $\mathbf{S}_{ff}(\mathbf{k}) = \sum_{s,s'} \boldsymbol{\sigma}_{ss'} f^{\dagger}_{\mathbf{k},s} f_{\mathbf{k},s'}$ and $\mathbf{S}_{cf}(\mathbf{k}) = \sum_{s,s'} \boldsymbol{\sigma}_{ss'} f^{\dagger}_{\mathbf{k},s} c_{\mathbf{k},s'}$. $\mathbf{g}_f(\mathbf{k}) = (\sin k_y, -\sin k_x, 0)$ and $\mathbf{g}_{cf}(\mathbf{k}) = (\sin^2 k_x \sin k_y, -\sin^2 k_y \sin k_x, 0)$ describe the g-vector for the intra- and inter-orbital Rashba SO interactions, respectively. The coefficients α_1 and $\alpha_2(\mathbf{k})$ have been obtained in Eqs. (6.10) and (6.18), respectively. The parameters ϵ and η vanish, and therefore α_1 and $\alpha_2(\mathbf{k})$ disappear in centrosymmetric systems.Thus, the Rashba SO interaction is generally derived in the tight-binding approximation by taking into account the parity mixing in the atomic states.

Note that the SO interaction arises from the combination of the atomic *LS* coupling and the parity mixing in the localized Ce 4*f* state. The inter-site hybridization between the *f* and admixed *d* (or *s*) orbitals gives rise to the intra-orbital SO interaction, while the inter-orbital SO interaction is induced by the hybridization between the conduction electrons and admixed *d*- (or *s*-)orbitals. Note that the cubic term $\propto \sin k_x \sin k_y \sin k_z (\sin^2 k_x - \sin^2 k_y)\sigma_z$ [22] does not appear in the above derivation.

The periodic Anderson model for the localized $|\Gamma_6'\pm\rangle = |\pm\frac{1}{2}\rangle$ state is derived in the same way. Then, we obtain the Hamiltonian that is similar to Eq. (6.19), but the g-vector for the inter-orbital SO interaction is replaced with $\mathbf{g}_{cf}(\mathbf{k}) = (\sin k_y, -\sin k_x, 0) = \mathbf{g}_f(\mathbf{k})$. Thus, the momentum dependence of g-vector depends on the symmetry of low-lying localized state and conduction electrons.

The kinetic energy term H_k in the periodic Anderson model is diagonalized by the unitary transformation $(a^{\dagger}_{1,\mathbf{k}\pm}, a^{\dagger}_{2,\mathbf{k}\pm}) = (f^{\dagger}_{\mathbf{k}}\pm, c^{\dagger}_{\mathbf{k}\pm})\hat{U}_{cf}(\mathbf{k})$ with

$$\hat{U}_{cf}(\mathbf{k}) = \begin{pmatrix} a_1(\mathbf{k}) & a_2^*(\mathbf{k}) \\ a_2(\mathbf{k}) & -a_1(\mathbf{k}) \end{pmatrix}. \tag{6.23}$$

Applying this unitary transformation to the periodic Anderson model in Eq. (6.19) and dropping the upper band described by $a^{\dagger}_{2,\mathbf{k}\pm}$, we obtain the single-orbital model with a Rashba SO interaction. Then, the g-vector is obtained as

$$\alpha \mathbf{g}(\mathbf{k}) = \alpha_1 a_1(\mathbf{k})^2 \mathbf{g}_{\mathrm{f}}(\mathbf{k}) + \alpha_2(\mathbf{k}) a_1(\mathbf{k})(a_2(\mathbf{k}) + a_2^*(\mathbf{k}))\mathbf{g}_{\mathrm{cf}}(\mathbf{k}). \tag{6.24}$$

The unitary transformation described by $\hat{U}_{\mathrm{cf}}(\mathbf{k})$ leads to the momentum dependence of the two-body interaction term H_{I}, as in the case of the multi-orbital Hubbard model [23]. By neglecting this momentum dependence for simplicity, we obtain the single-orbital Hubbard model

$$H = \sum_{k,s} \varepsilon(\mathbf{k}) a^{\dagger}_{\mathbf{k},s} a_{\mathbf{k},s} + \alpha \sum_{k} \mathbf{g}(\mathbf{k}) \cdot \mathbf{S}(\mathbf{k}) + U \sum_{i} n_{i,\uparrow} n_{i,\downarrow}. \tag{6.25}$$

As shown in Eqs. (6.18) and (6.24), the g-vector of SO interaction has a complicated momentum dependence, and therefore, an often-used assumption $g(\mathbf{k}) = (k_{\mathrm{y}}, -k_{\mathrm{x}}, 0)$ or $g(\mathbf{k}) = (\sin k_{\mathrm{y}}, -\sin k_{\mathrm{x}}, 0)$ is not justified. In subsection 6.2.3.3, we show a phenomenon which is induced by the complicated momentum dependence of Rashba SO interaction.

6.2.2 Microscopic Model and Approach

In this section we discuss the pairing state of $\mathrm{CePt_3Si}$ on the basis of the minimal model. We here introduce a single-orbital Hubbard model including the molecular field arising from the AF order

$$H = \sum_{k,s} \varepsilon(\mathbf{k}) c^{\dagger}_{\mathbf{k},s} c_{\mathbf{k},s} + \alpha \sum_{k} \mathbf{g}(\mathbf{k}) \cdot \mathbf{S}(\mathbf{k}) - \sum_{k} \mathbf{h}_{\mathrm{Q}} \cdot \mathbf{S}_{\mathrm{Q}}(\mathbf{k}) + U \sum_{i} n_{i,\uparrow} n_{i,\downarrow}, \tag{6.26}$$

where $\mathbf{S}(\mathbf{k}) = \sum_{s,s'} \boldsymbol{\sigma}_{ss'} c^{\dagger}_{\mathbf{k},s} c_{\mathbf{k},s'}$ and $\mathbf{S}_{\mathrm{Q}}(\mathbf{k}) = \sum_{s,s'} \boldsymbol{\sigma}_{ss'} c^{\dagger}_{\mathbf{k}+\mathbf{Q},s} c_{\mathbf{k},s'}$.

We consider a simple tetragonal lattice and assume the dispersion relation

$$\begin{aligned}\varepsilon(\mathbf{k}) = {} & 2t_1(\cos k_{\mathrm{x}} + \cos k_{\mathrm{y}}) + 4t_2 \cos k_{\mathrm{x}} \cos k_{\mathrm{y}} + 2t_3(\cos 2k_{\mathrm{x}} + \cos 2k_{\mathrm{y}}) \\ & + [2t_4 + 4t_5(\cos k_{\mathrm{x}} + \cos k_{\mathrm{y}}) + 4t_6(\cos 2k_{\mathrm{x}} + \cos 2k_{\mathrm{y}})] \cos k_{\mathrm{z}} \\ & + 2t_7 \cos 2k_{\mathrm{z}} - \mu,\end{aligned} \tag{6.27}$$

where the chemical potential μ is included. We determine the chemical potential μ so that the electron density per site is n. By choosing the parameters as $(t_1, t_2, t_3, t_4, t_5, t_6, t_7, n) = (1, -0.15, -0.5, -0.3, -0.1, -0.09, -0.2, 1.75)$, the dispersion relation Eq. (6.27) reproduces the Fermi surface of β-band in $\mathrm{CePt_3Si}$, which has been reported by band structure calculation without the AF order [24–26]. Fermi surfaces of this tight-binding model are depicted in Fig. 6.1. Thus, we assume that the superconductivity in $\mathrm{CePt_3Si}$ is mainly induced by the β-band because the

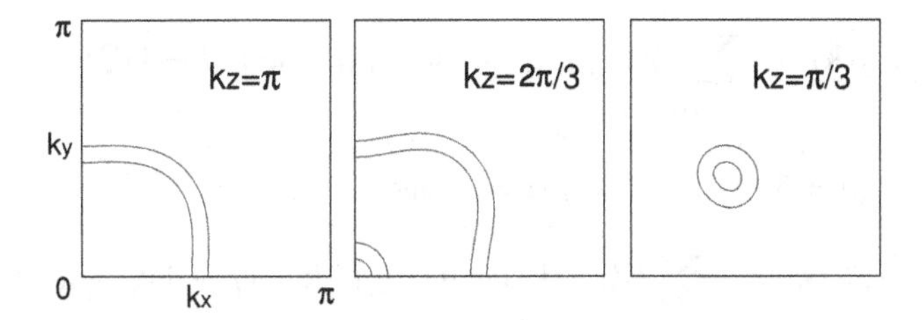

Fig. 6.1 Fermi surfaces of the model Eq. (6.26). We assume $\alpha = 0.3$, $U = 0$ and $h_{\rm Q} = 0$. The cross sections at $k_z = \pi$, $k_z = \frac{2\pi}{3}$ and $k_z = \frac{\pi}{3}$ are shown from the *left* to the *right*. (Ref. [18])

β-band has a substantial Ce 4*f*-electron character [24] and the largest density of states (DOS), namely 70% of the total DOS [25].

For the Rashba SO interaction, we assume a simple model for the g-vector $\mathbf{g}(\mathbf{k}) = (-v_{\rm y}(\mathbf{k}), v_{\rm x}(\mathbf{k}), 0)/\bar{v}$, where $v_{\rm x,y}(\mathbf{k}) = \partial\varepsilon(\mathbf{k})/\partial k_{\rm x,y}$, to study the deviation from the often-used form $\mathbf{g}(\mathbf{k}) = (\sin k_{\rm y}, -\sin k_{\rm x}, 0)$. The form of the g-vector hardly affects the microscopic mechanism of superconductivity and the pairing symmetry. However, some intriguing properties can be induced by the g-vector adopted here. We choose the coupling constant $\alpha = 0.3$ so that the band splitting due to Rashba SO interaction is consistent with the band structure calculations [25].

The AF order enters in our model through the staggered field $\mathbf{h}_{\rm Q}$ without discussing its microscopic origin. The experimentally determined AF order corresponds to $\mathbf{h}_{\rm Q} = h_{\rm Q}\hat{x}$ pointing in the [100] direction with a wave vector $\mathbf{Q} = (0, 0, \pi)$ [10]. For the magnitude, we choose $|h_{\rm Q}| \ll W$ where W is the bandwidth since the observed AF moment $\sim 0.16\mu_{\rm B}$ [10] is considerably less than the full moment of the $J = 5/2$ manifold in the Ce ion.

The undressed Green functions for $U = 0$ are represented by the matrix form $\hat{G}(\mathbf{k}, {\rm i}\omega_n) = ({\rm i}\omega_n \hat{1} - \hat{H}(\mathbf{k}))^{-1}$, where

$$\hat{G}(\mathbf{k}, {\rm i}\omega_n) = \begin{pmatrix} \hat{G}^1(\mathbf{k}, {\rm i}\omega_n) & \hat{G}^2(\mathbf{k}, {\rm i}\omega_n) \\ \hat{G}^2(\mathbf{k}+\mathbf{Q}, {\rm i}\omega_n) & \hat{G}^1(\mathbf{k}+\mathbf{Q}, {\rm i}\omega_n) \end{pmatrix}, \tag{6.28}$$

and

$$\hat{H}(\mathbf{k}) = \begin{pmatrix} \hat{e}(\mathbf{k}) & -h_{\rm Q}\hat{\sigma}^{(\rm x)} \\ -h_{\rm Q}\hat{\sigma}^{(\rm x)} & \hat{e}(\mathbf{k}+\mathbf{Q}) \end{pmatrix}, \tag{6.29}$$

with $\hat{e}(\mathbf{k}) = \varepsilon(\mathbf{k})\hat{\sigma}^{(0)} + \alpha\mathbf{g}(\mathbf{k})\boldsymbol{\sigma}$. The normal and anomalous Green functions $\hat{G}^i(\mathbf{k}, {\rm i}\omega_n)$ are the 2×2 matrix in spin space, where $\omega_n = (2n+1)\pi T$ and T is the temperature.

We study the superconducting instability which we assume to arise through electron–electron interaction incorporated in the effective on-site repulsion U. The linearized Éliashberg equation is obtained by the standard procedure:

$$\lambda\Delta_{p,s_1,s_2}(\mathbf{k}) = -\sum_{\mathrm{k}',\mathrm{q},\mathrm{s}_3,\mathrm{s}_4} V_{p,q,s_1,s_2,s_3,s_4}(\mathbf{k}, \mathbf{k}')\psi_{q,s_3,s_4}(\mathbf{k}'), \tag{6.30}$$

$$\psi_{p,s_1,s_2}(\mathbf{k}) = \sum_{i,j,s_3,s_4} \phi_{p,i,j,s_1,s_3,s_2,s_4}(\mathbf{k}) \Delta_{q,s_3,s_4}(\mathbf{k} + (i-1)\mathbf{Q}), \tag{6.31}$$

where $q = p$ ($q = 3 - p$) for $i = j$ ($i \neq j$) and

$$\phi_{p,i,j,s_1,s_2,s_3,s_4}(\mathbf{k}) = T \sum_n G^i_{s_1,s_2}(\mathbf{k}, \mathrm{i}\omega_n) G^j_{s_3,s_4}(-\mathbf{k} + (p-1)\mathbf{Q}, -\mathrm{i}\omega_n), \tag{6.32}$$

for $p = 1, 2$. Here, we adopt the so-called weak-coupling theory of superconductivity and ignore self-energy corrections and the frequency dependence of effective interaction [27, 28]. This simplification strongly affects the resulting transition temperature but hardly affects the pairing symmetry[29]. We optimize the momentum dependence of order parameters $\Delta_{p,s_1,s_2}(\mathbf{k})$ to study the mixing of singlet and triplet pairings due to Rashba SO interaction. Here, $\Delta_{1,s_1,s_2}(\mathbf{k})$ and $\Delta_{2,s_1,s_2}(\mathbf{k})$ describe the Cooper pairing with the total momenta $(0, 0, 0)$ and $(0, 0, \pi)$, respectively. The former is the order parameter for ordinary Cooper pairs, while the latter is that for π-singlet and π-triplet pairs [30–35].

We assume that the paring interaction $V_{p,q,s_1,s_2,s_3,s_4}(\mathbf{k}, \mathbf{k}')$ originates from spin fluctuations that we describe within the RPA [27, 28] according to the diagrammatic expression shown in Fig. 6.2. We obtain

$$V_{1,1,s1,s2,s3,s4}(\mathbf{k}, \mathbf{k}') = -[\hat{U}' \hat{\chi}_1(\mathbf{k}' - \mathbf{k}) \hat{U}']_{s3,s1,s4,s2} + \hat{U}_{s1,s2,s3,s4}, \tag{6.33}$$

$$V_{1,2,s1,s2,s3,s4}(\mathbf{k}, \mathbf{k}') = -[\hat{U}' \hat{\chi}_2(\mathbf{k}' - \mathbf{k}) \hat{U}']_{s3,s1,s4,s2}, \tag{6.34}$$

$\hat{V}_{2,2}(\mathbf{k}, \mathbf{k}') = \hat{V}_{1,1}(\mathbf{k}, \mathbf{k}')$ and $\hat{V}_{2,1}(\mathbf{k}, \mathbf{k}') = \hat{V}_{1,2}(\mathbf{k}, \mathbf{k}')$. The generalized susceptibility is expressed in the matrix form,

$$\begin{pmatrix} \hat{\chi}_1(\mathbf{q}) \\ \hat{\chi}_2(\mathbf{q}_+) \end{pmatrix} = \begin{pmatrix} \hat{1} - \hat{\chi}_1^{(0)}(\mathbf{q})\hat{U} & -\hat{\chi}_2^{(0)}(\mathbf{q})\hat{U} \\ -\hat{\chi}_2^{(0)}(\mathbf{q}_+)\hat{U} & \hat{1} - \hat{\chi}_1^{(0)}(\mathbf{q}_+)\hat{U} \end{pmatrix}^{-1} \begin{pmatrix} \hat{\chi}_1^{(0)}(\mathbf{q}) \\ \hat{\chi}_2^{(0)}(\mathbf{q}_+) \end{pmatrix}, \tag{6.35}$$

where $\mathbf{q}_+ = \mathbf{q} + \mathbf{Q}$. The matrix element of the bare susceptibility $\hat{\chi}_i^{(0)}(\mathbf{q})$ is expressed as

$$\begin{aligned} \chi^{(0)}_{1,s1,s2,s3,s4}(\mathbf{q}) = -T \sum_{\mathbf{k},\omega_n} [&G^1_{s4,s1}(\mathbf{k} + \mathbf{q}, \mathrm{i}\omega_n) G^1_{s2,s3}(\mathbf{k}, \mathrm{i}\omega_n) \\ &+ G^2_{s4,s1}(\mathbf{k} + \mathbf{q}, \mathrm{i}\omega_n) G^2_{s2,s3}(\mathbf{k} + \mathbf{Q}, \mathrm{i}\omega_n)], \end{aligned} \tag{6.36}$$

$$\begin{aligned} \chi^{(0)}_{2,s1,s2,s3,s4}(\mathbf{q}) = -T \sum_{\mathbf{k},\omega_n} [&G^1_{s4,s1}(\mathbf{k} + \mathbf{q}, \mathrm{i}\omega_n) G^2_{s2,s3}(\mathbf{k}, \mathrm{i}\omega_n) \\ &+ G^2_{s4,s1}(\mathbf{k} + \mathbf{q}, \mathrm{i}\omega_n) G^1_{s2,s3}(\mathbf{k} + \mathbf{Q}, \mathrm{i}\omega_n)]. \end{aligned} \tag{6.37}$$

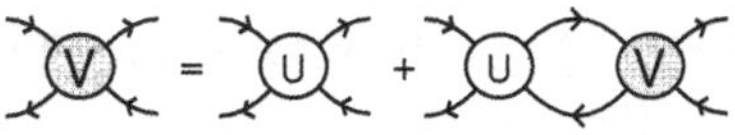

Fig. 6.2 Diagrammatic representation of the pairing interaction. The *white* circle represents the on-site interaction U. (Ref. [20])

We denote the element of 4×4 matrix, such as $\hat{\chi}_i^{(0)}(\mathbf{q})$ and $\hat{\chi}_i(\mathbf{q})$, using the spin indices as

$$\hat{A} = \begin{pmatrix} A_{\uparrow\uparrow\uparrow\uparrow} & A_{\uparrow\uparrow\uparrow\downarrow} & A_{\uparrow\uparrow\downarrow\uparrow} & A_{\uparrow\uparrow\downarrow\downarrow} \\ A_{\uparrow\downarrow\uparrow\uparrow} & A_{\uparrow\downarrow\uparrow\downarrow} & A_{\uparrow\downarrow\downarrow\uparrow} & A_{\uparrow\downarrow\downarrow\downarrow} \\ A_{\downarrow\uparrow\uparrow\uparrow} & A_{\downarrow\uparrow\uparrow\downarrow} & A_{\downarrow\uparrow\downarrow\uparrow} & A_{\downarrow\uparrow\downarrow\downarrow} \\ A_{\downarrow\downarrow\uparrow\uparrow} & A_{\downarrow\downarrow\uparrow\downarrow} & A_{\downarrow\downarrow\downarrow\uparrow} & A_{\downarrow\downarrow\downarrow\downarrow} \end{pmatrix}. \tag{6.38}$$

The matrix $\hat{U}$ is expressed as

$$\hat{U} = \begin{pmatrix} 0 & 0 & 0 & -U \\ 0 & 0 & U & 0 \\ 0 & U & 0 & 0 \\ -U & 0 & 0 & 0 \end{pmatrix}. \tag{6.39}$$

The superconducting T_c is determined by the criterion $\lambda = 1$. We here determine the leading instability to the superconductivity at $T = 0.02$, where the maximum eigenvalue is $\lambda = 0.3 \sim 0.6$. The pairing state below T_c is captured by this simplified procedure because the momentum and spin dependences of order parameter $\Delta_{p,s_1,s_2}(\mathbf{k})$ are nearly independent of the temperature.

6.2.3 Microscopic Theory for Magnetism and Superconductivity

In this subsection, we study the magnetism and superconductivity in $CePt_3Si$ from the microscopic point of view [18, 20]. To illustrate the basic properties, we ignore the AF order and assume $h_Q = 0$ in this section. While some properties are affected by the AF order as shown in the next Sect. 6.2.4, the mechanism and symmetry of superconductivity are not altered.

6.2.3.1 Helical Spin Fluctuations

Helical magnetism is one of the characteristic properties of non-centrosymmetric systems [36]. The RPA adopted in our calculation captures the helical anisotropy of spin fluctuations. The spin susceptibility tensor $\hat{\chi}(\mathbf{q})$ is expressed by the generalized susceptibility in Eq. (6.35) as

$$\chi^{\mu\nu}(\mathbf{q}) = \sum_{s1,s2,s3,s4} \sigma^{\mu}_{s1,s2}\chi_{1,s1,s2,s3,s4}(\mathbf{q})\sigma^{\nu}_{s3,s4}. \tag{6.40}$$

In our model Eq. (6.26), the spin susceptibility shows a peak around $\mathbf{q} = \mathbf{Q} = (0, 0, \pi)$. The helical spin fluctuation is described by the spin susceptibility tensor around $\mathbf{q} = \mathbf{Q}$, which is approximated to first order of SO interaction α as

$$\hat{\chi}(\mathbf{q}) = \begin{pmatrix} \chi^{\mathrm{d}}(\mathbf{q}) & 0 & \mathrm{i}\alpha B q_{\mathrm{x}} \\ 0 & \chi^{\mathrm{d}}(\mathbf{q}) & \mathrm{i}\alpha B q_{\mathrm{y}} \\ -\mathrm{i}\alpha B q_{\mathrm{x}} & -\mathrm{i}\alpha B q_{\mathrm{y}} & \chi^{\mathrm{d}}(\mathbf{q}) \end{pmatrix}, \tag{6.41}$$

where $\chi^{\mathrm{d}}(\mathbf{q}) = \chi(\mathbf{Q}) - a|\mathbf{q}_{\parallel}|^2 - c q_{\mathrm{z}}^2$ with $(\mathbf{q}_{\parallel}, q_{\mathrm{z}}) = \mathbf{q} - \mathbf{Q}$. This form can be viewed as a result of the Dzyaloshinski–Moriya-type interaction $H_{\mathrm{DM}} = \sum_{\mathbf{q}} \mathrm{i} D(\mathbf{q}) \cdot S(\mathbf{q}) \times S(-\mathbf{q})$ [37, 38] with $D(\mathbf{q}) \propto \alpha \hat{z} \times \mathbf{q}$. Owing to the Rashba SO interaction, the spin susceptibility tensor has the maximum eigenvalue at $\mathbf{q}_{\parallel} \neq 0$ with an eigenvector $\mathbf{S}(\mathbf{q}) = \frac{1}{\sqrt{2}}(\tilde{q}_{\mathrm{x}}, \tilde{q}_{\mathrm{y}}, \pm \mathrm{i})$, where $\tilde{q}_{\mathrm{x,y}} = q_{\mathrm{x,y}}/|\mathbf{q}_{\parallel}|$. This mode describes the fluctuations of cycloid-type helical magnetism along $\mathbf{q}_{\parallel}$. Thus, the fluctuation of helical magnetism is appropriately described by the RPA. More detailed discussion on the anisotropic spin fluctuations is given in Ref. [39].

6.2.3.2 Pairing States

We examine here the pairing state realized in the model Eq. (6.26). We do not adopt the approximated form of the spin susceptibility in Eq. (6.41), but numerically calculate Eqs. (6.35–6.37) in the following part.

Two pairing states are stabilized by the pairing interaction mediated by the helical spin fluctuation discussed in the previous subsection. One is the $s + P$-wave state, in which the order parameter has the leading odd-parity component $\mathbf{d}(\mathbf{k}) \sim (-\sin k_{\mathrm{y}}, \sin k_{\mathrm{x}}, 0)$ and the admixed even-parity part $\Phi(\mathbf{k}) \sim \delta + \cos k_{\mathrm{x}} + \cos k_{\mathrm{y}}$ with $\delta \sim 0.2$. Here and in the following, the leading angular momentum contribution is written in the capital letter. The $s + P$-wave symmetry has been proposed in Refs. [3, 40, 41]. In contrast to these literature, the spin-singlet component is viewed as an extended s-wave pairing in our calculation, where the order parameter changes its sign along the radial direction in order to avoid the local repulsive interaction U. The extended s-wave pairing in the admixed singlet component can be distinguished from the conventional s-wave pairing by the NMR $1/T_1T$. The coherence peak appears in $1/T_1T$ just below T_{c} for the latter [40, 42]. On the other hand, the coherence peak almost vanishes in case of the extended s-wave pairing [20]. Recent NMR measurements show no coherence peak in $1/T_1T$ [15] and therefore consistent with the extended s-wave pairing.

Note that the often-assumed relation $\mathbf{d}(\mathbf{k}) \parallel \mathbf{g}(\mathbf{k})$ is not satisfied in the entire Brillouin zone. This is a general consequence since the g-vector has a complicated momentum dependence as shown in Sect. 6.2.1. The momentum dependence of d-vector is mainly determined by the paring interaction, which favors the short-range

Cooper pairing and leads to the simple form $\mathbf{d}(\mathbf{k}) \sim (-\sin k_y, \sin k_x, 0)$ in our calculation. In the next section, we show that the mismatch of d-vector and g-vector gives rise to the line node of superconducting gap.

The other stable solution is the predominantly d-wave state that can be viewed as an interlayer Cooper pairing state: $\Phi(\mathbf{k}) \sim \{\sin k_x \sin k_z, \sin k_y \sin k_z\}$ (two-fold degenerate) admixed with an odd-parity component $\mathbf{d}(\mathbf{k}) \sim \Phi(\mathbf{k})(\sin k_y, \sin k_x, 0)$. In the paramagnetic phase, the most stable combination of the two degenerate states is chiral: $\Phi_{\pm}(\mathbf{k}) \sim (\sin k_x \pm i \sin k_y) \sin k_z$ which gains the maximal condensation energy in the weak-coupling approach. Since the spin-triplet order parameter has both the p-wave and f-wave components, we denote this state as the $p+D+f$-wave state.

Among these states, the $s+P$-wave state is stable for a small U, while the $p+D+f$-wave state is favored by a large U [20]. This is because the dimensionality of spin fluctuation changes as increasing U. For a small (large) U, the spin fluctuation has a two-dimensional (three-dimensional) nature. The intra-plane nearly ferromagnetic correlation induces the $s+P$-wave superconductivity, while the AF inter-plane coupling induced by a large U stabilizes the $p+D+f$-wave state.

Next we discuss the roles of Rashba SO interaction on the pairing state. It has been argued that the SO interaction leads to the parity mixing. In our case, the mixing of spin-singlet and -triplet components is closely related to the helical spin fluctuation discussed in Sect. 6.2.3.1. Since the helical anisotropy of spin fluctuations is pronounced around $\mathbf{q} = \mathbf{Q}$ and the spin susceptibility is enhanced by the electron correlation there, a significant mixing of order parameters occurs. We obtain $|\Phi(\mathbf{k})|_{max}/|\vec{d}(\mathbf{k})|_{max} = 0.2 \sim 0.3$ in the $s+P$-wave state, and $|\mathbf{d}(\mathbf{k})|_{max}/|\Phi(\mathbf{k})|_{max} = 0.2 \sim 0.3$ in the $p+D+f$-wave state for $\alpha = 0.3$, where the suffix $||_{max}$ means the maximum in the Brillouin zone.

The Rashba SO interaction also affects the stability of pairing states. Figure 6.3 shows the α dependence of eigenvalues λ for the $s+P$-wave and $p+D+f$-wave states. A large eigenvalue λ of Éliashberg equation indicates a high T_c. We see the concave structure of λ for the $s+P$-wave state. A small Rashba SO interaction destabilizes the $s+P$-wave state. Although the depairing effect due to the Rashba SO interaction is almost avoided in the $s+P$-wave state with $\mathbf{d}(\mathbf{k}) \sim (-\sin k_y, \sin k_x, 0)$, it does not vanish owing to the mismatch of d-vector and g-vector. This is the reason why λ decreases with increasing the SO interaction for $\alpha < 0.4$. On the other hand, λ for the $s+P$-wave state increases with α for $\alpha > 0.4$. This is because the helical anisotropy of spin fluctuations enhances the predominantly spin-triplet pairing state. If we assume an isotropic spin fluctuation, λ for the $s+P$-wave state monotonically decreases with increasing α, as shown by the dashed line in Fig. 6.3. This means that the enhancement of λ for $\alpha > 0.4$ is due to the helical spin fluctuation induced by the SO interaction. In contrast to the $s+P$-wave state, the $p+D+f$-wave state is not favored by the helical spin fluctuations as shown in Fig. 6.3. Thus, the predominantly spin-triplet pairing state can be stabilized by a large SO interaction near the AF critical point.

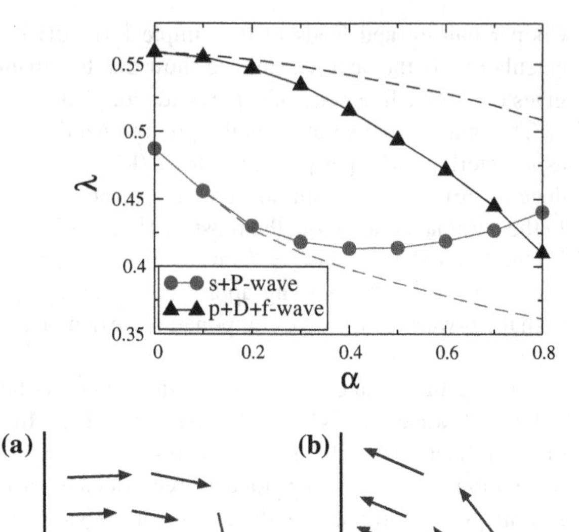

Fig. 6.3 Eigenvalues of Éliashberg equation λ for the $s+P$-wave (*circles*) and $p+D+f$-wave states (*triangles*). We assume $U=4$ and $h_{\mathrm{Q}}=0$. Éliashberg equation is solved in the $128 \times 128 \times 32$ lattice. The *dashed lines* show λ which is estimated by using the generalized susceptibility $\hat{\chi}_i(\mathbf{q})$ for $\alpha=0$. (Ref. [20])

Fig. 6.4 Schematic sketch of **a** the g-vector and **b** the d-vector around $\mathbf{k}=\mathbf{k}_0$. The *arrows* show the direction of vectors. A topological defect of g-vector is shown by the *red circle* in **a**

6.2.3.3 Topologically Protected Accidental Line Node

We have pointed out that the often assumed relation $\mathbf{d}(\mathbf{k}) \parallel \mathbf{g}(\mathbf{k})$ [3, 43, 44] is not satisfied in the entire Brillouin zone. While the g-vector generally has a complicated momentum dependence as shown in Sect. 6.2.1, the d-vector generally has a simple momentum dependence dominated by the short range Cooper pairing. Such a mismatch occurs because the momentum dependence of the vector in an irreducible representation of the point group is not unique. As a consequence, some intriguing properties appear. For instance, we here show that the line node of the superconducting gap is induced by the topological properties of the d-vector and g-vector . Strictly speaking, the excitation gap is not zero on this line node, but is negligible since the minimum of the gap is in the order of 10^{-5} of the maximum gap. This line node is not protected by symmetry, but robust against any perturbations because of the topological properties.

We here assume that the g-vector has some topological defects in the k-space, where $\mathbf{g}(\mathbf{k}_0)=0$ with $\mathbf{k}_0 \neq 0$, and the d-vector has no nontrivial topological defect except for $(k_{\mathrm{x}}, k_{\mathrm{y}})=(0,0)$, $(0,\pi)$, $(\pi,0)$, and (π,π). This is the case of our calculation, where $\mathbf{g}(\mathbf{k})=(-v_{\mathrm{y}}(\mathbf{k}), v_{\mathrm{x}}(\mathbf{k}), 0)/\bar{v}$ is assumed. A schematic sketch of the d-vector and g-vector in our case is shown in Fig. 6.4. Since the difference of winding number around $\mathbf{k}=\mathbf{k}_0$ between the g-vector and d-vector, these vectors have to be orthogonal along the line passing through $\mathbf{k}=\mathbf{k}_0$.

We assume the $|\alpha| \gg |\mathbf{d}(\mathbf{k})|_{\max}$ for simplicity. Then, the superconducting gaps in Fermi surfaces split by the SO interaction are expressed as $\Delta_{\pm}(\mathbf{k})=\pm\Phi(\mathbf{k})+\mathbf{d}(\mathbf{k})\cdot$

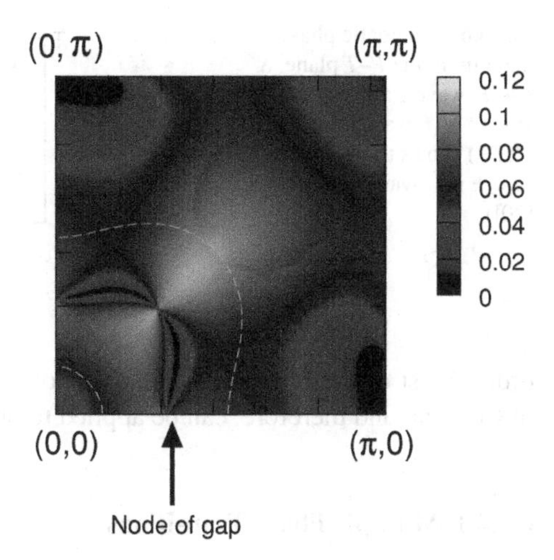

Fig. 6.5 Superconducting gap $|\Delta_+(\mathbf{k})|$ in the $s+P$-wave state at $k_z = \frac{2\pi}{3}$. Node of the gap arising from the topological defect of g-vector is shown by the *arrow*. Although the zeros of superconducting gap do not intersect with the Fermi surface at $k_z = \frac{2\pi}{3}$ (*thin dashed lines*), the line nodes exist on the three-dimensional Fermi surface at another k_z. (Ref. [20])

$\mathbf{g}(\mathbf{k})/|\mathbf{g}(\mathbf{k})|$. Both gaps have a line node around the line passing through $\mathbf{k} = \mathbf{k}_0$, when the leading order parameter is the spin-triplet component and the admixed part $|\Phi(\mathbf{k})|$ is much less than $|\mathbf{d}(\mathbf{k})|$. For instance, we show the superconducting gap $|\Delta_+(\mathbf{k})|$ at $k_z = \frac{2\pi}{3}$ in Fig. 6.5. We see the line node of the gap around a topological defect of g-vector at $(k_x, k_y) \sim (\pi/3, \pi/3)$.

For a finite $|\mathbf{d}(\mathbf{k})|_{\max}/\alpha$, the combination of the admixed spin-singlet component $|\Phi(\mathbf{k})|$ and the d-vector perpendicular to the g-vector gives rise to a tiny excitation gap. The amplitude of the gap is in the order of $\Delta_{\min} \sim |\Phi(\mathbf{k})||d_\perp(\mathbf{k})|^2/\alpha^2$, where $d_\perp(\mathbf{k}) = \mathbf{d}(\mathbf{k}) \times \tilde{g}(\mathbf{k})|_z$ with $\tilde{g}(\mathbf{k}) = \mathbf{g}(\mathbf{k})/|\mathbf{g}(\mathbf{k})|$. This gap is much smaller than the maximum gap. If we assume $|\mathbf{d}(\mathbf{k})|_{\max}/\alpha \sim 0.01$, we obtain $\Delta_{\min}/|\mathbf{d}(\mathbf{k})|_{\max} \sim 10^{-5}$.

Note that this node of the superconducting gap is not protected by the symmetry and, therefore, classified as an accidental line node. However, this node is protected by the topological property of g-vector and therefore robust for the perturbation. This is a general mechanism for the line node in the NCS dominated by the spin-triplet pairing. The topologically protected accidental line node in NCS should be contrasted to the accidental line node in spin-triplet superconductors with inversion symmetry, such as Sr_2RuO_4 [45]. In the latter, the accidental line node disappears for an infinitesimal perturbation.

6.2.4 NCS in AF State

In this section, we discuss the superconducting properties of NCS in the presence of AF order and study the pairing state in $CePt_3Si$ at ambient pressure [18–20]. In all of the presently known NCS in heavy fermion systems, namely, $CePt_3Si$, $CeRhSi_3$, $CeIrSi_3$, and UIr, superconductivity coexists with the magnetic

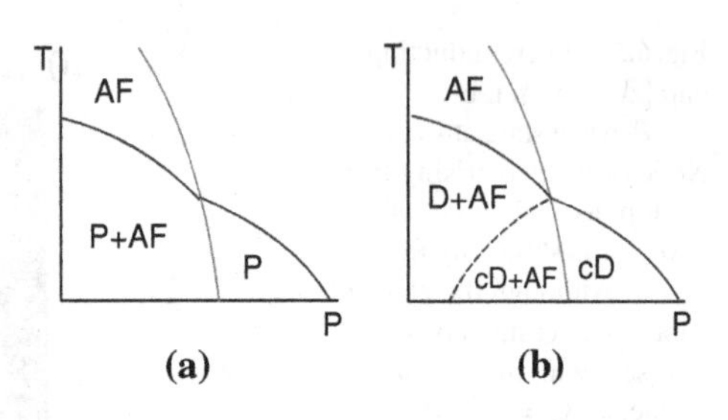

Fig. 6.6 Schematic phase diagram in the *P*–*T* plane. **a** $s + P$-wave state and **b** $p + D + f$-wave state. "D" ("cD") shows the d_{xz}-wave ($d_{xz} \pm i d_{yz}$-wave) state. (Ref. [20])

order. Most of the following results do not rely on the specific electronic structure of $CePt_3Si$, and therefore, can be applied to other compounds as well.

6.2.4.1 Multiple Phase Transitions

We first discuss the possibility of multiple phase transitions. To illustrate our results, we show the phase diagrams in the pressure-temperature (P–T) plane in Fig. 6.6.

Multiple superconducting transitions may occur in the superconductors having multi-component order parameters. The spin-triplet superfluid ^3He [46] and spin-triplet superconductors UPt_3, UBe_{13} [47], and Sr_2RuO_4 [45] actually show multiple phase transitions. On the other hand, multiple phases do not appear in the predominantly spin-triplet pairing state ($s + P$-wave state) in NCS, as shown in Fig. 6.6(a). This is because the six-fold degeneracy in the p-wave state in the absence of Rashba SO interaction is strongly split by the SO interaction. The $s + P$-wave state with $\mathbf{d}(\mathbf{k}) \sim (-\sin k_y, \sin k_x, 0)$ is the most stable, while the other predominantly p-wave state is strongly suppressed. Then, the phase transition between those states does not occur. This feature of NCS should be contrasted with the centrosymmetric spin triplet superconductors. For instance, the splitting of the degeneracy is very small in transition-metal oxides with inversion symmetry [48].

The $p + D + f$-wave state has two-component order parameters. Then, multiple phase transitions occur, as shown in Fig. 6.6(b). The phase transition from the chiral $d_{xz} \pm i d_{yz}$-wave state to the d_{xz}-wave state must occur around the critical pressure $P = P_c$. Thus, the observation of a multiple phase transition in the P–T plane might provide clear evidence of the $p + D + f$-wave state. Although the second superconducting transition has been observed in $CePt_3Si$ [14, 49], that may have an extrinsic origin. There are two superconducting phases with $T_c \sim 0.75$ K and $T_c \sim 0.45$ K owing to the sample inhomogeneity [6, 15, 49, 50]. Thus, the double transition in $CePt_3Si$ may be caused by the sample inhomogeneity.

6.2.4.2 Superconducting Gap

We here investigate the influence of AF order on the gap structure of both $s + P$-wave and $p + D + f$-wave states and discuss the experimental results for $CePt_3Si$ at ambient pressure [6, 13–15].

The quasiparticle spectrum in magnetic superconductor without inversion symmetry is obtained by diagonalizing the 8×8 matrix

$$\hat{H}_{\mathrm{s}}(\mathbf{k}) = \begin{pmatrix} \hat{H}(\mathbf{k}) & -\Delta_0 \hat{\Delta}(\mathbf{k}) \\ -\Delta_0 \hat{\Delta}^{\dagger}(\mathbf{k}) & -\hat{H}(-\mathbf{k})^{\mathrm{T}} \end{pmatrix}, \tag{6.42}$$

where $\hat{\Delta}(\mathbf{k})$ is the order parameter in the spin basis expressed as

$$\hat{\Delta}(\mathbf{k}) = \begin{pmatrix} \Delta_{1,s,s'}(\mathbf{k}) & \Delta_{2,s,s'}(\mathbf{k}) \\ \Delta_{2,s,s'}(\mathbf{k}+\mathbf{Q}) & \Delta_{1,s,s'}(\mathbf{k}+\mathbf{Q}) \end{pmatrix}. \tag{6.43}$$

The matrix element of $\hat{\Delta}(\mathbf{k})$ is determined from the linearized Éliashberg equation Eq. (6.30) by assuming that the momentum and spin dependences of the order parameter are only weakly dependent on temperature for $T \leq T_{\mathrm{c}}$. We choose Δ_0 so that the magnitude of the maximal gap is $\Delta_{\mathrm{g}} = 0.1$ in our energy units. We define the quasiparticle DOS as $\rho(\varepsilon) = \frac{1}{4N} \sum_i \sum'_{\mathbf{k}} \Delta(\varepsilon - E_i(\mathbf{k}))$, where $\sum'_k$ denotes the summation within $|k_{\mathrm{z}}| < \frac{\pi}{2}$. The eigenvalues of Eq. (6.42) $E_8(\mathbf{k}) > E_7(\mathbf{k}) > \cdots > E_1(\mathbf{k})$ satisfy the relation $E_i(\mathbf{k}) = -E_{9-i}(\mathbf{k})$.

To study the superconducting gap, it is more transparent to describe the superconducting order parameter in the band basis, which is obtained by unitary transformation using

$$\hat{\Delta}_{\mathrm{band}}(\mathbf{k}) = \hat{U}^{\dagger}(\mathbf{k}) \hat{\Delta}(\mathbf{k}) \hat{U}^{*}(-\mathbf{k}). \tag{6.44}$$

The unitary matrix $\hat{U}(\mathbf{k})$ diagonalizes the unperturbed Hamiltonian as

$$\hat{U}^{\dagger}(\mathbf{k}) \hat{H}(\mathbf{k}) \hat{U}(\mathbf{k}) = (e_i(\mathbf{k}) \Delta_{ij}). \tag{6.45}$$

The superconducting gap in the γth band is obtained as

$$\Delta_{\gamma}(\mathbf{k}) = \Delta_0 \Psi_{\gamma}(\mathbf{k}), \tag{6.46}$$

where $\Psi_{\gamma}(\mathbf{k})$ is the $(\gamma\gamma)$ component of the matrix $\hat{\Delta}_{\mathrm{band}}(\mathbf{k})$. Since the relation $T_{\mathrm{c}} \ll |\alpha|$ is satisfied in most of the NCS, the relation $|\Delta^{ij}_{\mathrm{band}}(\mathbf{k})| \ll |\alpha \hat{g}(\mathbf{k})|, \max\{h_{\mathrm{Q}}, |\varepsilon(\mathbf{k}) - \varepsilon(\mathbf{k}+\mathbf{q})|\}$ is valid for each matrix element of $\hat{\Delta}_{\mathrm{band}}(\mathbf{k})$ except for the special momentum such as $\mathbf{k} = (0, 0, k_{\mathrm{z}})$. Then, the off-diagonal components of $\hat{\Delta}_{\mathrm{band}}(\mathbf{k})$ hardly affect the electronic state, and the quasiparticle excitations $E_i(\mathbf{k})$ are expressed as $\pm E^{\mathrm{band}}_{\gamma}(\mathbf{k})$ with $E^{\mathrm{band}}_{\gamma}(\mathbf{k})^2 = e_{\gamma}(\mathbf{k})^2 + |\Delta_{\gamma}(\mathbf{k})|^2$. Thus, $|\Delta_{\gamma}(\mathbf{k})|$ is the superconducting gap in the γ-th band.

It is clear that the $p + D + f$-wave state has a horizontal line node because all of the matrix elements of $\hat{\Delta}(\mathbf{k})$ are zero at $k_{\mathrm{z}} = 0$. Since the pairing state changes from the chiral $d_{\mathrm{xz}} \pm \mathrm{i}d_{\mathrm{yz}}$-wave state in the paramagnetic state to the d_{xz}-wave state in the AF state (see Fig. 6.6b), another line node appears at $k_{\mathrm{x}} = 0$ in the AF state. These line nodes are protected by the symmetry.

An accidental line node of the gap is induced by the AF order in the $s + P$-wave state. The quasiparticle DOS $\rho(\varepsilon)$ is shown in Fig. 6.7, and the superconducting

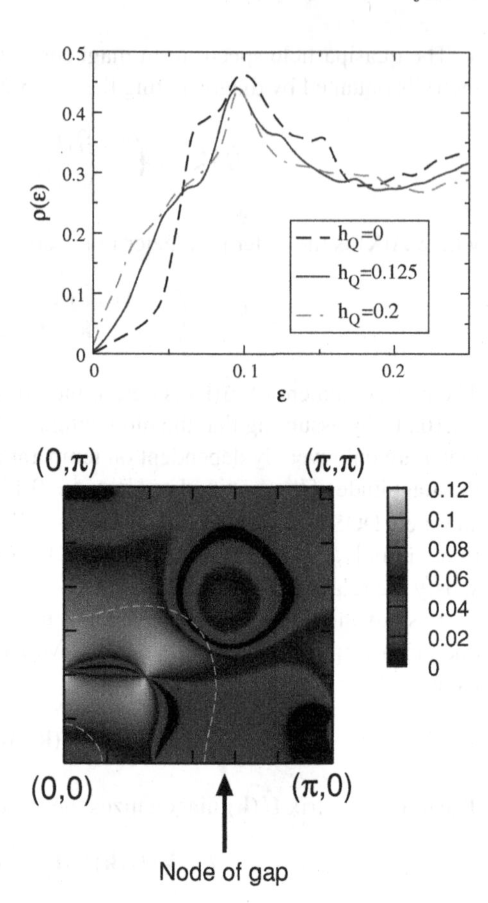

Fig. 6.7 DOS $\rho(\varepsilon)$ in the $s+P$-wave state. Paramagnetic state ($h_{\mathrm{Q}}=0$) and AF state ($h_{\mathrm{Q}}=0.125$ and $h_{\mathrm{Q}}=0.2$) are assumed. We fix the other parameters as $\alpha=0.3$ and $U=4$. We show the results for $\varepsilon>0$ because $\rho(\varepsilon)$ is particle-hole symmetric owing to our definition. (Ref. [20])

Fig. 6.8 Superconducting gap in the $s+P$-wave state in the presence of AF order. We show $|\Delta_4(\mathbf{k})|$ at $k_z=\frac{2\pi}{3}$ for $\alpha=0.3$, $U=4$, and $h_{\mathrm{Q}}=0.125$. A node of the gap induced by the AF order is shown by the *arrow*. (Ref. [20])

gap $|\Delta_\gamma(\mathbf{k})|$ for $\gamma=4$ is shown in Fig. 6.8. We see that the DOS at low energies is enhanced markedly by the AF order. This is mainly due to the mixing of the p-wave order parameter between the leading part $\mathbf{d}(\mathbf{k})\sim(-\sin k_{\mathrm{y}},\sin k_{\mathrm{x}},0)$ and the admixed part $\mathbf{d}(\mathbf{k})=(\sin k_{\mathrm{y}},\sin k_{\mathrm{x}},0)$. Because the a- and b-axes in the tetragonal lattice are no longer equivalent in the presence of the AF order, the p-wave order parameter is modified to $\mathbf{d}(\mathbf{k})=(-\sin k_{\mathrm{y}},\beta\sin k_{\mathrm{x}},0)$ with $\beta\neq1$. This change can be viewed as a rotation of the d-vector. According to the RPA theory, β decreases with increasing h_{Q}. Then, many low-energy excitations are induced around $k_{\mathrm{y}}=\pi/6$, as shown in Fig. 6.8. The superconducting gap in the 4th band (Fig. 6.8) is further decreased by the admixture of an extended s-wave order parameter.

Another line node is induced by the folding of Brillouin zone, as investigated in ref. [51]. However, the contribution to DOS from this line node is negligible in our results because the superconducting gap shows a steep increase around the node.

The DOS in the $s+P$-wave state clearly shows a linear dependence at low energies (Fig. 6.7), which is consistent with the experimental results of $\mathrm{CePt_3Si}$ at ambient pressure [6, 13–15]. We have calculated the temperature dependence of NMR

$1/T_1T$ and specific heat [20] and confirmed the existence of a line node observed in these experiments. Both $s + P$-wave state and $p + D + f$-wave state show the T-square law of $1/T_1T$ and the T-linear law of the specific heat coefficient C/T, when the superconductivity coexists with the AF order. No coherence peak appears in both states. Thus, the experimental results on the low-energy excitations are consistent with both states, and therefore, another experiment is needed to distinguish between these states. In the next section, we point out that the predominantly spin-singlet state and spin-triplet state can be distinguished by the non-linear response to the magnetic field. Other experimental tests have been discussed in ref. [20].

6.2.4.3 Helical Superconducting State

We study the NCS in the magnetic field. The helical superconductivity with modulation vector $\mathbf{q}_{\mathrm{H}} \perp \mathbf{H}$ is realized in NCS with Rashba SO interaction under the magnetic field along the *ab*-plane [52, 53]. In the helical state, the phase of superconducting order parameter is modulated as $\Delta(\mathbf{r}) = \Delta e^{\mathrm{i}\mathbf{q}_{\mathrm{H}}\mathbf{r}}$. This state can be viewed as a superconducting state with finite total momentum of Cooper pairs $\mathbf{q}_{\mathrm{H}}$. (This is the same as in the Fulde–Ferrell state in centrosymmetric superconductors [54–56].)

For the discussion of the helical superconducting phase, we investigate the following effective model using the mean-field approximation

$$H = H_0 + H_Z + H_I, \tag{6.47}$$

$$H_0 = \sum_{\mathbf{k},s} \varepsilon(\mathbf{k}) c^{\dagger}_{\mathbf{k},s} c_{\mathbf{k},s} + \alpha \sum_{\mathbf{k}} \mathbf{g}(\mathbf{k}) \cdot \mathbf{S}(\mathbf{k}) - \sum_{\mathbf{k}} \mathbf{h}_{\mathrm{Q}} \cdot \mathbf{S}_{\mathrm{Q}}(\mathbf{k}), \tag{6.48}$$

$$H_Z = -\sum_{\mathbf{k}} \mathbf{h} \cdot \mathbf{S}(\mathbf{k}), \tag{6.49}$$

$$H_{\mathrm{I}} = U \sum_i n_{i,\uparrow} n_{i,\downarrow} + (V - J/4) \sum_{\langle i,j \rangle} n_i n_j + J \sum_{\langle i,j \rangle} (\mathbf{S}_i \cdot \mathbf{S}_j - 2S_i^{\mathrm{x}} S_j^{\mathrm{x}}). \tag{6.50}$$

The first term is the same as in the Hubbard model in Eq. (6.26). We take into account the Zeeman coupling term H_Z due to the applied magnetic field to be oriented along the [010]-axis. We assume the AF moment along [100]-axis since the magnetization energy is maximal when $\mathbf{h}_{\mathrm{Q}} \perp \mathbf{h} = h\hat{y}$. Assuming the last term H_{I} with $U > 0$, $V = -0.8U$, and $J = 0.3V$, superconducting order parameters obtained by the RPA in Sect. 6.2.4.2 are reproduced. The coupling constant J describes the anisotropic spin–spin interaction arising from the AF order. This term plays an essential role to reproduce the results of the microscopic theory based on the Hubbard model. The details of the mean-field theory have been given in Ref. [19]. As we focus here on the paramagnetic limiting effect, we neglect the orbital depairing for simplicity.

First we show that the Pauli-limited critical magnetic field H_{P} along the *ab*-plane is significantly enhanced by the formation of helical superconducting phase. Figure 6.9 shows the H_{P} for the $s+P$-wave state. To illuminate the helical superconducting state,

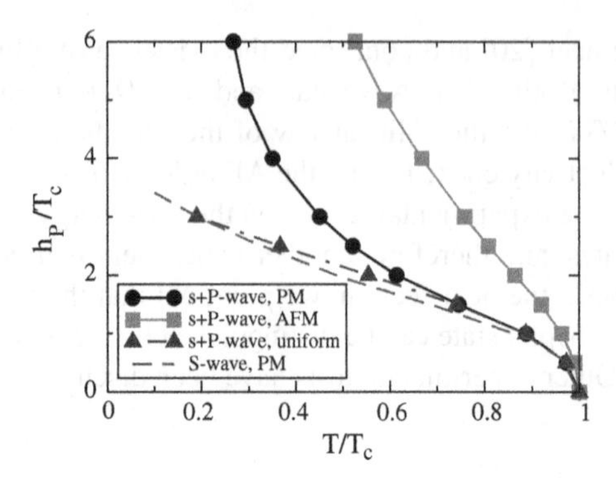

Fig. 6.9 Reduced pauli-limited critical magnetic field $h_P/T_c = \frac{1}{2}g\mu_B H_P/T_c$ against the reduced temperature T/T_c in the $s + P$-wave state. *Circles* and *squares* show the helical superconducting state in the paramagnetic state ($h_Q = 0$) and in the AF state ($h_Q = 0.125$), respectively. The *triangles* show the uniform BCS state at $h_Q = 0$. The s-wave state is also shown for a comparison (*dashed line*). (Ref. [19])

we first discuss the paramagnetic state with $h_Q = 0$. According to the linear response theory, the spin susceptibility in the NCS with Rashba SO interaction decreases to the half of the normal-state value for the magnetic field along the *ab*-plane [1, 2, 57], independent of the pairing symmetry [19]. When we assume a simple formula $H_P = \sqrt{N(0)\Delta^2/(\chi_N - \chi_S)}$, the Pauli-limited critical magnetic field H_P is enhanced by $\sqrt{2}$. By taking into account the non-linear response, the H_P is furthermore enhanced, as shown in Fig. 6.9. This is because the paramagnetic depairing effect is substantially avoided in the helical superconducting phase at high fields. Figure 6.9 shows that the H_P in the helical state (solid line, circles) is much larger than that in the uniform BCS state (dash-dotted line, triangles) for $T < T_c/2$.

The enhancement of H_P for the $s + P$-wave state below $T = T_c/2$ coincides with two distinct changes of the superconducting order parameters. First there is a significant increase of the helicity $|\mathbf{q}_H|$. Figure 6.10 shows the magnetic field dependence of the helicity $|\mathbf{q}_H|$ just below T_c. We see the non-linear increase from $|\mathbf{q}_H| \sim (\alpha/\varepsilon_F)h/v_F$ in the low-field region to $|\mathbf{q}_H| \sim h/v_F$ in the high-field region with a rapid crossover around $h \sim 1.5T_c$. This crossover can be viewed as a crossover from the helical superconducting state to the Fulde–Ferrell–Larkin–Ovchinnikov (FFLO) state. While a small helicity at low fields is essentially due to the inversion-symmetry-breaking, a large helicity $|\mathbf{q}_H| \sim h/v_F$ at high fields is an analogue of the FFLO state [58–60]. Although the FFLO state is restricted to a small region of the phase diagram in centrosymmetric superconductors [61], the high-field phase of NCS is robust in the $s + P$-wave state, as shown in Fig. 6.9. This is because the crossover from the helical superconducting state to the FFLO state coincides with the rotation of d-vector, as discussed below.

The enhancement of H_P in the $s + P$-wave state also coincides with the mixing of p-wave order parameters. Owing to the magnetic field along the [010]-axis the p-wave state $\mathbf{d}(\mathbf{k}) = (-\sin k_y, \sin k_x, 0)$ mixes with another p-wave state $\mathbf{d}(\mathbf{k}) =$

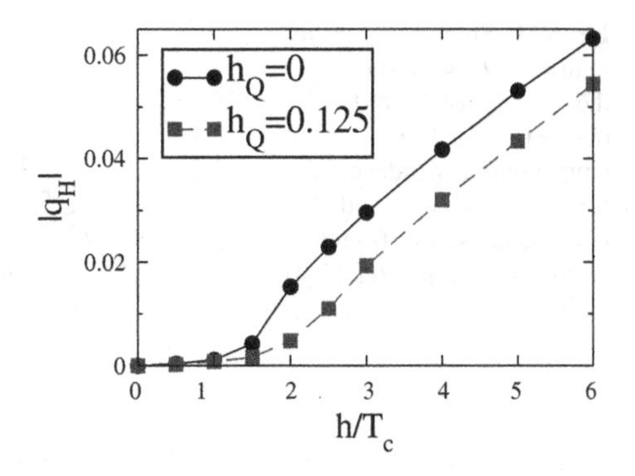

Fig. 6.10 Helicity $|\mathbf{q}_{\mathrm{H}}|$ just below T_{c} in the $s+P$-wave state. *Circles* and *squares* show the paramagnetic state and AF state, respectively. (Ref. [19])

$(\sin k_{\mathrm{y}}, \sin k_{\mathrm{x}}, 0)$ and leads to the d-vector $\mathbf{d}(\mathbf{k}) = (-\sin k_{\mathrm{y}}, \beta \sin k_{\mathrm{x}}, 0)$ with $\beta < 1$. The paramagnetic depairing effect is partly avoided by rotating the d-vector, and thus, the superconducting state is stabilized. The enhancement of H_{P} due to these effects, namely (i) the crossover from the helical superconducting state to the FFLO state and (ii) the rotation of p-wave order parameter is pronounced when the Fermi surfaces are anisotropic as in the model for $\mathrm{CePt_3Si}$ (Eq. (6.27)) [19].

We now turn to the AF state. Figure 6.9 shows that the AF order furthermore boosts H_{P} for the $s+P$-wave state. This is mainly because the coefficient β of the d-vector $\mathbf{d}(\mathbf{k}) = (-\sin k_{\mathrm{y}}, \beta \sin k_{\mathrm{x}}, 0)$ is decreased in the AF state, as discussed in Sect. 6.2.4.2. Thus, both magnetic field and AF order lead to a decrease of β. Since the small energy scale βT_{c} is relevant for the magnetic properties, the non-linear response to the magnetic field is pronounced for small β. The non-linearity with respect to the magnetic field generally enhances H_{P}. Actually, we see that the paramagnetic depairing effect is almost avoided in the AF state (squares in Fig. 6.9).

In the helical superconducting state with large helicity analogous to the FFLO state, the spin susceptibility remains nearly constant through T_{c}, as shown in Fig. 6.11. This is also a consequence of the pronounced non-linear response to the magnetic field. Even when T_{c} is not significantly suppressed by the magnetic field, the magnetization shows a significant non-linearity and leads to no decrease of spin susceptibility below T_{c}. These results on the H_{P} and spin susceptibility in the $s+P$-wave state are consistent with the experiments for $\mathrm{CePt_3Si}$ at ambient pressure [11, 16, 17]. However, they seem to be incompatible with the predominantly spin-singlet pairing state [19].

In contrast to the $s+P$-wave state, the influence of AF order on H_{P} is negligible for the predominantly spin-singlet pairing state [19]. Thus, the response to the AF order is quit different at high fields between the predominantly spin-singlet and spin-triplet pairing states. This is contrasted to the linear response to the magnetic field, which is *universal* in the sense that it is independent of the pairing symmetry. The influence of the AF order can be tested by using the fact that AF order is suppressed by pressure in $\mathrm{CePt_3Si}$, $\mathrm{CeRhSi_3}$, and $\mathrm{CeIrSi_3}$.

As the last issue of this section, we would like to point that $\mathrm{CePt_3Si}$ is a good candidate for an experimental observation of the helical superconducting phase and/or

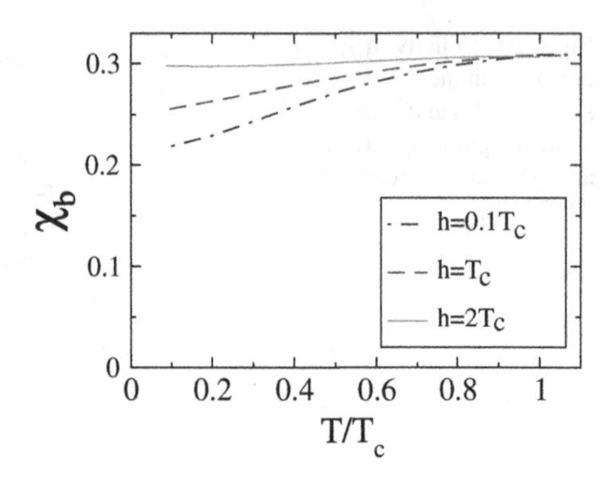

Fig. 6.11 Spin susceptibility in the $s+P$-wave state in the presence of AF order ($h_Q = 0.125$). The temperature dependences for $h = 0.1T_c$, $h = T_c$, and $h = 2T_c$ are shown from the *bottom* to the *top*. (Ref. [19])

FFLO phase. It seems to be difficult to detect the helical superconducting phase with a small helicity ($|\mathbf{q}_H| \sim (\alpha/\varepsilon_F)h/v_F$) because the wave length is much longer than the coherence length. Thus, the high-field phase with $|\mathbf{q}_H| \sim h/v_F$ is more promising for the experimental observation. The high-field phase realizes in a large part of the H-T plane for the $s+P$-wave state, as shown in Fig. 6.9. Comparing our theoretical results with experiments, the $s+P$-wave state is most likely realized in $CePt_3Si$. Therefore, the experimental search for the helical superconducting phase in $CePt_3Si$ is highly desirable. For the Rashba-type NCS, the helicity is perpendicular to the magnetic field, and therefore, the helical modulation is coupled to the vortices. Even then, a fingerprint of the helical phase appears [62].

6.2.5 Summary of $CePt_3Si$

We studied the pairing state in $CePt_3Si$ from the microscopic point of view. According to the microscopic analysis of the minimal Hubbard model in Eq. (6.26) based on RPA, two pairing states are stabilized. One is the predominantly spin-triplet pairing state ($s+P$-wave state) and the other is the predominantly spin-singlet pairing state ($p+D+f$-wave state). We investigated the properties of these states theoretically and compared with the experimental results. Although line-node behavior observed in thermal conductivity [13], superfluid density [14], specific heat [6], and NMR $1/T_1T$ [15] is consistent with both pairing states, magnetic properties clarified by the Knight shift and H_{c2} measurements are consistent with the $s+P$-wave state. We have elucidated some intriguing properties of NCS with predominantly spin-triplet pairing and proposed future experimental tests.

6.3 The Cases of $CeRhSi_3$ and $CeIrSi_3$

In this section, we consider the pairing mechanism of heavy fermion noncentrosymmetric superconductors $CeRhSi_3$ [63] and $CeIrSi_3$ [64]. Both of them possess the same tetragonal structure without inversion center along the c-direction as depicted

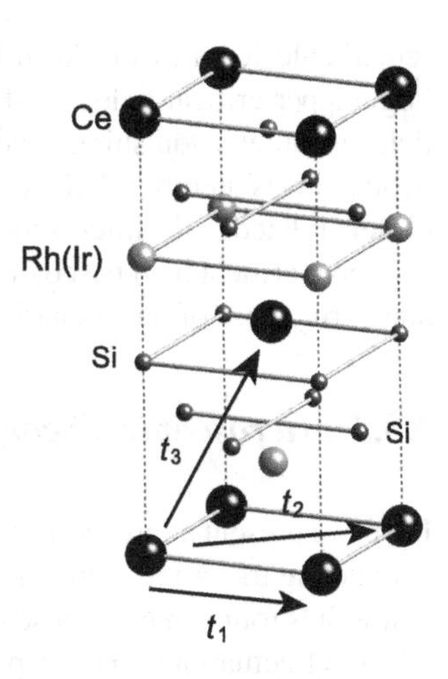

Fig. 6.12 Crystal structure of $CeRh(Ir)Si_3$. t_1, t_2, and t_3 denote the hopping integrals between Ce sites

in Fig. 6.12. $CeRhSi_3$ and $CeIrSi_3$ are antiferromagnets with the Neel temperatures $T_N = 1.6$ K and 5.0 K, respectively, at ambient pressure. When applied pressure increases, these systems exhibit the transition to superconductivity. The maximum of the superconducting transition temperature T_c is achieved in the paramagnetic phase where the antiferromagnetic order is completely destroyed by the applied pressure P. The highest superconducting transition temperatures are $T_c = 1.0$ K for $CeRhSi_3$ at $P = 2.63$ GPa and $T_c = 1.6$ K for $CeIrSi_3$ at $P = 2.6$ GPa. We refer to the pressure at which the highest T_c is realized as the "critical pressure" P_c^* in this section. The validity of this assignment will become clear later. Several experimental results of transport measurements [64] and NMR study [65] for these systems revealed that in the vicinity of the critical pressure P_c^*, there exist strong antiferromagnetic (AF) fluctuations interacting with itinerant electrons, which affect drastically the Fermi-liquid properties; e.g. the nuclear relaxation rate $1/T_1$ for $CeIrSi_3$ above T_c does not obey the Korringa law, but is proportional to $\sqrt{T}$ [65]. This temperature dependence is consistent with the prediction of Moriya's SCR theory for three-dimensional AF spin fluctuations [66]. Furthermore, just below T_c, the nuclear relaxation rate $1/T_1$ exhibits no coherence peak, showing a power-law behavior proportional to T^3 at low temperatures, which indicates that unconventional superconductivity with line nodes of the superconducting gap is realized. These observations strongly suggest that P_c^* is associated with an AF quantum critical point, and the AF fluctuations play an important role for the realization of superconductivity in these compounds. Thus, it is plausible to examine a scenario of spin-fluctuation-mediated Cooper pairing for these NCS. In this section, we develop a microscopic theory for the spin-fluctuation-mediated pairing mechanism applied to $CeRhSi_3$ and $CeIrSi_3$.

This section consists of two parts: in the first part, we analyse the pairing states and the superconducting transition temperatures realized for these systems employing microscopic calculations, and in the second part, we discuss the upper critical fields of these systems on the basis of the spin-fluctuation scenario. One of the most remarkable features of CeRh(Ir)Si_3 is the experimental observation of extremely large upper critical fields ~30–40 T [67, 68]. This magnitude of the upper critical fields is rather astonishing, since the transition temperature in the case without magnetic fields is merely ~1 K. As will be shown in the following, this notable feature is closely related to the microscopic origin of the pairing interaction. The calculation of the upper critical field based on the spin-fluctuation mechanism confirms the validity and correctness of our scenario for the pairing glue.

6.3.1 Microscopic Theory for the Pairing State

Here, we present microscopic theoretical analysis for the pairing states and the superconducting transition temperatures realized in $CeRhSi_3$ and $CeIrSi_3$ [44]. As shown below, it is found that, in these systems, the formation of Cooper pairs is mediated by AF spin fluctuations, and the pairing state is dominated by the spin-singlet extended s-wave state with a weak admixture of a spin-triplet component.

6.3.1.1 Microscopic Model and Approach

The maximum T_c of $CeRhSi_3$ ($CeIrSi_3$) is achieved at the pressure for which the antiferromagnetic order is completely destroyed. Thus, for the clarification of the pairing mechanism, it is sufficient to consider only the paramagnetic phase. The band structure of $CeRhSi_3$ in the paramagnetic state was calculated by Harima on the basis of the LDA calculation [69]. The LDA result of the Fermi surface is consistent with the de Haas–van Alphen experiments under applied pressures [70, 71]. In Fig. 6.13, a pair of the SO split Fermi surfaces obtained by the LDA method is shown. The spectral weight on these Fermi surfaces is dominated by that of f-electrons of Ce atoms, characterizing the heavy fermion states. We first construct an effective low-energy model for $CeRhSi_3$ which properly reproduces the essential features of the LDA results. As a first approximation, we consider a tight-binding model with a single band which consists of f-electrons of Ce atoms on the body-centered tetragonal lattice structure shown in Fig. 6.12. We take account of the hopping of electrons between the nearest-neighbor sites, the second nearest-neighbor sites, and the third nearest-neighbor sites. The energy band of the tight-binding model is given by

$$\varepsilon(\boldsymbol{k}) = -2t_1(\cos k_x + \cos k_y) + 4t_2 \cos k_x \cos k_y - 8t_3 \cos(k_x/2)\cos(k_y/2)\cos k_z - \mu \tag{6.51}$$

We choose the hopping parameters as $(t_1, t_2, t_3) = (1.0, 0.475, 0.3)$ and the chemical potential μ so that the electron density is near half-filling. Then, in spite of the drastic simplification, the model (6.51) well reproduces the LDA results of the Fermi surfaces which mainly consist of itinerant f-electrons. This implies that f-electrons in these systems possess notably itinerant character, which allows the description in terms of the single-band Hubbard model rather than the periodic Anderson model. This picture is also consistent with the experimental observation that the AF order

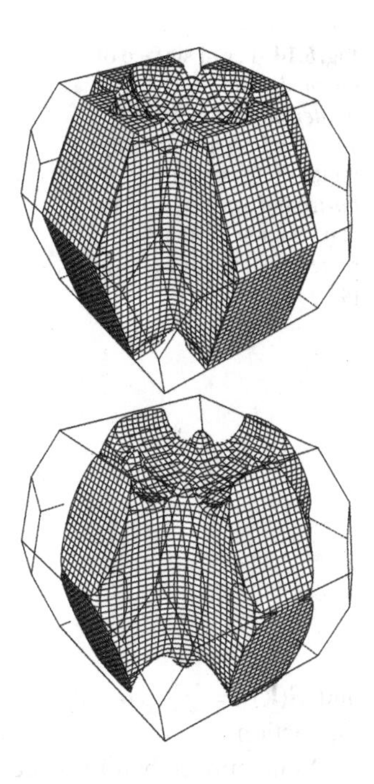

Fig. 6.13 Fermi surfaces of $CeRhSi_3$ obtained by the LDA calculation. *Top* and *bottom* figures are a pair of the SO split Fermi surfaces that have the largest weight of *f*-electrons among all Fermi surfaces. Quoted by courtesy of H. Harima

realized at ambient pressure in these systems is regarded as spin-density waves [72]. In Fig. 6.14, we show the Fermi surface of the tight-binding model (6.51). Since the crystal structure lacks mirror symmetry with respect to the *ab*-plane, we assume the existence of the Rashba SO interaction. Actually, the ground state of the *f*-electron level for $CeRhSi_3$ and $CeIrSi_3$ is the Γ_7 Kramers doublet. Transforming the basis of spin to the basis for the Kramers doublet, one can easily find that the antisymmetric SO interaction represented in terms of the Kramers doublet has the same form as that expressed in terms of a spin basis apart from a constant prefactor. Thus, even in the case of heavy fermion systems, as long as the Bloch states of *f*-electrons are labeled by pseudospins that constitute the Kramers doublet, we can use the same expression for the antisymmetric SO interaction as that for electron spins. Since the spectral weight in the vicinity of the Fermi level for $CeRh(Ir)Si_3$ is largely dominated by that of *f*-electrons, it is expected that this approximation works well.

We also consider the onsite Coulomb repulsion between *f*-electrons. Hence, the total Hamiltonian of the minimum model for the analysis of the superconducting transition is similar to Eq. (6.25):

$$H = \sum_{k,s} \varepsilon(\mathbf{k}) c^{\dagger}_{\mathbf{k},s} c_{\mathbf{k},s} + \alpha \sum_{k} \mathbf{g}(\mathbf{k}) \cdot \mathbf{S}(\mathbf{k}) + U \sum_{i} n_{i,\uparrow} n_{i,\downarrow}, \tag{6.52}$$

where $c_{\mathbf{k},s}$ with $s = \uparrow, \downarrow$ is the annihilation operator of the Kramers doublet. The second term of Eq. (6.52) is the Rashba SO interaction with $\mathbf{g}(\mathbf{k}) = (\sin k_y, -\sin k_x, 0)$

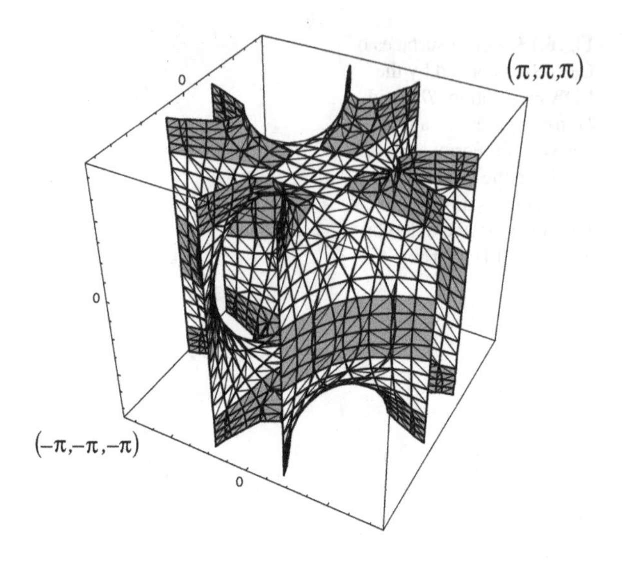

Fig. 6.14 Fermi surface of the model (6.51). The *borderlines* between the white regions and the gray regions represent the *horizontal line* nodes of the gap function for the A_1 representation $\cos 2k_z$. (Ref. [44])

and $\mathbf{S}(\mathbf{k}) = \sum_{s,s'} \boldsymbol{\sigma}_{ss'} c^{\dagger}_{\mathbf{k},s} c_{\mathbf{k},s'}$. The parameter α controls the strength of the SO interaction.

As mentioned before, since the highest T_c is realized in the vicinity of the critical pressure at which the AF spin fluctuations develop, it is natural to investigate the possibility that the Cooper pair is mediated by AF spin fluctuations [27, 28, 73, 74]. In general, the antisymmetric SO interaction gives rise to scattering processes with spin flip. To simplify the following analysis, we exploit an approximation scheme which captures the essential features of effects of the SO interaction, but omits complexity arising from spin-non-conserving scattering processes. One important and non-perturbative effect of the SO interaction is the SO splitting of the Fermi surface. Another important effect associated with pairing mechanisms is parity mixing. We take into account these two effects exactly. However, other aspects such as effects on AF fluctuations and the Fermi-liquid properties can be treated by perturbative approximations. For $CeRh(Ir)Si_3$, the magnitude of the SO interaction is much smaller than the Fermi energy; α/E_F is less than 0.1. Thus, corrections due to the SO interaction to AF fluctuations and the Fermi liquid properties are almost negligible at least qualitatively. Then, the expression for the single-electron Green function is considerably simplified,

$$G_{\alpha\beta}(k) = \sum_{\tau=\pm} \frac{1 + \tau \hat{\mathbf{g}}(\mathbf{k}) \cdot \boldsymbol{\sigma}_{\alpha\beta}}{2} G_\tau(k), \tag{6.53}$$

$$G_\tau^{-1}(k) = G_{0\tau}^{-1}(k) - \Sigma(k), \tag{6.54}$$

$$G_{0\tau}(k) = \frac{1}{i\omega_n - \varepsilon_\tau(\mathbf{k})}, \tag{6.55}$$

where $k = (\mathbf{k}, i\omega_n)$, $\alpha, \beta = \uparrow, \downarrow$, and $\hat{\mathbf{g}}(\mathbf{k}) = \mathbf{g}(\mathbf{k})/|\mathbf{g}(\mathbf{k})|$. The single-electron energy for $U = 0$ is given by $\varepsilon_\tau(\mathbf{k}) = \varepsilon(\mathbf{k}) + \tau\alpha|\mathbf{g}(\mathbf{k})|$. $\Sigma(k)$ is the single-electron self-energy. Because of the reason mentioned above, we have neglected the self-energy that is off-diagonal with respect to spin indices, and also scattering processes with spin flip for the diagonal self-energy $\Sigma(k)$. Furthermore, we neglect the anisotropy in the spin space due to the SO interaction. Based on these approximations, we exploit the standard random phase approximation (RPA) for the pairing interaction mediated by spin fluctuations; vertex corrections to the effective coupling between an electron and AF spin fluctuations are neglected. As is well-known, in contrast to the phonon-mediated pairing mechanism, it is generally not justified to neglect these corrections. However, according to several studies on effects of the vertex corrections beyond RPA, this inclusion does not change the most stable pairing state calculated within RPA. On the contrary, it raises the transition temperature for this pairing state [75, 76]. Thus, the RPA method used here is reliable for our purpose. The RPA calculations of the spin correlation function for the parameters mentioned above well reproduce important features of the momentum dependence obtained by neutron scattering measurements for $CeRhSi_3$ at ambient pressure [72]. It exhibits a prominent peak at the wave number vector $\boldsymbol{Q}_1 = (\pm 0.43\pi, 0, 0.5\pi)$ $\boldsymbol{Q}_2 = (0, \pm 0.43\pi, 0.5\pi)$. In the following analysis, we assume that this momentum dependence of the spin correlation function is almost unchanged by applying pressure, though the applied pressure increases the band width, resulting in the destruction of the AF order.

The superconducting transition temperature T_c is calculated by solving the linearized Éliashberg equation recast into an eigenvalue problem, which is similar to Eq. (6.30) with $p = q = 1$. However, an important difference is that, in this case, we need to take into account frequency-dependence of the pairing interaction and the gap function, and also the self-energy corrections to the single-electron Green function given by (6.54). Then, the linearized Éliashberg equation is

$$\lambda \Delta_{s_1 s_2}(k) = -\frac{T}{N} \sum_{k' s_3 s_4 \sigma_1 \sigma_2} V_{s_1 s_2 s_3 s_4}(k, k') G_{\sigma_1 s_3}(k') G_{\sigma_2 s_4}(-k') \Delta_{\sigma_1 \sigma_2}(k'). \quad (6.56)$$

Here, $k = (\mathbf{k}, i\omega_n)$ etc., $\sum_{k'} = \sum_{n'} \sum_{\mathbf{k}'}$, and we have omitted the subscripts p, q in Eq. (6.30). The maximum eigenvalue $\lambda_{\max} = 1$ corresponds to the superconducting transition. The pairing interaction $V_{s_1 s_2 s_3 s_4}(k, k')$ calculated within RPA is

$$V_{s_1 s_2 s_3 s_4}(k, k') = U \delta_{s_1 s_3} \delta_{s_2 s_4} \delta_{s_1 \bar{s}_2} + V^{\mathrm{bub}}_{s_1 s_2 s_3 s_4}(k, k') + V^{\mathrm{lad}}_{s_1 s_2 s_3 s_4}(k, k'), \quad (6.57)$$

where the contributions from bubble diagrams $V^{\mathrm{bub}}_{s_1 s_2 s_3 s_4}$ and ladder diagrams $V^{\mathrm{lad}}_{s_1 s_2 s_3 s_4}$ are, respectively given by $V^{\mathrm{bub}}_{ssss}(k, k') = v^{\mathrm{bub}}_{ssss}(k - k') - v^{\mathrm{bub}}_{ssss}(k + k')$, $V^{\mathrm{bub}}_{s\bar{s}s\bar{s}}(k, k') = v^{\mathrm{bub}}_{s\bar{s}s\bar{s}}(k - k')$, and $V^{\mathrm{lad}}_{ss\bar{s}\bar{s}}(k, k') = v^{\mathrm{lad}}_{ss\bar{s}\bar{s}}(k - k') - v^{\mathrm{lad}}_{ss\bar{s}\bar{s}}(k + k')$, $V^{\mathrm{lad}}_{s\bar{s}s\bar{s}}(k, k') = v^{\mathrm{lad}}_{s\bar{s}s\bar{s}}(k - k')$ with,

$$
\begin{aligned}
v_{ssss}^{\mathrm{bub}}(q) &= -U^2 \chi_{\bar{s}\bar{s}\bar{s}\bar{s}}^{0}(q)/D_{\mathrm{bub}}(q), \\
v_{s\bar{s}s\bar{s}}^{\mathrm{bub}}(q) &= U^2\left(-\chi_{\bar{s}ss\bar{s}}^{0}(q) - U\chi_{\bar{s}ss\bar{s}}^{0}(q)\chi_{s\bar{s}\bar{s}s}^{0}(q) + U\chi_{\bar{s}\bar{s}\bar{s}\bar{s}}^{0}(q)\chi_{ssss}^{0}(q)\right)/D_{\mathrm{bub}}(q), \\
D_{\mathrm{bub}}(q) &= \left(1 + U\chi_{\downarrow\uparrow\uparrow\downarrow}^{0}(q)\right)\left(1 + U\chi_{\uparrow\downarrow\downarrow\uparrow}^{0}(q)\right) - U^2\chi_{\uparrow\uparrow\uparrow\uparrow}^{0}(q)\chi_{\downarrow\downarrow\downarrow\downarrow}^{0}(q), \\
v_{ss\bar{s}\bar{s}}^{\mathrm{lad}}(q) &= U^2 \chi_{s\bar{s}s\bar{s}}^{0}(q)/D_{\mathrm{lad}}(q), \\
v_{s\bar{s}s\bar{s}}^{\mathrm{lad}}(q) &= U^2\left(\chi_{ss\bar{s}\bar{s}}^{0}(q) - U\chi_{\bar{s}\bar{s}ss}^{0}(q)\chi_{ss\bar{s}\bar{s}}^{0}(q) + U\chi_{s\bar{s}s\bar{s}}^{0}(q)\chi_{\bar{s}s\bar{s}s}^{0}(q)\right)/D_{\mathrm{lad}}(q), \\
D_{\mathrm{lad}}(q) &= \left(1 - U\chi_{\downarrow\downarrow\uparrow\uparrow}^{0}(q)\right)\left(1 - U\chi_{\uparrow\uparrow\downarrow\downarrow}^{0}(q)\right) - U^2\chi_{\uparrow\downarrow\uparrow\downarrow}^{0}(q)\chi_{\downarrow\uparrow\downarrow\uparrow}^{0}(q).
\end{aligned}
$$

The bare spin susceptibility $\chi_0(q)$ is $\chi_0(q) = -T\sum_m \sum_{\mathbf{k}} \sigma G_{0\uparrow\uparrow}(k+q)G_{0\downarrow\downarrow}(k)$. Here $G_{0\alpha\beta}$ is the electron Green function (6.53) without the selfenergy correction $\Sigma(k)$. This approximation scheme is sufficient to determine which pairing symmetry is favored by the spin-fluctuation-mediated pairing interaction.

6.3.1.2 Pairing States

Because of the parity mixing due to the antisymmetric SO interaction, the pairing gap function is generally expressed as,

$$
\Delta_{s_1 s_2}(k) = \Delta_s(i\varepsilon_n) d_0(\mathbf{k}) i(\sigma_2)_{s_1 s_2} + \Delta_t(i\varepsilon_n)\mathbf{d}(\mathbf{k}) \cdot i(\sigma\sigma_2)_{s_1 s_2}. \tag{6.58}
$$

The momentum dependence of the spin-singlet part $d_0(\mathbf{k})$ obeys the group theoretical classification of the pairing symmetry. For the tetragonal systems with the Rashba SO interaction such as CeRh(Ir)Si_3, $d_0(\mathbf{k})$ is given by one of basis functions of the irreducible representations for C_{4v}, which are listed in Table 6.1. In this table, the basis functions are expressed in terms of the second order harmonics $\cos 2k_x$, $\cos 2k_y$, $\cos 2k_z$ etc. rather than the first order harmonics $\cos k_x$, $\cos k_y$, $\cos k_z$ etc., because of the following reason. In our systems, the pairing interaction mediated by AF spin fluctuations (6.57) has a prominent peak for $\mathbf{k} - \mathbf{k}' \approx \mathbf{Q}$ with $\mathbf{Q} \approx (\pi/2, 0, \pi/2)$ or $(0, \pi/2, \pi/2)$. This ordering vector $\mathbf{Q}$ is consistent with experimental observations of neutron scattering measurements for $CeRhSi_3$ [72]. This momentum dependence stabilizes the gap functions that satisfy $d_0(\mathbf{k}+\mathbf{Q}) = -d_0(\mathbf{k})$. Thus, we use the basis functions listed in Table 6.1 in the following calculations. On the other hand, for the spin-triplet channel, $\mathbf{d}(\mathbf{k})$ is chosen so as to be consistent with the Rashba SO interaction, i.e. $\mathbf{d}(\mathbf{k}) = d_0(\mathbf{k})\mathbf{d}(\mathbf{k})$. For this choice of $\mathbf{d}(\mathbf{k})$, the interband pairings which give rise to pair-breaking effects are suppressed, and thus the optimum T_c is expected.

To examine which pairing state is the most stable in our model, we calculate the maximum eigenvalues $\lambda_{\max}$ of the lineraized Éliashberg equation (6.56) for five irreducible representations listed in Table 6.1. In this calculation, we neglect effects of the normal selfenergy, putting $\Sigma(k) = 0$ in (6.56). In Fig. 6.15, the $\lambda_{\max}$ values calculated for $\alpha = 0$ are shown. In the vicinity of the AF phase boundary determined by the condition $U\chi_0(0, \mathbf{Q}) = 1$, the highest T_c is achieved for the A_1 representation in the spin-singlet channel with the gap function $d_0^{A_1}(\mathbf{k}) = \cos 2k_z$, which

Table 6.1 Pairing symmetry for tetragonal systems with the Rashba SO interaction. The irreducible representations are given by those of C_{4v}. $\mathbf{g}(\mathrm{k}) = (\sin k_y, -\sin k_x, 0)$

Spin-singlet/triplet	Irreducible representation		Basis function
singlet	A_1	(extended s)	$d_0^{\mathrm{A}_1}(\mathbf{k}) = \cos 2k_z$
singlet	A_2	$(g_{xy(x^2-y^2)})$	$d_0^{\mathrm{A}_2}(\mathbf{k}) = \sin 2k_x \sin 2k_y \left(\cos 2k_x - \cos 2k_y\right)$
singlet	B_1	$(d_{x^2-y^2})$	$d_0^{\mathrm{B}_1}(\mathbf{k}) = \left(\cos 2k_x - \cos 2k_y\right)$
singlet	B_2	(d_{xy})	$d_0^{\mathrm{B}_2}(\mathbf{k}) = \sin 2k_x \sin 2k_y$
singlet	E	(d_{zx})	$d_0^{\mathrm{E}}(\mathbf{k}) = \sin k_x \sin 2k_z$
triplet			$d^{\Gamma}(\mathbf{k}) = d_0^{\Gamma}(\mathbf{k})\mathbf{g}(\mathbf{k})$

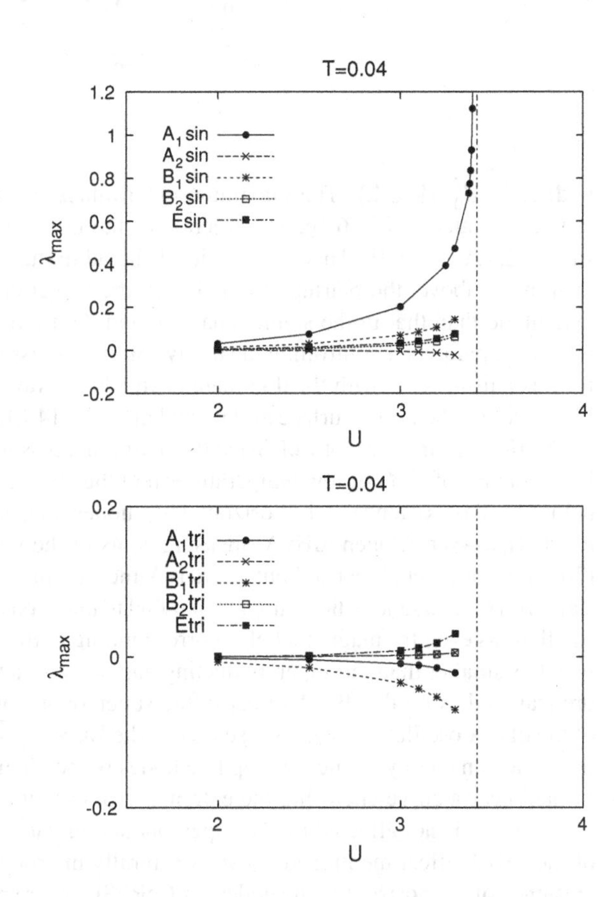

Fig. 6.15 Maximum eigen values of the linearized Éliashberg equation versus U for five irreducible representations with $\alpha = 0$ and $T = 0.04$. The *upper* (*lower*) panel is for the spin singlet (triplet) channel. The *dashed line* is the AF phase boundary. (Ref. [44])

corresponds to the extended s-wave state. In contrast, $\lambda_{\max}$'s for all the spin-triplet pairings are much smaller than unity. This is because that for the pairing interaction mediated by AF spin fluctuations (6.57), the spin-triplet channels are repulsive, or very weakly attractive. When the antisymmetric SO interaction is switched on, i.e. $\alpha \neq 0$, the pairing state with the highest T_c is the extended s-wave state with an admixture of the p-wave state. The gap function of the mixed p-wave component

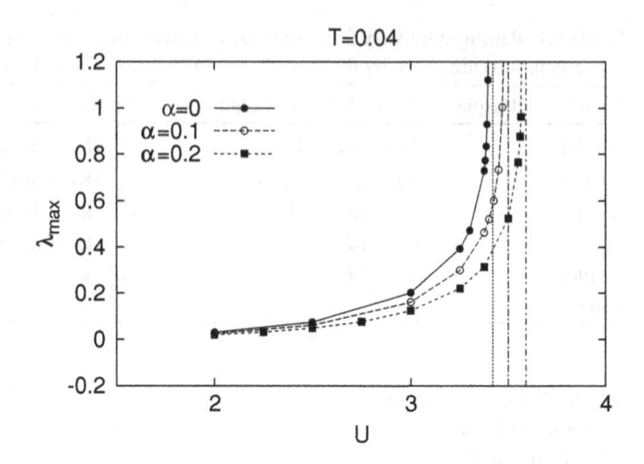

Fig. 6.16 Maximum eigen values of the linearized Éliashberg equation versus U for the extended $S + p$-wave state. $T = 0.04$. (Ref. [44])

is $\mathbf{d}(\mathbf{k}) = d_0^{A_1}(\mathbf{k})\mathbf{g}(\mathbf{k})$. The maximum eigenvalues for the extended $S + p$-wave state are plotted in Fig. 6.16. The fraction of the mixed p-wave component is very small; $\Delta_t/\Delta_s \sim 0.01$. This small ratio of the admixture is due to the fact that, as mentioned above, the pairing interaction in the triplet channel is much smaller in magnitude than that in the singlet channel, and the ratio $E_{\rm SO}/E_{\rm F}$ for CeRh(Ir)Si_3, which is a parameter controlling the parity-mixing, is less than 0.1. The gap function for this pairing state with the dominated extended s-wave symmetry has horizontal line nodes on the Fermi surface as depicted in Fig. 6.14 which is well consistent with the NMR measurement for $CeIrSi_3$ at the optimum pressure P_c^* indicating the power law behavior of $1/T_1$ at low temperatures and the absence of a coherence peak just below T_c. The existence of the horizontal line nodes for $CeRhSi_3$ is also suggested by the de Haas–van Alphen (dHvA) measurements of the Fermi surfaces for $CeRhSi_3$ [70]. Terashima et al. carried out the dHvA measurements for the superconducting state with high magnetic fields applied parallel to the c-axis. They observed the dHvA oscillations even for magnetic fields corresponding to the Landau level spacing $\hbar\omega_c$ which is smaller than the superconducting gap Δ by a factor 1/3. According to theoretical studies on the dHvA effect in the superconductors [77–79], the amplitude of the dHvA oscillation is suppressed when the BCS gap Δ is isotropic and exceeds the cyclotron energy, while, for gap line nodes perpendicular to magnetic fields and located near a curve enclosing the extremal cross section of the Fermi surface, the dHvA effect is not affected by the superconducting gap. Therefore, the observation of the dHvA effect mentioned above is naturally interpreted as an evidence for the existence of the horizontal line nodes for $CeRhSi_3$. More precisely, the line nodes of the $\cos 2k_z$ state do not lie on the curve enclosing the extremal cross section of the Fermi surface. However, as shown in Fig. 6.14, the Fermi surface has portions which are almost flat along the z-direction and crosses the line nodes of the gap. For this specific shape of the Fermi surface, it is plausible that the dHvA oscillation is not suppressed even though the line nodes are not located exactly on the extremal part of the Fermi surface.

Another experimental evidence which supports the scenario of the AF spin-fluctuation-mediated pairing mechanism for $CeRhSi_3$ and $CeIrSi_3$ is provided by the measurement of the upper critical field of these systems. We would like to address this issue in the next section.

6.3.2 Upper Critical Fields

One of the most remarkable experimental observations for $CeRhSi_3$ and $CeIrSi_3$ is the extremely large upper critical field H_{c2} for a magnetic field parallel to the c-axis in the vicinity of the critical pressure P_c^* [67, 68]. The observed $H_{c2} \sim 30-40$ T at P_c^* is almost comparable to those of high-T_c systems, though the transition temperature at zero field is merely $\sim$1 K. Also, the H–T curve exhibits upward curvature in a wide range of temperatures; i.e. the increase in the upper critical field is enormously enhanced as the temperature T is decreased, in sharp contrast to any other superconductors in which the increase of H_{c2} becomes weaker as T is decreased. Furthermore, the pressure dependence of H_{c2} is significantly strong in the vicinity of P_c^*, while, by contrast, the pressure dependence of T_c is moderate. If the SO interaction for these systems is assumed to be of the Rashba type, the absence of the Pauli depairing is understood as follows. The Rashba SO interaction splits the Fermi surface into two parts. For magnetic fields applied parallel to the c-axis (or the z-axis in the notation of Eq. (6.52)), the uniform spin susceptibility in the normal state χ_N is dominated by the van-Vleck-like susceptibility χ_{VV} governed by contributions from the transition between the two SO split bands, and there is no Pauli spin susceptibility dominated by contributions from the Fermi surface. When the SO-splitting of the Fermi surface E_{SO} is much larger than the superconducting gap Δ, which is the case of the heavy fermion NCS, the spin susceptibility $\chi_N \approx \chi_{VV}$ is not affected by the superconducting transition, leading to the infinite Pauli limiting field H_P; i.e. from the energy balance between the Zeeman energy and the condensation energy, $\frac{1}{2}\chi_N H_P^2 = \frac{1}{2}\chi_{VV} H_P^2 + \frac{1}{2}N(0)\Delta^2$, we have $H_P \to \infty$. In contrast to the Pauli depairing effect, however, the orbital depairing effect should always exist for NCS. It is important to clarify why the orbital limiting fields of superconductivity in $CeRhSi_3$ and $CeIrSi_3$ are so large as indicated by these experiments. In this section, we examine whether these remarkable features can be explained by the spin-fluctuation-mediated pairing mechanism considered in the previous section [80]. A key factor for understanding these experimental results is the existence of the quantum critical point (QCP) associated with AF order at sufficiently low temperatures where the huge H_{c2} is observed. The pairing interaction mediated by spin fluctuations should strongly increase just at the QCP where the correlation length of the AF order diverges. On the other hand, at sufficiently low temperatures near quantum criticality, pair-breaking effects of inelastic scattering due to spin fluctuations which are important for the determination of the transition temperature are suppressed. The interplay between the enhancement of the pairing interaction and the suppression of the pair-breaking effect in the vicinity of the QCP yields naturally the remarkable enhancement of H_{c2}. We examine this scenario in the following.

6.3.2.1 Quasiclassical Éliashberg Equation

For the calculation of the orbital limiting field in the vicinity of a QCP, it is important to take account of pair-breaking effects due to the interaction with spin fluctuations. This task is rather involved. To simplify the following analysis, we use a phenomenological model for the AF spin fluctuations, instead of calculating the spin fluctuation propagator microscopically as done in the previous section. We assume the following form of an effective action for the interaction between electrons and spin fluctuations,

$$S_{SF} = -\sum_q g^2 \chi(q) \mathbf{S}_q \cdot \mathbf{S}_{-q}. \tag{6.59}$$

Here, $\mathbf{S}_q = \sum_k c^\dagger_{k+q\alpha} \sigma_{\alpha\beta} c_{k\beta}$, $q = (\mathbf{q}, i\varepsilon_n)$ with $\varepsilon_n = 2\pi n T$, and,

$$\chi(q) = \sum_a \frac{\chi_0 \xi^2}{1 + \xi^2 (\mathbf{q} - \mathbf{Q}_a)^2 + |\varepsilon_n| / (\Gamma_0 \xi^{-2})}. \tag{6.60}$$

We choose the propagating vectors of spin fluctuations as $\mathbf{Q}_1 = (\pm 0.43\pi, 0, 0.5\pi)$ and $\mathbf{Q}_2 = (0, \pm 0.43\pi, 0.5\pi)$ in accordance with neutron scattering measurements for $CeRhSi_3$ [72]. We also assume that the temperature dependence of the AF correlation length $\xi(T)$ is given by $\xi(T) = \frac{\bar{\xi}}{\sqrt{T+\theta}}$, which implies that the critical exponent of $\xi(T)$ is equal to the mean field value $1/2$. This assumption for $\xi(T)$ is consistent with the NMR experiment for $CeIrSi_3$ at $P \sim P_c^*$, which indicates $1/T_1 \propto \sqrt{T}$ above T_c [65]. The action (6.59) replaces the onsite repulsion term in (6.52). We interpret that the interaction (6.59) is generated by renormalizing high-energy scattering processes due to the onsite Coulomb repulsion U.

To investigate the orbital depairing effect, we need to deal with a spatially varying vector potential $\mathbf{A}(\mathbf{r})$. For this purpose, we use a standard quasiclassical approximation, which is justified for $k_F l_H \ll 1$, where k_F is the Fermi momentum, and l_H the magnetic length. The single-electron Green function within this approximation is

$$G_{\alpha\beta}(\mathbf{x}, \mathbf{y}, i\omega_n; \mathbf{A}) = e^{ie \int_{\mathbf{y}}^{\mathbf{x}} \mathbf{A}(\mathbf{s}) ds} G_{\alpha\beta}(\mathbf{x} - \mathbf{y}, i\omega_n; \mathbf{A} = 0). \tag{6.61}$$

Here, $G_{\alpha\beta}(i\omega_n, \mathbf{x} - \mathbf{y}; \mathbf{A} = 0)$ is the Fourier transform of (6.53) with respect to momentum $\mathbf{k}$. The interaction with critical AF fluctuations strongly modifies the Fermi-liquid properties of electrons. This effect is incorporated into the normal self-energy $\Sigma(k)$. Up to the first order in g^2, it is given by

$$\Sigma(k) = \frac{T}{N} \sum_{k',\alpha} g^2 \chi(k - k') G_{0\alpha\alpha}(k'). \tag{6.62}$$

As in the previous section, we neglect scattering processes with spin flip. We also add the Zeeman term $-\mu_B H_z \sum_k c^\dagger_k \sigma_z c_k$ to the Hamiltonian to take account of the Pauli depairing effect.

The upper critical field H_{c2} in the strong-coupling regime is obtained by solving the linearized Éliashberg equation,

$$\Delta_{\alpha\alpha'}(\mathbf{k}, i\omega_m; \mathbf{R}) = -\frac{T}{N} \sum_{k',\beta\beta'\gamma\gamma'} V_{\alpha\alpha',\beta\beta'}(k,k') \times G_{\beta\gamma}(\mathbf{k}' + \mathbf{\Pi}, i\omega_n) G_{\beta'\gamma'}(-\mathbf{k}', -i\omega_n) \Delta_{\gamma\gamma'}(\mathbf{k}', i\omega_n; \mathbf{R}), \tag{6.63}$$

where $\mathbf{R}$ is the center of mass coordinate of Cooper pairs, and $\mathbf{\Pi}(\mathbf{R}) = -i\nabla_R + 2e\mathbf{A}(\mathbf{R})$ with $\nabla_R \times \mathbf{A}(\mathbf{R}) = (0, 0, H)$. The pairing interaction mediated by AF spin fluctuations is

$$V_{ss,ss}(k,k') = -\frac{1}{2} g^2 \chi(k-k') + \frac{1}{2} g^2 \chi(k+k'), \tag{6.64}$$

$$V_{s\bar{s},s\bar{s}}(k,k') = -V_{s\bar{s},\bar{s}s}(k,k') \tag{6.65}$$

$$= \frac{1}{2} g^2 \chi(k-k') + g^2 \chi(k+k'), \tag{6.66}$$

Here, we have assumed that the energy scale of the AF spin fluctuations is sufficiently larger than the Zeeman energy introduced by the applied magnetic field, and that the propagator for the spin fluctuations is not affected by the magnetic field. Since the magnetic field is parallel to the c-direction and the SO interaction is the Rashba-type, we do not need to consider nonuniform states like the helical vortex phase or the FFLO state; i.e. under this situation, the nonuniform pairing states with center-of-mass momentum require interband pairing between electrons of the two SO-split bands, which is strongly suppressed when the SO splitting is much larger than the superconducting gap. Then, it is sufficient to take only the lowest Landau level in the evaluation of the kernel of the gap equation (6.63), and use the expression for the gap function up to the lowest Landau level, $\Delta_{\alpha\alpha'}(k; \mathbf{R}) = \Delta_{\alpha\alpha'}(k)\phi_0(R_x, R_y)$, where ϕ_0 is the basis function for the lowest Landau level. As a result, the Éliashberg equation is rewritten as,

$$\Delta_{\alpha\alpha'}(k) = \frac{T}{N} \sum_{k',\beta\beta'\gamma\gamma'} V_{\alpha\alpha',\beta\beta'}(k,k') \sum_{\tau=\pm} \left(\frac{1 + \tau \hat{\mathbf{g}}^H(\mathbf{k}) \cdot \boldsymbol{\sigma}}{2} \right)_{\beta\gamma} i\,\mathrm{sgn}(\tilde{\omega}') \times \left(\frac{2}{a_\tau(\mathbf{k}')} \right)^{1/2} f\left(\frac{b_\tau(k')}{\sqrt{2a_\tau(\mathbf{k}')}} \right) G_{\beta'\gamma'}(-k') \Delta_{\gamma\gamma'}(k'), \tag{6.67}$$

where $\hat{\mathbf{g}}^H(\mathbf{k}) = \mathbf{g}^H(\mathbf{k})/|\mathbf{g}^H(\mathbf{k})|$ with $\mathbf{g}^H(\mathbf{k}) = (\sin k_y, -\sin k_x, -\mu_B H_z/\alpha)$, and $\tilde{\omega}(k) = \omega_n - \mathrm{Im}\Sigma_0(k)$, $a_\tau(\mathbf{k}) = \sqrt{|e|H}(v_{\tau x}(\mathbf{k})^2 + v_{\tau y}(\mathbf{k})^2)$, $b_\tau(k) = |\tilde{\omega}| + i\,\mathrm{sgn}(\tilde{\omega})\varepsilon_\tau(\mathbf{k})$, with the single-particle energy $\varepsilon_\tau(\mathbf{k}) = \varepsilon(\mathbf{k}) + \tau\alpha|\vec{g}^H(\mathbf{k})|$ and the velocity $\vec{v}_\tau(\mathbf{k}) = \nabla\varepsilon_\tau(\mathbf{k})$. $f(z)$ is defined as $f(z) = \frac{\sqrt{\pi}}{2} e^{z^2} \mathrm{erfc}(z)$. Eq. (6.67) is the basis for our analysis of the upper critical field in the strong-coupling regime.

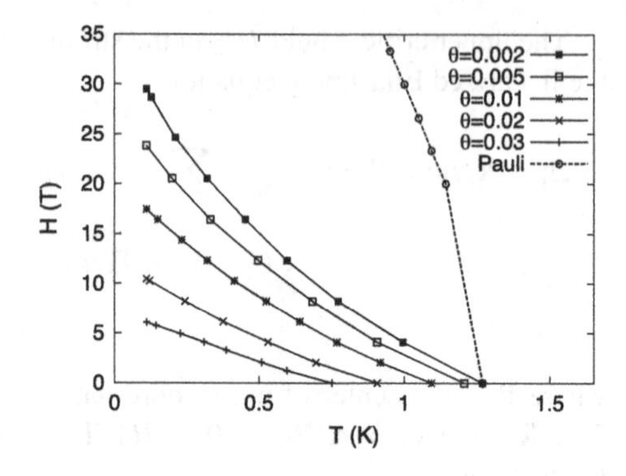

Fig. 6.17 *Upper* critical fields plotted as functions of temperatures. The Pauli limiting field for $\theta = 0.002$ (a *dotted line* with open *circles*), and the orbital limiting fields for several θ (*solid lines*) are shown. The unit of θ is $t_1 = 113$ K. (Ref. [80])

6.3.2.2 Extremely Large H_{c2} Due to Critical Spin Fluctuation

The transition temperature $T_c(H_z)$ for a finite magnetic field H_z and the upper critical field H_{c2} can be calculated from Eq. (6.67). Since the phenomenological form of spin correlation function (6.60) possesses the momentum dependence quite similar to that obtained by the RPA calculations for the model presented in the previous section, the extended s-wave state with the small fraction of the admixture of p-wave state has the highest T_c at zero field. This pairing symmetry is not changed by a magnetic field along the z-direction within our calculations. As mentioned before, the ratio of the spin-triplet gap to the spin-singlet gap obtained by our calculation is very small, $\sim$0.01. Thus, we can safely neglect the admixture of the triplet component for the discussion on H_{c2} in the following. In Fig. 6.17, we show the calculated upper critical field for this pairing state as a function of temperatures T for several values of the parameter θ. In this calculation we set the coupling constant as $g^2\chi_0 = 13$, and the lattice constant $a = 4$ Å, $t_1 = 113$ K. The value of t_1 is determined by identifying the calculated maximum transition temperature $T_c = 0.0115$ as 1.3 K. As shown in Fig. 6.17, because of the existence of the large van-Vleck-like susceptibility due to the antisymmetric SO interaction, the Pauli limiting field for the magnetic field parallel to the z-axis is so large that the upper critical field is solely determined by the orbital limit. As the parameter θ decreases, and the system approaches the AF QCP, H_{c2} increases remarkably reaching to $H_{c2} \sim 30$ T, even though T_c for zero field is merely $\sim$1 K. These results successfully explain the experimental observations for CeRh(Ir)Si_3 mentioned before. The main features of the upper critical fields stem from the combination of the strong enhancement of the pairing interaction due to the AF spin fluctuation in the vicinity of the QCP and the significant suppression of the pair-breaking effect caused by inelastic scattering with the spin fluctuations at sufficiently low temperatures. The enhancement of H_{c2} is prominent especially for small θ, giving rise to the huge values of H_{c2}^{orb}, as seen in Fig. 6.17.

Another intriguing feature observed for $CeRhSi_3$ and $CeIrSi_3$ is the strong pressure dependence of H_{c2} which is in sharp contrast to the quite weak dependence of T_c on pressures in the vicinity of the QCP. This remarkable pressure dependence is also reproduced from Eq. (6.67). To see this, we plot H_{c2} and T_c at zero field as a function

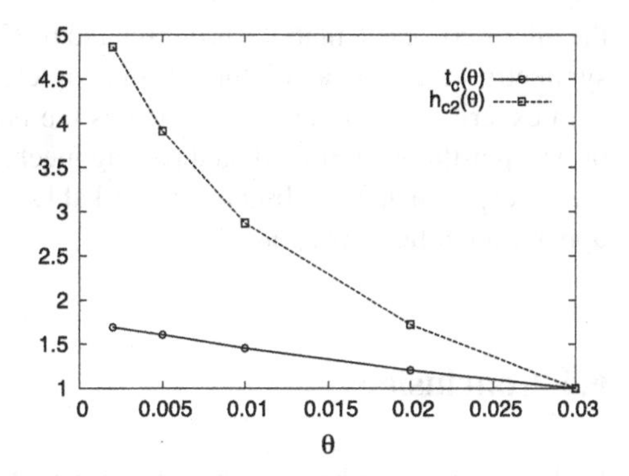

Fig. 6.18 The normalized transition temperature $t_c(\theta) \equiv T_c(H = 0, \theta)/T_c(H = 0, \theta = 0.03)$ and the normalized *upper* critical field $h_{c2}(\theta) \equiv H_{c2}(T = 0.001, \theta)/H_{c2}(T = 0.001, \theta = 0.03)$ versus θ. The unit of θ is $t_1 = 113$ K. (Ref. [80])

of θ in Fig. 6.18. As the system under consideration approaches the QCP ($\theta \to 0$), H_{c2} increases very rapidly while, by contrast, the change of T_c is moderate. These behaviors are also understood as a result of the strongly enhanced pairing interaction at low temperatures in the vicinity of the QCP. Although the increase of $T_c(H = 0)$ near the QCP (i.e. $\theta = 0$) is considerably suppressed by the pair breaking effects due to inelastic scattering by spin fluctuations, H_{c2} at low temperatures is not seriously affected by them, resulting in the strong enhancement of H_{c2} for $\theta \sim 0$.

We would like to stress that the large enhancement of H_{c2} near a magnetic QCP for a spin-fluctuation-mediated pairing mechanism is not specific to the NCS considered here, but rather expected to occur more generally for orbital-limited superconductors caused by spin fluctuations, as long as the Pauli depairing effect is suppressed by some other mechanisms such as the lack of inversion symmetry as in the case of $CeRh(Ir)Si_3$. For instance, $CeCoIn_5$ is also believed to be an unconventional superconductor located in the vicinity of AF QCP. However, because of the Pauli depairing effect, the strong enhancement of H_{c2} near the QCP is not realized for this system. It would be intriguing to search for the huge enhancement of H_{c2} for triplet superconductors in the vicinity of a ferromagnetic QCP.

As clarified above, the coincidence between the theoretical results presented in this section and the experimental observations for the upper critical fields strongly supports the scenario that superconductivity realized in $CeRhSi_3$ and $CeIrSi_3$ is caused by AF spin fluctuations in the vicinity of AF criticality.

6.3.3 Summary of $CeRhSi_3$ and $CeIrSi_3$

Superconductivity in $CeRhSi_3$ and $CeIrSi_3$ under applied pressure is realized in the vicinity of AF criticality. There are several pieces of experimental evidence which support the existence of strong AF spin fluctuations in these systems, as indicated by the measurements of the NMR signal, the specific heat, and the transport phenomena [63–65]. It is natural to expect that the AF spin fluctuations mediate Cooper pairing in these superconductors. The superconducting state obtained by the microscopic calculations is dominated by Cooper pairs with the extended *s*-wave symmetry with small admixture of spin-triplet pairs. This gap symmetry is consistent with experimental observations such as the NMR measurement [65]. The scenario of AF spin-fluctuation-mediated pairing mechanisms for $CeRh(Ir)Si_3$ explains quite nicely experimentally observed remarkable features such as extraordinarily large upper critical fields [67, 68].

6.4 Conclusions

In this chapter, we have overviewed our understanding of superconductivity realized in two classes of heavy fermion NCS, $CePt_3Si$ and $CeRh(Ir)Si_3$, based upon microscopic theories of pairing mechanisms. In both of these systems, non-phonon mechanisms play a crucial role, stabilizing unconventional superconductivity. In particular, the spin-fluctuation-mediated pairing mechanism is a promising candidate for the microscopic origin of superconductivity in these systems. There are some important differences of superconducting states between $CePt_3Si$ and $CeRh(Ir)Si_3$. In $CePt_3Si$, the $s + p$ wave state with the dominant *p*-wave component is stabilized by helical spin fluctuations. On the other hand, in $CeRh(Ir)Si_3$, the dominant extended *s*-wave state with small admixture of the *p*-wave pairs is realized by strong AF spin fluctuations in the vicinity of the magnetic QCP. This difference of the character of spin fluctuations is due to the difference of the band structures and magnetic properties between these systems. Furthermore, the node structures of the superconducting gap functions are also different. For $CePt_3Si$, there are accidental line nodes of the gap caused by the parity mixing of the pairing states and by the AF order, while, for $CeRh(Ir)Si_3$, there are horizontal line nodes associated with the extended *s*-wave gap symmetry.

In addition to these heavy fermion NCS, there are other classes of NCS, which are also expected to be unconventional superconductors: e.g. UIr [81], and $Li_2(Pt_{1-x}Pd_x)_3B$ [82–84]. In particular, for Li_2Pt_3B, the NMR study and the measurement of the penetration depth suggest that there are line nodes of the superconducting gap [83, 84]. A model of the parity-mixed $s + p$ wave state was proposed for understanding these experimental observations [83]. In this model, the *p*-wave pairing dominates the superconducting state. However, this superconductor is regarded as a weakly correlated system in which effects of electron–electron interaction are negligible. It is an unresolved important issue to identify the microscopic origin of a pairing interaction in the *p*-wave channel, if it does indeed exist in Li_2Pt_3B.

Finally, we mention some other theoretical studies on microscopic pairing mechanisms of NCS. In ref. [85, 86], a spin-fluctuation mechanism is investigated for the two-dimensional Hubbard model with the Rashba SO interaction on the basis of the RPA method. The $d + f$-wave state with a dominant $d_{x^2-y^2}$-wave component is stabilized for electron density close to half-filling because of AF spin fluctuations. The gap function for this state has the form given by Eq. (6.1). However, the weight of the mixed *f*-wave component is very small for $E_{SO}/E_F < 0.1$, as in the case of

$CeRh(Ir)Si_3$ discussed in Sect. 6.3.3. In ref. [87], the Kohn–Luttinger mechanism for pairing interactions due to the Coulomb repulsion U is examined for a single-band model with the Rashba SO interaction by using perturbative expansions in terms of U. In this model, there are substantially strong attractive interactions in both the p-wave channel and the d-wave channel. It is found that, in some parameter regions, strong parity mixing between p-wave pairing and d-wave pairing occurs. However, in this state, as in the case of $CePt_3Si$ discussed in Sect. 6.3.2, the **d**-vector for p-wave pairing is not parallel to $\mathbf{g}(\mathbf{k})$ vector because of frustration between the Rashba SO interaction and pairing interactions, and interband pairs between the SO-split bands are induced.

The results presented in Sects. 6.3.2 and 6.3.3 and the above-mentioned related works imply that pairing states realized in NCS crucially depend on both the detailed structure of pairing interactions and the character of the antisymmetric SO interaction. In general, when there are substantial spin-triplet pairing correlations, it may be possible that the momentum dependence of the dominant spin-triplet pairing interaction is not compatible with that of the antisymmetric SO interaction, because these two kinds of interactions usually have different microscopic origins. In this case, there are interband pairs between the SO-split bands, and the gap function can not be expressed in terms of Eq. (6.1). The strong parity mixing with the gap function of the form (6.1) satisfying $\mathbf{d}(\mathbf{k}) \parallel \mathbf{g}(\mathbf{k})$ is realized only when the dominant spin-triplet pairing interaction is matched with the antisymmetric SO interaction. Thus, for the exploration of pairing states in NCS, it is particularly important to clarify the microscopic nature of pairing interactions.

Acknowledgements The authors are grateful to M. Sigrist, Y. Tada, and N. Kawakami for fruitful collaborations on the studies of noncentrosymmetric superconductors, upon which the content of this chapter is largely based. The authors also thank E. Bauer, Y. Onuki, M. Sigrist, N. Kimura, R. Settai, H. Mukuda, H. Harima, H. Yamagami, T. Terashima, T. Takeuchi, Y. Kitaoka, J. Flouquet, R. Ikeda, K. Izawa, Y. Matsuda, M.-A. Measson, G. Motoyama, T. Takimoto, M. Yogi, T. Tateiwa for valuable discussions. The authors are in particular grateful to H. Harima for his kindly providing figures of the Fermi surfaces of $CeRhSi_3$ obtained by the LDA calculation (Fig. 6.13).

References

1. Edelstein, V. M.: Sov. Phys. JETP **68**, 1244 (1989)
2. Gor'kov, L. P., Rashba, E.: Phys. Rev. Lett. **87**, 037004 (2001)
3. Frigeri, P. A., Agterberg, D. F., Koga, A., Sigrist, M.: Phys. Rev. Lett. **92**, 097001 (2004)
4. Sergienko, I. A., Curnoe, S. H.: Phys. Rev. B **70**, 214510 (2004)
5. Bauer, E., Hilscher, G., Michor, H., Paul Ch., Scheidt, E. W., Gribanov, A., Seropegin Yu., Noel, H., Sigrist, M., Rogl, P.: Phys. Rev. Lett **92**, 027003 (2004)
6. Takeuchi, T., Yasuda, T., Tsujino, M., Shishido, H., Settai, R., Harima, H., Onuki, Y.: J. Phys. Soc. Jpn. **76**, 014702 (2007)
7. Bauer, E., Kaldarar, H., Prokofiev, A., Royanian, E., Amato, A., Sereni, J., Bramer-Escamilla, W., Bonalde, I.: J. Phys. Soc. Jpn. **76**, 051009 (2007)
8. Amato, A., Bauer, E., Baines, C.: Phys. Rev. B **71**, 092501 (2005)

9. Yogi, M., Kitaoka, Y., Hashimoto, S., Yasuda, T., Settai, R., Matsuda, T. D., Haga, Y., Onuki, Y., Rogl, P., Bauer, E.: Phys. Rev. Lett **93**, 027003 (2004)
10. Metoki, N., Kaneko, K., Matsuda, T. D., Galatanu, A., Takeuchi, T., Hashimoto, S., Ueda, T., Settai, R., Onuki, Y., Bernhoeft, N.: J. Phys. Condens. Matter **16**, L207 (2004)
11. Yasuda, T., Shishido, H., Ueda, T., Hashimoto, S., Settai, R., Takeuchi, T., Matsuda, T. D., Haga, Y., Onuki, Y.: J. Phys. Soc. Jpn. **73**, 1657 (2004)
12. Tateiwa, T., Haga, Y., Matsuda, T. D., Ikeda, S., Yasuda, T., Takeuchi, T., Settai, R., Onuki, Y.: J. Phys. Soc. Jpn. **74**, 1903 (2005)
13. Izawa, K., Kasahara, Y., Matsuda, Y., Behnia, K., Yasuda, T., Settai, R., Onuki, Y.: Phys. Rev. Lett **94**, 197002 (2005)
14. Bonalde, I., Bramer-Escamilla, W., Bauer, E.: Phys. Rev. Lett **94**, 207002 (2005)
15. Mukuda, H., Nishide, S., Harada, A., Iwasaki, K., Yogi, M., Yashima, M., Kitaoka, Y., Tsujino, M., Takeuchi, T., Settai, R., Onuki, Y., Bauer, E., Itoh, K. M., Haller, E. E.: J. Phys. Soc. Jpn. **78**, 014705 (2009)
16. Yogi, M., Mukuda, H., Kitaoka, Y., Hashimoto, S., Yasuda, T., Settai, R., Matsuda, T. D., Haga, Y., Onuki, Y., Rogl, P., Bauer, E.: J. Phys. Soc. Jpn. **75**, 013709 (2006)
17. Higemoto, W., Haga, Y., Matsuda, T. D., Onuki, Y., Ohishi, K., Ito, T. U., Koda, A., Saha, S. R., Kadono, R.: J. Phys. Soc. Jpn. **75**, 124713 (2006)
18. Yanase, Y., Sigrist, M.: J. Phys. Soc. Jpn. **76**, 043712 (2007)
19. Yanase, Y., Sigrist, M.: J. Phys. Soc. Jpn. **76**, 124709 (2007)
20. Yanase, Y., Sigrist, M.: J. Phys. Soc. Jpn. **77**, 124711 (2008)
21. Motoyama, G., Watanabe, M., Sumiyama, A., Oda, Y.: J. Phys.: Conf. Ser. **150**, 052173 (2009)
22. Samokhin, K. V.: Phys. Rev. Lett **94**, 027004 (2005)
23. Yanase, Y., Mochizuki, M., Ogata, M.: J. Phys. Soc. Jpn. **74**, 430 (2005)
24. A. Kozhevnikov and V. Ansimov, private communication.
25. Samokhin, K. V., Zijlstra, E. S., Bose, S. K.: Phys. Rev. B **69**, 094514 (2004)
26. Hashimoto, S., Yasuda, T., Kubo, T., Shishido, H., Ueda, T., Settai, R., Matsuda, T. D., Haga, Y., Harima, H., Onuki, Y.: J. Phys. Condens. Matter **16**, L287 (2004)
27. Miyake, K., Schmitt-Rink, S., Varma, C. M.: Phys. Rev. B **34**, 6554 (1986)
28. Scalapino, D. J., Loh Jr. E., Hirsch, J. E.: Phys. Rev. B **34**, 8190 (1986)
29. Yanase, Y., Jujo, T., Nomura, T., Ikeda, H., Hotta, T., Yamada, K.: Phys. Rep. **387**, 1 (2003)
30. Machida, K.: J. Phys. Soc. Jpn. **50**, 2195 (1981)
31. Psaltakis, G. C., Fenton, E. W.: J. Phys. C **16**, 3913 (1983)
32. Kato, M., Machida, K.: Phys. Rev. B **37**, 1510 (1988)
33. Murakami, M., Fukuyama, H.: J. Phys. Soc. Jpn. **67**, 2784 (1998)
34. Kyung, B.: Phys. Rev. B **62**, 9083 (2000)
35. Aperis, A., Varelogiannis, G., Littlewood, P. B., Simon, B. D.: . J. Phys. Condens. Matter **20**, 434235 (2008)
36. Pfleiderer, C., Reznik, D., Pintschovius, L., Lohneysen, H. v., Garst, M., Rosch, A.: Nature **427**, 227 (2004)
37. Dzyaloshinsky, I.: J. Phys. Chem. Solids **4**, 241 (1959)
38. Moriya, T.: Phys. Rev. **120**, 91 (1960)
39. Takimoto, T.: J. Phys. Soc. Jpn. **77**, 113706 (2008)
40. Hayashi, N., Wakabayashi, K., Frigeri, P. A., Sigrist, M.: Phys. Rev. B **73**, 024504 (2006)
41. Hayashi, N., Wakabayashi, K., Frigeri, P. A., Sigrist, M.: Phys. Rev. B **73**, 092508 (2006)
42. Fujimoto, S.: Phys. Rev. B **74**, 024515 (2005)
43. Fujimoto, S.: J. Phys. Soc. Jpn. **76**, 034712 (2007)
44. Tada, Y., Kawakami, N., Fujimoto, S.: J. Phys. Soc. Jpn. **77**, 054707 (2008)
45. Mackenzie, A. P., Maeno, Y.: Rev. Mod. Phys. **75**, 657 (2003)
46. Leggett, A. J.: Rev. Mod. Phys. **47**, 331 (1975)
47. Sigrist, M., Ueda, K.: Rev. Mod. Phys. **63**, 239 (1991)
48. Yanase, Y., Ogata, M.: J. Phys. Soc. Jpn. **72**, 673 (2003)

49. Nakatsuji, K., Sumiyama, A., Oda, Y., Yasuda, T., Settai, R., Onuki, Y.: J. Phys. Soc. Jpn. **75**, 084717 (2006)
50. Motoyama, G., Maeda, K., Oda, Y.: J. Phys. Soc. Jpn. **77**, 044710 (2008)
51. Fujimoto, S.: J. Phys. Soc. Jpn. **75**, 083704 (2006)
52. Samokhin, K. V.: Phys. Rev. B **70**, 104521 (2004)
53. Kaur, R. P., Agterberg, D. F., Sigrist, M.: Phys. Rev. Lett **94**, 137002 (2005)
54. Fulde, P., Ferrel, R. A.: Phys. Rev. **135**, A550 (1964)
55. Larkin, A. I., Ovchinnikov Yu. N.: Zh. Eksp. Teor. Fiz. **47**, 1136 (1964)
56. Larkin, A. I., Ovchinnikov Yu. N.: Sov. Phys. JETP **20**, 762 (1965)
57. Frigeri, P. A., Agterberg, D. F., Sigrist, M.: New. J. Phys. **6**, 115 (2004)
58. Dimitrova, O. V., Feigel'man, M. V.: JETP Lett. **78**, 637 (2003)
59. Dimitrova, O. V., Feigel'man, M. V.: Phys. Rev. B **76**, 014522 (2007)
60. Agterberg, D. F., Kaur, R. P.: Phys. Rev. B **75**, 064511 (2007)
61. Matsuda, Y., Shimahara, H.: J. Phys. Soc. Jpn. **76**, 051005 (2007)
62. Matsunaga, Y., Hiasa, N., Ikeda, R.: Phys. Rev. B **78**, 220508(R) (2008)
63. Kimura, N., Ito, K., Saitoh, K., Umeda, Y., Aoki, H., Terashima, T.: Phys. Rev. Lett. **95**, 247004 (2005)
64. Sugitani, I., Okuda, Y., Shishido, H., Yamada, T., Thamizhavel, A., Yamamoto, E., Matsuda, T. D., Haga, Y., Takeuchi, T., Settai, R., Onuki, Y.: J. Phys. Soc. Jpn. **75**, 043703 (2006)
65. Mukuda, H., Fujii, T., Ohara, T., Harada, A., Yashima, M., Kitaoka, Y., Okuda, Y., Settai, R, Ōnuki, Y.:Phys. Rev. Lett. **100**, 107003 (2008)
66. Moriya, T.: Spin Fluctuations in Itinerant Electron Magnetism. Springer, New York (1985)
67. Kimura, N., Ito, K., Aoki, H., Uji, S., Terashima, T.: Phys. Rev. Lett. **98**, 197001 (2007)
68. Settai, R., Miyauchi, Y., Takeuchi, T., Lévy, F., Siieikin, I., Ōnuki, Y.:J. Phys. Soc. Jpn. **74**, 073705 (2008)
69. H. Harima, unpublished.
70. Terashima, T., Takahide, Y., Matsumoto, T., Uji, S., Kimura, N., Aoki, H., Harima, H.: Phys. Rev. B **76**, 054506 (2007)
71. Terashima, T., Kimata, M., Uji, S., Sugawara, T., Kimura, N., Aoki, H., Harima, H.: Phys. Rev. B **78**, 205107 (2008)
72. Aso, N., Miyano, H., Yoshizawa, H., Kimura, N., Komatsubara, T., Aoki, H.: J. Magn. Magn. Mater. **310**, 602 (2007)
73. Monthoux, P., Pines, D.: Phys. Rev. Lett. **69**, 961 (1992)
74. Monthoux, P., Lonzarich, G. G.: Phys. Rev. B **59**, 14598 (1999)
75. Yonemitsu, K.: J. Phys. Soc. Jpn. **58**, 4576 (1989)
76. Fujimoto, S.: J. Phys. Soc. Jpn. **73**, 2061 (2004)
77. Vavilov, M. G., Mineev, V. P.: JETP **86**, 1191 (1998)
78. Norman, M. R., MacDonald, A. H.: Phys. Rev. B **54**, 4239 (1996)
79. Yasui, K., Kita, T.: Phys. Rev. B **66**, 184516 (2002)
80. Tada, Y., Kawakami, N., Fujimoto, S.: Phys. Rev. Lett. **101**, 267006 (2008)
81. Akazawa, T., Hidaka, H., Kotegawa, H., Kobayashi, T., Fujiwara, T., Yamamoto, E., Haga, Y., Settai, R., Ōnuki, Y.: J. Phys. Soc. Jpn. **73**, 3129 (2004)
82. Badica, P., Kondo, T., Togano, K.: J. Phys. Soc. Jpn. **74**, 1014 (2005)
83. Yuan, H. Q., Agterberg, D. F., Hayashi, N., Badica, P., Vandervelde, D., Togano, K., Sigrist, M., Salamon, M. B.: Phys. Rev. Lett. **97**, 017006 (2006)
84. Nishiyama, M., Inada, Y., Zheng, G. -q.: Phys. Rev. Lett. **98**, 047002 (2007)
85. Yokoyama, T., Onari, S., Tanaka, Y.: Phys. Rev. B **75**, 172511 (2007)
86. Yokoyama, T., Onari, S., Tanaka, Y.: Phys. Rev. B **78**, 029902 (2008)
87. Tada, Y., Kawakami, N., Fujimoto, S.: New J. Phys. **11**, 055070 (2009)

Chapter 7
Kinetic Theory for Response and Transport in Non-centrosymmetric Superconductors

Ludwig Klam, Dirk Manske and Dietrich Einzel

Abstract We formulate a kinetic theory for non-centrosymmetric superconductors at low temperatures in the clean limit. The transport equations are solved quite generally in spin- and particle-hole (Nambu) space by performing first a transformation into the band basis and second a Bogoliubov transformation to the quasiparticle-quasihole phase space. Our result is a particle-hole-symmetric, gauge-invariant and charge conserving description, which is valid in the whole quasiclassical regime ($|\boldsymbol{q}| \ll k_F$ and $\hbar\omega \ll E_F$). We calculate the current response, the specific heat capacity, and the Raman response function. For the Raman case, we investigate within this framework the polarization dependence of the electronic (pair-breaking) Raman response for the recently discovered non-centrosymmetric superconductors at zero temperature. Possible applications include the systems $CePt_3$ Si and Li_2 Pd_x Pt_{3-x} B, which reflect the two important classes of the involved spin-orbit coupling. We provide analytical expressions for the Raman vertices for these two classes and calculate the polarization dependence of the electronic spectra. We predict a two-peak structure and different power laws with respect to the unknown relative magnitude of the singlet and triplet contributions to the superconducting order parameter, revealing a large variety of characteristic fingerprints of the underlying condensate.

L. Klam (✉) · D. Manske
Max-Planck-Institut für Festkörperforschung,
Heisenbergstrasse 1,
70569 Stuttgart
e-mail: L.Klam@fkf.mpg.de

e-mail: D. Manske@fkf.mpg.de

D. Einzel
Walther-Meissner-Institut,
Bayerische Akademie der Wissenschaften,
85748 Garching
e-mail: Dietrich.Einzel@wmi.badw.de

E. Bauer and M. Sigrist (eds.), *Non-centrosymmetric Superconductors*,
Lecture Notes in Physics 847, DOI: 10.1007/978-3-642-24624-1_7,

7.1 Introduction

In a large class of conventional and in particular unconventional superconductors a classification of the order parameter with respect to spin singlet/even parity and spin triplet/odd parity is possible, using the Pauli exclusion principle. A necessary prerequisite for such a classification is, however, the existence of an inversion center. Something of a stir has been caused by the discovery of the bulk superconductor $CePt_3Si$ without inversion symmetry [3], which initiated extensive theoretical [12, 26] and experimental studies [2, 11]. In such systems the existence of an antisymmetric potential gradient causes a parity-breaking antisymmetric spin-orbit coupling (ASOC) that gives rise to the possibility of having admixtures of spin-singlet and spin-triplet pairing states. Such parity-violated, non-centrosymmetric superconductors (NCS) are the topic of this chapter, which is dedicated particularly to a theoretical study of response and transport properties at low temperatures. We will use the framework of a kinetic theory described by a set of generalized Boltzmann equations, successfully used before in [10], to derive various response and transport functions such as the normal and superfluid densities, the specific heat capacity (i. e. normal fraction and condensate properties that are native close to the long wavelength, stationary limit) and in particular the electronic Raman response in NCS (which involves frequencies $\hbar\omega$ comparable to the energy gap $\Delta_{\mathbf{k}}$ of the superconductor).

A few general remarks about the connection between response and transport phenomena are appropriate at this stage. Traditionally, the notion of transport implies that the theoretical description takes into account the effects of quasiparticle scattering processes, represented, say, by a scattering rate Γ. Therefore, we would like to demonstrate with a simple example, how response and transport are intimately connected: consider the density response of normal metal electrons to the presence of the two electromagnetic potentials Φ^{ext} and $\mathbf{A}^{\mathrm{ext}}$, which generate the gauge-invariant form of the electric field $\mathbf{E} = -\nabla\Phi^{\mathrm{ext}} - \partial\mathbf{A}/c\partial t$. In Fourier space ($\nabla \to i\mathbf{q}$, $\partial/\partial t \to -i\omega$) one may write for the response of the charge density:

$$\delta n_e = e^2 i\mathbf{q}\cdot\mathbf{M}_0(\mathbf{q},\omega)\cdot\mathbf{E}$$

with $\mathbf{M}_0$ the Lindhard tensor and $\mathbf{q}\cdot\mathbf{M}_0\cdot\mathbf{q} \equiv M_0$ the Lindhard function, appropriately renormalized by collision effects [22]:

$$M_0(\mathbf{q},\omega) = \frac{\mathscr{L}_0(\mathbf{q},\omega+i\Gamma)}{1 - \dfrac{i\Gamma}{\omega+i\Gamma}\left[1 - \dfrac{\mathscr{L}_0(\mathbf{q},\omega+i\Gamma)}{\mathscr{L}_0(\mathbf{q},0)}\right]}$$

Here $\mathscr{L}_0(q,\omega)$ denotes the unrenormalized Lindhard function in the collisionless limit $\Gamma \to 0$:

$$\mathscr{L}_0(\mathbf{q},\omega) = \frac{1}{V}\sum_{\mathbf{p}\sigma}\frac{n^0_{\mathbf{p}+\mathbf{q}/2} - n^0_{\mathbf{p}-\mathbf{q}/2}}{\varepsilon_{\mathbf{p}+\mathbf{q}/2} - \varepsilon_{\mathbf{p}-\mathbf{q}/2} - \hbar\omega}.$$

In this definition of the Lindhard function, $n^0_{\mathbf{k}}$ denotes the equilibrium Fermi-Dirac distribution function and $\varepsilon_{\mathbf{k}} = \xi_{\mathbf{k}} + \mu$ represents the band structure with the chemical potential μ. Now the aspect of transport comes into play by the observation that $M_0(\mathbf{q}, \omega + i\Gamma)$ may be expressed through the full dynamic conductivity tensor $\sigma(\mathbf{q}, \omega) = e^2(\partial n/\partial\mu)\mathbf{D}(\mathbf{q}, \omega)$ of the electron system as follows:

$$M_0(\mathbf{q}, \omega) \equiv \frac{\mathbf{q} \cdot \sigma(\mathbf{q}, \omega) \cdot \mathbf{q}}{i\omega - \mathbf{q} \cdot \mathbf{D}(\mathbf{q}, \omega) \cdot \mathbf{q}/(1 - i\omega\tau)}$$

with $\mathbf{q}\cdot\boldsymbol{\sigma}\cdot\mathbf{q} \overset{\Gamma\to 0}{\equiv} i\omega e^2 \mathscr{L}_0(\mathbf{q}, \omega)$, the particle density n, and with the so-called diffusion pole including the diffusion tensor $\mathbf{D}(\mathbf{q}, \omega)$ in the denominator of $M_0(\mathbf{q}, \omega)$ reflecting the charge conservation law. This expression for the Lindhard response function M_0 clearly demonstrates the connection between response (represented by M_0 itself) and transport (represented by the conductivity σ), which can be evaluated both in the clean limit $\Gamma \to 0$ and in the presence of collisions $\Gamma \neq 0$. In this sense, the notions of response and transport are closely connected and therefore equitable. In this whole chapter we shall limit or considerations to the collisionless case.

An important example for a response phenomenon involving finite frequencies is the electronic Raman effect. Of particular interest is the so-called pair-breaking Raman effect, in which an incoming photon breaks a Cooper pair of energy $2\Delta_{\mathrm{k}}$ on the Fermi surface, and a scattered photon leaves the sample with a frequency reduced by $2\Delta_{\mathrm{k}}/\hbar$, has turned out to be a very effective tool to study unconventional superconductors with gap nodes. This is because various choices of the photon polarization with respect to the location of the nodes on the Fermi surface allow one to draw conclusions about the node topology and hence the pairing symmetry. An example for the success of such an analysis is the important work by Devereaux et al. [6] in which the $d_{x^2-y^2}$ symmetry of the order parameter in cuprate superconductors could be traced back to the frequency dependence of the electronic Raman spectra, that directly measured the pair-breaking effect. Various theoretical studies of NCS have revealed a very rich and complex node structure in parity-mixed order parameters, which can give rise to qualitatively very different shapes, i. e. frequency dependencies, of the Raman intensities, ranging from threshold- and cusp- to singularity-like behavior. Therefore the study of the polarization dependence of Raman spectra enables one to draw conclusions about the internal structure of the parity-mixed gap parameter in a given NCS.

This chapter is organized as follows: In Sect. 7.2 we introduce our model for the ASOC, the two order parameters on the spin-orbit split bands and the pairing interaction. Then, in Sect. 7.3 we derive the kinetic transport equations for NCS at low temperatures in the clean limit and transform these equations into the more convenient band basis. In Sect. 7.4, the transport equations are solved quite generally in band- and particle-hole (Nambu) space by first performing a Bogoliubov transformation to the quasiparticle-quasihole phase space and then performing the inverse Bogoliubov transformation to recover the original distribution functions. We demonstrate gauge invariance of our theory in Sect. 7.5 by taking the fluctuations of the order parameter into account. Within this framework, we calculate the normal and superfluid densities in Sect. 7.6 and the specific heat capacity in Sect. 7.7. In Sect. 7.8, our particular interest is focused on the electronic Raman response. We investigate the polarization dependence of the pair-breaking Raman response at zero temperature for two important classes of the involved spin-orbit coupling. Finally, in Sect. 7.9 we summarize our results and draw our conclusions.

7.2 Antisymmetric Spin-Orbit Coupling

We start from a model Hamiltonian for noninteracting electrons in a non-centrosymmetric crystal [25]

$$\hat{H} = \sum_{\mathbf{k}\sigma\sigma'} \hat{c}^{\dagger}_{\mathbf{k}\sigma} \left[\xi_{\mathbf{k}}\delta_{\sigma\sigma'} + \boldsymbol{\gamma}_{\mathbf{k}} \cdot \boldsymbol{\tau}_{\sigma\sigma'}\right] \hat{c}_{k\sigma'}, \tag{7.1}$$

where $\xi_{\mathbf{k}}$ represents the bare band dispersion assuming time reversal symmetry ($\xi_{-\mathbf{k}} = \xi_{\mathbf{k}}$), $\sigma, \sigma' = \uparrow, \downarrow$ label the spin state, and $\boldsymbol{\tau}$ are the Pauli matrices. The second term describes an antisymmetric spin-orbit coupling (ASOC) with a (vectorial) coupling constant $\boldsymbol{\gamma}_{\mathbf{k}}$. The pseudovector function $\boldsymbol{\gamma}_{\mathbf{k}}$ has the following symmetry properties: $\gamma_{-\mathbf{k}} = -\boldsymbol{\gamma}_{\mathbf{k}}$ and $g\gamma_{g^{-1}\mathbf{k}} = \boldsymbol{\gamma}_{\mathbf{k}}$. Here g denotes any symmetry operation of the point group $\mathcal{G}$ of the crystal under consideration. In NCSs two important classes of ASOCs are realized, reflecting the underlying point group $\mathcal{G}$ of the crystal. In particular, we shall be interested in the tetragonal point group C_{4v} (applicable to the heavy fermion compound $CePt_3Si$ with $T_c = 0.75$K[3] for example) and the cubic point group O (applicable to the system Li_2 Pd_x Pt_{3-x} B with T_c =2.2–2.8 K for x=0 and T_c =7.2–8 K for x=3 [1]). For $\mathcal{G} = C_{4v}$ the ASOC reads [14, 25]

$$\boldsymbol{\gamma}_{\mathbf{k}} = g_{\perp}(\hat{\mathbf{k}} \times \hat{\mathbf{e}}_z) + g_{\parallel}\hat{k}_x\hat{k}_y\hat{k}_z(\hat{k}_x^2 - \hat{k}_y^2)\hat{\mathbf{e}}_z. \tag{7.2}$$

In the purely two-dimensional case ($g_{\parallel} = 0$) one recovers, what is known as the Rashba interaction [7, 8, 15]. We will choose for simplicity $g_{\parallel} = 0$ for our Raman results. For the cubic point group $\mathcal{G} = O$, $\boldsymbol{\gamma}_{\mathbf{k}}$ reads [30]

$$\boldsymbol{\gamma}_{\mathbf{k}} = g_1\hat{\mathbf{k}} - g_3\left[\hat{k}_x(\hat{k}_y^2 + \hat{k}_z^2)\hat{\mathbf{e}}_x + \hat{k}_y(\hat{k}_z^2 + \hat{k}_x^2)\hat{\mathbf{e}}_y + \hat{k}_z(\hat{k}_x^2 + \hat{k}_y^2)\hat{\mathbf{e}}_z\right], \tag{7.3}$$

where the ratio $g_3/g_1 \simeq 3/2$ is estimated by Ref. [30]. Because of the larger prefactor $g_3 > g_1$ we will keep the higher order term for our further considerations. Thus, in terms of spherical angles, $\hat{\mathbf{k}} = (\cos\phi\sin\theta, \sin\phi\sin\theta, \cos\theta)$, the absolute value of the $\boldsymbol{\gamma}_{\mathbf{k}}$-vectors for both point groups, illustrated in Fig. 7.1, reads

$$|\boldsymbol{\gamma}_{\mathbf{k}}| = \sin\theta \qquad \text{for } C_{4v} \tag{7.4}$$

$$|\boldsymbol{\gamma}_{\mathbf{k}}| = \sqrt{1 - \frac{15}{16}\sin^2 2\theta - \frac{3}{16}\sin^4\theta\sin^2 2\phi\left(9\sin^2\theta - 4\right)} \quad \text{for } O \tag{7.5}$$

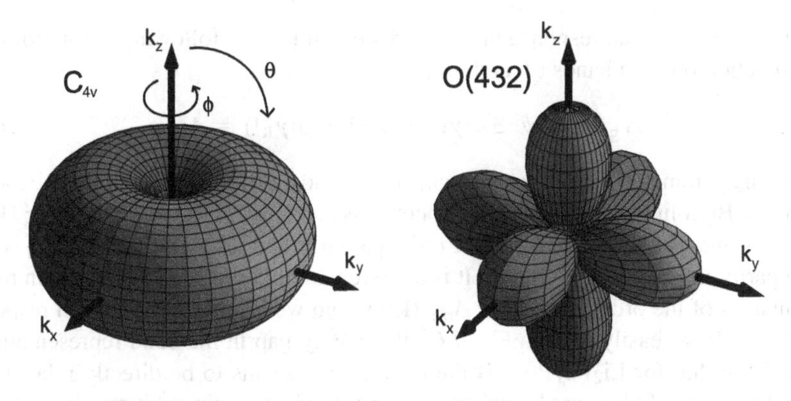

Fig. 7.1 The angular dependence of $|\boldsymbol{\gamma}_{\mathbf{k}}|$ for the point groups C_{4v} and O. Since $\mathbf{d}_{\mathbf{k}} || \boldsymbol{\gamma}_{\mathbf{k}}$, these plots show also the magnitude of the gap function in the pure triplet case for both point groups

By diagonalizing the Hamiltonian of Eq. (7.1), one finds the eigenvalues $\xi_\lambda(\mathbf{k}) = \xi_{\mathbf{k}} + \lambda|\boldsymbol{\gamma}_{\mathbf{k}}|$, which physically correspond to the lifting of the Kramers degeneracy between the two spin states at a given $\mathbf{k}$ in the presence of ASOC. The basis in which the band is diagonal can be referred to as the *band basis* where the Fermi surface defined by $\xi_\pm(\mathbf{k}) = 0$ is split into two parts labeled $\pm$. Sigrist and co-workers have shown that the presence of the ASOC generally allows for an admixture of a spin-triplet component to the otherwise spin-singlet pairing gap [12]. This implies that we may write down the following ansatz for the energy gap matrix in spin space:

$$\Delta_{\sigma\sigma'}(\mathbf{k}) = \{[\psi_{\mathbf{k}}(T)\mathbf{1} + \mathbf{d}_{\mathbf{k}}(T) \cdot \boldsymbol{\tau}] i\tau^y\}_{\sigma\sigma'}, \tag{7.6}$$

where $\psi_{\mathbf{k}}(T)$ and $\mathbf{d}_{\mathbf{k}}(T)$ reflect the singlet and triplet part of the pair potential, respectively. In the band basis we find immediately

$$\Delta_\pm(\mathbf{k}) = \psi_{\mathbf{k}}(T) \pm |\mathbf{d}_{\mathbf{k}}(T)|. \tag{7.7}$$

It has been demonstrated that a large ASOC compared to $k_B T_c$ is not detrimental to triplet pairing if one assumes $\mathbf{d}_{\mathbf{k}} \| \boldsymbol{\gamma}_{\mathbf{k}}$ [12, 26]:

$$\mathbf{d}_{\mathbf{k}}(T) = d(T)\hat{\boldsymbol{\gamma}}_{\mathbf{k}}, \tag{7.8}$$

whereas the temperature-dependent magnitudes $\psi(T)$ and $d(T)$ of the spin-singlet and triplet energy gaps are solutions of coupled self-consistency equations and $\hat{\boldsymbol{\gamma}}_{\mathbf{k}}$ is defined by

$$\hat{\boldsymbol{\gamma}}_{\mathbf{k}} = \frac{\boldsymbol{\gamma}_{\mathbf{k}}}{\sqrt{\langle|\boldsymbol{\gamma}_{\mathbf{k}}|^2\rangle_{FS}}}. \tag{7.9}$$

Thus the energy gap of Eq. (7.7) can be written as:

$$\Delta_\pm(\mathbf{k}) = \psi(T) \pm d(T)|\hat{\boldsymbol{\gamma}}_{\mathbf{k}}|. \tag{7.10}$$

For the $T = 0$ Raman response in Sect. 7.8 we will use the following ansatz for the gap function on both bands (+ and −) [13]:

$$\Delta_{\pm}(\mathbf{k}) = \psi \pm d|\boldsymbol{\gamma}_{\mathbf{k}}| = \psi\,(1 \pm p|\boldsymbol{\gamma}_{\mathbf{k}}|) \equiv \Delta_{\pm}, \tag{7.11}$$

where the parameter $p = d/\psi$ represents the unknown triplet-singlet ratio. Accordingly, the Bogoliubov quasiparticle dispersion is given by $E_{\pm}^2(\mathbf{k}) = \xi_{\pm}^2(\mathbf{k}) + \Delta_{\pm}^2(\mathbf{k})$. If we assume no $\mathbf{q}$-dependence of the order parameter, $\Delta_{\lambda}(\mathbf{k})$ [and also $E_{\lambda}(\mathbf{k})$] is of even parity i.e. $\Delta_{\lambda}(-\mathbf{k}) = \Delta_{\lambda}(\mathbf{k})$. It is quite remarkable that although the spin representation of the order parameter $\Delta_{\sigma\sigma'}(\mathbf{k})$ has no well-defined parity with respect to $\mathbf{k} \to -\mathbf{k}$, as easily seen in Eq. (7.6), the energy gap in the band representation does. Note that for $Li_2Pd_xPt_{3-x}B$ the parameter p seems to be directly related to the substitution of platinum by palladium, since the larger spin-orbit coupling of the heavier platinum is expected to enhance the triplet contribution [20]. This seems to be confirmed by penetration-depth experiments [31, 30].

The corresponding weak-coupling gap equation reads

$$\Delta_{\lambda}(\mathbf{k}, T) = -\sum_{\mathbf{k}',\mu} V_{\mathbf{k}\mathbf{k}'}^{\lambda\mu} \Delta_{\mu}(\mathbf{k}', T)\theta_{\mu}(\mathbf{k}') \tag{7.12}$$

with

$$\theta_{\lambda}(\mathbf{k}) = \frac{1}{2E_{\lambda}(\mathbf{k})} \tanh \frac{E_{\lambda}(\mathbf{k})}{2k_B T} \tag{7.13}$$

and its solution are extensively discussed in Ref. [13]. Here and in the following we choose a separable ansatz for the pairing-interaction (cf. Ref. [13] with $e_m = 0$, i.e. without Dzyaloshinskii-Moriya interaction):

$$V_{\mathbf{k}\mathbf{k}'}^{\lambda\mu} = \Gamma_s + \lambda\mu\Gamma_t|\hat{\boldsymbol{\gamma}}_{\mathbf{k}}||\hat{\boldsymbol{\gamma}}_{\mathbf{k}'}|, \tag{7.14}$$

where Γ_s and Γ_t represent the singlet and triplet contribution, respectively. Although an exact numerical solution of Eqs. (7.12)–(7.14) with a microscopic pairing interaction would be desirable, we restrict ourselves in this work to a phenomenological description which allows for an analytical treatment of response and transport in NCS.

7.3 Derivation of the Transport Equations

In this section, we study the linear response of the superconducting system to an effective external perturbation potential of the form

$$\delta\xi_{\mathbf{k}\sigma\sigma'}^{\text{ext}} = \left[e\Phi(\mathbf{q}, \omega) - \frac{e}{c}\mathbf{v}_{\mathbf{k}} \cdot \mathbf{A}(\mathbf{q}, \omega)\right]\delta_{\sigma\sigma'} + \frac{e^2}{c^2}\mathbf{A}_i^I(\mathbf{q}, \omega)\frac{\partial^2\varepsilon_{\mathbf{k}}}{\hbar^2\partial\mathbf{k}_i\partial\mathbf{k}_j}\mathbf{A}_j^S(\mathbf{q}, \omega)\delta_{\sigma\sigma'}. \tag{7.15}$$

Table 7.1 External perturbations can be decomposed into a vertex function and a potential

vertex $a_{\mathbf{k}}$	(fictive) potential $\delta\xi_a$	parity	dimension	response
e	Φ^{ext}	even	scalar	charge and
$e\mathbf{v_k}$	$\mathbf{A}^{\text{ext}}$	odd	vector	current
$\left(\frac{E_\lambda(\mathbf{k})}{T} - \frac{\partial E_\lambda(\mathbf{k})}{\partial T}\right)$	ΔT	even	scalar	specific heat capacity
$m(\mathbf{M}_{\mathbf{k}}^{-1})_{i,j}$	$r_0 A_i^I A_j^S$	even	tensor	Raman

The vertex function is characteristic for each response function and can be classified according to parity (with respect to $\mathbf{k} \to -\mathbf{k}$) and dimension

Here Φ and $\mathbf{A}$ denote the electromagnetic scalar and vector potential. Electronic Raman scattering is described by the third term in Eq. (7.15). It describes a Raman process where an incoming photon with vector potential $\mathbf{A}^I$, polarization $\hat{\mathbf{e}}^I$ and frequency ω_I is scattered off an electronic excitation. The scattered photon with vector potential $\mathbf{A}^S$, polarization $\hat{\mathbf{e}}^S$ and frequency $\omega_S = \omega_I - \omega$ gives rise to a Raman signal (Stokes process) and creates an electronic excitation with momentum transfer $\mathbf{q}$. Further, the Raman vertex in the so-called effective-mass approximation reads

$$\gamma_{\mathbf{k}}^{(R)} = m \sum_{i,j} \hat{\mathbf{e}}_i^S \frac{\partial^2 \varepsilon(\mathbf{k})}{\hbar^2 \partial k_i \partial k_j} \hat{\mathbf{e}}_j^I. \tag{7.16}$$

In general, an external perturbation can be decomposed into a vertex function $a_{\mathbf{k}}$ and a related potential $\delta\xi_a$:

$$\delta\xi_{\mathbf{k}}^{\text{ext}} = \sum_a a_{\mathbf{k}} \delta\xi_a. \tag{7.17}$$

A list of all relevant vertex functions and potentials that will be discussed in this chapter is given in Table 7.1. The charge density response to the electric field $\mathbf{E} = -\nabla\Phi^{\text{ext}} - \partial\mathbf{A}/c\partial t$ is characterized by a constant vertex $a_{\mathbf{k}} = e$ (electron charge) and therefore of even parity (with respect to $\mathbf{k} \to -\mathbf{k}$), whereas the current response to the vector potential $\mathbf{A}$ depends on the odd vertex function $a_{\mathbf{k}} = e\mathbf{v_k}$ (electron velocity). In case of the specific heat capacity $C_V(T)$, the role of the fictive potential is played by the temperature change δT, which couples to the energy variable $\xi_{\mathbf{k}}$. For the Raman response, this fictive potential depends essentially on the vector potential of the incoming and scattered light. The response and transport functions will be obtained as moments of the momentum distribution functions with the corresponding vertex (see Sects. 7.6–7.8).

In addition to the external perturbation potentials, molecular potentials can be taken into account within a mean-field approximation:

$$\delta\xi_{\mathbf{k}} = \delta\xi_{\mathbf{k}}^{\text{ext}} + \sum_{\mathbf{p}\sigma} \left(f_{\mathbf{kp}}^s + V_{\mathbf{q}}\right) \delta n_{\mathbf{p}} = \sum_a a_{\mathbf{k}} \delta\xi_a + V_{\mathbf{q}} \delta n_1 + \sum_{\mathbf{p}\sigma} f_{\mathbf{kp}}^s \delta n_{\mathbf{p}}. \tag{7.18}$$

The short-range Fermi-liquid interaction $f^s_{\mathbf{kp}}$ leads to a renormalization of the electron mass [24] and the long-ranged Coulomb interaction with $V_{\mathbf{q}} = 4\pi e^2/\mathbf{q}^2$ is included self-consistently through the macroscopic density fluctuations $\delta n_1 = \sum_{\mathbf{p}\sigma} \delta n_{\mathbf{p}}$ with the non-equilibrium momentum distribution function $\delta n_{\mathbf{p}}$.

The potentials $\delta\xi^{\text{ext}}_{\mathbf{k}}$ are assumed to vary in time and space $\propto \exp(i\mathbf{q}\cdot\mathbf{r} - i\omega t)$. Then the response to the perturbation potentials can generally be described by a non-equilibrium momentum distribution function $\underline{n}_{\mathbf{pp}'}$, which is a matrix in Nambu, momentum and spin space with $\mathbf{p} = \mathbf{k} + \mathbf{q}/2$ and $\mathbf{p}' = \mathbf{k} - \mathbf{q}/2$. The evolution of the non-equilibrium matrix distribution function in time and space is governed by the matrix kinetic (von Neumann) equation [4, 29]

$$\omega \underline{n}_{\mathbf{pp}'} + \sum_{\mathbf{p}''} \left[\underline{n}_{\mathbf{pp}''}, \underline{\xi}_{\mathbf{p}''\mathbf{p}'}\right]_- = 0 \tag{7.19}$$

in which the full quasiparticle energy $\underline{\xi}_{\mathbf{pp}'}$ plays the role of the Hamiltonian of the system. This equation holds for $\hbar\omega \ll E_{\text{F}}$ and $|\mathbf{q}| \ll k_{\text{F}}$. In general, a collision integral (see e.g. [29]) could be inserted on the right-hand side of Eq. (7.19) that accounts for the relaxation of the system into local-equilibrium through collisions. In the following we will assume the absence of collisions.[1] After linearization according to

$$\underline{n}_{\mathbf{pp}'} = \underline{n}_{\mathbf{k}}(\mathbf{q},\omega) = \underline{n}^0_{\mathbf{k}}\delta_{\mathbf{q},0} + \delta\underline{n}_{\mathbf{k}}(\mathbf{q},\omega) \tag{7.20}$$

$$\underline{\xi}_{\mathbf{pp}'} = \underline{\xi}_{\mathbf{k}}(\mathbf{q},\omega) = \underline{\xi}^0_{\mathbf{k}}\delta_{\mathbf{q},0} + \delta\underline{\xi}_{\mathbf{k}}(\mathbf{q},\omega), \tag{7.21}$$

the matrix kinetic equation assumes the following form in $\omega - \mathbf{q}$- and spin-space:

$$\omega\delta\underline{n}_{\mathbf{k}} + \delta\underline{n}_{\mathbf{k}}\underline{\xi}^0_{\mathbf{k}-} - \underline{\xi}^0_{\mathbf{k}+}\delta\underline{n}_{\mathbf{k}} = \delta\underline{\xi}_{\mathbf{k}}\underline{n}^0_{\mathbf{k}-} - \underline{n}^0_{\mathbf{k}+}\delta\underline{\xi}_{\mathbf{k}}. \tag{7.22}$$

Here, ω is the frequency and $\mathbf{k}\pm = \mathbf{k} \pm \mathbf{q}/2$, with $\mathbf{q}$ representing the wave number of the external perturbation. The equilibrium distribution function $\underline{n}^0_{\mathbf{k}}$ and quasiparticle energy $\underline{\xi}^0_{\mathbf{k}}$ are matrices in Nambu and spin space:

$$\underline{n}^0_{\mathbf{k}} = \begin{pmatrix} \mathbf{n}_{\mathbf{k}} & \mathbf{g}_{\mathbf{k}} \\ \mathbf{g}^\dagger_{\mathbf{k}} & \mathbf{1} - \mathbf{n}_{-\mathbf{k}} \end{pmatrix} \tag{7.23}$$

$$\underline{\xi}^0_{\mathbf{k}} = \begin{pmatrix} \xi_{\mathbf{k}} + \boldsymbol{\gamma}_{\mathbf{k}}\cdot\tau & \Delta_{\mathbf{k}} \\ \Delta^\dagger_{\mathbf{k}} & -\left[\xi_{\mathbf{k}} - \boldsymbol{\gamma}_{\mathbf{k}}\cdot\tau\right]^T \end{pmatrix}. \tag{7.24}$$

---

[1] An example for collision integrals in the Raman response can be found in Ref. [10]

The momentum and frequency-dependent deviations from equilibrium are defined as

$$\delta \underline{n}_{\mathbf{k}}^{0} = \begin{pmatrix} \delta \mathbf{n}_{\mathbf{k}} & \delta \mathbf{g}_{\mathbf{k}} \\ \delta \mathbf{g}_{\mathbf{k}}^{\dagger} & -\delta \mathbf{n}_{-\mathbf{k}} \end{pmatrix} \tag{7.25}$$

and

$$\delta \underline{\xi}_{\mathbf{k}}^{0} = \begin{pmatrix} \delta \xi_{\mathbf{k}} & \delta \Delta_{\mathbf{k}} \\ \delta \Delta_{\mathbf{k}}^{\dagger} & -\delta \xi_{-\mathbf{k}} \end{pmatrix}, \tag{7.26}$$

respectively. In the spin basis, the matrix kinetic equation Eq. (7.22) represents a set of 16 equations, which can be reduced to a set of 8 equations by an unitary transformation into the band basis (also referred to as helicity basis). This SU(2) rotation is given by [28]

$$\underline{U}_{\mathbf{k}} = \begin{pmatrix} \mathbf{U}_{\mathbf{k}} & 0 \\ 0 & \mathbf{U}_{\mathbf{k}}^{*} \end{pmatrix} \tag{7.27}$$

$$\mathbf{U}_{\mathbf{k}} = \exp\left(-i \frac{\theta_{\gamma}}{2} \hat{\mathbf{n}}_{\gamma} \cdot \tau \right) = \cos \frac{\theta_{\gamma}}{2} - i \hat{\mathbf{n}}_{\gamma} \cdot \tau \sin \frac{\theta_{\gamma}}{2} \tag{7.28}$$

$$\mathbf{n}_{\gamma} = \frac{\boldsymbol{\gamma}_{\mathbf{k}} \times \hat{\mathbf{z}}}{|\boldsymbol{\gamma}_{\mathbf{k}} \times \hat{\mathbf{z}}|}, \tag{7.29}$$

which corresponds to a rotation in spin space into the $\hat{\mathbf{z}}$-direction about the polar angle θ_{γ} between $\boldsymbol{\gamma}_{\mathbf{k}}$ and $\hat{\mathbf{z}}$, Then Eq. (7.22) may be written as

$$\begin{aligned} \omega \underline{U}_{\mathbf{k}+}^{\dagger} \delta \underline{n}_{\mathbf{k}} \underline{U}_{\mathbf{k}-} & + \underline{U}_{\mathbf{k}+}^{\dagger} \delta \underline{n}_{\mathbf{k}} \underline{U}_{\mathbf{k}-} \underline{U}_{\mathbf{k}-}^{\dagger} \underline{\xi}_{\mathbf{k}-}^{0} \underline{U}_{\mathbf{k}-} \quad - \underline{U}_{\mathbf{k}+}^{\dagger} \underline{\xi}_{\mathbf{k}+}^{0} \underline{U}_{\mathbf{k}+} \underline{U}_{\mathbf{k}+}^{\dagger} \delta \underline{n}_{\mathbf{k}} \underline{U}_{\mathbf{k}-} \\ & = \underline{U}_{\mathbf{k}+}^{\dagger} \delta \underline{\xi}_{\mathbf{k}} \underline{U}_{\mathbf{k}-} \underline{U}_{\mathbf{k}-}^{\dagger} \underline{n}_{\mathbf{k}-}^{0} \underline{U}_{\mathbf{k}-} \quad - \underline{U}_{\mathbf{k}+}^{\dagger} \underline{n}_{\mathbf{k}+}^{0} \underline{U}_{\mathbf{k}+} \underline{U}_{\mathbf{k}+}^{\dagger} \delta \underline{\xi}_{\mathbf{k}} \underline{U}_{\mathbf{k}-} \end{aligned} \tag{7.30}$$

or, more simply

$$\omega \delta \underline{n}_{\mathbf{k}}^{b} + \delta \underline{n}_{\mathbf{k}}^{b} \underline{\xi}_{\mathbf{k}-}^{b} - \underline{\xi}_{\mathbf{k}+}^{b} \delta \underline{n}_{\mathbf{k}}^{b} = \delta \underline{\xi}_{\mathbf{k}}^{b} \underline{n}_{\mathbf{k}-}^{b} - \underline{n}_{\mathbf{k}+}^{b} \delta \underline{\xi}_{\mathbf{k}}^{b}, \tag{7.31}$$

where the equilibrium distribution function and energy shifts in the band basis are given by

$$\underline{n}_{\mathbf{k}}^{b} = \begin{pmatrix} \frac{1}{2}(1 - \xi_{+}\theta_{+}) & 0 & 0 & -\Delta_{+}\theta_{+} \\ 0 & \frac{1}{2}(1 - \xi_{-}\theta_{-}) & \Delta_{-}\theta_{-} & 0 \\ 0 & \Delta_{-}^{*}\theta_{-} & \frac{1}{2}(1 + \xi_{-}\theta_{-}) & 0 \\ -\Delta_{+}^{*}\theta_{+} & 0 & 0 & \frac{1}{2}(1 + \xi_{+}\theta_{+}) \end{pmatrix} \tag{7.32}$$

and

$$\underline{\xi}_{\mathbf{k}}^{b} = \begin{pmatrix} \xi_+ & 0 & 0 & \Delta_+ \\ 0 & \xi_- & -\Delta_- & 0 \\ 0 & -\Delta_-^* & -\xi_- & 0 \\ \Delta_+^* & 0 & 0 & -\xi_+ \end{pmatrix}. \tag{7.33}$$

The deviations from equilibrium can be parameterized as follows:

$$\delta \underline{n}_{\mathbf{k}}^{b} = \underline{U}_{\mathbf{k}+}^{\dagger} \delta \underline{n}_{\mathbf{k}} \underline{U}_{\mathbf{k}-} = \begin{pmatrix} \delta n_+^b & 0 & 0 & \delta g_+^b \\ 0 & \delta n_-^b & -\delta g_-^b & 0 \\ 0 & -\delta g_-^{b*} & -\delta n_-^b & 0 \\ \delta g_+^{b*} & 0 & 0 & -\delta n_+^b \end{pmatrix} \tag{7.34}$$

$$\delta \underline{\xi}_{\mathbf{k}}^{b} = \underline{U}_{\mathbf{k}+}^{\dagger} \delta \underline{\xi}_{\mathbf{k}} \underline{U}_{\mathbf{k}-} = \begin{pmatrix} \delta \xi_+^b & 0 & 0 & \delta \Delta_+^b \\ 0 & \delta \xi_-^b & -\delta \Delta_-^b & 0 \\ 0 & -\delta \Delta_-^{b*} & -\delta \xi_-^b & 0 \\ \delta \Delta_+^{b*} & 0 & 0 & -\delta \xi_+^b \end{pmatrix}. \tag{7.35}$$

Thus, we have now derived a set of equations in spin and band basis [Eqs. (7.22) and (7.31)] that allow us to determine the diagonal and off-diagonal non-equilibrium momentum distribution functions. In Sects. 7.6–7.8 we will use these distribution functions to determine the normal and superfluid densities, specific heat capacity, and the Raman response of NCS. From now on, we will omit the index "b" indicating the band basis, since all further considerations will be made in the band picture.

7.4 Solution by Bogoliubov Transformation

In what follows we will solve the kinetic equation (7.31), derived in the previous section. For this purpose, we perform first a Bogoliubov transformation into quasiparticle space, where the kinetic equations are easily decoupled and then solved. For the subsequent inverse Bogoliubov transformation we will introduce parity projected quantities to obtain finally a relation between the diagonal and off-diagonal energy shifts on the one side and the non-equilibrium distribution functions on the other. As a fist step towards the solution of the kinetic equations, the momentum distribution matrix $\underline{n}_{\mathbf{k}}$ and the energy matrix $\underline{\xi}_{\mathbf{k}}$ (both in band basis) are diagonalized through the following Bogoliubov transformation

$$\underline{\nu}_{\mathbf{k}} = \underline{B}_{\mathbf{k}}^{\dagger} \underline{n}_{\mathbf{k}} \underline{B}_{\mathbf{k}} = \begin{pmatrix} f(E_+) & 0 & 0 & 0 \\ 0 & f(E_-) & 0 & 0 \\ 0 & 0 & f(-E_-) & 0 \\ 0 & 0 & 0 & f(-E_+) \end{pmatrix} \tag{7.36}$$

$$\underline{E}_{\mathbf{k}} = \underline{B}_{\mathbf{k}}^{\dagger} \underline{\xi}_{\mathbf{k}} \underline{B}_{\mathbf{k}} = \begin{pmatrix} E_+ & 0 & 0 & 0 \\ 0 & E_- & 0 & 0 \\ 0 & 0 & -E_- & 0 \\ 0 & 0 & 0 & -E_+ \end{pmatrix} \tag{7.37}$$

with the Fermi-Dirac distribution function $f(E_\lambda) = [\exp(E_\lambda / k_{\mathrm{B}} T) + 1]^{-1}$. The Bogoliubov matrix has been found to read in the band basis

$$\underline{B}_{\mathbf{k}} = \begin{pmatrix} u_+ & 0 & 0 & v_+ \\ 0 & u_- & -v_- & 0 \\ 0 & v_-^* & u_- & 0 \\ -v_+^* & 0 & 0 & u_+ \end{pmatrix} \tag{7.38}$$

with the coherence factors

$$u_\lambda(\mathbf{k}) = \sqrt{\frac{1}{2}\left(1 + \frac{\xi_\lambda(\mathbf{k})}{E_\lambda(\mathbf{k})}\right)} \tag{7.39}$$

$$v_\lambda(\mathbf{k}) = -\sqrt{\frac{1}{2}\left(1 - \frac{\xi_\lambda(\mathbf{k})}{E_\lambda(\mathbf{k})}\right)} \frac{\Delta_\lambda(\mathbf{k})}{|\Delta_\lambda(\mathbf{k})|} \tag{7.40}$$

satisfying the condition $|u_\lambda|^2 + |v_\lambda|^2 = 1$, by which the fermionic character of the Bogoliubov quasiparticles is established. In order to solve the transport equation in the band basis (7.31), one may multiply from the left with the Bogoliubov matrix $\underline{B}_{\mathbf{k}+}^{\dagger}$ and from the right with $\underline{B}_{\mathbf{k}-}$. The result is

$$\begin{aligned} \omega \underline{B}_{\mathbf{k}+}^{\dagger} \delta \underline{n}_{\mathbf{k}} \underline{B}_{\mathbf{k}-} &+ \underline{B}_{\mathbf{k}+}^{\dagger} \delta \underline{n}_{\mathbf{k}} \underline{B}_{\mathbf{k}-} \underline{B}_{\mathbf{k}-}^{\dagger} \underline{\xi}_{\mathbf{k}-}^{0} \underline{B}_{\mathbf{k}-} - \underline{B}_{\mathbf{k}+}^{\dagger} \underline{\xi}_{\mathbf{k}+}^{0} \underline{B}_{\mathbf{k}+} \underline{B}_{\mathbf{k}+}^{\dagger} \delta \underline{n}_{\mathbf{k}} \underline{B}_{\mathbf{k}-} \\ &= \underline{B}_{\mathbf{k}+}^{\dagger} \delta \underline{\xi}_{\mathbf{k}} \underline{B}_{\mathbf{k}-} \underline{B}_{\mathbf{k}-}^{\dagger} \underline{n}_{\mathbf{k}-}^{0} \underline{B}_{\mathbf{k}-} - \underline{B}_{\mathbf{k}+}^{\dagger} \underline{n}_{\mathbf{k}+}^{0} \underline{B}_{\mathbf{k}+} \underline{B}_{\mathbf{k}+}^{\dagger} \delta \underline{\xi}_{\mathbf{k}} \underline{B}_{\mathbf{k}-} \end{aligned} \tag{7.41}$$

or, more simply

$$\omega \delta \underline{\nu}_{\mathbf{k}} + \delta \underline{\nu}_{\mathbf{k}} \underline{E}_{\mathbf{k}-} - \underline{E}_{\mathbf{k}+} \delta \underline{\nu}_{\mathbf{k}} = \delta \underline{E}_{\mathbf{k}} \underline{\nu}_{\mathbf{k}-} - \underline{\nu}_{\mathbf{k}+} \delta \underline{E}_{\mathbf{k}}. \tag{7.42}$$

The new Bogoliubov-transformed quantities describing the deviation from equilibrium are identified from the preceding equations and labeled as follows:

$$\delta \underline{\nu}(\mathbf{k}) = \underline{B}_{\mathbf{k}+}^{\dagger} \delta \underline{n}_{\mathbf{k}} \underline{B}_{\mathbf{k}-} = \begin{pmatrix} \delta\nu_+(\mathbf{k}) & 0 & 0 & \delta\gamma_+(\mathbf{k}) \\ 0 & \delta\nu_-(\mathbf{k}) & -\delta\gamma_-(\mathbf{k}) & 0 \\ 0 & -\delta\gamma_-^*(\mathbf{k}) & -\delta\nu_-(-\mathbf{k}) & 0 \\ \delta\gamma_+^*(\mathbf{k}) & 0 & 0 & -\delta\nu_+(-\mathbf{k}) \end{pmatrix} \tag{7.43}$$

$$\delta \underline{E}(\mathbf{k}) = \underline{B}^{\dagger}_{\mathbf{k}+} \delta \underline{\xi}_{\mathbf{k}} \underline{B}_{\mathbf{k}-} = \begin{pmatrix} \delta E_+(\mathbf{k}) & 0 & 0 & \delta D_+(\mathbf{k}) \\ 0 & \delta E_-(\mathbf{k}) & -\delta D_-(\mathbf{k}) & 0 \\ 0 & -\delta D^*_-(\mathbf{k}) & -\delta E_-(-\mathbf{k}) & 0 \\ \delta D^*_+(\mathbf{k}) & 0 & 0 & -\delta E_+(-\mathbf{k}) \end{pmatrix}. \tag{7.44}$$

The solution of Eq. (7.42) for the quasiparticle distribution functions is the set of the following eight equations ($\lambda = \pm$):

$$\delta \nu_\lambda(\mathbf{k}) = \frac{\eta^-_\lambda(\mathbf{k})}{\omega - \eta^-_\lambda(\mathbf{k})} \tilde{y}_\lambda(\mathbf{k}) \delta E_\lambda(\mathbf{k}), \tag{7.45a}$$

$$\delta \nu_\lambda(-\mathbf{k}) = -\frac{\eta^-_\lambda(\mathbf{k})}{\omega + \eta^-_\lambda(\mathbf{k})} \tilde{y}_\lambda(\mathbf{k}) \delta E_\lambda(-\mathbf{k}), \tag{7.45b}$$

$$\delta \gamma_\lambda(\mathbf{k}) = \frac{\eta^+_\lambda(\mathbf{k})}{\omega - \eta^+_\lambda(\mathbf{k})} \Theta_\lambda(\mathbf{k}) \delta D_\lambda(\mathbf{k}), \tag{7.45c}$$

$$\delta \gamma^*_\lambda(\mathbf{k}) = -\frac{\eta^+_\lambda(\mathbf{k})}{\omega + \eta^+_\lambda(\mathbf{k})} \Theta_\lambda(\mathbf{k}) \delta D^*_\lambda(\mathbf{k}), \tag{7.45d}$$

where we have introduced the following abbreviations:

$$\eta^\pm_\lambda(\mathbf{k}) = E_\lambda(\mathbf{k}+) \pm E_\lambda(\mathbf{k}-), \tag{7.46}$$

$$\tilde{y}_\lambda(\mathbf{k}) = -\frac{f[E_\lambda(\mathbf{k}+)] - f[E_\lambda(\mathbf{k}-)]}{E_\lambda(\mathbf{k}+) - E_\lambda(\mathbf{k}-)}, \tag{7.47}$$

and

$$\Theta_\lambda(\mathbf{k}) = \frac{1 - f[E_\lambda(\mathbf{k}+)] - f[E_\lambda(\mathbf{k}-)]}{E_\lambda(\mathbf{k}+) + E_\lambda(\mathbf{k}-)}. \tag{7.48}$$

The expressions for these quantities in the long-wavelength limit can be found in appendix 1. In this limit, the difference quotient $\tilde{y}_\lambda(\mathbf{k})$ is equal to the Yosida kernel $y_\lambda(\mathbf{k})$ which is given by the derivative of the quasiparticle distribution function

$$y_\lambda(\mathbf{k}) = -\frac{\partial f[E_\lambda(\mathbf{k})]}{\partial E_\lambda(\mathbf{k})} = \frac{1}{4k_{\mathrm{B}}T} \frac{1}{\cosh^2\left(\frac{E_\lambda(\mathbf{k})}{2k_{\mathrm{B}}T}\right)} \tag{7.49}$$

and is crucial for the temperature dependence of all response and transport functions. Accordingly, $\Theta_\lambda(\mathbf{k}) \overset{\mathbf{q}\to 0}{\to} \theta_\lambda(\mathbf{k})$ represents the kernel of the self-consistency Eq. (7.12). It is instructive to note that the distribution functions $\delta \nu_\lambda(\mathbf{k})$ and $\delta \gamma_\lambda(\mathbf{k})$ have a clear physical meaning: The diagonal component $\delta \nu_\lambda(\mathbf{k}) = \delta \langle \hat{\alpha}^\dagger_\lambda \hat{\alpha}_\lambda \rangle(\mathbf{k})$ describes the response of the Bogoliubov quasiparticles (with the quasiparticle creation and annihilation operators $\hat{\alpha}^\dagger_\lambda$, $\hat{\alpha}_\lambda$ in the band λ). The off-diagonal component

$\delta\gamma_\lambda(\mathbf{k}) = \delta\langle\hat{\alpha}_\lambda\hat{\alpha}_\lambda\rangle(\mathbf{k})$ describes the pair-response. Note that the abbreviations $\eta_\lambda^\pm(\mathbf{k})$ are of even (+) and odd (−) parity with respect to $\mathbf{k} \to -\mathbf{k}$ and become very simple expressions in the small wavelength limit (see Appendix 1).

For the inverse Bogoliubov transformation it is convenient to introduce parity-projected quantities which are labeled by $s = \pm 1$:

$$\delta n_\lambda^{(s)}(\mathbf{k}) = \frac{1}{2}\left[\delta n_\lambda(\mathbf{k}) + s\delta n_\lambda(-\mathbf{k})\right], \tag{7.50}$$

$$\delta\xi_\lambda^{(s)}(\mathbf{k}) = \frac{1}{2}\left[\delta\xi_\lambda(\mathbf{k}) + s\delta\xi_\lambda(-\mathbf{k})\right]. \tag{7.51}$$

In almost the same manner also the off-diagonal components are decomposed by

$$\delta g_\lambda^{(s)}(\mathbf{k}) = \frac{1}{2}\left[\delta g_\lambda(\mathbf{k})\frac{\Delta_\lambda^*(\mathbf{k})}{|\Delta_\lambda(\mathbf{k})|} + s\frac{\Delta_\lambda(\mathbf{k})}{|\Delta_\lambda(\mathbf{k})|}\delta g_\lambda{}^*(-\mathbf{k})\right], \tag{7.52}$$

and

$$\delta\Delta_\lambda^{(s)}(\mathbf{k}) = \frac{1}{2}\left[\delta\Delta_\lambda(\mathbf{k})\frac{\Delta_\lambda^*(\mathbf{k})}{|\Delta_\lambda(\mathbf{k})|} + s\frac{\Delta_\lambda(\mathbf{k})}{|\Delta_\lambda(\mathbf{k})|}\delta\Delta_\lambda^*(-\mathbf{k})\right]. \tag{7.53}$$

We use the same symmetry classification for the Bogoliubov transformed quantities. The physical meaning of $\delta\Delta_\lambda(\mathbf{k},\mathbf{q},\omega)$ becomes clear after a decomposition into its real and imaginary part

$$\begin{aligned}\delta\Delta_\lambda(\mathbf{k},\mathbf{q},\omega) &= a_\lambda(\mathbf{k},\mathbf{q},\omega)e^{i\varphi_\lambda(\mathbf{q},\omega)} - \Delta_\lambda(\mathbf{k}) \\ &= \left[\delta a_\lambda(\mathbf{k},\mathbf{q},\omega) + i\delta\varphi_\lambda(\mathbf{q},\omega)|\Delta_\lambda(\mathbf{k})|\right]\frac{\Delta_\lambda(\mathbf{k})}{|\Delta_\lambda(\mathbf{k})|}.\end{aligned} \tag{7.54}$$

With Eq. (7.53) we can identify $\delta\Delta_\lambda^{(+)}(\mathbf{k},\mathbf{q},\omega) = \delta a_\lambda(\mathbf{k},\mathbf{q},\omega)$ as the amplitude fluctuations and $\delta\Delta_\lambda^{(-)}(\mathbf{k},\mathbf{q},\omega)/\Delta_\lambda(\mathbf{k}) = i\delta\varphi_\lambda(\mathbf{q},\omega)$ as the phase fluctuations of the order parameter.

The off-diagonal energy shift $\delta\Delta_\lambda^{(s)}(\mathbf{k})$ can be determined from a straightforward variation of the self-consistency Eq. (7.12):

$$\delta\Delta_\lambda^{(s)}(\mathbf{k}) = \sum_{\mathbf{k}'\mu} V_{\mathbf{kk}'}^{\lambda\mu}\delta g_\mu^{(s)}(\mathbf{k}') \tag{7.55}$$

with $\delta g_\lambda^{(s)}(\mathbf{k}) = -\theta_\lambda(\mathbf{k})\delta\Delta_\lambda^{(s)}(\mathbf{k})$. This off-diagonal self-consistency equation will play an important role for the gauge invariance of the theory, as will be discussed in Sect. 7.5.

From the symmetry classification we can assign to each transport and response function (see Table 7.1.) the corresponding momentum distribution function $\delta n_\lambda^{(+)}(\mathbf{k})$ or $\delta n_\lambda^{(-)}(\mathbf{k})$: The vertex function of the (charge) density and Raman responses is even in $\mathbf{k}$, thus only the even distribution function $\delta n_\lambda^{(+)}(\mathbf{k})$ contributes to those response functions. For the current response (dynamic conductivity), the vertex function ($a_\sigma(\mathbf{k}) = e\mathbf{v}_\mathbf{k}$) is odd in momentum. Thus, only $\delta n_\lambda^{(-)}(\mathbf{k})$ contributes to the conductivity upon summation over $\mathbf{k}$. Furthermore, the Bogoliubov transformation can now be written in this simple form

$$\begin{pmatrix} \delta v_\lambda^{(s)}(\mathbf{k}) \\ \delta \gamma_\lambda^{(s)}(\mathbf{k}) \end{pmatrix} = \begin{pmatrix} q_\lambda^{(s)}(\mathbf{k}) & p_\lambda^{(s)}(\mathbf{k}) \\ -p_\lambda^{(s)}(\mathbf{k}) & q_\lambda^{(s)}(\mathbf{k}) \end{pmatrix} \cdot \begin{pmatrix} \delta n_\lambda^{(s)}(\mathbf{k}) \\ \delta g_\lambda^{(s)}(\mathbf{k}) \end{pmatrix} \tag{7.56}$$

$$\begin{pmatrix} \delta E_\lambda^{(s)}(\mathbf{k}) \\ \delta D_\lambda^{(s)}(\mathbf{k}) \end{pmatrix} = \begin{pmatrix} q_\lambda^{(s)}(\mathbf{k}) & p_\lambda^{(s)}(\mathbf{k}) \\ -p_\lambda^{(s)}(\mathbf{k}) & q_\lambda^{(s)}(\mathbf{k}) \end{pmatrix} \cdot \begin{pmatrix} \delta \xi_\lambda^{(s)}(\mathbf{k}) \\ \delta \Delta_\lambda^{(s)}(\mathbf{k}) \end{pmatrix} \tag{7.57}$$

which might easily be inverted by using the sum rule

$$\left[q_\lambda^{(s)}(\mathbf{k})\right]^2 + \left[p_\lambda^{(s)}(\mathbf{k})\right]^2 = 1. \tag{7.58}$$

Here, we have defined the real-valued coherence factors

$$q_\lambda^{(s)}(\mathbf{k}) = |u_\lambda(\mathbf{k}+)u_\lambda(\mathbf{k}-)| - s|v_\lambda(\mathbf{k}+)v_\lambda(\mathbf{k}-)| \tag{7.59}$$

and

$$p_\lambda^{(s)}(\mathbf{k}) = |u_\lambda(\mathbf{k}+)v_\lambda(\mathbf{k}-)| + s|u_\lambda(\mathbf{k}-)v_\lambda(\mathbf{k}+)| \tag{7.60}$$

with the explicit form

$$q_\lambda^{(s)}(\mathbf{k}) = \sqrt{\frac{1}{2} + \frac{\xi_\lambda(\mathbf{k}+)\xi_\lambda(\mathbf{k}-) - s|\Delta_\lambda(\mathbf{k})|^2}{2E_\lambda(\mathbf{k}+)E_\lambda(\mathbf{k}-)}} \tag{7.61}$$

and

$$p_\lambda^{(s)}(\mathbf{k}) = \sqrt{\frac{1}{2} - \frac{\xi_\lambda(\mathbf{k}+)\xi_\lambda(\mathbf{k}-) - s|\Delta_\lambda(\mathbf{k})|^2}{2E_\lambda(\mathbf{k}+)E_\lambda(\mathbf{k}-)}}. \tag{7.62}$$

From Eqs. (7.57) and (7.42) we finally obtain the following solution of the matrix kinetic equation

$$\begin{pmatrix} \delta n_\lambda^+(\mathbf{k}) \\ \delta n_\lambda^-(\mathbf{k}) \\ \delta g_\lambda^+(\mathbf{k}) \\ \delta g_\lambda^-(\mathbf{k}) \end{pmatrix} = \begin{pmatrix} N_{11} & N_{12} & N_{13} & N_{14} \\ N_{21} & N_{22} & N_{23} & N_{24} \\ N_{31} & N_{32} & N_{33} & N_{34} \\ N_{41} & N_{42} & N_{43} & N_{44} \end{pmatrix} \cdot \begin{pmatrix} \delta \xi_\lambda^+(\mathbf{k}) \\ \delta \xi_\lambda^-(\mathbf{k}) \\ \delta \Delta_\lambda^+(\mathbf{k}) \\ \delta \Delta_\lambda^-(\mathbf{k}) \end{pmatrix} \tag{7.63}$$

The vector on the left-hand side contains the non-equilibrium momentum distribution functions [defined in Eq. (7.34)] which can be expressed in terms of the diagonal and off-diagonal energy shifts [defined in Eq. (7.35) and obtained from Table 7.1 and Eq. (7.55)]. The matrix elements N_{ij} read in detail:

$$N_{11} = q_\lambda^{(+)2}(\mathbf{k})\tilde{y}_\lambda^{(+)}(\mathbf{k}) + p_\lambda^{(+)2}(\mathbf{k})\Theta_\lambda^{(+)}(\mathbf{k}) \tag{7.64a}$$

$$N_{12} = q_\lambda^{(+)}(\mathbf{k})q_\lambda^{(-)}(\mathbf{k})\tilde{y}_\lambda^{(-)}(\mathbf{k}) + p_\lambda^{(+)}(\mathbf{k})p_\lambda^{(-)}(\mathbf{k})\Theta_\lambda^{(-)}(\mathbf{k}) \tag{7.64b}$$

$$N_{13} = q_\lambda^{(+)}(\mathbf{k})p_\lambda^{(+)}(\mathbf{k})\left[\tilde{y}_\lambda^{(+)}(\mathbf{k}) - \Theta_\lambda^{(+)}(\mathbf{k})\right] \tag{7.64c}$$

$$N_{14} = q_\lambda^{(+)}(\mathbf{k})p_\lambda^{(-)}(\mathbf{k})\tilde{y}_\lambda^{(-)}(\mathbf{k}) - q_\lambda^{(-)}(\mathbf{k})p_\lambda^{(+)}(\mathbf{k})\Theta_\lambda^{(-)}(\mathbf{k}) \tag{7.64d}$$

$$N_{22} = q_\lambda^{(-)2}(\mathbf{k})\tilde{y}_\lambda^{(+)}(\mathbf{k}) + p_\lambda^{(-)2}(\mathbf{k})\Theta_\lambda^{(+)}(\mathbf{k}) \tag{7.64e}$$

$$N_{23} = q_\lambda^{(-)}(\mathbf{k})p_\lambda^{(+)}(\mathbf{k})\tilde{y}_\lambda^{(-)}(\mathbf{k}) - q_\lambda^{(+)}(\mathbf{k})p_\lambda^{(-)}(\mathbf{k})\Theta_\lambda^{(-)}(\mathbf{k}) \tag{7.64f}$$

$$N_{24} = q_\lambda^{(-)}(\mathbf{k})p_\lambda^{(-)}(\mathbf{k})\left[\tilde{y}_\lambda^{(+)}(\mathbf{k}) - \Theta_\lambda^{(+)}(\mathbf{k})\right] \tag{7.64g}$$

$$N_{33} = p_\lambda^{(+)2}(\mathbf{k})\tilde{y}_\lambda^{(+)}(\mathbf{k}) + q_\lambda^{(+)2}(\mathbf{k})\Theta_\lambda^{(+)}(\mathbf{k}) \tag{7.64h}$$

$$N_{34} = p_\lambda^{(+)}(\mathbf{k})p_\lambda^{(-)}(\mathbf{k})\tilde{y}_\lambda^{(-)}(\mathbf{k}) + q_\lambda^{(+)}(\mathbf{k})q_\lambda^{(-)}(\mathbf{k})\Theta_\lambda^{(-)}(\mathbf{k}) \tag{7.64i}$$

$$N_{44} = p_\lambda^{(-)2}(\mathbf{k})\tilde{y}_\lambda^{(+)}(\mathbf{k}) + q_\lambda^{(-)2}(\mathbf{k})\Theta_\lambda^{(+)}(\mathbf{k}). \tag{7.64j}$$

The matrix elements N_{ij} are symmetric, i.e. $N_{ij} = N_{ji}$ and the occurring products of coherence factors can be found in the appendix 1. Above, we have introduced the following abbreviations:

$$\tilde{y}_\lambda^{(s)}(\mathbf{k}) = \frac{\eta_\lambda^{(s)2}(\mathbf{k})}{\omega^2 - \eta_\lambda^{(s)2}(\mathbf{k})}\tilde{y}_\lambda(\mathbf{k}) \tag{7.65}$$

$$\Theta_\lambda^{(s)}(\mathbf{k}) = \frac{\eta_\lambda^{(s)2}(\mathbf{k})}{\omega^2 - \eta_\lambda^{(s)2}(\mathbf{k})}\Theta_\lambda(\mathbf{k}).$$

The matrix elements N_{13}, N_{23} and N_{34} are shown to be odd with respect to $\xi_\lambda(\mathbf{k}) \to -\xi_\lambda(\mathbf{k})$. Thus in a particle-hole symmetric theory, these terms will vanish upon integration over $\xi_\lambda(\mathbf{k})$ and are labeled $O(\text{pha})$ which stands for "particle-hole asymmetric". It is convenient to rewrite these matrix elements in terms of the functions

$$\lambda_\lambda(\mathbf{k}) = \left[p_\lambda^{(+)2}(\mathbf{k}) - q_\lambda^{(-)2}(\mathbf{k})\right]\left[\tilde{y}_\lambda^{(+)}(\mathbf{k}) - \Theta_\lambda^{(+)}(\mathbf{k})\right] \tag{7.66}$$

$$\Phi_\lambda(\mathbf{k}) = q_\lambda^{(+)2}\tilde{y}_\lambda(\mathbf{k}) + p_\lambda^{(+)2}\tilde{y}_\lambda(\mathbf{k})\Theta_\lambda(\mathbf{k}) \tag{7.67}$$

$$\frac{\Theta_\lambda^{(+)}(\mathbf{k})}{2} = \frac{\eta_\lambda^{(+)2}(\mathbf{k})\Theta_\lambda(\mathbf{k}) - \eta_\lambda^{(-)2}(\mathbf{k})\tilde{y}_\lambda(\mathbf{k})}{\eta_\lambda^{(+)2}(\mathbf{k}) - \eta_\lambda^{(-)2}(\mathbf{k})} \tag{7.68}$$

where the first one, $\lambda_\lambda(\mathbf{k})$ is referred to as the Tsuneto function [27]. A straightforward but lengthy calculation yields

$$N_{11} = \frac{\eta^2\Phi_\lambda(\mathbf{k}) - \omega^2\lambda_\lambda(\mathbf{k})}{\omega^2 - \eta^2} \tag{7.69a}$$

$$N_{12} = \frac{\omega\eta[\Phi_\lambda(\mathbf{k}) - \lambda_\lambda(\mathbf{k})]}{\omega^2 - \eta^2} \tag{7.69b}$$

$$N_{13} = O(\text{pha}) \tag{7.69c}$$

$$N_{14} = \frac{\omega}{2\Delta_\lambda(\mathbf{k})}\lambda_\lambda(\mathbf{k}) \tag{7.69d}$$

$$N_{22} = \frac{\eta^2[\Phi_\lambda(\mathbf{k}) - \lambda_\lambda(\mathbf{k})]}{\omega^2 - \eta^2} \tag{7.69e}$$

$$N_{23} = O(\text{pha}) \tag{7.69f}$$

$$N_{24} = \frac{\eta}{2\Delta_\lambda(\mathbf{k})}\lambda_\lambda(\mathbf{k}) \tag{7.69g}$$

$$N_{33} = -\frac{\theta_\lambda^{(+)}(\mathbf{k})}{2} - \frac{\omega^2 - \eta^2 - 4\Delta_\lambda^2(\mathbf{k})}{4\Delta_\lambda^2(\mathbf{k})}\lambda_\lambda(\mathbf{k}) \tag{7.69h}$$

$$N_{34} = O(\text{pha}) \tag{7.69i}$$

$$N_{44} = -\frac{\theta_\lambda^{(+)}(\mathbf{k})}{2} - \frac{\omega^2 - \eta^2}{4\Delta_\lambda^2(\mathbf{k})}\lambda_\lambda(\mathbf{k}), \tag{7.69j}$$

where $\eta = \mathbf{v_k}\cdot\mathbf{q}$. Note, that all expressions are valid in the whole quasiclassical limit, i.e. for $\mathbf{q} \ll k_\mathrm{F}$ and $\hbar\omega \ll E_\mathrm{F}$. For small wave numbers, as required e.g. in the Raman case, the Tsuneto and related functions $\lambda_\lambda(\mathbf{k})$, $\Phi_\lambda(\mathbf{k})$ and $\theta_\lambda^{(+)}$ simplify considerably. The results for such a small-$\mathbf{q}$ expansion can be found in Appendix 1. Our further considerations for response and transport properties require both main results of this section: The solution of the transport equation in quasiparticle space, given by Eq. (7.45), will be used directly in Sect. 7.7 to derive the specific heat capacity in NCS (see Table 7.1.). While for the discussion of the gauge mode (Sect. 7.5), the normal and superfluid densities (Sect. 7.6) and the Raman response (Sect. 7.8) the non-equilibrium distribution functions after an inverse Bogoliubov transformation, given in Eq. (7.63) and Eq. (7.69), are necessary.

7.5 Gauge Invariance

The gauge invariance of our theory is an important issue that will be discussed in this section. For this purpose, it is very instructive to rebuild the original distribution function by combining δn_λ^+ and δn_λ^- from Eq. (7.63) and Eq. (7.69):

$$\omega\delta n_\lambda - \eta\left[\delta n_\lambda + \Phi_\lambda \delta\xi_\lambda\right] = -\lambda_\lambda\left[\omega\delta\xi_\lambda^+ + \eta\delta\xi_\lambda^-\right] + \lambda_\lambda\left(\omega^2-\eta^2\right)\frac{\delta\Delta_\lambda^-}{2\Delta_\lambda}. \quad (7.70)$$

The left-hand side of this equation is of the same structure as the linearized Landau-Boltzmann equation of the normal state. In what follows, we want to discuss the right-hand side of the above equation. Note that all terms coupling to $\delta\Delta_\lambda^+$ have vanished because of particle-hole symmetry. This means that the amplitude fluctuations of the order parameter do not contribute to the response in a particle-hole symmetric theory. The phase fluctuations are also given by Eq. (7.63):

$$\delta g_\lambda^- + \left[\frac{\theta_\lambda^+}{2} + \frac{\omega^2-\eta^2}{4\Delta_\lambda^2}\lambda_\lambda\right]\delta\Delta_\lambda^- = \frac{\omega\delta\xi_\lambda^+ + \eta\delta\xi_\lambda^-}{2\Delta_\lambda}\lambda_\lambda. \quad (7.71)$$

Multiplication with the pairing-interaction $V_{\mathbf{kk}'}^{\lambda\mu}$ and summation over $\mathbf{k}'$ and the band-index μ yields

$$\delta\Delta_\lambda^-(\mathbf{k}) + \sum_{\mathbf{k}'\mu} V_{\mathbf{kk}'}^{\lambda\mu}\left[\theta_\mu + \delta\theta_\mu + \frac{\omega^2-\eta^2}{4\Delta_\lambda^2}\lambda_\lambda\right]\delta\Delta_\mu^-(\mathbf{k}') \quad (7.72)$$
$$= \sum_{\mathbf{k}'\mu} V_{\mathbf{kk}'}^{\lambda\mu}\frac{\omega\delta\xi_\lambda^+(\mathbf{k}') + \eta\delta\xi_\lambda^-(\mathbf{k}')}{2\Delta_\lambda(\mathbf{k}')}\lambda_\lambda(\mathbf{k}'),$$

where we have introduced $\delta\theta_\lambda = \theta_\lambda^+/2 - \theta_\lambda$. It can be shown, that the $\xi_\mu(\mathbf{k})$-integral over $\delta\theta_\mu$ vanishes identically for all $\mathbf{q}$. Using the equilibrium gap-equation [Eq. (7.12)] we arrive at

$$\sum_\mu \frac{\delta\Delta_\mu^-}{\Delta_\mu}\sum_{\mathbf{k}'\mu} V_{\mathbf{kk}'}^{\lambda\mu}\frac{\omega^2-\eta^2}{4\Delta_\lambda^2}\lambda_\lambda = \sum_{\mathbf{k}'\mu} V_{\mathbf{kk}'}^{\lambda\mu}\frac{\omega\delta\xi_\lambda^+(\mathbf{k}') + \eta\delta\xi_\lambda^-(\mathbf{k}')}{2\Delta_\lambda(\mathbf{k}')}\lambda_\lambda(\mathbf{k}'). \quad (7.73)$$

These are two coupled equations (for $\mu = \pm$) which determine the phase fluctuations of the order parameter (gauge mode). Note that in the weak-coupling BCS theory, there are only two collective excitations possible: the Anderson-Bogoliubov and 2Δ mode. In NCS, there exist two gauge modes due to the band splitting, which can be connected with the particle number conservation law. In addition, due to existence of a triplet fraction, there could be further collective excitation analogous to Leggett's spontaneously broken spin-orbit symmetry modes [21] predicted for the superfluid phases of ^{3}He. The latter should be connectable with the spin conservation law in NCS. Finally, massive collective modes with frequencies below $2\Delta/\hbar$ may exist in

NCS. It can be shown, that the right-hand side of Eq. (7.70) vanishes upon $\mathbf{k}$ and λ (band) summation when inserting the above expressions for the gauge mode. This leads us to the following continuity equation for the electron density:

$$\omega \sum_{\mathbf{k},\lambda} \delta n_\lambda(\mathbf{k}) - \mathbf{q} \cdot \sum_{\mathbf{k},\lambda} \mathbf{v_k} \left[\delta n_\lambda(\mathbf{k}) + \Phi_\lambda(\mathbf{k})\delta\xi_\lambda(\mathbf{k})\right] = 0. \tag{7.74}$$

For a conserved quantity such as the particle ($a_\mathbf{k} = 1$) or charge density ($a_\mathbf{k} = e$), we can identify the corresponding generalized density and current density

$$\delta n_a = \sum_{\mathbf{k},\lambda} a_\mathbf{k} \delta n_\lambda(\mathbf{k}) \tag{7.75}$$

$$\mathbf{j}_a = \sum_{\mathbf{k},\lambda} a_\mathbf{k} \mathbf{v_k} \left[\delta n_\lambda(\mathbf{k}) + \Phi_\lambda(\mathbf{k})\delta\xi_\lambda(\mathbf{k})\right], \tag{7.76}$$

obeying the continuity equation

$$\omega \delta n_a - \mathbf{q} \cdot \mathbf{j}_a = 0. \tag{7.77}$$

Therefore, we have demonstrated charge conservation and gauge invariance of the theory for $\hbar\omega \ll E_\mathrm{F}$ and $\mathbf{q} \ll k_\mathrm{F}$.

7.6 Normal and Superfluid Densities

The normal and superfluid densities are derived in the static and long-wavelength limit ($\omega \to 0$ and $\mathbf{q} \to 0$). In order to preserve gauge invariance, gradient terms of the order $O(\mathbf{q})$ are still taken into account. The parity-projected distribution functions are obtained from Eq. (7.63) and from Eq. (7.69):

$$\delta n_\lambda^+(\mathbf{k}) = -\phi_\lambda(\mathbf{k})\delta\xi_\lambda^+(\mathbf{k}) \tag{7.78}$$

$$\delta n_\lambda^-(\mathbf{k}) = -\left[\phi_\lambda(\mathbf{k}) - \lambda_\lambda(\mathbf{k})\right]\delta\xi_\lambda^-(\mathbf{k}) + \eta\lambda_\lambda(\mathbf{k})\frac{\delta\Delta_\lambda^-(\mathbf{k})}{2\Delta_\lambda(\mathbf{k})}, \tag{7.79}$$

where we made use of the $\mathbf{q} \to 0$ limit with the coherence factors $q_\lambda^-(\mathbf{k}) \to 1$, $p_\lambda^-(\mathbf{k}) \to 0$, and $\Phi_\lambda(\mathbf{k}) \to \phi_\lambda(\mathbf{k})$, $\tilde{y}_\lambda(\mathbf{k}) \to y_\lambda(\mathbf{k})$, as well as the Tsuneto function $\lambda_\lambda(\mathbf{k}) \to \phi_\lambda(\mathbf{k}) - y_\lambda(\mathbf{k})$ (see appendix 1). The combined expression for $\delta n_\lambda^+(\mathbf{k})$ and $\delta n_\lambda^-(\mathbf{k})$ are now inserted in Eq. (7.76) to derive the supercurrent density (vertex function $a_\mathbf{k} = e$):

$$\begin{aligned} \mathbf{j}_i^s &= \sum_{\mathbf{p}\lambda} e\mathbf{v}_{\mathbf{p}i} \left[\delta n_\lambda(\mathbf{p}) + \phi_\lambda(\mathbf{p})\delta\xi_\lambda^-(\mathbf{p})\right] \\ &= e\sum_{\mathbf{p}\lambda} \mathbf{v}_{\mathbf{p}i}\mathbf{v}_{\mathbf{p}j}\lambda_\lambda(\mathbf{p}) \left(-\frac{e}{c}\mathbf{A} + \frac{\hbar}{2}\nabla\delta\varphi_\lambda\right). \end{aligned} \tag{7.80}$$

Here we used the result from Sect. 7.4 that $\delta\Delta_\lambda^-(\mathbf{k})/\Delta_\lambda(\mathbf{k}) = i\delta\varphi_\lambda \equiv i\delta\varphi$ represents the phase fluctuations of the order parameter, assumed to be independent of the band index λ. These phase fluctuations ensure gauge invariance in the above expression for the supercurrent. By rewriting the supercurrent as product of the superfluid density and the corresponding velocity $\mathbf{v}^s$, we can easily identify

$$\mathbf{j}^s = e\,\mathbf{n}^s \cdot \mathbf{v}^s \tag{7.81}$$

$$\mathbf{v}^s = \frac{e}{m}\left(-\frac{e}{c}\mathbf{A} + \frac{\hbar}{2}\nabla\delta\varphi_\lambda\right). \tag{7.82}$$

Therefore, the superfluid and normal fluid density tensor read

$$n_{ij}^s = \sum_{\mathbf{p}\lambda} \mathbf{p}_i \mathbf{v}_{\mathbf{p}j} \lambda_\lambda(\mathbf{p}) \tag{7.83}$$

$$n_{ij}^n = n\delta_{ij} - \mathbf{n}_{ij}^s = \sum_{\mathbf{p}\lambda} \mathbf{p}_i \mathbf{v}_{\mathbf{p}j} y_\lambda(\mathbf{p}). \tag{7.84}$$

Thus, in this static and small-$\mathbf{q}$ limit we obtain a very clear picture: The Yosida kernel $y_\lambda(\mathbf{k}) = -\partial f[E_\lambda(\mathbf{k})]/\partial E_\lambda(\mathbf{k})$ generates the normal fluid density and the Tsuneto function $\lambda_\lambda(\mathbf{k})$ gives rise to the superfluid density.

It is important to realize that this result can be derived in the following alternative simple way from local-equilibrium considerations. In terms of the Fermi-Dirac distribution function on both bands $f[E_\lambda(\mathbf{p})]$ for the Bogoliubov quasiparticles, the supercurrent can be written in the standard quantum-mechanical form:

$$\begin{aligned} j_i^{\mathrm{s}} &= n v_i^{\mathrm{s}} + \frac{1}{V}\sum_{\mathbf{p}\lambda} v_{\mathbf{p}i}(\mathbf{p}) f(E_\lambda(\mathbf{p}) + \mathbf{p}\cdot\mathbf{v}^{\mathrm{s}}) \\ &= n v_i^{\mathrm{s}} + \frac{1}{V}\sum_{\mathbf{p}\lambda} \mathbf{v}_{\mathbf{p}i}\left\{ f(E_\lambda(\mathbf{p})) + \frac{\partial f(E_\lambda(\mathbf{p}))}{\partial E_\lambda(\mathbf{p})} p_j v_j^{\mathrm{s}} \right\} \\ &= \left\{ n\delta_{ij} - \frac{1}{V}\sum_{\mathbf{p}\lambda} \frac{p_i}{m}\left(-\frac{\partial f(E_\lambda(\mathbf{p}))}{\partial E_\lambda(\mathbf{p})}\right) p_j \right\} v_j^{\mathrm{s}}. \end{aligned} \tag{7.85}$$

This immediately implies the definition of the normal fluid density in the form

$$n_{ij}^{\mathrm{n}} = \frac{1}{V}\sum_{\mathbf{p}\lambda} p_i v_j y_\lambda(\mathbf{p}). \tag{7.86}$$

Thus, the results obtained with our simple local-equilibrium picture are in agreement with the results in Ref. [9].

7.7 The Specific Heat Capacity

In order to derive the specific heat capacity, we start from an expression for the entropy of a NCS, which has to be written in the general form

$$\begin{aligned} T\sigma(T) &= -\frac{k_B}{V}\sum_{\mathbf{p}\lambda} f[E_\lambda(\mathbf{p})]\ln f[E_\lambda(\mathbf{p})] + \{1 - f[E_\lambda(\mathbf{p})]\}\ln\{1 - f[E_\lambda(\mathbf{p})]\} \\ &= \frac{1}{V}\sum_{\mathbf{p}\lambda} \xi_\lambda^2(\mathbf{p}) y_{\mathbf{p}}^{(\lambda)}. \end{aligned} \tag{7.87}$$

The change of the entropy as a consequence of a temperature change δT can then be written in the form [9]

$$T\delta\sigma(T) = \frac{1}{V}\sum_{\mathbf{p}\lambda} E_\lambda(\mathbf{p})\delta\nu_\lambda(\mathbf{p}), \tag{7.88}$$

where the quasiparticle distribution function is given by Eq. (7.45). In the static and homogeneous limit, i.e. $\omega \to 0$ and $\mathbf{q} \to 0$, this expression simplifies considerably to $\delta\nu_\lambda(\mathbf{k}) = y_\lambda(\mathbf{k})\delta E_\lambda(\mathbf{k})$. The quasiparticle energy shift for a temperature change is $\delta E_\lambda(\mathbf{k}) = (E_\lambda(\mathbf{k})/T - \partial E_\lambda(\mathbf{k})/\partial T)\delta T$ for each band [9]. Therefore, our result for the entropy change reads

$$\begin{aligned} T\delta\sigma(T) &= \frac{1}{V}\sum_{\mathbf{p}\lambda} y_\lambda(\mathbf{p}) E_\lambda(\mathbf{p}) \left[E_\lambda(\mathbf{p}) - T\frac{\partial E_\lambda(\mathbf{p})}{\partial T}\right]\delta T \\ &= C_V(T)\delta T \end{aligned} \tag{7.89}$$

and one may easily identify the specific heat capacity as

$$C_V(T) = \frac{1}{V}\sum_{\mathbf{p}\lambda} y_\lambda(\mathbf{p}) \left[E_\lambda^2(\mathbf{p}) - \frac{T}{2}\frac{\partial \Delta_\lambda^2(\mathbf{p})}{\partial T}\right]. \tag{7.90}$$

An alternative way to derive the specific heat capacity employs again the concept of local-equilibrium:

$$T\delta\sigma(T) = \frac{1}{V}\sum_{\mathbf{p}\lambda} E_\lambda(\mathbf{p})\delta f(E_\lambda(\mathbf{p}))\,. \tag{7.91}$$

The change of the Bogoliubov quasiparticle (Fermi-Dirac) distribution function with temperature has two causes: first the direct change $T \to T + \delta T$ and second the change of the quasiparticle energy with temperature through the T dependence of the energy gap:

$$\delta f(E_\lambda(\mathbf{p})) = f\left(\frac{E_\lambda(\mathbf{p}) + \frac{\partial E_\lambda(\mathbf{p})}{\partial T}\delta T}{k_B[T+\delta T]}\right) - f\left(\frac{E_\lambda(\mathbf{p})}{k_B T}\right) \tag{7.92}$$

$$= \underbrace{\left(-\frac{\partial f(E_\lambda(\mathbf{p}))}{\partial E_\lambda(\mathbf{p})}\right)}_{y_\lambda(\mathbf{p})}\left(\frac{E^0_{\mathbf{p}\lambda}}{T} - \frac{\partial E_\lambda(\mathbf{p})}{\partial T}\right)\delta T.$$

Hence we arrive at the same result for the entropy change

$$T\delta\sigma(T) = \frac{1}{V}\sum_{\mathbf{p}\lambda} y_\lambda(\mathbf{p})E_\lambda(\mathbf{p})\left[E_\lambda(\mathbf{p}) - T\frac{\partial E_\lambda(\mathbf{p})}{\partial T}\right]\delta T \tag{7.93}$$

$$= C_V(T)\delta T$$

and the result for the specific heat capacity is confirmed. Again, as in the case of the normal and superfluid densities, the result for the specific heat capacity can be viewed to consist of contributions from the two bands, in the sense that the sum over the spin projections $\sigma = \pm 1$ is replaced by a sum over the pseudospin variable $\lambda = \pm$.

7.8 A Case Study: Raman Response

In the following section we will discuss in detail the electronic Raman response for $T = 0$ in NCS [18]. An extensive description of the electronic Raman effect in unconventional superconductors can be found in Ref. [5]. A Raman experiment detects the intensity of the scattered light with frequency-shift $\omega = \omega_I - \omega_S$, where the incoming photon of frequency ω_I is scattered on an elementary excitation and gives rise to a scattered photon with frequency ω_S and a momentum transfer $\mathbf{q}$. The differential photon scattering cross section of this process is given by Ref. [19]

$$\frac{\partial^2\sigma}{\partial\omega\partial\Omega} = \frac{\omega_S}{\omega_I} r_0^2 S_{\gamma\gamma}(\mathbf{q},\omega) \tag{7.94}$$

with the solid angle Ω and the Thompson radius $r_0 = e^2/mc^2$. The generalized structure function $S_{\gamma\gamma}(\mathbf{q},\omega)$ is connected through the fluctuation-dissipation theorem to the imaginary part of the Raman response function $\chi_{\gamma\gamma}(\mathbf{q},\omega)$:

$$S_{\gamma\gamma}(\mathbf{q},\omega) = -\frac{\hbar}{\pi}\left[1 + n(\omega)\right]\chi''_{\gamma\gamma}(\mathbf{q},\omega). \tag{7.95}$$

Here, $n(\omega) = \left[\exp(\hbar\omega/k_B T) - 1\right]$ denotes the Bose distribution function. After Coulomb renormalisation and in the long-wavelength limit ($\mathbf{q} = 0$), the Raman response function is given by the imaginary part of (see also Ref. [23])

$$\chi_{\gamma\gamma}(\omega) = \chi_{\gamma\gamma}^{(0)}(\omega) - \frac{\left[\chi_{\gamma 1}^{(0)}(\omega)\right]^2}{\chi_{11}^{(0)}(\omega)}. \tag{7.96}$$

Within our notation, the unscreened Raman response is given by

$$\chi_{ab}^{(0)}(\omega) = \frac{1}{V}\sum_{\mathbf{p},\sigma} a_{\mathbf{p}} b_{\mathbf{p}} \lambda_{\mathbf{p}}(\omega), \tag{7.97}$$

where the vertex functions $a_{\mathbf{p}}, b_{\mathbf{p}}$ are either 1 or the corresponding momentum-dependent Raman vertex $\gamma \equiv \gamma_{\mathbf{k}}^{(R)}$ that describes the coupling of polarized light to the sample. The long-wavelength limit of the Tsuneto function $\lambda_{\mathbf{p}}(\mathbf{q} = 0) = 4\Delta_{\mathbf{p}}^2\theta_{\mathbf{p}}/\ (4E_{\mathbf{p}}^2 - \omega^2)$ is given in appendix 1. Since we are interested in the $T = 0$ Raman response, it is possible to perform the integration on the energy variable $\xi_{\mathbf{k}}$ (see e.g. [5]). Note that the second term in Eq. (7.96) is often referred to as the screening contribution that originates from gauge invariance. Since the ASOC leads to a splitting of the Fermi surface, the total Raman response is given by $\chi_{\gamma\gamma}^{\text{total}} = \sum_{\lambda=\pm}\chi_{\gamma\gamma}^{\lambda}$ with $\chi_{\gamma\gamma}^{\pm} = \chi_{\gamma\gamma}(\Delta_{\pm})$, in which the usual summation over the spin variable σ is replaced by a summation over the pseudo-spin (band) index λ. With Eq. (7.11) the unscreened Raman response for both bands in the clean limit [$l \gg \xi(0)$ with the mean free path l and the coherence length $\xi(T = 0)$] can be analytically expressed as

$$\Im\chi_{\gamma\gamma}^{(0)\pm} = \frac{\pi N_{\mathrm{F}}^{\pm}\psi}{\omega}\Re\left\langle \gamma_{\mathbf{k}}^{(R)2}\frac{|1 \pm p|\gamma_{\mathbf{k}}||^2}{\sqrt{(\frac{\omega}{2\psi})^2 - |1 \pm p|\gamma_{\mathbf{k}}||^2}}\right\rangle_{\mathrm{FS}}. \tag{7.98}$$

Here, $N_{\mathrm{F}}^{\pm}$ reflect the different densities of states on both bands and $\langle\ldots\rangle_{\mathrm{FS}}$ denotes an average over the Fermi surface. We consider small momentum transfers ($\mathbf{q} \to 0$) and neglect interband scattering processes, assuming non-resonant scattering. Then, the Raman tensor is approximately given by

$$\gamma_{\mathbf{k}}^{(R)} = m\sum_{i,j}\hat{\mathbf{e}}_i^S\frac{\partial^2\varepsilon(\mathbf{k})}{\hbar^2\partial k_i\partial k_j}\hat{\mathbf{e}}_j^I, \tag{7.99}$$

where $\hat{\mathbf{e}}^{S,I}$ denote the unit vectors of scattered and incident polarization light, respectively. The light polarization selects elements of this Raman tensor, where $\gamma_{\mathbf{k}}^{(R)}$ can be decomposed into its symmetry components and, after a straight forward calculation (see appendix 2), expanded into a set of basis functions on a spherical Fermi surface. Our results for the tetragonal group C_{4v} are

$$\gamma_{A_1}^{(R)} = \sum_{k=0}^{\infty}\sum_{l=0}^{l\le k/2}\gamma_{k,l}^{(R)}\cos 4l\phi\sin^{2k}\theta, \tag{7.100a}$$

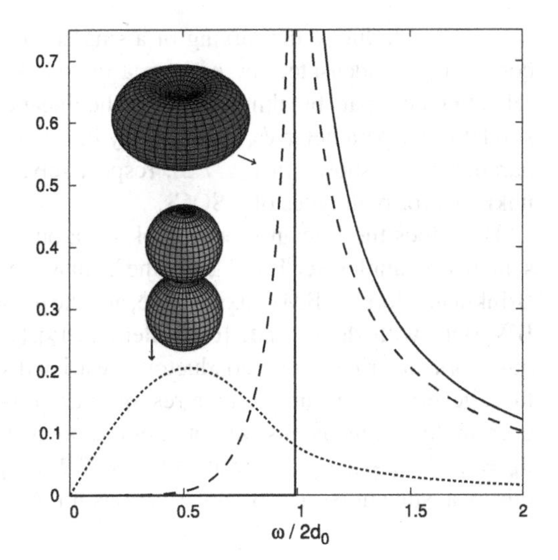

Fig. 7.2 Calculated Raman spectra for a pure triplet order parameter (i.e. $\psi = 0$) for $B_{1,2}$ polarization of the point group C_{4v} in backscattering geometry $(z\bar{z})$. The ABM (axial) state with $|\mathbf{d_k}| = d_0 \sin\theta$ is displayed as a *dashed line* and the polar state with $|\mathbf{d_k}| = d_0 |\cos\theta|$ as a *dotted line*. For a comparison, also the threshold behavior of the Raman response for the BW state (*solid line*) with $|\mathbf{d_k}| = d_0$ is shown

$$\gamma_{B_1}^{(R)} = \sum_{k=1}^{\infty} \sum_{l=1}^{l \le (k+1)/2} \gamma_{k,l}^{(R)} \cos(4l-2)\phi \sin^{2k}\theta, \qquad (7.100b)$$

$$\gamma_{B_2}^{(R)} = \sum_{k=1}^{\infty} \sum_{l=1}^{l \le (k+1)/2} \gamma_{k,l}^{(R)} \sin(4l-2)\phi \sin^{2k}\theta, \qquad (7.100c)$$

and for the cubic group O we obtain

$$\gamma_{A_1}^{(R)} = \sum_{k=0}^{\infty} \sum_{l=0}^{l \le k/2} \gamma_{k,l}^{(R)} \cos 4l\phi \sin^{2k}\theta, \qquad (7.101a)$$

$$\gamma_{E^{(1)}}^{(R)} = \gamma_0^{(R)} (2 - 3\sin^2\theta) + \cdots, \qquad (7.101b)$$

$$\gamma_{E^{(2)}}^{(R)} = \sum_{k=1}^{\infty} \sum_{l=1}^{l \le (k+1)/2} \gamma_{k,l}^{(R)} \cos(4l-2)\phi \sin^{2k}\theta, \qquad (7.101c)$$

$$\gamma_{T_2}^{(R)} = \sum_{k=1}^{\infty} \sum_{l=1}^{l \le (k+1)/2} \gamma_{k,l}^{(R)} \sin(4l-2)\phi \sin^{2k}\theta \qquad (7.101d)$$

in a backscattering-geometry experiment $(z\bar{z})$[2]. In what follows, we neglect higher harmonics and thus use only the leading term in the expansions of $\gamma_{\mathbf{k}}^{(R)}$[3].

---

[2] The vertices $E^{(1)}$ and $E^{(2)}$ seem to be quite different, but it turns out that the Raman response is exactly the same because $E^{(1)}$ and $E^{(2)}$ are both elements of the same symmetry class.

[3] Due to screening, the constant term $(k = 0, l = 0)$ in the A_1 vertex generates no Raman response, thus we used $(k = 1, l = 0)$. For all the other vertices the leading term is given by $(k = 1, l = 1)$.

In general, due to the mixing of a singlet and a triplet component to the superconducting condensate, one expects a two-peak structure in parity-violated NCS, reflecting both pair-breaking peaks for the linear combination [see Eq. (7.11)] of the singlet order parameter $\psi_{\mathbf{k}}$ (extensively discussed in Ref. [5]) and the triplet order parameter $\mathbf{d}_{\mathbf{k}}$ (shown in Fig. 7.2), respectively. The ratio $p = d/\psi$, however, is unknown for both types of ASOCs.

How does the Raman spectra look for a pure triplet p-wave state? Some representative examples, see Fig. 7.2, are the Balian-Werthamer (BW) state, the Anderson-Brinkman-Morel (ABM or axial) state, and the polar state. The simple pseudoisotropic BW state with $\mathbf{d}_{\mathbf{k}} = d_0\hat{\mathbf{k}}$ [equivalent to Eq. (7.3) for $g_{3=0}$], as well as previous work on triplet superconductors, restricted on a (cylindrical) 2D Fermi surface, generates the same Raman response as an s-wave superconductor [17]. However, in three dimensions we obtain more interesting results for the axial state with $\mathbf{d}_{\mathbf{k}} = d_0(\hat{k}_y\hat{\mathbf{e}}_x - \hat{k}_x\hat{\mathbf{e}}_y)$ [equivalent to Eq. (7.2) for $g_{\parallel} = 0$]. The Raman response for this axial state in B_1 and B_2 polarizations for $\mathcal{G} = C_{4v}$ is given by

$$\chi''_{B_{1,2}}(x) = \frac{\pi N_F \gamma_0^{(R)2}}{128} \times \left(-10 - \frac{28}{3}x^2 - 10x^4 + \frac{5+3x^2+3x^4+5x^6}{x}\ln\left|\frac{x+1}{x-1}\right|\right) \tag{7.102}$$

with the dimensionless frequency $x = \omega/2d_0$. An expansion for low frequencies reveals a characteristic exponent [$\chi''_{B_{1,2}} \propto (\omega/2d_0)^6$], due to the overlap between the gap and the vertex function. Moreover, we calculate the Raman response for the polar state with $\mathbf{d}_{\mathbf{k}} = d_0\hat{k}_z\hat{\mathbf{e}}_x$; in this case one equatorial line node crosses the Fermi surface and we obtain:

$$\chi''_{B_{1,2}}(x) = \frac{\pi N_F \gamma_0^{(R)2}}{8x}\begin{cases} \frac{\pi}{2}x^2 - \frac{3\pi}{4}x^4 + \frac{5\pi}{16}x^6 & x \le 1 \\ \left(x^2 - \frac{3}{2}x^4 + \frac{5}{8}x^6\right)\arcsin\frac{1}{x} & \\ -\left(\frac{1}{3} - \frac{13}{12}x^2 + \frac{5}{8}x^4\right)\sqrt{x^2-1} & x > 1 \end{cases} \tag{7.103}$$

with the trivial low-frequency expansion $\chi''_{B_{1,2}} \propto \omega/2d_0$. While the pair-breaking peaks for the BW and ABM state were both located at $\omega = 2d_0$ (similar to the B_{1g} polarization in the singlet d-wave case, which is peaked at $2\Delta_0$), for the polar state this peak is significantly shifted to lower frequencies ($\omega = 1.38d_0$).

Let's turn to the Raman spectra predicted for the tetragonal point group $\mathcal{G} = C_{4v}$. In Fig. 7.3 we show the calculated Raman response using Eq. (7.2) with $g_{\parallel} = 0$. This Rashba-type ASOC splits the Fermi surface into two bands; while on the one band the gap function is $\Delta_{\mathbf{k}} = \psi(1 + p|\gamma_{\mathbf{k}}|) \equiv \Delta_+$, it is $\Delta_- \equiv \psi(1 - p|\gamma_{\mathbf{k}}|)$ on the other band. Thus, depending on the ratio $p = d/\psi$, four different cases (see polar diagrams in the insets) have to be considered: (a) no nodes; (b) one (equatorial) line node (Δ_- band); (c) two line nodes (Δ_- band); and (d) two point nodes on both bands. Since the Raman intensity in NCS is proportional to the imaginary part of

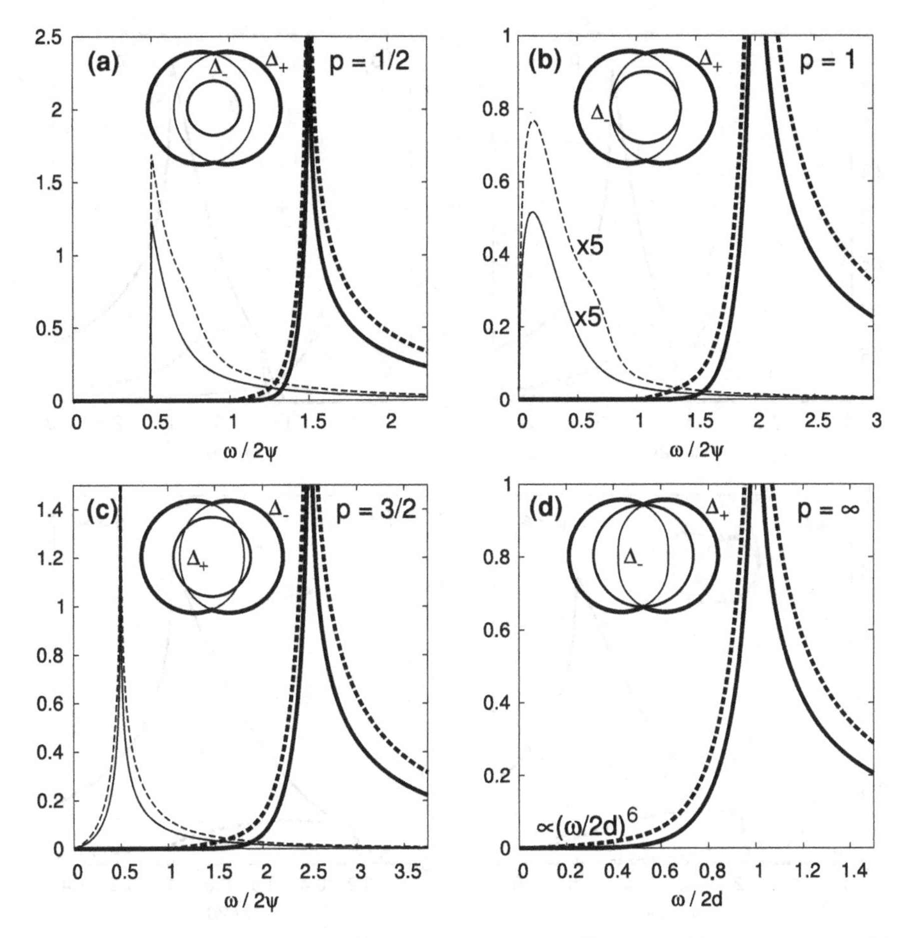

Fig. 7.3 Calculated Raman spectra $\chi''_{\gamma\gamma}(\Delta_-)$ (*thin lines*) and $\chi''_{\gamma\gamma}(\Delta_+)$ (*thick lines*) for $B_{1,2}$ (*solid lines*) and for A_1 (*dashed lines*) polarizations for the point group C_{4v}. We obtain the same spectra for the B_1 and B_2 symmetry. The polar diagrams in the insets demonstrate the four qualitative different cases for the unknown ratio $p = d/\psi$

$$\chi^{\text{total}}_{\gamma\gamma} = \chi_{\gamma\gamma}(\Delta_-) + \chi_{\gamma\gamma}(\Delta_+), \tag{7.104}$$

it is interesting to display both contributions separately (thick and thin lines, respectively). Although (except for $\psi = 0$) we always find two pair-breaking peaks at

$$\frac{\omega}{2\psi} = |1 \pm p| \tag{7.105}$$

we stress that our results for NCS are not just a superposition of a singlet and a triplet spectra. This is clearly demonstrated in Fig. 7.3(a), for example, where we show the results for a small triplet contribution ($p = 1/2$). For $\chi''_{\gamma\gamma}(\Delta_-)$

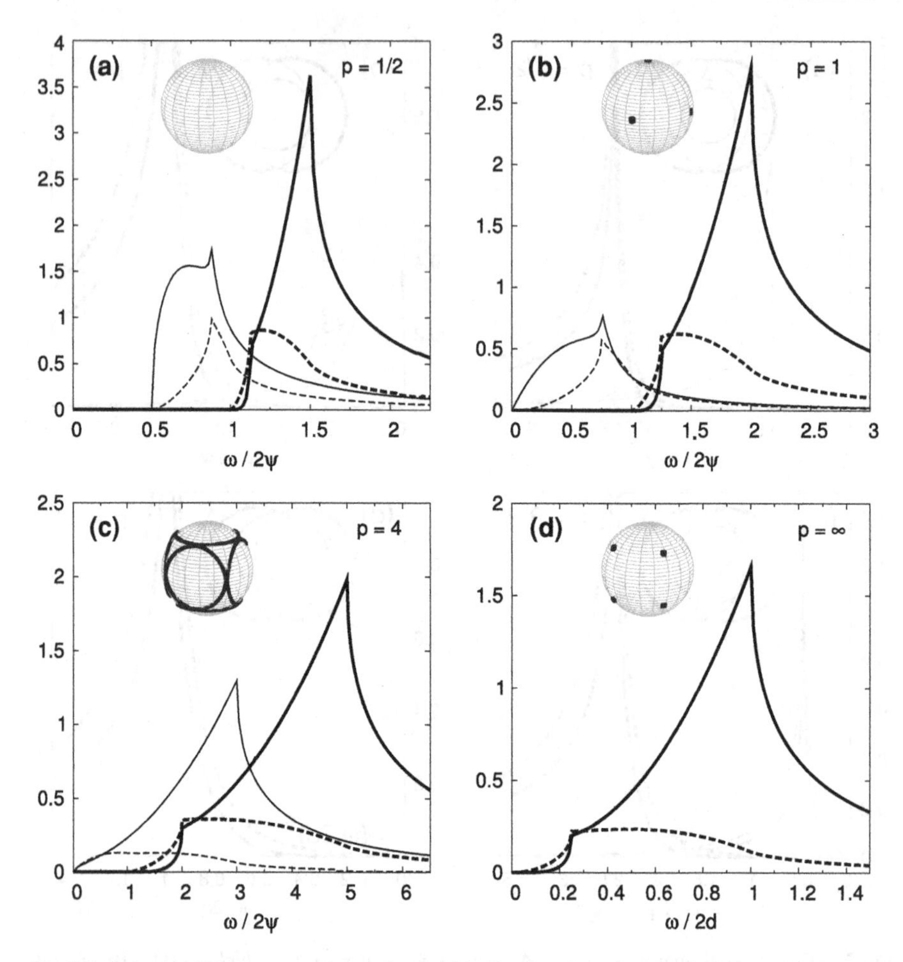

Fig. 7.4 Calculated Raman spectra $\chi_{\gamma\gamma}(\Delta_-)$ (*thin lines*) and $\chi_{\gamma\gamma}(\Delta_+)$ (*thick lines*) for E (*solid lines*), T_2 (*dashed lines*) polarizations for the point group O. The insets display the point and line nodes of the gap function Δ_-

we find a threshold behavior with an adjacent maximum value of $\chi''_{B_{1,2}}(\Delta_-) = N_F^- \gamma_0^{(R)2} \pi^2/8\sqrt{p^{-1}-1}$. In contrast for $\chi''_{\gamma\gamma}(\Delta_+)$ a zero Raman signal to twice the singlet contribution followed by a smooth increase and a singularity is obtained.[4] In the special case, where the singlet contribution equals the triplet one ($p = 1$), the gap function Δ_- displays an equatorial line node without sign change. This is displayed in Fig. 7.3b. Because of the nodal structure and strong weight from the vertex function ($\propto \sin^2\theta$), many low-energy quasiparticles can be excited, which leads to

[4] Note that even though the gap function does not depend on ϕ (see Fig. 7.1, we obtain a small polarization dependence. This unusual behavior only in A_1 symmetry is due to screening and leads to a small shoulder for $p \leq 1$.

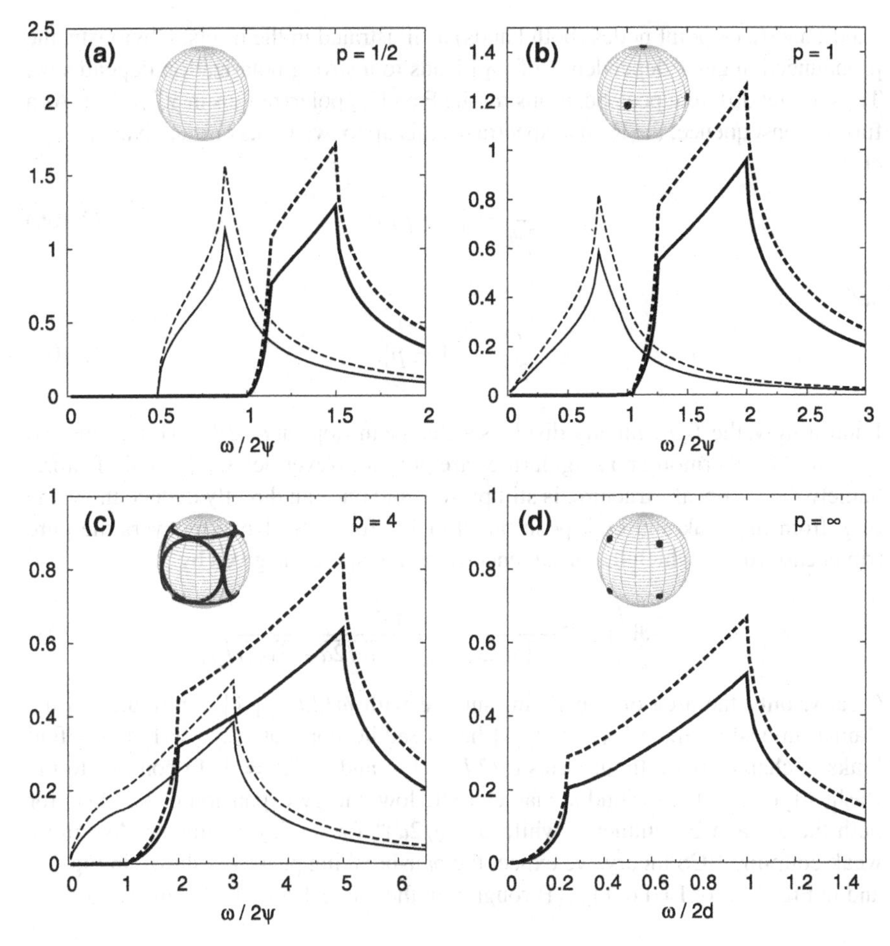

Fig. 7.5 Calculated Raman spectra $\chi_{\gamma\gamma}(\Delta_-)$ (*thin lines*) and $\chi_{\gamma\gamma}(\Delta_+)$ (*thick lines*) for A_1 polarization with screening (*solid lines*) and without screening (*dashed lines*) for the point group O. The insets display the point and line nodes of the gap function Δ_-

the square-root-like increase in the Raman intensity. In this special case the pair-breaking peak is located very close to elastic scattering ($\omega = 0.24\psi$). In Fig. 7.3(c) the gap function Δ_- displays two circular line nodes. The corresponding Raman response for $p > 1$ shows two singularities with different low-frequency power laws $[\chi''_{B_{1,2}}(\Delta_-) \propto \omega/2\psi$ and $\chi''_{B_{1,2}}(\Delta_+) \propto (\omega/2\psi - 1)^{11/2}]$. Finally, for $p \gg 1$ one recovers the pure triplet case (d) which is given analytically by Eq. (7.102).

The Raman response for the point group O, using Eq. (7.3), is shown in Fig. 7.4 for the E and T_2 symmetries and in Fig. 7.5 for the A_1 symmetry with and without screening. As in the previous (tetragonal) case, there is only little difference between the unscreened and the screened Raman response. We again consider four different cases: (a) no nodes; (b) six point nodes (Δ_- band); (c) six connected line nodes (Δ_-

band); and (d) 8 point nodes (both bands) as illustrated in the insets. Obviously, the pronounced angular dependence of $|\gamma_{\mathbf{k}}|$ leads to a strong polarization dependence. Thus we get different peak positions for the E and T_2 polarizations in $\chi''_{\gamma\gamma}(\Delta_+)$. As a further consequence, the Raman spectra reveals up to two kinks on each band (+, −) at

$$\frac{\omega}{2\psi} = |1 \pm p/4| \tag{7.106}$$

and

$$\frac{\omega}{2\psi} = |1 \pm p|. \tag{7.107}$$

Interestingly, the T_2 symmetry displays a change in slope at $\omega/2\psi = |1 + p|$ instead of a kink. Furthermore, no singularities are present. Nevertheless, the main feature, namely the two-peak structure, is still present and one can directly deduce the value of p from the peak and kink positions. Finally, for $p \gg 1$ one recovers the pure triplet case (d), in which the unscreened Raman response is given by

$$\chi''_{\gamma\gamma}(\omega) \propto \frac{2d}{\omega}\Re\left\langle \gamma_{\mathbf{k}}^{(R)2} \frac{|\gamma_{\mathbf{k}}|^2}{\sqrt{(\omega/2d + |\gamma_{\mathbf{k}}|)(\omega/2d - |\gamma_{\mathbf{k}}|)}} \right\rangle_{FS}. \tag{7.108}$$

Clearly, only the area on the Fermi surface with $\omega/2d > |\gamma_{\mathbf{k}}|$ contributes to the Raman intensity. Since $|\gamma_{\mathbf{k}}| \in [0, 1]$ has a saddle point at $|\gamma_{\mathbf{k}}| = 1/4$, we find kinks at characteristic frequencies $\omega/2d = 1/4$ and $\omega/2d = 1$. In contrast to the Rashba-type ASOC, we find a characteristic low-energy expansion $\propto (\omega/2d)^2$ for both the A_1 and E symmetry, while $\propto (\omega/2d)^4$ for the T_2 symmetry. Assuming weak-coupling BCS theory, we expect the pair-breaking peaks (as shown in Fig. 7.4 and in Fig. 7.5) for $Li_2Pd_xPt_{3-x}B$ roughly in the range $4\,cm^{-1}$ to $30\,cm^{-1}$.

7.9 Conclusion

In this chapter, we derived response and transport functions for non-centrosymmetric superconductors from a kinetic theory with particular emphasis on the Raman response. We started from the generalized von Neumann equation which describes the evolution of the momentum distribution function in time and space and derived a linearized matrix kinetic (Boltzmann) equation in ω-$\mathbf{q}$ space. This kinetic equation is a 4×4 matrix equation in both particle-hole (Nambu) and spin space. We explored the Nambu structure and solved the kinetic equation quite generally by first performing an SU(2) rotation into the band basis and second applying a Bogoliubov transformation into quasiparticle space. Our theory is particle-hole symmetric, applies to any kind of antisymmetric spin-orbit coupling, and holds for arbitrary quasiclassical frequency and momentum with $\hbar\omega \ll E_F$ and $|\mathbf{q}| \ll k_F$. Furthermore, assuming a separable ansatz in the pairing interaction, we demonstrated gauge invariance and charge conservation for our theory. Within this framework, we derived expressions for the normal and superfluid densities and compared the results in the static and long-wavelength limit with those from a local-equilibrium analysis. The same investigations were done for the specific heat capacity. In both cases we recover the same results, which validates our theory.

Finally, we presented analytic and numeric results for the electronic (pair-breaking) Raman response in noncentrosymmetric superconductors for zero temperature. For this purpose we analyzed the two most interesting classes of tetragonal and cubic symmetry, applying for example to $CePt_3Si$($\mathcal{G} = C_{4v}$) and $Li_2Pd_xPt_{3-x}B$ $\mathcal{G} = O$. Accounting for the antisymmetric spin-orbit coupling, we provide various analytic results such as the Raman vertices for both point groups, the Raman response for several pure triplet states, and power laws and kink positions for mixed-parity states. Our numerical results cover all relevant cases from weak to strong triplet-singlet ratio and demonstrate a characteristic two-peak structure for Raman spectra of non-centrosymmetric superconductors. Our theoretical predictions can be used to analyze the underlying condensate in parity-violated noncentrosymmetric superconductors and allow the determination of the unknown triplet-singlet ratio.

Acknowledgements We thank M. Sigrist for helpful discussions.

Appendix 1: Small q-Expansion

For small wave numbers, i.e. $\mathbf{q} \to 0$, the Tsuneto and related functions, which play an important role in the matrix elements N_{ij} [see Eq. (7.69)], will simplify considerably. Taking into account terms to the order $O(\eta_{\mathbf{k}}^2)$ with $\eta_{\mathbf{k}} = \mathbf{v_k} \cdot \mathbf{q}$, we obtain the well-known expression for the Tsuneto function [16]

$$\lim_{\mathbf{q}\to 0} \lambda_\lambda(\mathbf{k}) = -4\Delta_\lambda^2(\mathbf{k}) \frac{(\omega^2 - \eta_{\mathbf{k}}^2)\theta_\lambda(\mathbf{k}) + \eta_{\mathbf{k}}^2 \phi_\lambda(\mathbf{k})}{\omega^2[\omega^2 - 4E_\lambda^2(\mathbf{k})] - \eta_{\mathbf{k}}^2[\omega^2 - 4\xi_\lambda^2(\mathbf{k})]} \tag{7.109}$$

where

$$\phi_\lambda(\mathbf{k}) = -\frac{\partial n_\lambda(\mathbf{k})}{\partial \xi_\lambda(\mathbf{k})} = \frac{\xi_\lambda^2(\mathbf{k})}{E_\lambda^2(\mathbf{k})} y_\lambda(\mathbf{k}) + \frac{\Delta_\lambda^2(\mathbf{k})}{E_\lambda^2(\mathbf{k})} \theta_\lambda(\mathbf{k}) \tag{7.110}$$

is the derivative of the electron distribution function in the band λ and

$$y_\lambda(\mathbf{k}) = -\frac{\partial f[E_\lambda(\mathbf{k})]}{\partial E_\lambda(\mathbf{k})} = \frac{1}{4k_{\mathrm{B}}T} \frac{1}{\cosh^2\left(\frac{E_\lambda(\mathbf{k})}{2k_{\mathrm{B}}T}\right)} \tag{7.111}$$

is the derivative of the quasiparticle distribution function.

The following limits are also of interest: the homogeneous limit ($\mathbf{q} = 0$), e.g. for the Raman response and the static limit ($\omega = 0$), used in local-equilibrium situation

$$\lambda_\lambda(\mathbf{k}, \mathbf{q} = 0) = \frac{4\Delta_\lambda^2(\mathbf{k})\theta_\lambda(\mathbf{k})}{4E_\lambda^2(\mathbf{k}) - \omega^2} \tag{7.112}$$

$$\lim_{\omega\to 0}\lim_{\mathbf{q}\to 0}\lambda_\lambda(\mathbf{k}) = \phi_\lambda(\mathbf{k}) - y_\lambda(\mathbf{k}). \tag{7.113}$$

For the following small **q**-expansion we omitted the band-label λ for better readability:

$$\lim_{\mathbf{q}\to 0}\theta^+_{\mathbf{k}} = 2\theta_{\mathbf{k}} + \frac{\eta^2_{\mathbf{k}}}{4E^2_{\mathbf{k}}}\left[\frac{\Delta^2_{\mathbf{k}} - 2\xi^2_{\mathbf{k}}}{E^2_{\mathbf{k}}}(y_{\mathbf{k}} - \theta_{\mathbf{k}}) - \frac{\xi^2_{\mathbf{k}}}{E_{\mathbf{k}}}f''_{\mathbf{k}}\right] \tag{7.114a}$$

$$\lim_{\mathbf{q}\to 0}\theta^-_{\mathbf{k}} = \frac{\eta_{\mathbf{k}}\xi_{\mathbf{k}}}{E^2_{\mathbf{k}}}(y_{\mathbf{k}} - \theta_{\mathbf{k}}) \tag{7.114b}$$

$$\lim_{\mathbf{q}\to 0}\Phi_{\mathbf{k}} = \phi_{\mathbf{k}} + \frac{\eta^2_{\mathbf{k}}}{4E^2_{\mathbf{k}}}\frac{\Delta^2_{\mathbf{k}}(\Delta^2_{\mathbf{k}} - 4\xi^2_{\mathbf{k}})}{2E^4_{\mathbf{k}}}(y_{\mathbf{k}} - \theta_{\mathbf{k}}) - \frac{\eta^2_{\mathbf{k}}\xi^2_{\mathbf{k}}}{4E^2_{\mathbf{k}}}\left[\frac{\Delta^2_{\mathbf{k}}}{E_{\mathbf{k}}}f''_{\mathbf{k}} + \frac{\xi^2_{\mathbf{k}}}{6E^3_{\mathbf{k}}}f'''_{\mathbf{k}}\right] \tag{7.114c}$$

$$\begin{aligned}\delta\theta_{\mathbf{k}} &= \frac{\theta^+_{\mathbf{k}}}{2} - \theta_{\mathbf{k}}\\ &= \frac{\eta^2_{\mathbf{k}}}{8E^2_{\mathbf{k}}}\left[\frac{\Delta^2_{\mathbf{k}} - 2\xi^2_{\mathbf{k}}}{E^2_{\mathbf{k}}}(y_{\mathbf{k}} - \theta_{\mathbf{k}}) - \frac{\xi^2_{\mathbf{k}}}{E_{\mathbf{k}}}f''_{\mathbf{k}}\right]\end{aligned} \tag{7.114d}$$

$$\begin{aligned}\delta\phi_{\mathbf{k}} &= \Phi_{\mathbf{k}} - \phi_{\mathbf{k}}\\ &= \frac{\eta^2_{\mathbf{k}}}{4E^2_{\mathbf{k}}}\frac{\Delta^2_{\mathbf{k}}(\Delta^2_{\mathbf{k}} - 4\xi^2_{\mathbf{k}})}{2E^4_{\mathbf{k}}}(y_{\mathbf{k}} - \theta_{\mathbf{k}}) - \frac{\eta^2_{\mathbf{k}}\xi^2_{\mathbf{k}}}{4E^2_{\mathbf{k}}}\left[\frac{\Delta^2_{\mathbf{k}}}{E_{\mathbf{k}}}f''_{\mathbf{k}} + \frac{\xi^2_{\mathbf{k}}}{6E^3_{\mathbf{k}}}f'''_{\mathbf{k}}\right],\end{aligned} \tag{7.114e}$$

where $f^{(n)}_{\mathbf{k}}$ denotes the nth derivative of $f(E_{\mathbf{k}})$ with respect to $E_{\mathbf{k}}$. Furthermore we find the following expansions:

$$\lim_{\mathbf{q}\to 0}\eta^+_{\mathbf{k}} = 2E_{\mathbf{k}}\left(1 + \frac{\eta^2_{\mathbf{k}}\Delta^2_{\mathbf{k}}}{8E^4_{\mathbf{k}}}\right) \tag{7.115a}$$

$$\lim_{\mathbf{q}\to 0}\eta^-_{\mathbf{k}} = \frac{\xi_{\mathbf{k}}}{E_{\mathbf{k}}}\eta_{\mathbf{k}}\left(1 - \frac{\eta^2_{\mathbf{k}}\Delta^2_{\mathbf{k}}}{8E^4_{\mathbf{k}}}\right) \tag{7.115b}$$

$$\lim_{\mathbf{q}\to 0}\tilde{y}_{\mathbf{k}} = y_{\mathbf{k}} - \frac{\eta^2_{\mathbf{k}}}{8E^2_{\mathbf{k}}}\left(\frac{\Delta^2_{\mathbf{k}}}{E_{\mathbf{k}}}v''_{\mathbf{k}} + \frac{\xi^2_{\mathbf{k}}}{3}v'''_{\mathbf{k}}\right) \tag{7.115c}$$

$$\lim_{\mathbf{q}\to 0}\Theta_{\mathbf{k}} = \theta_{\mathbf{k}} + \frac{\eta^2_{\mathbf{k}}}{8E^2_{\mathbf{k}}}\left[\frac{\Delta^2_{\mathbf{k}}}{E^2_{\mathbf{k}}}(y_{\mathbf{k}} - \theta_{\mathbf{k}}) - \frac{\xi^2_{\mathbf{k}}}{E_{\mathbf{k}}}v''_{\mathbf{k}}\right]. \tag{7.115d}$$

The ten products of coherence factors in Eq. (7.64) have the following explicit form:

$$\left[q_{\mathbf{k}}^{(s)}\right]^2 = \frac{1}{2}\frac{E_{\mathbf{k}+}E_{\mathbf{k}-} + \xi_{\mathbf{k}+}\xi_{\mathbf{k}-} - s\Delta_{\mathbf{k}}^2}{E_{\mathbf{k}+}E_{\mathbf{k}-}} \tag{7.116a}$$

$$\left[p_{\mathbf{k}}^{(s)}\right]^2 = \frac{1}{2}\frac{E_{\mathbf{k}+}E_{\mathbf{k}-} - \xi_{\mathbf{k}+}\xi_{\mathbf{k}-} + s\Delta_{\mathbf{k}}^2}{E_{\mathbf{k}+}E_{\mathbf{k}-}} \tag{7.116b}$$

$$q_{\mathbf{k}}^{(+)}q_{\mathbf{k}}^{(-)} = \frac{1}{2}\frac{E_{\mathbf{k}-}\xi_{\mathbf{k}+} + E_{\mathbf{k}+}\xi_{\mathbf{k}-}}{E_{\mathbf{k}+}E_{\mathbf{k}-}} \tag{7.116c}$$

$$p_{\mathbf{k}}^{(+)}p_{\mathbf{k}}^{(-)} = \frac{1}{2}\frac{E_{\mathbf{k}-}\xi_{\mathbf{k}+} - E_{\mathbf{k}+}\xi_{\mathbf{k}-}}{E_{\mathbf{k}+}E_{\mathbf{k}-}} \tag{7.116d}$$

$$q_{\mathbf{k}}^{(s)}p_{\mathbf{k}}^{(s)} = \frac{\Delta_{\mathbf{k}}}{2}\frac{\xi_{\mathbf{k}+} + s\xi_{\mathbf{k}-}}{E_{\mathbf{k}+}E_{\mathbf{k}-}} \tag{7.116e}$$

$$q_{\mathbf{k}}^{(-)}p_{\mathbf{k}}^{(+)} = \frac{\Delta_{\mathbf{k}}}{2}\frac{E_{\mathbf{k}+} + E_{\mathbf{k}-}}{E_{\mathbf{k}+}E_{\mathbf{k}-}} \tag{7.116f}$$

and the small-$\mathbf{q}$ limit of each coherence factors reads:

$$\lim_{\mathbf{q}\to 0} q_{\mathbf{k}}^{(+)} = \frac{\xi_{\mathbf{k}}}{E_{\mathbf{k}}}\left(1 - \frac{\eta_{\mathbf{k}}^2\Delta_{\mathbf{k}}^2}{4E_{\mathbf{k}}^4}\right) \tag{7.117a}$$

$$\lim_{\mathbf{q}\to 0} q_{\mathbf{k}}^{(-)} = 1 - \frac{\eta_{\mathbf{k}}^2\Delta_{\mathbf{k}}^2}{8E_{\mathbf{k}}^4} \tag{7.117b}$$

$$\lim_{\mathbf{q}\to 0} p_{\mathbf{k}}^{(+)} = \frac{\Delta_{\mathbf{k}}}{E_{\mathbf{k}}}\left(1 + \frac{\eta_{\mathbf{k}}^2\xi_{\mathbf{k}}^2}{4E_{\mathbf{k}}^4}\right) \tag{7.117c}$$

$$\lim_{\mathbf{q}\to 0} p_{\mathbf{k}}^{(-)} = \frac{\eta_{\mathbf{k}}\Delta_{\mathbf{k}}}{2E_{\mathbf{k}}^2}. \tag{7.117d}$$

Appendix 2: Derivation of the Raman Vertices

In order to derive the relevant expressions for the polarization-dependent Raman vertices, we start from a general dispersion relation for tetragonal symmetry (C_{4v})

$$\begin{aligned}\varepsilon_{\mathbf{k}} = & \sum_{n=1}^{\infty}\sum_{r=0}^{\infty} a_{n,r}^{C_{4v}} \left[\cos(nk_x a) + \cos(nk_y a)\right] \cos(rk_z c) \\ & + \sum_{n=0}^{\infty}\sum_{r=0}^{\infty} b_{n,r}^{C_{4v}} \cos(nk_x a) \cos(nk_y a) \cos(rk_z c) \\ & + \sum_{n=1}^{\infty}\sum_{m=1}^{\infty}\sum_{r=0}^{\infty} c_{n,m,r}^{C_{4v}} \left[\cos(nk_x a)\cos(mk_y a) + \cos(mk_x a)\cos(nk_y a)\right] \cos(rk_z c)\end{aligned} \tag{7.118}$$

and for the cubic symmetry (O)

$$\begin{aligned}\varepsilon_{\mathbf{k}} = & \sum_{n=1}^{\infty} a_n^O \left[\cos(nk_x a) + \cos(nk_y a) + \cos(nk_z c)\right] \\ & + \sum_{n=0}^{\infty} b_n^O \cos(nk_x a)\cos(nk_y a)\cos(rk_z a) \\ & + \sum_{n=1}^{\infty}\sum_{m=1}^{n-1} c_{n,m}^O \left[\cos(mk_x a)\cos(mk_y a)\cos(nk_z a)\right. \\ & + \left.\cos(mk_x a)\cos(nk_y a)\cos(mk_z a) + \cos(nk_x a)\cos(mk_y a)\cos(mk_z a)\right] \\ & + \sum_{n=2}^{\infty}\sum_{m=1}^{n-1}\sum_{r=0}^{m-1} d_{n,m,r}^O \left[\cos(nk_x a)\cos(mk_y a)\cos(rk_z a)\right. \\ & + \cos(nk_x a)\cos(rk_y a)\cos(mk_z a) + \cos(mk_x a)\cos(nk_y a)\cos(rk_z a) \\ & + \cos(rk_x a)\cos(nk_y a)\cos(mk_z a) + \cos(mk_x a)\cos(rk_y a)\cos(nk_z a) \\ & + \left.\cos(rk_x a)\cos(mk_y a)\cos(nk_z a)\right].\end{aligned} \tag{7.119}$$

Time reversal symmetry allows only for even functions of momentum $\mathbf{k}$ in the energy dispersion. Furthermore the dispersion must be invariant under all symmetry elements of the point group $\mathcal{G}$ of the crystal. For small momentum transfers and non-resonant scattering, the Raman tensor is given by the effective-mass approximation

$$\gamma(\mathbf{k}) = m \sum_{i,j} \hat{\mathbf{e}}_i^S \frac{\partial^2 \varepsilon(\mathbf{k})}{\hbar^2 \partial k_i \partial k_j} \hat{\mathbf{e}}_j^I. \tag{7.120}$$

where $\hat{\mathbf{e}}^{S,I}$ denote the unit vectors of the scattered and incident polarization light, respectively.

The light polarization vectors select elements of the Raman tensor according to

$$\gamma_{\mathbf{k}}^{IS} = \mathbf{e}^I \cdot \gamma_{\mathbf{k}}^{(R)} \cdot \mathbf{e}^S , \tag{7.121}$$

where the Raman tensor $\gamma_{\mathbf{k}}$ can be decomposed into its symmetry components and later expanded into Fermi surface harmonics:

$$\gamma_{\mathbf{k}}^{C_{4v}} = \begin{pmatrix} \gamma_{A_1^{(1)}} + \gamma_{B_1} & \gamma_{B_2} & \gamma_{E^{(1)}} \\ \gamma_{B_2} & \gamma_{A_1^{(1)}} - \gamma_{B_1} & \gamma_{E^{(2)}} \\ \gamma_{E^{(1)}} & \gamma_{E^{(2)}} & \gamma_{A_1^{(2)}} \end{pmatrix} \tag{7.122}$$

$$\gamma_{\mathbf{k}}^{O} = \begin{pmatrix} \gamma_{A_1} + \gamma_{E^{(1)}} - \sqrt{3}\gamma_{E^{(2)}} & \gamma_{T_2^{(1)}} & \gamma_{T_2^{(2)}} \\ \gamma_{T_2^{(1)}} & \gamma_{A_1} + \gamma_{E^{(1)}} + \sqrt{3}\gamma_{E^{(2)}} & \gamma_{T_2^{(3)}} \\ \gamma_{T_2^{(2)}} & \gamma_{T_2^{(3)}} & \gamma_{A_1} - 2\gamma_{E^{(1)}} \end{pmatrix}. \tag{7.123}$$

Here we have omitted all non-Raman active symmetries such as A_{2g}. The vertices $A_1^{(1)}$ and $A_1^{(2)}$ are equal up to some constants determined by the band structure, and the vertices for $E^{(1)}$ and $E^{(2)}$ in C_{4v} differ only by a rotation of the azimuthal angle ϕ by $\pi/2$. Since this rotation is an element of the corresponding point groups, these vertices are identical, too. The same holds for $T_2^{(1)}$, $T_2^{(2)}$ and $T_2^{(3)}$. Therefore the upper indices will be omitted in the following (whenever possible). For the tetragonal group C_{4v} the A_1, B_1, B_2 and E symmetries are Raman active in backscattering geometry. Relevant polarizations for this group are:

$$\begin{aligned} \gamma_{\mathbf{k}}^{xx} &= \gamma_{\mathbf{k}}^{A_1} + \gamma_{\mathbf{k}}^{B_1} & \gamma_{\mathbf{k}}^{x'x'} &= \gamma_{\mathbf{k}}^{A_1} + \gamma_{\mathbf{k}}^{B_2} \\ \gamma_{\mathbf{k}}^{yy} &= \gamma_{\mathbf{k}}^{A_1} - \gamma_{\mathbf{k}}^{B_1} & \gamma_{\mathbf{k}}^{y'y'} &= \gamma_{\mathbf{k}}^{A_1} - \gamma_{\mathbf{k}}^{B_2} \\ \gamma_{\mathbf{k}}^{xy} &= \gamma_{\mathbf{k}}^{B_2} & \gamma_{\mathbf{k}}^{x'y'} &= \gamma_{\mathbf{k}}^{B_1} \\ \gamma_{\mathbf{k}}^{xz} &= \gamma_{\mathbf{k}}^{E} & \gamma_{\mathbf{k}}^{RR} &= \gamma_{\mathbf{k}}^{A_1} \\ \gamma_{\mathbf{k}}^{yz} &= \gamma_{\mathbf{k}}^{E} & \gamma_{\mathbf{k}}^{LL} &= \gamma_{\mathbf{k}}^{A_1} \\ \gamma_{\mathbf{k}}^{zz} &= \gamma_{\mathbf{k}}^{A_1} & \gamma_{\mathbf{k}}^{RL} &= \gamma_{\mathbf{k}}^{B_1} - i\gamma_{\mathbf{k}}^{B_2}. \end{aligned} \tag{7.124}$$

The cubic group O reveals three Raman active symmetries, namely A_1, $(E^{(1)}, E^{(2)})$, and T_2 (still assuming backscattering geometry). The relevant polarizations are:

$$\begin{aligned} \gamma_{\mathbf{k}}^{xx} &= \gamma_{\mathbf{k}}^{A_1} + \gamma_{\mathbf{k}}^{E^{(1)}} - \sqrt{3}\gamma_{\mathbf{k}}^{E^{(2)}} & \gamma_{\mathbf{k}}^{x'x'} &= \gamma_{\mathbf{k}}^{A_1} + \gamma_{\mathbf{k}}^{E^{(1)}} + \gamma_{\mathbf{k}}^{T_2} \\ \gamma_{\mathbf{k}}^{yy} &= \gamma_{\mathbf{k}}^{A_1} + \gamma_{\mathbf{k}}^{E^{(1)}} + \sqrt{3}\gamma_{\mathbf{k}}^{E^{(2)}} & \gamma_{\mathbf{k}}^{y'y'} &= \gamma_{\mathbf{k}}^{A_1} + \gamma_{\mathbf{k}}^{E^{(1)}} - \gamma_{\mathbf{k}}^{T_2} \\ \gamma_{\mathbf{k}}^{xy} &= \gamma_{\mathbf{k}}^{T_2} & \gamma_{\mathbf{k}}^{x'y'} &= -\sqrt{3}\gamma_{\mathbf{k}}^{E^{(2)}} \\ \gamma_{\mathbf{k}}^{xz} &= \gamma_{\mathbf{k}}^{T_2} & \gamma_{\mathbf{k}}^{RR} &= \gamma_{\mathbf{k}}^{A_1} + \gamma_{\mathbf{k}}^{E^{(1)}} \\ \gamma_{\mathbf{k}}^{yz} &= \gamma_{\mathbf{k}}^{T_2} & \gamma_{\mathbf{k}}^{LL} &= \gamma_{\mathbf{k}}^{A_1} + \gamma_{\mathbf{k}}^{E^{(1)}} \\ \gamma_{\mathbf{k}}^{zz} &= \gamma_{\mathbf{k}}^{A_1} - 2\gamma_{\mathbf{k}}^{E^{(1)}} & \gamma_{\mathbf{k}}^{RL} &= -\sqrt{3}\gamma_{\mathbf{k}}^{E^{(2)}} - i\gamma_{\mathbf{k}}^{T_2}. \end{aligned} \tag{7.125}$$

Here, we have defined the unit polarization vectors $\hat{\mathbf{x}}' = (\hat{\mathbf{x}} + \hat{\mathbf{y}})/\sqrt{2}$ and $\hat{\mathbf{y}}' = (\hat{\mathbf{x}} - \hat{\mathbf{y}})/\sqrt{2}$. L and R denote left and right circularly polarized light with positive and negative helicity, respectively ($\mathbf{e}^L = (\hat{\mathbf{x}} + i\hat{\mathbf{y}})/\sqrt{2}$, $\mathbf{e}^R = (\hat{\mathbf{x}} - i\hat{\mathbf{y}})/\sqrt{2}$). Note that in a backscattering configuration the polarization vectors $\mathbf{e}^{I,S}$ are pinned to the coordinate system of the crystal axes. Therefore some caution is advised when choosing the proper helicity for the scattered polarization vector $\mathbf{e}^S$. Although the Raman vertices $E^{(1)}$ and $E^{(2)}$ seem to look completely different, the Raman response turns out to be exactly the same. From a tight-binding analysis we obtain the same (band-structure) prefactors for both vertices, thus $\gamma_{\mathbf{k}}^{E^{(1)}}$ and $\sqrt{3}\gamma_{\mathbf{k}}^{E^{(2)}}$ generate both the same Raman response. Note that it is not possible to measure A_1 and $E^{(1)}$ independently in backscattering geometry with the crystal c-axis aligned parallel to the laser beam.

The Raman vertices are extracted from the band structure by comparing the symmetry components of the Raman tensor with the second derivative of the energy dispersion. This can be done by solving a set of 6 coupled linear equations—the 6 equations correspond exactly to the 6 free components of the symmetric tensor of inverse effective-mass and to the 6 symmetry elements (vertices) to be determined. Finally we make a series expansion in $\mathbf{k}$, in order to get the angular dependence of the vertices on the Fermi surface. Our results for the tetragonal point group C_{4v} are

$$\gamma_{A_1}^{(R)} = \sum_{k=0}^{\infty} \sum_{l=0}^{l\le k/2} \gamma_{k,l}^{(R)} \cos 4l\phi \sin^{2k}\theta \tag{7.126a}$$

$$\gamma_{B_1}^{(R)} = \sum_{k=1}^{\infty} \sum_{l=1}^{l\le (k+1)/2} \gamma_{k,l}^{(R)} \cos(4l-2)\phi \sin^{2k}\theta \tag{7.126b}$$

$$\gamma_{B_2}^{(R)} = \sum_{k=1}^{\infty} \sum_{l=1}^{l\le (k+1)/2} \gamma_{k,l}^{(R)} \sin(4l-2)\phi \sin^{2k}\theta \tag{7.126c}$$

$$\gamma_{E}^{(R)} = \sum_{k=1}^{\infty} \sum_{l=1}^{\infty} \gamma_{k,l}^{(R)} \sin(2l-1)\phi \sin 2k\theta \tag{7.126d}$$

and for the cubic point group O we obtain

$$\gamma_{A_1}^{(R)} = \sum_{k=0}^{\infty} \sum_{l=0}^{l\le k/2} \gamma_{k,l}^{(R)} \cos 4l\phi \sin^{2k}\theta \tag{7.127a}$$

$$\gamma_{E^{(1)}}^{(R)} = \gamma_0^{(R)}(2-3\sin^2\theta) + \cdots \tag{7.127b}$$

$$\gamma_{E^{(2)}}^{(R)} = \sum_{k=1}^{\infty} \sum_{l=1}^{l\le (k+1)/2} \gamma_{k,l}^{(R)} \cos(4l-2)\phi \sin^{2k}\theta \tag{7.127c}$$

$$\gamma_{T_2}^{(R)} = \sum_{k=1}^{\infty} \sum_{l=1}^{l \le (k+1)/2} \gamma_{k,l}^{(R)} \sin(4l-2)\phi \sin^{2k}\theta \tag{7.127d}$$

in a backscattering-geometry experiment ($z\bar{z}$).

References

1. Badica, P., Kondo, T., Togano, K. (2005) J. Phys. Soc. Jpn. **74**:1014
2. Bauer, E., Bonalde, I., Sigrist, M. (2005) Low Temp. Phys. **31**:748
3. Bauer, E., Hilscher, G., Michor, H., Paul, C., Scheidt, E. W., Gribanov, A., Seropegin Yu., Noël, H., Sigrist, M., Rogl, P. (2004) Phys. Rev. Lett. **92**:027003
4. Betbeder-Matibet, O., Nozières, P. (1969) Ann. Phys. **51**:392
5. Devereaux, T. P., Einzel, D. (1995) Phys. Rev. B **51**:16336
6. Devereaux, T. P., Einzel, D. (1996) Phys. Rev. B **54**:15547
7. Dresselhaus, G. (1955) Phys. Rev. **100**:580
8. Edelstein, V. M. (1989) Zh. Eksp. Teor. Fiz. **95**:2151
9. Einzel, D. (2003) J. Low Temp. Phys. **131**:1
10. Einzel, D., Klam, L. (2008) J. Low Temp. Phys. **150**:57
11. Fak, B., Raymond, S., Braithwaite, D., Lapertot, G., Mignot, J.-M. (2008) Phys. Rev. B **78**:184518
12. Frigeri, P. A., Agterberg, D. F., Koga, A., Sigrist, M. (2004) Phys. Rev. Lett. **92**:097001
13. Frigeri, P. A., Agterberg, D. F., Milat, I., Sigrist, M. (2006) Eur. Phys. J. B **54**:435
14. Frigeri, P. A., Agterberg, D. F., Sigrist, M. (2004) New J. Phys. **6**:115
15. Gor'kov, L. P., Rashba, E. I. (2001) Phys. Rev. Lett. **87**:037004
16. Hirschfeld, P. J., Wölfle, P., Sauls, J. A., Einzel, D., Putikka, W. O. (1989) Phys. Rev. B **40**:6695
17. Kee, H.-Y., Maki, K., Chung, C. H. (2003) Phys. Rev. B, **67**:180504
18. Klam, L., Einzel, D., Manske, D. (2009) Phys. Rev. Lett. **102**:027004
19. Klein, M. V., Dierker, S. B. (1984) Phys. Rev. B **29**:4976
20. Lee, K.-W., Pickett, W. E. (2005) Phys. Rev. B **72**:174505
21. Leggett, A. J. (1975) Rev. Mod. Phys. **47**:331
22. Mermin, N. D. (1970) Phys. Rev. B **1**:2362
23. Monien, H., Zawadowski, A. (1990) Phys. Rev. B **41**:8798
24. Pines, D., Nozières, P., Benjamin, W. A. (1966) New York
25. Samokhin, K. V. (2007) Phys. Rev. B **76**:094516
26. Samokhin, K. V., Mineev, V. P. (2008) Phys. Rev. B **77**:104520
27. Tsuneto, T. (1960) Phys. Rev. **118**:1029
28. Vorontsov, A. B., Vekhter, I., Eschrig M. (2008) Phys. Rev. Lett. **101**:127003
29. Wölfle, P. (1976) J. Low Temp. Phys. **22**:157
30. Yuan, H. Q., Agterberg, D. F., Hayashi, N., Badica, P., Vandervelde, D., Togano, K., Sigrist, M., Salamon, M. B. (2006) Phys. Rev. Lett. **97**:017006 and references therein.
31. Yuan, H. Q., Vandervelde, D., Salamon, M. B., Badica, P., Togano, K. arXiv:cond-mat/0506771 (2005).

Chapter 8
Aspects of Spintronics

S. Fujimoto and S. K. Yip

Abstract In this chapter, transport properties raised by antisymmetric spin-orbit interactions in noncentrosymmetric systems are discussed. We consider magneto-electric effects, the anomalous Hall effect, the spin Hall effect, and topological transport phenomena which are in analogy with the quantum spin Hall effect realized in Z_2 topological insulators. These topics are supposed to be relevant to potential applications to spintronics.

8.1 Introduction

Spin-orbit (SO) interactions in electron systems generally induce the coupling between charge degrees of freedom and spin degrees of freedom, giving rise to distinct transport phenomena involving both charge and spin of electrons. A well-known example is the anomalous Hall effect for which a charge Hall current is raised not by the Lorentz force, but by the coupling between a momentum of an electron and a spin moment through the SO interaction [1, 2, 3, 4, 5]. A closely related phenomenon also caused by the SO interactions is the spin Hall effect: a spin Hall current is generated by an applied longitudinal electric field in the absence of a magnetic field [6, 7, 8, 9]. The spin Hall effect opens a possibility of manipulating electron spins coherently, and may be utilized for potential applications to spintronics devices. In systems with noncentrosymmetric crystal structures, in addition to spherical SO interactions, there is an antisymmetric SO interaction,

S. Fujimoto (✉)
Department of Physics, Kyoto University
e-mail: fuji@scphys.kyoto-u.ac.jp

S. K. Yip
Academia Sinica, Institute of Physics
e-mail: yip@phys.sinica.edu.tw

E. Bauer and M. Sigrist (eds.), *Non-centrosymmetric Superconductors*,
Lecture Notes in Physics 847, DOI: 10.1007/978-3-642-24624-1_8,

$$\mathcal{H}_{\rm SO} = \alpha(\boldsymbol{k} \times \nabla V) \cdot \boldsymbol{\sigma}. \tag{8.1}$$

Here ∇V is an asymmetric potential gradient due to atoms, the locations of which break inversion symmetry. The antisymmetric SO interaction (8.1) introduces another nontrivial coupling between charge degrees of freedom and spin degrees of freedom, which associates parity-violation in momentum space with broken spin-rotational symmetry. This leads to unique transport phenomena such as magneto-electric effects [10, 11, 12, 13, 14, 15, 17, 18, 19, 20, 21]. For instance, a magnetic field coupled to electron spins controls charge current dynamics of electrons, and conversely, a charge current flow induces and affects magnetic moment of electron spins, which implies potential applications to spintronics. In this chapter, we overview the present theoretical understanding on these phenomena associated with the SO interaction in noncentrosymmetric systems. In the Sects. 8.2, 8.3, 8.4, our main concern are focused on bulk transport phenomena. Some of the above-mentioned effects are related to paramagnetic effects, and hence, drastically influenced by electron correlation effects. Furthermore, some noncentrosymmetric superconductors (NCS) discovered so far are heavy fermion systems, which are regarded as strongly correlated electron systems. Thus, we will present discussions about electron correlation effects on these transport phenomena, examining feasibility of experimental observations of them in heavy fermion NCS. In the Sect. 8.5, we will discuss a transport phenomenon analogous to the quantum spin Hall effect: spin currents carried by edge excitations which appear on open boundaries of systems. This phenomenon has been extensively studied for a certain class of band insulators. We demonstrate that a similar effect also occurs in NCS under a particular circumstance.

8.2 Model Systems

In the following, our argument for the case of normal states is largely based on the Hamiltonian,

$$\mathcal{H} = \mathcal{H}_0 + \mathcal{H}_{\rm SO}, \tag{8.2}$$

$$\mathcal{H}_0 = \sum_{k,\sigma} \varepsilon_k c_k^\dagger c_k + U \sum_i c_{\uparrow i}^\dagger c_{\uparrow i} c_{\downarrow i}^\dagger c_{\downarrow i}, \tag{8.3}$$

$$\mathcal{H}_{\rm SO} = \alpha \sum_k c_k^\dagger \boldsymbol{\mathcal{L}}_0(k) \cdot \boldsymbol{\sigma} c_k, \tag{8.4}$$

where $c_k^\dagger = (c_{\uparrow k}^\dagger, c_{\downarrow k}^\dagger)$ is the two-component spinor field for an electron with spin $\uparrow$, $\downarrow$, and momentum k. $\boldsymbol{\sigma} = (\sigma_x, \sigma_y, \sigma_z)$ with σ_ν, $\nu = x, y, z$, the Pauli matrices. $\mathcal{H}_{\rm SO}$ is an antisymmetric SO interaction with a coupling constant α. $\boldsymbol{\mathcal{L}}_0(k)$ is given by an average of the operator $(\boldsymbol{k} \times \nabla V)$ over Bloch wave functions. For tetragonal lattice structures and small k, $\boldsymbol{\mathcal{L}}_0(k) = (k_y, -k_x, 0)$, which is the Rashba interaction [22]. We also include an onsite Coulomb repulsion U in $\mathcal{H}_0$ to discuss electron correlation effects on transport properties, which may be important for heavy fermion NCS. For the discussion on superconducting states, which is presented in the Sect. 8.3 and in the Sect. 8.7, we add the BCS mean field pairing term

$$\mathcal{H}_{\text{BCS}} = -\frac{1}{2}\sum_{k}[\Delta_{\sigma\sigma'}(k)c^{\dagger}_{\sigma k}c^{\dagger}_{\sigma' -k} + h.c.] \tag{8.5}$$

to the Hamiltonian (8.2). Here the gap function is [11]

$$\Delta(k) = \Delta_s(k)i\sigma_2 + \Delta_t(k)\mathcal{L}_0(k)\cdot\boldsymbol{\sigma}i\sigma_2. \tag{8.6}$$

The first (second) term of Eq. (8.6) is the superconducting gap for a spin-singlet (spin-triplet) component. The $\boldsymbol{d}$-vector for the spin-triplet component of (8.6) is chosen soo as to optimize the antisymmetric SO interaction.

8.3 Magnetoelectric Effect

The existence of the antisymmetric SO interaction $\alpha(\boldsymbol{k}\times\nabla V)\cdot\boldsymbol{\sigma}$ yields nontrivial coupling between charge and spin degrees of freedom, giving rise to magnetoelectric effect. Magnetoelectric effects have been studied extensively for multiferroic systems, i.e. insulators. However, our argument here is focused on itinerant electron systems. This effect is possible in both the normal state and the superconducting state. In particular, in the superconducting state, the magnetoelectric effect involves dissipationless supercurrent, and, in fact, is related to static and thermodynamic properties rather than non-equilibrium transport.

8.3.1 Normal State

The magnetoelectric effect in the normal metal was originally discussed by Levitov et al. [13, 14]. We explain this effect in the case of cubic systems without mirror symmetry. When an electric field is applied, the antisymmetric SO interaction generates the magnetization,

$$\boldsymbol{M} = \hat{\Upsilon}\boldsymbol{E}. \tag{8.7}$$

Here the magnetoelectric-effect coefficient $\hat{\Upsilon}$ is a tensor. For cubic symmetry, $\hat{\Upsilon}$ is a pseudoscalar, $(\hat{\Upsilon})_{\mu\nu} = \Upsilon\delta_{\mu\nu}$ with $\mu, \nu = x, y, z$. As an inverse effect, an AC magnetic field gives rise to the charge current flow,

$$\boldsymbol{J} = -\Upsilon\frac{d\boldsymbol{B}}{dt}. \tag{8.8}$$

Equations (8.7) and (8.8) have opposite signs because the entropy generations $dS = (\boldsymbol{J} \cdot \boldsymbol{E} - \boldsymbol{M} \cdot \dot{\boldsymbol{B}})dt/T$ under these equilibrium processes must be nonzero. Note that the inverse effect (8.8) involves dissipation due to current flows, and thus requires dynamical magnetic fields which supply the system with energy.

The physical origin of these effects are easily understood as follows. When the charge current flows along the μ-axis ($\mu = x, y, z$), the Fermi surface is deformed into asymmetric shape, and because of the SO interaction which couples momentum of electrons with spins, the deformation of the Fermi surface yields imbalance of distributions of up-spins and down-spins, giving rise to magnetization along the same axis. Conversely, an applied magnetic field in the μ-direction changes the distribution of spins, which also deforms the Fermi surface asymmetrically, leading to the charge current.

It should be noticed that Eq. (8.8) does not include contributions from magnetization current $J^M = c\nabla \times \boldsymbol{M}$. The magnetization $\boldsymbol{M} = \Upsilon \boldsymbol{E}$ is related to the AC magnetic field via $\nabla \times \boldsymbol{E} = -c^{-1}\partial \boldsymbol{B}/\partial t$. Then, we obtain $J^M = c\nabla \times \boldsymbol{M} = -\Upsilon d\boldsymbol{B}/dt$. The total current induced by the AC magnetic field is

$$J_{\text{tot}} = J + J^M = -2\Upsilon \frac{d\boldsymbol{B}}{dt}. \tag{8.9}$$

The charge current is doubled by the magnetization current.

The magnetoelectric effect is possible also for the case of tetragonal systems with the Rashba-type SO interaction. However, in this case, only the off-diagonal components of the magnetoelectric-effect coefficient $\Upsilon_{\mu\nu}$ with $(\mu, \nu) = (x, y)$ or (y, x) are nonzero. Because of broken inversion symmetry along the z-direction, the Onsager relation for $\Upsilon_{\mu\nu}$ is $\Upsilon_{xy} = -\Upsilon_{yx}$. Thus, Eqs. (8.7) and (8.8) are changed to

$$M_\mu = -\Upsilon_{\mu\nu} E_\nu, \tag{8.10}$$

$$J_\mu = -\Upsilon_{\mu\nu} \frac{dB_\nu}{dt}. \tag{8.11}$$

For this definition of $\Upsilon_{\mu\nu}$, the entropy generations dS is nonzero. Since the sign of Eq. (8.10) is negative in contrast to the positive sign in the case of cubic systems (8.7), the magnetization current partially cancels the magnetoelectric-effect current; i.e. $\boldsymbol{J} + \boldsymbol{J}^M = -c\Upsilon_{xy}(\partial_x E_z, \partial_y E_z, -\partial_x E_x - \partial_y E_y)$. Thus, when these gradients of an electric field are zero, the current induced by the magnetoelectric effect vanishes.[1]

The magnetoelectric-effect coefficient in the normal state $\Upsilon_{\mu\nu}$ is calculated by using the standard linear response theory based on the Kubo formula,

$$\Upsilon_{\mu\nu}(\omega) = \frac{1}{i\omega} K^{\text{ME}}_{\mu\nu}(i\omega_n)|_{i\omega_n \to \omega + i0}, \tag{8.12}$$

$$K^{\text{ME}}_{\mu\nu}(i\omega_n) = \int_0^{1/T} d\tau \langle T_\tau \{S_\mu(\tau) J_\nu(0)\}\rangle e^{i\omega_n \tau}. \tag{8.13}$$

[1] The argument in Ref. [20] on the additive contribution of magnetization currents for the case of the Rashba model (Eq. (8.86) in Ref. [20]) is not correct.

Here S_μ and J_ν are, respectively, the total spin and the total current defined by

$$S_\mu = \mu_B \sum_k c_k^\dagger \sigma_\mu c_k, \tag{8.14}$$

$$J_\mu = e \sum_k c^\dagger \hat{v}_{k\mu} c_k, \tag{8.15}$$

with

$$\hat{v}_{k\mu} = \partial_{k_\mu}(\varepsilon_k + \alpha \boldsymbol{\sigma} \cdot \boldsymbol{\mathcal{L}}_0(\boldsymbol{k})). \tag{8.16}$$

Here we put the g factor equal to 2. In the case that electron-electron interaction is negligible, and mean free path is determined by scattering due to impurities, $\Upsilon_{\mu\nu}$ is easily calculated by using the Green function formalism;

$$K_{\mu\nu}^{\mathrm{ME}}(i\omega_n) = e\mu_B T \sum_{\varepsilon_m} \sum_k \mathrm{tr}[\sigma_\mu \hat{G}(k, \varepsilon_m + \omega_n) \hat{v}_{k\nu} \hat{G}(k, \varepsilon_m)], \tag{8.17}$$

where the single-electron Green function in the absence of the Coulomb repulsion U is

$$\hat{G}(k, \varepsilon_m) = \sum_{\tau=\pm} \frac{1 + \tau \hat{\mathcal{L}}_0(k) \cdot \boldsymbol{\sigma}}{2} G_\tau(k, \varepsilon_m), \tag{8.18}$$

$$G_\tau(k, \varepsilon_m) = \frac{1}{i\varepsilon_m - \varepsilon_{k\tau} + i\,\mathrm{sgn}(\varepsilon_m)\gamma_k}, \tag{8.19}$$

with $\varepsilon_{k\tau} = \varepsilon_k + \tau\alpha|\boldsymbol{\mathcal{L}}_0(k)|$, $\hat{\boldsymbol{\mathcal{L}}}_0(k) = \boldsymbol{\mathcal{L}}_0(k)/|\boldsymbol{\mathcal{L}}_0(k)|$, and γ_k the quasiparticle damping. ε_n and ω_n are, respectively, fermionic and bosonic Matsubara frequencies. If we assume a spherical Fermi surface, $\Upsilon_{\mu\nu}$ is $\sim e\mu_B m\alpha\ell/v_F$ where ℓ is a mean free path of an electron.

By using more elaborated analysis based on the Fermi liquid theory, we can take account of electron correlation effects on $\Upsilon_{\mu\nu}$, which may be important for heavy fermion NCS. According to this analysis, we obtain a simple relation among $\Upsilon_{\mu\nu}$, the specific heat coefficient γ, and the resistivity ρ: [19]

$$\Upsilon_{\mu\nu} \sim \frac{\mu_B}{e v_F^* \rho} \cdot \frac{\alpha k_F}{E_F} \propto \frac{\gamma}{\rho} \cdot \frac{\alpha k_F}{E_F}. \tag{8.20}$$

In general, for heavy fermion systems, the resistivity is given by $\rho \sim \rho_0 + AT^2$, with ρ_0 a residual resistivity and A a constant factor $\propto \gamma^2$. At sufficiently low temperatures, for clean systems, $\Upsilon_{\mu\nu}$ can become large.

We now estimate the order of the magnitude of these effects. We assume that the Fermi velocity is $v_F^* \sim 10^5$ cm/s, which corresponds to the mass enhancement of order ~ 100, i.e. a typical value for heavy fermion systems, and that the SO splitting is sufficiently large, e.g. $\alpha k_F/E_F \sim 0.1$. To consider the magnetization induced by an electric field, we assume that the charge current density is $J \sim 1$ A/cm^2. Then, the induced magnetization is estimated as, $M = \Upsilon E \approx \mu_B(\alpha k_F/E_F)(J/ev_F^*)$ ~ 1 Gauss, which is experimentally measurable. To evaluate the charge current induced by an AC magnetic field we assume that an AC magnetic field $B = B_0\cos(\omega t)$ with $B_0 \sim 100$ Gauss, and $\omega \sim 100$ kHz is applied, and the normal resistivity is $\rho \sim 10\,\mu\Omega \cdot$ cm. Then we obtain the charge current, $J = -2\Upsilon(dB/dt) \approx \mu_B(\alpha k_F/E_F)(dB/dt)/(ev_F^*\rho) \sim 1$ mA/cm^2. This magnitude is also experimentally accessible. However, in this case, it is required to discriminate between the current due to the magnetoelectric effect and the usual eddy current induced by the time-dependent magnetic field. For cubic systems, the current induced by the magnetoelectric effect is parallel to the direction of the applied magnetic field, and hence perpendicular to the eddy current. These two currents are distinguished by this directional dependence.

8.3.2 Superconducting State

Magnetoelectric effects in the superconducting state we shall discuss involve equilibrium dissipationless supercurrent in contrast to the non-equilibrium transport in the normal state discussed in the previous section. We shall see that there exist an extra contribution to the supercurrent induced by the Zeeman magnetic field, and conversely, an extra bulk magnetization induced by the supercurrent flow. These phenomena were originally predicted by Levitov et al. and Edelstein [10, 11, 12, 13, 17, 18, 19, 20, 21, 22]. This mentioned supercurrent is an additional contribution to the ordinary one which is due to finite phase gradients. Its physical origin is also the asymmetric deformation of the Fermi surface due to an applied Zeeman magnetic field as in the case of the normal state. However, in the present case of a static Zeeman field, no net current can arise in the normal state because of the cancellation between the contributions due to the changes in the occupation numbers versus the quasiparticle dispersion. A net finite contribution arises only within the superconducting state where this cancellation is no longer perfect [11, 16]. For the realization of this supercurrent flow induced by Zeeman magnetic fields, a system must allow a bulk current flow without dissipation. One example of such a system may be realized by attaching leads made of superconductors to the sample. Without leads to the outside, the current from phase gradient and the magnetoelectric effect must sum to be zero in the ground state and the system must develop instead a finite phase gradient and therefore be in the "helical state" [24].

To explain the magnetoelectric effects, we first exploit the Ginzburg-Landau (GL) theory, and later, we will present microscopic analysis. The GL free energy for superconductors without inversion symmetry was derived by Edelstein, Samokhin, and Kaur et al. [15, 26, 24], which reads,

$$
\begin{aligned}
F_s - F_n = & a|\Psi|^2 + \frac{\beta}{2}|\Psi|^4 + \frac{1}{2m_\mu}|D_\mu \Psi|^2 \\
& + \frac{\mathcal{K}_{\mu\nu}}{2en_s} B_\mu (\Psi (D_\nu \Psi)^* + \Psi^* D_\nu \Psi) \\
& + \frac{\boldsymbol{B}^2}{8\pi} - \frac{\chi_{\mu\mu} B_\mu^2}{2},
\end{aligned} \tag{8.21}
$$

where $a = a_0(T - T_{c0})$, $D_\mu = -\hbar \nabla_\mu - 2eA_\mu/c$, A_μ is a vector potential, $\boldsymbol{B} = \nabla \times \boldsymbol{A}$, $\boldsymbol{M}$ is a magnetization density, and n_s is a superfluid density. The forth term of Eq. (8.21) with the coefficient $\mathcal{K}_{\mu\nu}$ stems from the antisymmetric SO interaction, and is the origin of the magnetoelectric effects. Differentiating the free energy with respect to $\boldsymbol{A}$ and $\boldsymbol{B}$, we obtain the following relations for the supercurrent density J^s and the magnetization density $\boldsymbol{M}$, [17]

$$
J_\mu^s = J_\mu^{\text{dia}} + \mathcal{K}_{\nu\mu} B_\nu, \tag{8.22}
$$

$$
M_\mu = -\mathcal{K}_{\mu\nu} \Lambda J_\nu^{\text{dia}} + M_\mu^{\text{Zee}}, \tag{8.23}
$$

where J_μ^{dia} is the usual diamagnetic supercurrent given by $J_\mu^{\text{dia}} = (\hbar \nabla_\mu \phi - 2eA_\mu/c)/(2e\Lambda)$ with ϕ the phase of the order parameter Ψ, and $\Lambda^{-1} = 4e^2|\Psi|^2/m$. The last term of Eq. (8.23), M_μ^{Zee}, is magnetization due to the usual Zeeman effect. Also, we have put $|\Psi|^2 = n_s$. The second term of the right-hand side of Eq. (8.22) is the supercurrent due to the magnetoelectric effect, and the first term of the right-hand side of (8.23) is the magnetization induced by the supercurrent flow.

The structure of $\mathcal{K}_{\mu\nu}$ is constrained by the symmetry requirement as described below:

1. *Tetragonal systems with C_{4v} symmetry*—In this case, the systems are invariant with respect to the reflection $x \to -x$ (or $y \to -y$). Under this reflection, the current J_x (J_y) changes its sign, while B_x (B_y) does not. Thus, this symmetry and Eq. (8.22) imply that $-\mathcal{K}_{xx} = \mathcal{K}_{xx} = 0$. (Also, $\mathcal{K}_{yy} = 0$.) Also, under the reflection $x \to -x$, J_z is invariant, while B_z changes its sign. This implies $\mathcal{K}_{zz} = 0$. Furthermore, the systems are invariant under the $\pi/2$-rotation around the z-axis, i.e. $x \to -y$, $y \to x$. This leads to $\mathcal{K}_{xy} \to -\mathcal{K}_{yx} = \mathcal{K}_{xy}$, and $\mathcal{K}_{xz} \to -\mathcal{K}_{yz} = -\mathcal{K}_{xz} = \mathcal{K}_{xz} = 0$. Only $\mathcal{K}_{xy} = -\mathcal{K}_{yx}$ is nonzero. This case is relevant to $CePt_3Si$ (space group $P4mm$), and $CeRhSi_3$, $CeIrSi_3$ ($I4mm$).
2. *Cubic systems without mirror symmetry*—For cubic systems, if we take μ and ν as the principal axes of the crystal structure, $\mathcal{K}_{\mu,\nu} = \mathcal{K}\delta_{\mu\nu}$ holds; i.e. the supercurrent induced by the magnetoelectric effect is

$$
\boldsymbol{J} = \mathcal{K}\boldsymbol{B}. \tag{8.24}
$$

Since the left-hand side is a polar vector whereas the right-hand side is an axial vector, $\mathcal{K}$ is a pseudoscalar. For cubic systems without mirror symmetry (O symmetry), $\mathcal{K}$ is nonzero. This case is realized in $Li_2(Pd_{1-x}Pt_x)_3B$ (space group $P4_332$).

3. *Cubic systems with mirror symmetry*—An example of the crystal structure for this case is that with T_d symmetry. The SO interaction is the Dresselhaus type [25]. Equation (8.24) is still applicable to this case. However, since the pseudoscalar should vanish in the presence of mirror symmetry, $\mathcal{K} = 0$ and thus the magnetoelectric effect is absent.

It is noted that the above lists are by no means exhaustive. The same symmetry constraint is also applicable to the magnetoelectric-effect coefficient in the normal state $\Upsilon_{\mu\nu}$ discussed in the previous section.

In the case of the Rashba SO interaction, the paramagnetic supercurrent induced by Zeeman fields is partially canceled with magnetization current $\boldsymbol{J}^M = c\nabla \times \boldsymbol{M}$. This was first pointed out by Yip in the case of the Rashba interaction [18]. To see this, using Eqs. (8.22), (8.23), and the relation $\nabla \times \boldsymbol{J}^{\text{dia}} = -\boldsymbol{B}/c\Lambda$, we write down the total current,

$$\begin{aligned} \boldsymbol{J}_s + \boldsymbol{J}_M = \boldsymbol{J}^{\text{dia}} + c\nabla \times \boldsymbol{M}_{\text{Zee}} \\ + c\mathcal{K}\Lambda(-\partial_x J_z^{\text{dia}}, -\partial_y J_z^{\text{dia}}, \partial_x J_x^{\text{dia}} + \partial_y J_y^{\text{dia}}). \end{aligned} \tag{8.25}$$

The last term of the right-hand side of (8.25) is the paramagnetic supercurrent. In the complete Meissner state and in the thermodynamic limit, this term vanishes, and thus there is no paramagnetic supercurrent. Yip pointed out that because of this cancellation, the penetration depth is symmetric under the transformation $z \to -z$ [18]. However, in finite systems, or in the mixed state, the last term of (8.25) gives nonzero contributions to the magnetoelectric effect.

In the case of cubic systems without mirror symmetry, the cancellation between the paramagnetic supercurrent and the magnetization current does not occur, and instead, these currents contribute additively for the magnetoelectric effect, as elucidated in [20, 21, 22]. This is due to the fact that the magnetoelectric-effect coefficient is a pseudo scalar $\mathcal{K}_{\mu\nu} = \mathcal{K}\delta_{\mu\nu}$, as explained in the previous section for the case of the normal state.

The magnetoelectric effect coefficient $\mathcal{K}_{\mu\nu}$ can be calculated by using a linear response theory as in the case of the normal state. Since the magnetoelectric effect in the superconducting state is a static and thermodynamic phenomenon, the coefficient $\mathcal{K}_{\mu\nu}$ is given by a static correlation function,

$$\mathcal{K}_{\mu\nu} = K_{\mu\nu}^{\text{ME}}(0). \tag{8.26}$$

Here the expression of $K_{\mu\nu}^{\text{ME}}$ is the same as Eq. (8.13), but is evaluated in the superconducting state. For the case without electron-electron interaction, $\mathcal{K}_{\mu\nu}$ is calculated from

$$\mathcal{K}_{\mu\nu} = -e\mu_{\text{B}}T\sum_{n,k}\frac{1}{2}\text{tr}[\hat{S}_\mu\hat{\mathcal{G}}(k,\varepsilon_n)\hat{V}_{k\nu}\hat{\mathcal{G}}(k,\varepsilon_n)], \tag{8.27}$$

where

$$\hat{S}_\mu = \begin{pmatrix} \sigma_\mu & 0 \\ 0 & -\sigma^t_\mu \end{pmatrix}, \tag{8.28}$$

$$\hat{V}_{k\nu} = \begin{pmatrix} \hat{v}_{k\nu} & 0 \\ 0 & -\hat{v}^t_{-k\nu} \end{pmatrix}, \tag{8.29}$$

with $\hat{v}_{ky}$ defined by Eq. (8.16), and $\hat{\mathcal{G}}(k, \varepsilon_n)$ is the single-electron Green function for the model (8.2) with the pairing term (8.5), defined by

$$\hat{\mathcal{G}}(k, \varepsilon_n) = \begin{pmatrix} \hat{G}_s(k, \varepsilon_n) & \hat{F}(k, \varepsilon_n) \\ \hat{F}^\dagger(k, \varepsilon_n) & -\hat{G}^t_s(-k, -\varepsilon_n) \end{pmatrix}, \tag{8.30}$$

with

$$\hat{G}_s(k, \varepsilon_n) = \sum_{\tau=\pm 1} \frac{1 + \tau \hat{\mathcal{L}}_0(k) \cdot \boldsymbol{\sigma}}{2} G_{s\tau}(k, \varepsilon_n), \tag{8.31}$$

$$\hat{F}(k, \varepsilon_n) = \sum_{\tau=\pm 1} \frac{1 + \tau \hat{\mathcal{L}}_0(k) \cdot \sigma}{2} i\sigma_y F_\tau(k, \varepsilon_n), \tag{8.32}$$

$$G_{s\tau}(k, \varepsilon_n) = \frac{i\varepsilon + \varepsilon_{k\tau}}{(i\varepsilon + i\gamma_k \mathrm{sgn}\varepsilon)^2 - E^2_{k\tau}}, \tag{8.33}$$

$$F_\tau(k, \varepsilon_n) = \frac{\Delta_{k\tau}}{(i\varepsilon + i\gamma_k \mathrm{sgn}\varepsilon)^2 - E^2_{k\tau}}, \tag{8.34}$$

Here $E_{k\tau} = \sqrt{\varepsilon^2_{k\tau} + \Delta^2_{k\tau}}$, $\Delta_{k\tau} = \Delta_s(k) + \tau |\mathcal{L}_0(k)| \Delta_t(k)$, and $\Delta_{s(t)}(k)$ is the BCS gap for spin-singlet (spin-triplet) pairs.

Using the Fermi liquid theory, we can take account of electron correlation effects in Eq. (8.26). The most important electron correlation effect appears in the response to a Zeeman magnetic field, i.e. the renormalization of g-factor by effective mass enhancement. In the case with a spherical Fermi surface, up to the first order in $\alpha k_F / E_F$, the magnetoelectric coefficient is simplified as,

$$\mathcal{K}_{\mu\nu} = \frac{e\mu_{\mathrm{B}} n_s \alpha}{8\pi^3 z E_F}, \tag{8.35}$$

where n_s is the superfluid density. $\mathcal{K}_{\mu\nu}$ is amplified by the mass enhancement factor $1/z$. This feature is in contrast to the electron correlation effect on a conventional diamagnetic supercurrent which is suppressed by the factor z. As a result, the magnetoelectric effect in the superconducting state is much more enhanced in heavy fermion systems with large effective mass than in weakly correlated metals. In the derivation of Eq. (8.35), we assumed that there is no strong ferromagnetic spin fluctuations. If the system is in the vicinity of ferromagnetic criticality, there is additional enhancement of $\mathcal{K}_{\mu\nu}$ due to spin fluctuations. the magnetoelectric effect is enhanced by spin fluctuations.

We now discuss the feasibility of experimental observations of these effects. We use material parameters suitable for heavy fermion systems. Then, assuming $\alpha k_F/E_F \sim 0.1$, the electron density $n \sim 10^{22}$ cm^{-3}, the mass enhancement factor $1/z \sim 100$, and $v_s/v_F^* \sim \Delta/E_F \sim 0.01$, we estimate the magnitude of the bulk magnetization induced by the supercurrent as $M \approx \mu_B n(\alpha k_F/E_F)(v_s/v_F^*)/(8\pi^3 z) \sim 0.1$ Gauss. The experimental detection of this internal field may be possible. For the above conditions, the magnitude of the paramagnetic supercurrent is also accessible to usual experimental measurements. It should be emphasized again that to detect the paramagnetic supercurrent, one needs to prepare a circuit in which the bulk current flow without dissipation is possible.

8.4 Anomalous Hall Effect

In this section and the next section, we mainly consider transport phenomena in the normal state. The anomalous Hall effect is the Hall effect that is not due to Lorentz force but caused by SO interactions combined with spin polarization raised by an external magnetic field or a spontaneous magnetization in ferromagnets [1]. This effect has been explained in terms of two different mechanisms; (1) intrinsic mechanism due to bulk SO interaction, (2) extrinsic mechanism due to impurity SO scattering. Here, we are concerned with the former effect due to antisymmetric SO interactions. In the case of the Rashba SO interaction, this effect is intuitively understood as follows. When an electric field applied along the y-axis induces the current flow along this direction, deforming the Fermi surface into an asymmetric shape, the distribution of spins, which is constrained to be perpendicular to the Fermi momentum by the Rashba SO interaction, becomes anisotropic. A magnetic field H_z applied along the z-axis gives rise to torque which rotates spins around the z-axis. Because of the anisotropic distribution of spins and the SO interaction, the rotation of spins accompanies the rotation of the asymmetrically deformed Fermi surfaces on the xy-plane. As a consequence, the net current along the x-axis occurs. The Hall current in this situation is carried by electrons with anomalous velocity associated with the SO interaction. The origin of the anomalous velocity is also understood in terms of Berry-phase effects due to the SO interaction [4]; i.e. the modulation of the Bloch wave function $|u_{\boldsymbol{k}}(\boldsymbol{r})\rangle$ due to the SO coupling gives rise to the Berry curvature defined by $\frac{i}{2}\epsilon_{\alpha\beta\gamma}[\langle\frac{\partial u}{\partial k_\alpha}|\frac{\partial u}{\partial k_\beta}\rangle - \langle\frac{\partial u}{\partial k_\beta}|\frac{\partial u}{\partial k_\alpha}\rangle]$, which plays a role similar to a magnetic field, i.e. a curvature of the gauge field, yielding the transverse force on moving electrons. In the case of the Rashba interaction, the anomalous velocity, $\boldsymbol{v}_{\mathrm{A}} = \alpha(\boldsymbol{n} \times \boldsymbol{\sigma})$ with $\boldsymbol{n} = (0, 0, 1)$, is perpendicular to the z-axis. Thus, the anomalous Hall effect is possible only for magnetic fields along the z-axis.

The anomalous Hall conductivity $\sigma_{xy}^{\rm AHE}$ can be computed from the Kubo formula for an anomalous-current correlation function. For general forms of SO interactions, the expression for $\sigma_{xy}^{\rm AHE}$ is quite involved. However, in the case that, for all $\boldsymbol{k}$ on the Fermi surfaces, the SO split of electron bands $\alpha|\boldsymbol{\mathcal{L}}_0(\boldsymbol{k})|$ is nonzero and sufficiently larger than the magnitude of quasiparticle damping, the expression is much simplified. We also ignore the Coulomb repulsion between electrons, $U=0$, for simplicity. Then, the anomalous Hall conductivity for the model (8.2) with the Rashba SO interaction and a magnetic field H_z parallel to the z-axis is given by, [19]

$$\frac{\mathrm{Re}\,\sigma_{xy}^{\rm AHE}}{H_z}=e^2\mu_{\rm B}\sum_{\tau=\pm}\sum_{k}\frac{-\tau f(\varepsilon_{k\tau})}{2\alpha|\boldsymbol{\mathcal{L}}_0(\boldsymbol{k})|^3}\left(\frac{\partial\mathcal{L}_{0x}}{\partial k_x}\frac{\partial\mathcal{L}_{0y}}{\partial k_y}-\frac{\partial\mathcal{L}_{0y}}{\partial k_x}\frac{\partial\mathcal{L}_{0x}}{\partial k_y}\right). \tag{8.36}$$

In this derivation, we have ignored orbital motions of electrons due to the coupling with a vector potential, which are not important for the anomalous Hall effect. In Eq. (8.36), the quasiparticle damping γ_k does not appear, and thus, the Hall current is dissipationless in the sense that it does not involve any relaxation mechanisms. Equation (8.36) is derived assuming $\alpha|\boldsymbol{\mathcal{L}}_0(\boldsymbol{k})|\gg\gamma_k$. In the case that for a certain $\boldsymbol{k}$, the SO split vanishes, the factor $\alpha|\boldsymbol{\mathcal{L}}_0(\boldsymbol{k})|$ in the denominator of Eq. (8.36) for this wave number $\boldsymbol{k}$ is replaced with quasiparticle damping γ_k, regularizing possible divergences of (8.36). In this case, the Hall effect is dissipative in the sense that momentum dissipation mechanisms play an important role.

According to an analysis based on the Fermi liquid theory, in the case with Coulomb repulsion $U\neq 0$, $\sigma_{xy}^{\rm AHE}$ is enhanced by the mass renormalization factor $1/z$. This is because that the magnetic field H_z couples to the anomalous velocity through the Zeeman effect, and the paramagnetic effect is enhanced by the mass renormalization effect due to electron correlation. More precisely, the enhancement of $\sigma_{xy}^{\rm AHE}$ due to electron correlation effects is associated to the enhancement of the van-Vleck-like spin susceptibility which is governed by transitions between the SO split bands [19]. For heavy fermion systems, this factor is of the same order as the mass enhancement factor $1/z\approx 100\sim 1000$, and thus, the anomalous Hall conductivity can be significantly large. For instance, let us assume the resistivity $\rho\sim 10\,\mu\Omega\cdot$cm, the mass enhancement factor $1/z_{k\tau}\sim 100$, the Fermi velocity $v_F^*\sim 10^5$ cm/s, and the carrier density $n\sim 10^{22}$ cm^{-3}. Then, the ratio of $\sigma_{xy}^{\rm AHE}\sim e^2\mu_{\rm B}B/(h^2v_F^*z)$ to the normal Hall conductivity $\sigma_{xy}^{\rm NHE}$ is estimated as $\sigma_{xy}^{\rm AHE}/\sigma_{xy}^{NHE}\sim 40$. The anomalous Hall effect overwhelms the normal Hall effect.

An analogous Hall effect for heat current is also possible. The anomalous Hall conductivity for heat current is expressed as, [19]

$$\kappa_{xy}^{\rm AHE}=\frac{1}{T}(L_{xy}^{(2)}-\sum_{\mu\nu}L_{x\mu}^{(1)}L_{\mu\nu}^{(0)-1}L_{\nu y}^{(1)}), \tag{8.37}$$

where $L_{\mu\nu}^{(0)}$ is equal to the conductivity tensor $\sigma_{\mu\nu}$, and, in the absence of electron correlation, $U=0$, for the Rashba model,

$$\frac{L_{xy}^{(m)\mathrm{AHE}}}{H_z} = e^{2-m}\mu_{\mathrm{B}} \sum_{\tau=\pm} \sum_{k} \frac{-\tau(\varepsilon_{k\tau})^m f(\varepsilon_{k\tau})}{2\alpha|\mathcal{L}_0(\boldsymbol{k})|^3} \left(\frac{\partial \mathcal{L}_{0x}}{\partial k_x} \frac{\partial \mathcal{L}_{0y}}{\partial k_y} - \frac{\partial \mathcal{L}_{0y}}{\partial k_x} \frac{\partial \mathcal{L}_{0x}}{\partial k_y} \right), \tag{8.38}$$

with $m = 1, 2$. In the case with electron correlation effects, $U \neq 0$, $L_{xy}^{(m)\mathrm{AHE}}$ is enhanced by the mass renormalization factor $1/z$.

It should be noted that in Eqs. (8.36) and (8.38), electrons away from the Fermi surface give dominant contributions to the anomalous Hall conductivity. This feature is in accordance with the fact that the magnetic response against the magnetic field along the z-axis is governed by the van-Vleck-like term. This observation leads us to an interesting implication for the superconducting state. In the superconducting state, the Hall effect for heat current is possible at finite temperature, and when the superconducting gap is much smaller than the size of the SO splitting, the thermal anomalous Hall conductivity is not affected by the superconducting transition. Furthermore, even in the limit of zero temperature, $\kappa_{xy}^{\mathrm{AHE}}/(TH_z)$ is nonzero, and behaves like in the normal state, even though the quasiparticle density is vanishingly small. The experimental detection of this effect is an intriguing future issue.

8.5 Spin Hall Effect

The SO interaction gives rise to a transverse spin current under an applied longitudinal electric field even in the absence of external magnetic fields. This effect is called the spin Hall effect [8, 9, 10, 27, 28, 29, 30, 31]. The origin of the spin Hall effect is deeply related to the existence of the anomalous Hall effect. [7, 8] To explain this phenomenon, we consider the Rashba model again. Suppose that a longitudinal electric field E_x along the x-direction and a magnetic field H_z along the z-direction are applied to a system, and there is a nonzero anomalous Hall current; i.e. $J^{\mathrm{AHE}}/e = (n_\uparrow v_\uparrow + n_\downarrow v_\downarrow) = \frac{n_\uparrow + n_\downarrow}{2}(v_\uparrow + v_\downarrow) + \frac{n_\uparrow - n_\downarrow}{2}(v_\uparrow - v_\downarrow) \neq 0$ with $n_{\uparrow(\downarrow)}$ density of electrons with up (down) spin and $v_{\uparrow(\downarrow)}$ velocity of electrons with up (down) spin. On the other hand, in the absence of the magnetic field, J^{AHE} must be zero, and also there is no spin magnetization, i.e. $n_\uparrow - n_\downarrow = 0$, which leads to $v_\uparrow + v_\downarrow = 0$. As a result, for $H_z = 0$ and $E_x \neq 0$, there is a nonzero spin Hall current $J^{\mathrm{SHE}}/\mu_{\mathrm{B}} = n_\uparrow v_\uparrow - n_\downarrow v_\downarrow = \frac{n_\uparrow + n_\downarrow}{2}(v_\uparrow - v_\downarrow) \neq 0$, while the charge Hall current is zero, $J^{\mathrm{AHE}} = 0$. From a different point of view, the origin of the spin Hall effect is understood in terms of spin torque raised by the SO interaction [9]. The applied electric field $E_x \neq 0$ changes the x-component of momentum of electrons by $\Delta p_x = eE_x\Delta t$. This raises the change of the SO interaction, $\alpha\boldsymbol{\sigma} \cdot (\Delta p_x \times \nabla V)$. Since the SO interaction can be regarded as an effective Zeeman effect which depends on the direction of momentum, this change gives rise to torque of spins along $\Delta p_x \times \nabla V$. For the Rashba model, the x-component of electron spins for $p_y > 0$ is opposite to that for $p_y < 0$, and thus the spin torque yields the positive (negative) z-component of spins for $p_y > 0$ ($p_y < 0$), leading to the spin Hall current along the y-direction. Recently, the existence of the spin Hall effect in the Rashba model has been extensively investigated by several authors [9, 10]. For the Rashba model with broken inversion symmetry along the z-axis, the in-plane spin current with the magnetization in the z-direction is considered. Then, the spin Hall conductivity is defined as,

$$\sigma_{xy}^{\mathrm{SHE}} = \lim_{\omega \to 0} \frac{1}{i\omega} K^{\mathrm{SHE}}(i\omega_n)|_{i\omega_n \to \omega + i0}, \tag{8.39}$$

$$K^{\mathrm{SHE}}(i\omega_n) = \int_0^{1/T} d\tau \langle T_\tau \{J_x^{sz}(\tau) J_y(0)\}\rangle e^{i\omega_n \tau}. \tag{8.40}$$

Here the total spin current J_x^{sz} is,

$$J_x^{sz} = \frac{g\mu_{\mathrm{B}}}{4} \sum_k c_k^\dagger (\hat{v}_{kx}\sigma^z + \sigma^z \hat{v}_{kx}) c_k, \tag{8.41}$$

with g the g-factor.

To obtain an explicit formula for the spin Hall conductivity, we assume again that $\alpha|\mathcal{L}_0(\boldsymbol{k})| \gg \gamma_k$ is satisfied for all $\boldsymbol{k}$. Then, in the absence of electron correlation, $U = 0$, a straightforward calculation yields,

$$\sigma_{xy}^{\mathrm{SHE}} = \frac{eg\mu_{\mathrm{B}}}{2} \sum_k \sum_{\tau=\pm} \frac{\tau f(\varepsilon_{k\tau})}{2\alpha|\mathcal{L}_0(\boldsymbol{k})|^3} v_x \left(\mathcal{L}_{0y}(\boldsymbol{k}) \frac{\partial \mathcal{L}_{0x}}{\partial k_y} - \mathcal{L}_{0x}(\boldsymbol{k}) \frac{\partial \mathcal{L}_{0y}}{\partial k_y} \right), \tag{8.42}$$

where $v_\mu = \partial_{k_\mu} \varepsilon_k$. As in the case of the anomalous Hall conductivity (8.36), quasiparticle damping does not appear in the expression (8.42), which indicates that the effect is dissipationless. In the case of a two-dimensional electron gas model with the Rashba interaction $\mathcal{L}_0 = (k_y, -k_x, 0)$ and $\varepsilon_k = k^2/2m$, when the Fermi level crosses both of two SO split bands, the spin Hall conductivity calculated from (8.42) is

$$\sigma_{xy}^{\mathrm{SHE}} = \frac{eg\mu_{\mathrm{B}}}{8\pi}. \tag{8.43}$$

Remarkably, its value is universal, and independent of any parameters specific to the system such as the SO coupling α and electron density [9]. However, this by no means implies that $\sigma^{\mathrm{SHE}} \neq 0$ even for $\alpha \to 0$. It should be noted that the above result is obtained under the assumption that the SO split is much larger than the quasiparticle damping. As $\alpha \to 0$, the quasiparticle damping which should appear in the denominator of (8.42) becomes important, leading to $\sigma^{\mathrm{SHE}} \to 0$ [27]. In more general cases where the SO interaction is not the Rashba type and the Fermi surface is not spherical, the magnitude of $\sigma_{xy}^{\mathrm{SHE}}$ depends on the detail of the electronic structure and is not universal.

When the quasiparticle damping is governed by impurity scattering, $\sigma_{xy}^{\mathrm{SHE}}$ is partially cancelled with current vertex corrections due to impurity scattering which are related to the single-electron selfenergy (the quasiparticle damping) via the Ward-Takahashi identity. In particular, this cancellation is perfect, $\sigma_{xy}^{\mathrm{SHE}} = 0$, in the case of the Rashba model with $\mathcal{L}_0(\boldsymbol{k}) = (k_y, -k_x, 0)$ even when the SO split is much larger than the scattering rate. However, this complete cancellation is accidental, and does not hold for general forms of SO interactions [32]. Thus, to calculate $\sigma_{xy}^{\mathrm{SHE}}$ correctly, one needs to take account of both the detail band structure and scattering processes which govern the quasiparticle damping.

According to a precise analysis based on the Fermi liquid theory, the spin Hall conductivity $\sigma_{xy}^{\mathrm{SHE}}$ is not affected by electron correlation effects, but determined solely by the band structure, in contrast to the anomalous Hall conductivity discussed before. This is simply due to the absence of paramagnetic effects (Zeeman fields) for the spin Hall effect [19].

The experimental observations of the spin Hall effect were successfully achieved for semiconductors [30, 31]. In these experiments, spin polarization at the edges of samples due to spin currents under an applied electric field was detected by optical measurements. Unfortunately, experiments for NCS have not been achieved so far, partly because it is difficult, up to now, to synthesize single crystals of NCS with a size large enough to make the detection of the spin Hal effect feasible.

8.6 Quantum (Spin) Hall Effect in the Superconducting State: Topological Transport Phenomena

The subject in this section is conceptually different from the bulk transport phenomena considered in the previous sections. Here, we discuss transport phenomena raised by nontrivial topological structures of the many-body Hilbert space. As mentioned before, the anomalous Hall effect and the spin Hall effect are also related to a topological property: nonzero Berry curvature in momentum space. However, the topological transport phenomena discussed here are distinct from these Hall effect in that transport currents are carried not by bulk quasiparticles, but by edge excitations which exist on boundaries of systems. Such transport phenomena occur in the case that there are both a bulk excitation energy gap and gapless edge excitations. The studies on topological transport phenomena were initiated in the celebrated paper by Thouless et al., in which the topological explanation for the quantum Hall effect realized in two-dimensional electron gas in a strong magnetic field was presented [33]. In the quantum Hall state, there is a bulk energy gap due to the Landau quantization of the energy band, and Hall currents are mainly carried by gapless edge states, which propagate along one direction only, and are topologically protected from perturbations such as disorder [34, 35]. Here, the topological protection means that the existence of edge states is closely related to a nonzero topological number, i.e. the first Chern number n_{Ch} for the U(1) bundle corresponding to the wave function. That is, the U(1) phase of the wave function is not smooth in the entire (magnetic) Brillouin zone, and there is a jump of the phase somewhere in the $\boldsymbol{k}$-space, which leads to the nonzero Berry curvature, and the nonzero Chern number. As a result, the edge states are stable against any local perturbations which can not change the topology of the Hilbert space. The modulus of the Chern number represents the total number of the edge modes. The Hall conductivity is expressed in terms of the Chern number as $\sigma_{xy} = (e^2/h)n_{\text{Ch}}$ [33].

It was pointed out by several authors that a similar phenomenon is possible in chiral $p + ip$ superconductors, in which there is a gapless edge mode, which propagates along only one direction, reflecting broken time-reversal-symmetry in chiral superconductors [36, 37].

For a certain class of insulators with time-reversal symmetry, there exists another topological transport phenomenon, which is associated with spin currents, and called the quantum spin Hall effect; i.e. in a certain class of insulators with a bulk energy gap, a spin Hall current is induced by a longitudinal electric field [38, 39, 40, 41]. In this state, the Chern number is zero, because of time-reversal symmetry. However, instead, this state is characterized by another topological number called the Z_2 topological invariant [38]. These insulators are called the Z_2 topological insulators.

As there is the close relation between the quantum Hall state and chiral $p + ip$ superconductors mentioned above, there is also parallelism between Z_2 insulators and $s + p$-wave NCS [44, 45, 46, 47]. Moreover, in the case with a magnetic field, the $s + p$-wave NCS also exhibit a topological phase in analogy with the quantum Hall state characterized by the nonzero Chern number. In the following, we discuss these topological phenomena realized in NCS.

8.6.1 Z_2 Insulator and Quantum Spin Hall Effect

Before considering NCS, we briefly summarize the fundamental properties of the Z_2 insulator relevant to the discussion on NCS. The Z_2 insulator possesses a bulk excitation energy gap which separates the ground state from excited states. In contrast to the quantum Hall state where the bulk gap is due to the filled Landau level, the bulk gap of the Z_2 insulator is a band gap, or a gap generated by some symmetry-breaking of the system which preserves time-reversal symmetry. The most important feature of the Z_2 insulator is the existence of two gapless edge modes which propagate in the opposite directions, and carry, respectively, up-spin and down-spin. This leads to a nonzero spin current flowing on the edge without net charge current flow. As a result of it, the quantum spin Hall effect occurs; i.e. an applied electric field parallel to the edges gives rise to spin Hall current traverse.

The Z_2 insulator is regarded as a pair of two quantum Hall states in which magnetic fields are applied in the opposite directions, and time-reversal symmetry is preserved in the whole system. For a while, we assume that the total spin is conserved. Then, the two quantum Hall states are, respectively, associated with spin up and spin down states. In such a system, each of two quantum Hal states possesses nonzero Chern numbers with the same magnitude but different signs. Thus, the total Chern number is zero. However, there are another topological numbers which characterize the topological phase [38, 42, 43, 48, 49]. Let us consider the case that there are m gapless edge modes ($m > 1$) for each spin state (i.e. the total number of edge modes is $2m$), and that the spin-resolved Chern number (Chern number for each spin state) is m. All of these gapless edge modes are not necessarily topologically protected. For instance, two edge modes in the same spin state may propagate in the opposite directions. In this situation, the two gapless edge modes are backscattered by non-magnetic impurity, and become gapful. Thus, for the case of even m, the system is not topologically-protected. When m is odd, there is, at least, one gapless edge mode which is stable against disorder, characterizing the topological phase. This implies that as long as the topological nature is concerned, there are only two states; i.e. topologically trivial or non-trivial. These two states are classified by the parity of spin-resolved Chern number m. Originally, the topological number which characterizes this topological phase was introduced by Kane and Mele by using the Pfaffian of a matrix $M_{mn}(k) = \langle u_{k,m}|\Theta|u_{k,n}\rangle$ where $|u_{k,n}\rangle$ is the Bloch state with a wave vector k and a band index n ($n = 1, 2, \dots, N$), and Θ is the time reversal operator [38]. Note that each Bloch function $|u_{k,n}\rangle$ is two component spinor which consists of the Kramers doublet, and that $M_{mn}(k)$ is a $2N \times 2N$ matrix. The total number ν of zeros of the Pfaffian Pf$[M(k)]$ in half the Brillouin zone which includes only one of $\boldsymbol{k}$ and $-\mathbf{k}$ discriminates between the topological phase and trivial insulators. For the Z_2 topological insulator, $\nu = 1$ (mod 2), and for trivial insulators, $\nu = 0$ (mod 2). Later, it turned out that the Z_2 invariant is equivalent to the parity of the spin-resolved Chern number [50, 42].

In the above explanation, we consider the case that the spin projection S_z is a good quantum number. However, the concept of the Z_2 invariant is more general and applicable also to the case without spin conservation, as long as time-reversal symmetry is preserved and there is the Kramers degeneracy. Actually, in microscopic models for the Z_2 insulator proposed so far, SO interactions which violate spin conservation play important roles to stabilize the topological phase [38, 39, 40, 41]. When the total spin is not conserved but time reversal symmetry is still preserved, the above argument is valid if we replace the spin-up and spin-down states with the Kramers doublet. The stability of two gapless edge modes which form the Kramers doublet is ensured by the nonzero Z_2 invariant. More precise arguments on the relation between the gapless edge modes and topological numbers, and effects of electron-electron interaction are given in Refs. [50, 51].

8.6.2 Z_2 Topological Phase in Noncentrosymmetric Superconductors

As mentioned before, there is parallelism between Z_2 insulators and $s + p$-wave NCS in the absence of magnetic fields [44, 45, 46, 47]. To explain this point, we consider 2D NCS with the Rashba SO interaction defined on a square lattice. We assume the d-vector of the p-wave pairing is compatible with the Rashba interaction, i.e. $\boldsymbol{d} \propto (\sin k_y, -\sin k_x, 0)$. We also allows for the admixture of the s-wave pairing.

In two dimension, the superconducting gaps in the two SO split bands have no nodes provided that the p-wave gap $\Delta_p(\boldsymbol{k})$ is not equal to the s-wave gap $\Delta_s(\boldsymbol{k})$ for any $\boldsymbol{k}$ on the Fermi surfaces. To clarify the topological nature of this system, we consider the energy spectrum of edge states in the case that the geometry of the system is a cylinder with open boundaries at $x = 0$ and $x = L$ [44]. According to the numerical analysis for this system, when $\Delta_p(\boldsymbol{k}) > \Delta_s(\boldsymbol{k})$ is satisfied on the Fermi surfaces, two gapless edge modes on each boundary emerge [44, 46]. The two gapless edge modes on the same boundary are, respectively, associated with the two SO split bands which constitute the Kramers doublet, and propagate in the directions opposite to each other. This state is characterized by the Z_2 topological number, in analogy with the Z_2 insulators [44, 46]. In this phase, each of superconducting states realized in two SO split bands is similar to a chiral $p + ip$ superconducting state with different chirality. Actually, the Hilbert space of this phase can be deformed into a topologically equivalent one which is a product of the spaces of a chiral superconductor with $p_x + ip_y$ gap symmetry and that with $p_x - ip_y$ gap symmetry. The deformation into a topologically equivalent phase is possible when bulk excitation gaps are not closed by this deformation. In the case that $\Delta_p(\boldsymbol{k}) > \Delta_s(\boldsymbol{k})$ is fulfilled on the Fermi surfaces, we are able to change the magnitudes of $\Delta_s(\boldsymbol{k})$ and the SO coupling α continuously to zero without closing the bulk superconducting gap. In this state, because of time-reversal symmetry, the Chern number is zero, and there is no charge Hall current flowing on the edge. However, a spin current carried by the edge states exists, which gives rise to the spin Hall effect, in analogy with the Z_2 insulator.

8.6.3 Analogue of Quantum Hall State in the Case with Magnetic Fields

We consider again the 2D Rashba superconductors with $s + p$-wave pairing gaps satisfying the condition $\Delta_p(\boldsymbol{k}) > \Delta_s(\boldsymbol{k})$ on the Fermi surfaces. In the case with a magnetic field, a topological state similar to the quantum Hall state is realized for a particular electron density [44]. When the Fermi level crosses the Γ point in the Brillouin zone, and a magnetic field is applied to the system, a gap opens at the Γ point. If the magnetic field is smaller than an upper critical field of the superconducting state, one gapless edge mode associated with the band at the Γ point disappears, leaving only one gapless edge mode. This chiral edge state is analogous to the quantum Hall effect state. However, in contrast to the quantum Hall effect state, this gapless edge state does not carry a charge current, because the quasiparticles in the edge state are Majorana fermions; i.e. the antiparticles of them are equivalent to themselves. The Majorana edge state may be probed by thermal transport measurement.

The existence of the gapless edge mode is deeply related to the existence of a zero energy mode in a vortex core which is also described by a Majorana fermion, as clarified by analysing the Bogoliubov-de Gennes equations [52, 44]. In fact, when the geometry of the system is a disk with a closed boundary, and there is no vortex core in the system, i.e. the geometry of the system is simply-connected, the edge mode has an excitation gap of order $1/L$ where L is the perimeter of the closed system. In contrast, when there is a single vortex with odd vorticity in the bulk system, the edge mode becomes gapless, and simultaneously, a zero energy state in the vortex core appears. In this sense, the gapless edge mode is a concomitant of the zero energy vortex core state. A quasiparticle on the edge in the gapless case is also a Majorana fermion. This implies that a Majorana fermion can not exist in isolation, but should always accompany a Majorana partner, with which it forms a complex fermion.

The chiral Majorana edge state is also realizable even in a purely s-wave Rashba superconducor with $\Delta_s \neq 0$ and $\Delta_p = 0$, provided that the Zeeman energy due to a magnetic field H is larger than the s-wave gap Δ_s, i.e. $\mu_B H > \Delta_s$, and that the Fermi level is located within the energy gap around $\boldsymbol{k} \sim 0$ generated by the Zeeman effect [53, 54]. There are several proposals for the realization of this system, which utilize, e.g., ultracold fermionic atoms, heavy fermion superconductors, and semiconductor heterostructures [53, 54, 55, 56].

8.6.4 Accidentally Protected Spin Hall State Without Time-Reversal Symmetry

In the case with a magnetic field, because of broken time reversal symmetry, the topological characterization in terms of the Z_2 number is not applicable. However, even in such a situation, a pair of two gapless edge modes which carry a spin current is stable for the 2D Rashba superconductors with the condition $\Delta_p > \Delta_s$, provided that the magnetic field is perpendicular to the direction along which the edge modes propagate [44]. In this phase, both the Z_2 number and the Chern number are zero. However, there is another topological number which ensures the stability of this phase. This topological number is a winding number defined for particular symmetry points in the Brillouin zone inherent in the Rashba model [44]. In this sense, the stability of this phase is accidental, and fragile when there is a magnetic field component parallel to the propagating direction of the edge modes.

8.6.5 Topological Transport Phenomena

The transport phenomena associated with edge states can be experimentally detected by using the measurements for a system with a Hall bar geometry as considered before for the case of the quantum Hall effect in two-dimensional semiconductors [57]. In superconducting systems, instead of charge currents in semiconductors, the measurement of a heat current is useful for the detection of quasiparticle contributions to transport phenomena. In the topological phases mentioned above, heat currents are mainly carried by gapless edge states, and hence the thermal conductivity exhibits power law behavior $\propto T$ as a function of temperature, in contrast to the bulk contributions to the thermal conductivity which should decay exponentially at low temperatures $\sim \exp(-\Delta/T)$ in the superconducting state with full gap Δ. A more drastic effect characterizing the existence of edge states is a non-local transport phenomenon. In a Hall-bar geometry in which two terminals (1 and 2) are attached to one of two longer edges and another two terminals (3 and 4) are attached to the other longer edge, the temperature gradient between the terminals 1 and 3 gives rise to a heat current flowing between the terminals 2 and 4. This non-local transport can not be explained if one considers only the contributions from bulk quasiparticles, when the distance between the terminals 1, 3 and the terminals 2, 4 is sufficiently large. The detection of this effect may be a direct evidence for the existence of edge states governing low-energy transport. The experimental verification of these phenomena has not been achieved so far for NCS. The exploration for the topological superconducting state in NCS is an interesting and important future issue.

8.7 Conclusions

The antisymmetric SO interactions inherent in noncentrosymmetric systems are sources of remarkable transport phenomena both in the superconducting state and in the normal state, which are characterized by nontrivial coupling between charge and spin degrees of freedom. Although experimental verification of these phenomena in noncentrosymmetric superconductors is not yet achieved; it is naturally expected that some of them related to the paramagnetic effect such as the anomalous Hall effect and magnetoelectric effects are enhanced in strongly correlated electron systems, and their experimental detections may be feasible.

The antisymmetric SO interactions are also origins of topological order and topological transport phenomena such as the quantum spin Hall effect. In noncentrosymmetric superconductors under certain circumstances, topological phases akin to Z_2 topological insulators can be realized.

Acknowledgements The work of SF was supported by the Grant-in-Aids for Scientific Research from MEXT of Japan (Grants No.18540347 and No.19052003) The work of SKY was supported by the National Science Council of Taiwan under Grant number NSC95-2112-M-001-054-MY3.

References

1. Karplus, R., Luttinger, J.M.: Phys. Rev. **95**, 1154 (1954)
2. Luttinger, J.M.: Phys. Rev. **112**, 739 (1958)
3. Adams, E.N., Blount, E.I.: J. Chem. Phys. **10**, 286 (1959)
4. Sundaram, G., Niu, Q.: Phys. Rev. B **59**, 14915 (1999)
5. Jungwirth, T., Niu, Q., MacDonald, A.H.: Phys. Rev. Lett. **88**, 207208 (2002)
6. Lee, W.L., Watauchi, S., Miller, V.L., Cava, R.J., Ong, N.P.: Science **303**, 1647 (2004)
7. Dyakonov, M.I., Perel, V.I.: Phys. Lett. A **35**, 459 (1971)
8. Hirsch, J.E.: Phys. Rev. Lett. **83**, 1834 (1999)
9. Murakami, S., Nagaosa, N., Zhang, S.C.: Science **301**, 1348 (2003)

10. Sinova, J., Culcer, D., Niu, Q., Sinitsyn, N.A., Jungwirth, T., MacDonald, A.H.: Phys. Rev. Lett. **92**, 126603 (2004)
11. Levitov, L.S., Nazarov Yu. V., Eliashberg, G.M.: JETP Lett. **41**, 445 (1985)
12. Edelstein, V.M.: Sov. Phys. JETP **68**, 1244 (1989)
13. Edelstein, V.M.: Phys. Rev. Lett. **75**, 2004 (1995)
14. Levitov, L.S., Nazarov Yu. V., Eliashberg, G.M.: Sov. Phys. JETP **61**, 133 (1985)
15. Edelstein, V.M.: Solid State Commun. **73**, 233 (1990)
16. Edelstein, V.M.: J. Phys. Condens. Matter **8**, 339 (1996)
17. Yip, S.K.: Phys. Rev. B **65**, 144508 (2002)
18. Yip, S.K.: J. Low Temp. Phys. **140**, 67 (2005) cond-mat0502477
19. Fujimoto, S.: Phys. Rev. B **72**, 024515 (2005)
20. Fujimoto, S.: J. Phys. Soc. Jpn. **76**, 034712 (2007)
21. Lu, C.K., Yip, S.K.: Phys. Rev. B **77**, 054515 (2008)
22. Lu, C.K., Yip, S.K.: J. Low Temp. Phys. **155**, 160 (2009)
23. Rashba, E.I.: Sov. Phys. Solid State **2**, 1109 (1960)
24. Kaur, R.P., Agterberg, D.F., Sigrist, M.: Phys. Rev. Lett. **94**, 137002 (2005)
25. Dresselhaus, G.: Phys. Rev. **100**, 580 (1955)
26. Samokhin, K.: Phys. Rev. B **70**, 104521 (2004)
27. Schliemann, J., Loss, D.: Phys. Rev. B **69**, 165315 (2004)
28. Engel, H.A., Halperin, B.I., Rashba, E.I.: Phys. Rev. Lett. **95**, 166605 (2005)
29. Dyakonov, M.I.: Phys. Rev. Lett. **99**, 126601 (2007)
30. Kato, Y., Myers, R.C., Gossard, A.C., Awschalom, D.D.: Science **306**, 1910 (2004)
31. Wunderlich, J., Kaestner, B., Sinova, J., Jungwirth, T.: Phys. Rev. Lett. **94**, 047204 (2005)
32. Murakami, S.: Phys. Rev. B **69**, 241202 (2004)
33. Thouless, D.J., Kohmoto, M., Nightingale, M.P., den Nijs, M.: Phys. Rev. Lett. **49**, 405 (1982)
34. Hatsugai, Y.: Phys. Rev. Lett. **71**, 3697 (1993)
35. Wen, X.G.: Phys. Rev. Lett. **64**, 2206 (1990)
36. Stone, M., Roy, R.: Phys. Rev. B **69**, 184511 (2004)
37. Read, N., Green, D.: Phys. Rev. B **61**, 10267 (2000)
38. Kane, C.L., Mele, E.J.: Phys. Rev. Lett. **95**, 146802 (2005)
39. Kane, C.L., Mele, E.J.: Phys. Rev. Lett. **95**, 226801 (2005)
40. Bernevig, B.A., Zhang, S.C.: Phys. Rev. Lett. **96**, 106802 (2006)
41. Bernevig, B.A., Hughes, T.L., Zhang, S.C.: Science **314**, 1757 (2006)
42. Roy, R.: Phys. Rev. B **79**, 195322 (2009) cond-mat/0608064; arXiv: 0803.2868
43. Sheng, D.N., Weng, Z.Y., Sheng, L., Haldane, F.D.M.: Phys. Rev. Lett. **97**, 036808 (2006)
44. Sato, M., Fujimoto, S.: Phys. Rev. B **79**, 094504 (2009)
45. Vorontsov, A.B., Vekhter, I., Eschrig, M.: Phys. Rev. Lett. **101**, 127003 (2008)
46. Tanaka, Y., Yokoyama, T., Balatsky, A.V., Nagaosa, N.: Phys. Rev. B **79**, 060505(R) (2009)
47. Qi, X.-L., Hughes, T.L., Raghu, S., Zhang, S.C.: Phys. Rev. Lett. **102**, 187001 (2009)
48. Moore, J.E., Balents, L.: Phys. Rev. B **75**, 121306 (2007)
49. Fu, L., Kane, C.L.: Phys. Rev. B **74**, 195312 (2006)
50. Fu, L., Kane, C.L.: Phys. Rev. B **76**, 045302 (2007)
51. Qi, X.-L., Wu, Y.-S., Zhang, S.C.: Phys. Rev. B **74**, 045125 (2006)
52. Lu, C.K., Yip, S.K.: Phys. Rev. B **78**, 132502 (2008)
53. Sato, M., Takahashi, Y., Fujimoto, S.: Phys. Rev. Lett. **103**, 020401 (2009)
54. Sato, M., Takahashi, Y., Fujimoto, S.: Phys. Rev. B **82**, 134521 (2010)
55. Sau, J.D., Lutchyn, R.M., Tewari, S., Das Sarma, S.: Phys. Rev. Lett. **104**, 040502 (2010)
56. Alicea, J.: Phys. Rev. B **81**, 125318 (2010)
57. McEuen, P.L., Szafer, A., Richter, C.A., Alphenaar, B.W., Jain, J.K., Stone, A.D., Wheelerr, R.G.: Phys. Rev. Lett. **64**, 2062 (1990)

Part III
Special Topics of Non-centrosymmetric Superconductors

Chapter 9
Effects of Impurities in Non-centrosymmetric Superconductors

K. V. Samokhin

Abstract Effects of disorder on superconducting properties of noncentrosymmetric compounds are discussed. Elastic impurity scattering, even for scalar impurities, leads to a strongly anisotropic mixing of the electron states in the bands split by the spin-orbit coupling. We focus on the calculation of the critical temperature T_c, the upper critical field H_{c2}, and the spin susceptibility χ_{ij}. It is shown that the impurity effects on the critical temperature are similar to those in multi-band centrosymmetric superconductors. In particular, Anderson's theorem holds for isotropic singlet pairing. In contrast, scalar impurities affect the spin susceptibility in the same way as spin-orbit impurities do in centrosymmetric superconductors. Another peculiar feature is that in the absence of inversion symmetry scalar disorder can mix singlet and triplet pairing channels. This leads to significant deviations of the upper critical field from the predictions of the Werthamer-Helfand-Hohenberg theory in the conventional centrosymmetric case.

9.1 Introduction

The discovery of superconductivity in the heavy-fermion compound $CePt_3Si$ (Ref. [1]) has stimulated considerable interest, both experimental and theoretical, in the properties of superconductors whose crystal lattice lacks a center of inversion. The list of noncentrosymmetric superconductors has been steadily growing and now includes dozens of materials, such as UIr (Ref. [2]), $CeRhSi_3$ (Ref. [3]), $CeIrSi_3$ (Ref. [4]), Y_2C_3 (Ref. [5]), $Li_2(Pd_{1-x}, Pt_x)_3B$ (Ref. [6, 7]), and many others.

A peculiar property of noncentrosymmetric crystals is that the spin-orbit (SO) coupling of electrons with the crystal lattice qualitatively changes the nature of the

K. V. Samokhin (✉)
Department of Physics, Brock University, St. Catharines,
Ontario, L2S 3A1 Canada
e-mail: kirill.samokhin@brocku.ca

E. Bauer and M. Sigrist (eds.), *Non-centrosymmetric Superconductors*,
Lecture Notes in Physics 847, DOI: 10.1007/978-3-642-24624-1_9,

Bloch states, lifting the spin degeneracy of the electron bands almost everywhere in the Brillouin zone. The resulting nondegenerate bands are characterized by a complex spin texture and a nontrivial wavefunction topology in momentum space [8]. This has profound consequences for superconductivity, including unusual nonuniform superconducting phases, both with and without magnetic field [9–14], magnetoelectric effect [15–19], and a strongly anisotropic spin susceptibility with a large residual component at zero temperature [16, 20–22]. These and other properties are discussed in other chapters of this volume.

In this chapter we present a theoretical review of the effects of nonmagnetic impurities in superconductors without inversion symmetry. In Sect. 9.2, the disorder-averaged Green's functions in the normal and superconducting states are calculated. In Sect. 9.3, the equations for the superconducting gap functions renormalized by impurities are used to find the critical temperature T_c. In Sect. 9.4, the upper critical field H_{c2} is calculated for arbitrary temperatures. In Sect. 9.5, we calculate the spin susceptibility, focusing, in particular, on the effects of impurities on the residual susceptibility at $T = 0$. Section 9.6 contains a discussion of our results. Throughtout this chapter we use the units in which $k_B = \hbar = 1$.

9.2 Impurity Scattering in Normal and Superconducting State

Let us consider one spin-degenerate band with the dispersion given by $\varepsilon_0(\boldsymbol{k})$, and turn on the SO coupling. The Hamiltonian of noninteracting electrons in the presence of scalar impurities can be written in the form $H = H_0 + H_{imp}$, where

$$H_0 = \sum_{\boldsymbol{k},\alpha\beta} [\varepsilon_0(\boldsymbol{k})\delta_{\alpha\beta} + \boldsymbol{\gamma}(\boldsymbol{k})\boldsymbol{\sigma}_{\alpha\beta}] a^\dagger_{\boldsymbol{k}\alpha} a_{\boldsymbol{k}\beta}, \tag{9.1}$$

$\alpha, \beta = \uparrow, \downarrow$ is the spin projection on the z-axis, $\sum_{\boldsymbol{k}}$ stands for the summation over the first Brillouin zone, $\hat{\boldsymbol{\sigma}}$ are the Pauli matrices, and the chemical potential is included in $\varepsilon_0(\boldsymbol{k})$. The "bare" band dispersion satisfies $\varepsilon_0(-\boldsymbol{k}) = \varepsilon_0(\boldsymbol{k})$, $\varepsilon_0(g^{-1}\boldsymbol{k}) = \varepsilon_0(\boldsymbol{k})$, where g is any operation of the point group $\mathbf{G}$ of the crystal. The electron-lattice SO coupling is described by the pseudovector $\boldsymbol{\gamma}(\boldsymbol{k})$, which has the following symmetry properties: $\boldsymbol{\gamma}(\boldsymbol{k}) = -\boldsymbol{\gamma}(-\boldsymbol{k})$, $g\boldsymbol{\gamma}(g^{-1}\boldsymbol{k}) = \boldsymbol{\gamma}(\boldsymbol{k})$. Its momentum dependence crucially depends on $\mathbf{G}$, see Ref. [8]. For example, in the case of a tetragonal point group $\mathbf{G} = \mathbf{C}_{4v}$, which is realized, e.g., in $CePt_3Si$, $CeRhSi_3$, and $CeIrSi_3$, the simplest expression for the SO coupling is $\boldsymbol{\gamma}(\boldsymbol{k}) = \gamma_0(k_y\hat{x} - k_x\hat{y})$, which is also known as the Rashba model [23]. In contrast, in a cubic crystal with $\mathbf{G} = \mathbf{O}$, which describes $Li_2\,(Pd_{1-x},Pt_x)_3\,B$, we have $\boldsymbol{\gamma}(\boldsymbol{k}) = \gamma_0\boldsymbol{k}$.

The impurity scattering is described by the following Hamiltonian:

$$H_{imp} = \int d^3\boldsymbol{r} \sum_\alpha U(\boldsymbol{r})\psi^\dagger_\alpha(\boldsymbol{r})\psi_\alpha(\boldsymbol{r}). \tag{9.2}$$

The impurity potential $U(\boldsymbol{r})$ is a random function with zero mean and the correlator $\langle U(\boldsymbol{r}_1)U(\boldsymbol{r}_2)\rangle_{imp} = n_{imp}U_0^2\delta(\boldsymbol{r}_1 - \boldsymbol{r}_2)$, where n_{imp} is the impurity concentration, and U_0 is the strength of an individual point-like impurity.

The Hamiltonian (9.1) can be diagonalized by a unitary transformation $a_{\boldsymbol{k}\alpha} = \sum_\lambda u_{\alpha\lambda}(\boldsymbol{k})c_{\boldsymbol{k}\lambda}$, where $\lambda = \pm$ is the band index (helicity), and

$$u_{\uparrow\lambda} = \frac{1}{\sqrt{2}}\sqrt{1+\lambda\frac{\gamma_z}{|\boldsymbol{\gamma}|}}, \qquad u_{\downarrow\lambda} = \lambda\frac{1}{\sqrt{2}}\frac{\gamma_x + i\gamma_y}{\sqrt{\gamma_x^2+\gamma_y^2}}\sqrt{1-\lambda\frac{\gamma_z}{|\boldsymbol{\gamma}|}}, \tag{9.3}$$

with the following result:

$$H = \sum_{\boldsymbol{k}}\sum_{\lambda=\pm}\xi_\lambda(\boldsymbol{k})c^\dagger_{\boldsymbol{k}\lambda}c_{\boldsymbol{k}\lambda}. \tag{9.4}$$

The energy of the fermionic quasiparticles in the λth band is given by $\xi_\lambda(\boldsymbol{k}) = \varepsilon_0(\boldsymbol{k})+\lambda|\boldsymbol{\gamma}(\boldsymbol{k})|$. This expression is even in $\boldsymbol{k}$ despite the antisymmetry of the SO coupling, which is a manifestation of the Kramers degeneracy: the states $|\boldsymbol{k}\lambda\rangle$ and $|-\boldsymbol{k}\lambda\rangle$ are related by time reversal and therefore have the same energy. In real noncentrosymmetric materials, the SO splitting between the helicity bands is strongly anisotropic. Its magnitude can be characterized by $E_{SO} = 2\max_{\boldsymbol{k}}|\boldsymbol{\gamma}(\boldsymbol{k})|$. For instance, in $CePt_3Si$ E_{SO} ranges from 50 to 200 meV (Ref. [24]), while in Li_2Pd_3B it is 30 meV, reaching 200 meV in Li_2Pt_3B (Ref. [25]).

In the band representation the impurity Hamiltonian (9.2) becomes

$$H_{imp} = \frac{1}{\mathscr{V}}\sum_{\boldsymbol{k}\boldsymbol{k}'}\sum_{\lambda\lambda'}U(\boldsymbol{k}-\boldsymbol{k}')w_{\lambda\lambda'}(\boldsymbol{k},\boldsymbol{k}')c^\dagger_{\boldsymbol{k}\lambda}c_{\boldsymbol{k}'\lambda'}, \tag{9.5}$$

where $\mathscr{V}$ is the system volume, $U(\boldsymbol{q})$ is the Fourier transform of the impurity potential, and $\hat{w}(\boldsymbol{k},\boldsymbol{k}') = \hat{u}^\dagger(\boldsymbol{k})\hat{u}(\boldsymbol{k}')$. We see that the impurity scattering amplitude in the band representation is momentum-dependent, even for isotropic scalar impurities, and also acquires both intraband and interband components, the latter causing mixing of the helicity bands. In the case of a slowly-varying random potential, keeping only the forward-scattering contribution $U(\boldsymbol{q}) \sim \delta_{\boldsymbol{q},0}$, one obtains: $w_{\lambda\lambda'}(\boldsymbol{k},\boldsymbol{k}) = \delta_{\lambda\lambda'}$, i.e. the bands are not mixed.

The electron Green's function in the helicity band representation is introduced in the standard fashion: $G_{\lambda\lambda'}(\boldsymbol{k},\tau;\boldsymbol{k}',\tau') = -\langle T_\tau c_{\boldsymbol{k}\lambda}(\tau)c^\dagger_{\boldsymbol{k}'\lambda'}(\tau')\rangle$. In the absence of impurities, we have $G_{0,\lambda\lambda'}(\boldsymbol{k},\omega_n) = \delta_{\lambda\lambda'}/[i\omega_n - \xi_\lambda(\boldsymbol{k})]$, where $\omega_n = (2n+1)\pi T$ is the fermionic Matsubara frequency.

We will now show that the impurity-averaged Green's function remains band-diagonal. The disorder averaging with the Hamiltonian (9.5) can be performed using the standard methods [26], resulting in the Dyson equation of the form $\hat{G}^{-1} = \hat{G}_0^{-1} - \hat{\Sigma}$, where $\hat{G}$ is the average Green's function and $\hat{\Sigma}$ is the impurity self-energy, see Fig. 9.1. In the Born approximation, taking the thermodynamic limit $\mathscr{V}\to\infty$, we have

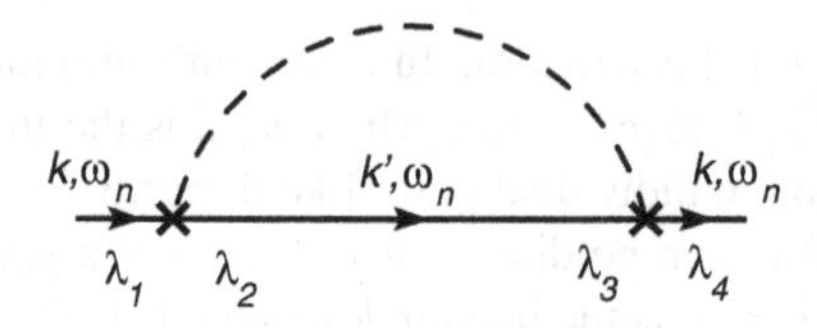

Fig. 9.1 The impurity self-energy in the band representation. The dashed line corresponds to $n_{imp}U_0^2$, the vertices include the anisotropy factors $\hat{w}(\boldsymbol{k},\boldsymbol{k}')$, and the solid line is the average Green's function of electrons in the normal state. It is shown in the text that the self-energy is nonzero only if $\lambda_1=\lambda_4$ and $\lambda_2=\lambda_3$

$$\hat{\Sigma}(\boldsymbol{k},\omega_n)=n_{imp}U_0^2\int\frac{d^3\boldsymbol{k}'}{(2\pi)^3}\hat{w}(\boldsymbol{k},\boldsymbol{k}')\hat{G}(\boldsymbol{k}',\omega_n)\hat{w}(\boldsymbol{k}',\boldsymbol{k}). \tag{9.6}$$

Seeking a solution of the Dyson equation in a band-diagonal form, $G_{\lambda\lambda'}=G_\lambda\delta_{\lambda\lambda'}$, the integrand on the right-hand side of Eq. (9.6) can be written as follows:

$$\hat{u}(\boldsymbol{k}')\hat{G}(\boldsymbol{k}',\omega_n)\hat{u}^\dagger(\boldsymbol{k}')=\frac{G_+(\boldsymbol{k}',\omega_n)+G_-(\boldsymbol{k}',\omega_n)}{2}\hat{\tau}_0 +\frac{G_+(\boldsymbol{k}',\omega_n)-G_-(\boldsymbol{k}',\omega_n)}{2}\hat{\boldsymbol{\gamma}}(\boldsymbol{k}')\hat{\boldsymbol{\tau}},$$

where $\hat{\tau}_i$ are the Pauli matrices, and $\hat{\boldsymbol{\gamma}}=\boldsymbol{\gamma}/|\boldsymbol{\gamma}|$. The second line in this expression vanishes after the momentum integration, therefore $\hat{\Sigma}(\boldsymbol{k},\omega_n)=\Sigma(\omega_n)\hat{\tau}_0$. The real part of the self-energy renormalizes the chemical potential, while for the imaginary part we obtain: $\mathrm{Im}\Sigma(\omega_n)=-\Gamma\,\mathrm{sign}\,\omega_n$. Here $\Gamma=\pi n_{imp}U_0^2N_F$ is the elastic scattering rate, with N_F defined as follows: $N_F=(N_++N_-)/2$, where $N_\lambda=\mathcal{V}^{-1}\sum_{\boldsymbol{k}}\delta[\xi_\lambda(\boldsymbol{k})]$ is the Fermi-level density of states in the λth band. Thus we arrive at the following expression for the average Green's function of the band electrons:

$$G_{\lambda\lambda'}(\boldsymbol{k},\omega_n)=\frac{\delta_{\lambda\lambda'}}{i\omega_n-\xi_\lambda(\boldsymbol{k})+i\Gamma\,\mathrm{sign}\,\omega_n}. \tag{9.7}$$

This derivation is valid under the assumption that the elastic scattering rate is small compared with the Fermi energy ε_F, which justifies neglecting the diagrams with crossed impurity lines in the self-energy in Fig. 9.1.

9.2.1 Impurity Averaging in Superconducting State

In the limit of strong SO coupling, i.e. when the band splitting exceeds all superconducting energy scales, the Cooper pairing between the electrons with opposite momenta occurs only if they are from the same nondegenerate band. The pairing interaction in the strong SO coupling case is most naturally introduced using the basis of the exact band states [20, 24, 27], which already incorporate the effects of the crystal lattice potential and the SO coupling. The total Hamiltonian including the pairing interaction is given by $H = H_0 + H_{imp} + H_{int}$, where the first two terms are given by Eqs. (9.1) and (9.5) respectively, and the last term has the following form:

$$H_{int} = \frac{1}{2\mathcal{V}} \sum_{\boldsymbol{k}\boldsymbol{k}'\boldsymbol{q}} \sum_{\lambda\lambda'} V_{\lambda\lambda'}(\boldsymbol{k}, \boldsymbol{k}') c^\dagger_{\boldsymbol{k}+\boldsymbol{q},\lambda} c^\dagger_{-\boldsymbol{k},\lambda} c_{-\boldsymbol{k}',\lambda'} c_{\boldsymbol{k}'+\boldsymbol{q},\lambda'}. \tag{9.8}$$

Physically, the pairing interaction is mediated by some bosonic excitations, e.g. phonons, and is effective only at frequencies smaller than a cutoff frequency ω_c, which has to be included in the appropriate Matsubara sums. Alternatively, the cutoff can be imposed on the momenta in Eq. (9.8), as in the original Bardeen-Cooper-Schrieffer (BCS) model. The diagonal elements of the pairing potential $V_{\lambda\lambda'}$ describe the intraband Cooper pairing, while the off-diagonal ones correspond to the pair scattering from one band to the other.

The pairing potential can be represented in the following form: $V_{\lambda\lambda'}(\boldsymbol{k}, \boldsymbol{k}') = t_\lambda(\boldsymbol{k}) t^*_{\lambda'}(\boldsymbol{k}') \tilde{V}_{\lambda\lambda'}(\boldsymbol{k}, \boldsymbol{k}')$, see Ref. [28]. Here $t_\lambda(\boldsymbol{k}) = -t_\lambda(-\boldsymbol{k})$ are non-trivial phase factors originating in the expression for the time reversal operation for electrons in the helicity bands: $K|\boldsymbol{k}\lambda\rangle = t_\lambda(\boldsymbol{k})|-\boldsymbol{k}\lambda\rangle$ [20, 27], while the components of $\tilde{V}_{\lambda\lambda'}$ are even in both $\boldsymbol{k}$ and $\boldsymbol{k}'$ and invariant under the point group operations: $\tilde{V}_{\lambda\lambda'}(g^{-1}\boldsymbol{k}, g^{-1}\boldsymbol{k}') = \tilde{V}_{\lambda\lambda'}(\boldsymbol{k}, \boldsymbol{k}')$. The latter can be expressed in terms of the basis functions of the irreducible representations of the point group [29]. In general, the basis functions are different for each matrix element. Neglecting this complication, and also considering only the one-dimensional representation corresponding to the pairing channel with the maximum critical temperature, one can write

$$\tilde{V}_{\lambda\lambda'}(\boldsymbol{k}, \boldsymbol{k}') = -V_{\lambda\lambda'} \phi_\lambda(\boldsymbol{k}) \phi^*_{\lambda'}(\boldsymbol{k}'), \tag{9.9}$$

where the coupling constants $V_{\lambda\lambda'}$ form a symmetric positive-definite 2×2 matrix, and $\phi_\lambda(\boldsymbol{k})$ are even basis functions. While $\phi_+(\boldsymbol{k})$ and $\phi_-(\boldsymbol{k})$ have the same symmetry, their momentum dependence does not have to be the same. The basis functions are assumed to be real and normalized: $\langle |\phi_\lambda(\boldsymbol{k})|^2 \rangle_\lambda = 1$, where the angular brackets denote the Fermi-surface averaging in the λth band: $\langle (\ldots) \rangle_\lambda = (1/N_\lambda \mathcal{V}) \sum_{\boldsymbol{k}} (\ldots) \delta[\xi_\lambda(\boldsymbol{k})]$.

Treating the pairing interaction (9.8) in the mean-field approximation, one introduces the superconducting order parameters in the helicity bands, which have the following form: $\Delta_\lambda(\boldsymbol{k}, \boldsymbol{q}) = t_\lambda(\boldsymbol{k}) \tilde{\Delta}_\lambda(\boldsymbol{k}, \boldsymbol{q})$. The superconducting order parameter is given by a set of complex gap functions, one for each band, which are coupled due to the interband scattering of the Cooper pairs and other mechanisms, e.g. impurity scattering. Thus the overall structure of the theory resembles that of multi-band superconductors [30, 31]. If the pairing corresponds to a one-dimensional representation, see Eq. (9.9), then we have $\tilde{\Delta}_\lambda(\boldsymbol{k}, \boldsymbol{q}) = \eta_\lambda(\boldsymbol{q}) \phi_\lambda(\boldsymbol{k})$.

An important particular case is a BCS-like model in which the pairing interaction is local in real space:

$$H_{int} = -V \int d^3\boldsymbol{r} \, \psi^\dagger_\uparrow(\boldsymbol{r}) \psi^\dagger_\downarrow(\boldsymbol{r}) \psi_\downarrow(\boldsymbol{r}) \psi_\uparrow(\boldsymbol{r}), \tag{9.10}$$

where $V > 0$ is the coupling constant. One can show [28] that in this model there is no interband pairing for any strength of the SO coupling, the order parameter has only one component η, the gap symmetry corresponds to the unity representation with $\phi_\lambda(\boldsymbol{k}) = 1$, and all coupling constants in Eq. (9.9) take the same value: $V_{\lambda\lambda'} = V/2$.

Let us calculate the impurity-averaged Green's functions in the superconducting state. To make notations compact, the normal and anomalous Green's functions [26] can be combined into a 4×4 matrix $\mathscr{G}(\boldsymbol{k}_1, \boldsymbol{k}_2; \tau) = -\langle T_\tau C_{\boldsymbol{k}_1}(\tau) C^\dagger_{\boldsymbol{k}_2}(0)\rangle$, where $C_{\boldsymbol{k}} = (c_{\boldsymbol{k}\lambda}, c^\dagger_{-\boldsymbol{k},\lambda})^T$ are four-component Nambu operators. Averaging with respect to the impurity positions restores translational invariance: $\langle \mathscr{G}(\boldsymbol{k}_1, \boldsymbol{k}_2; \omega_n)\rangle_{imp} = \delta_{\boldsymbol{k}_1,\boldsymbol{k}_2} \mathscr{G}(\boldsymbol{k}, \omega_n)$, where

$$\mathscr{G}(\boldsymbol{k}, \omega_n) = \begin{pmatrix} \hat{G}(\boldsymbol{k}, \omega_n) & -\hat{F}(\boldsymbol{k}, \omega_n) \\ -\hat{F}^\dagger(\boldsymbol{k}, \omega_n) & -\hat{G}^T(-\boldsymbol{k}, -\omega_n) \end{pmatrix}, \tag{9.11}$$

and the hats denote 2×2 matrices in the band space. The average matrix Green's function satisfies the Gor'kov equations, $(\mathscr{G}_0^{-1} - \Sigma_{imp})\mathscr{G} = 1$, where

$$\mathscr{G}_0^{-1}(\boldsymbol{k}, \omega_n) = \begin{pmatrix} i\omega_n - \hat{\xi}(\boldsymbol{k}) & -\hat{\Delta}(\boldsymbol{k}) \\ -\hat{\Delta}^\dagger(\boldsymbol{k}) & i\omega_n + \hat{\xi}(\boldsymbol{k}) \end{pmatrix}, \tag{9.12}$$

and the impurity self-energy in the self-consistent Born approximation is

$$\Sigma_{imp}(\boldsymbol{k}, \omega_n) = n_{imp} U_0^2 \int \frac{d^3\boldsymbol{k}'}{(2\pi)^3} W(\boldsymbol{k}, \boldsymbol{k}') \mathscr{G}(\boldsymbol{k}', \omega_n) W(\boldsymbol{k}', \boldsymbol{k}), \tag{9.13}$$

which is the Nambu-matrix generalization of Eq. (9.6). The 4×4 matrix W is defined as follows: $W(\boldsymbol{k}, \boldsymbol{k}') = \mathrm{diag}[\hat{w}(\boldsymbol{k}, \boldsymbol{k}'), -\hat{w}^T(-\boldsymbol{k}', -\boldsymbol{k})]$. It is straightforward to show that $[\hat{w}^T(-\boldsymbol{k}', -\boldsymbol{k})]_{\lambda\lambda'} = t^*_\lambda(\boldsymbol{k}) t_{\lambda'}(\boldsymbol{k}') w_{\lambda\lambda'}(\boldsymbol{k}, \boldsymbol{k}')$. We assume the disorder to be sufficiently weak, so that it is legitimate to use the Born approximation. Although there are some interesting qualitative effects in the opposite limit of strong disorder, such as the impurity resonance states [32], these are beyond the scope of our study.

In the absence of impurities, the Green's functions have the following form: $G_{0,\lambda\lambda'}(\boldsymbol{k}, \omega_n) = \delta_{\lambda\lambda'} G_{0,\lambda}(\boldsymbol{k}, \omega_n)$, and $F_{0,\lambda\lambda'}(\boldsymbol{k}, \omega_n) = \delta_{\lambda\lambda'} t_\lambda(\boldsymbol{k}) \tilde{F}_{0,\lambda}(\boldsymbol{k}, \omega_n)$, where

$$G_{0,\lambda} = -\frac{i\omega_n + \xi_\lambda}{\omega_n^2 + \xi_\lambda^2 + |\tilde{\Delta}_\lambda|^2}, \quad \tilde{F}_{0,\lambda} = \frac{\tilde{\Delta}_\lambda}{\omega_n^2 + \xi_\lambda^2 + |\tilde{\Delta}_\lambda|^2}. \tag{9.14}$$

In the presence of impurities, we seek solution of the Gor'kov equations in a band-diagonal form and require, for consistency, that the Nambu matrix components of the self-energy are also band-diagonal. Then,

$$\begin{pmatrix} \Sigma^{11}_{\lambda\lambda'}(\boldsymbol{k}, \omega_n) & \Sigma^{12}_{\lambda\lambda'}(\boldsymbol{k}, \omega_n) \\ \Sigma^{21}_{\lambda\lambda'}(\boldsymbol{k}, \omega_n) & \Sigma^{22}_{\lambda\lambda'}(\boldsymbol{k}, \omega_n) \end{pmatrix} = \delta_{\lambda\lambda'} \begin{pmatrix} \Sigma_1(\omega_n) & t_\lambda(\boldsymbol{k}) \Sigma_2(\omega_n) \\ t^*_\lambda(\boldsymbol{k}) \Sigma^*_2(\omega_n) & -\Sigma_1(-\omega_n) \end{pmatrix}, \tag{9.15}$$

where Σ_1 and Σ_2 satisfy the equations

$$\Sigma_1(\omega_n) = \frac{1}{2} n_{imp} U_0^2 \sum_\lambda \int \frac{d^3k}{(2\pi)^3} G_\lambda(\boldsymbol{k}, \omega_n),$$
$$\Sigma_2(\omega_n) = \frac{1}{2} n_{imp} U_0^2 \sum_\lambda \int \frac{d^3k}{(2\pi)^3} \tilde{F}_\lambda(\boldsymbol{k}, \omega_n). \tag{9.16}$$

Absorbing the real part of Σ_1 into the chemical potential, we have $\Sigma_1(\omega_n) = i\tilde{\Sigma}_1(\omega_n)$, where $\tilde{\Sigma}_1$ is odd in ω_n.

Solving the Gor'kov equations we obtain the following expressions for the disorder-averaged Green's functions:

$$G_\lambda(\boldsymbol{k}, \omega_n) = -\frac{i\tilde{\omega}_n + \xi_\lambda(\boldsymbol{k})}{\tilde{\omega}_n^2 + \xi_\lambda^2(\boldsymbol{k}) + |D_\lambda(\boldsymbol{k}, \omega_n)|^2},$$
$$\tilde{F}_\lambda(\boldsymbol{k}, \omega_n) = \frac{D_\lambda(\boldsymbol{k}, \omega_n)}{\tilde{\omega}_n^2 + \xi_\lambda^2(\boldsymbol{k}) + |D_\lambda(\boldsymbol{k}, \omega_n)|^2}, \tag{9.17}$$

where $\tilde{\omega}_n = \omega_n - \tilde{\Sigma}_1(\omega_n)$ and $D_\lambda(\boldsymbol{k}, \omega_n) = \tilde{\Delta}_\lambda(\boldsymbol{k}) + \Sigma_2(\omega_n)$. Substituting these into Eqs. (9.16), we arrive at the self-consistency equations for the Matsubara frequency and the gap functions renormalized by impurities:

$$\tilde{\omega}_n = \omega_n + \frac{\Gamma}{2} \sum_\lambda \rho_\lambda \left\langle \frac{\tilde{\omega}_n}{\sqrt{\tilde{\omega}_n^2 + |D_\lambda(\boldsymbol{k}, \omega_n)|^2}} \right\rangle_\lambda, \tag{9.18}$$

$$D_\lambda(\boldsymbol{k}, \omega_n) = \eta_\lambda \phi_\lambda(\boldsymbol{k}) + \frac{\Gamma}{2} \sum_{\lambda'} \rho_{\lambda'} \left\langle \frac{D_{\lambda'}(\boldsymbol{k}', \omega_n)}{\sqrt{\tilde{\omega}_n^2 + |D_{\lambda'}(\boldsymbol{k}', \omega_n)|^2}} \right\rangle_{\lambda'}, \tag{9.19}$$

where

$$\rho_\pm = \frac{N_\pm}{N_F} = 1 \pm \delta \tag{9.20}$$

are the fractional densities of states in the helicity bands. The parameter $\delta = (N_+ - N_-)/(N_+ + N_-)$ characterizes the strength of the SO coupling.

9.3 Gap Equations and the Critical Temperature

The Gor'kov equations must be supplemented by self-consistency equations for the order parameter components, which have the form usual for two-band superconductors. In particular, for a uniform order parameter, $\eta_\lambda(\boldsymbol{q}) = \eta_\lambda \delta(\boldsymbol{q})$, we have $\sum_{\lambda'} V_{\lambda\lambda'}^{-1} \eta_{\lambda'} = T \sum_n \int_{\boldsymbol{k}} \tilde{F}_\lambda(\boldsymbol{k}, \omega_n) \phi_\lambda(\boldsymbol{k})$ (recall that the basis functions are assumed to be real). Using Eq. (9.17), we obtain:

$$\sum_{\lambda'} V^{-1}_{\lambda\lambda'}\eta_{\lambda'} = \pi\rho_\lambda N_F T \sum_n{}' \left\langle \frac{D_\lambda(\boldsymbol{k},\omega_n)\phi_\lambda(\boldsymbol{k})}{\sqrt{\tilde{\omega}_n^2 + |D_\lambda(\boldsymbol{k},\omega_n)|^2}} \right\rangle_\lambda . \tag{9.21}$$

These equations are called the gap equations and, together with Eqs. (9.18) and (9.19), completely determine the properties of disordered noncentrosymmetric superconductors in the uniform state. The prime in the Matsubara sum means that the summation is limited to $\omega_n \le \omega_c$, where ω_c is the BCS frequency cutoff.

The superconducting critical temperature can be found from Eq. (9.21) after linearization with respect to the order parameter components. It follows from Eq. (9.18) that $\tilde{\omega}_n = \omega_n + \Gamma \operatorname{sign}\omega_n$ near T_c, and we obtain from Eq. (9.19) that $\Sigma_2(\omega_n) = (\Gamma/2|\omega_n|)\sum_\lambda \rho_\lambda \langle\phi_\lambda\rangle\eta_\lambda$ (here and below we omit, for brevity, the arguments of the basis functions and the subscripts λ in the Fermi-surface averages). Therefore the linearized gap equations take the form $\sum_{\lambda'} a_{\lambda\lambda'}\eta_{\lambda'} = 0$, where

$$a_{\lambda\lambda'} = \frac{1}{N_F} V^{-1}_{\lambda\lambda'} - \rho_\lambda \delta_{\lambda\lambda'} S_{01} - \frac{\Gamma}{2}\rho_\lambda\rho_{\lambda'}\langle\phi_\lambda\rangle\langle\phi_{\lambda'}\rangle S_{11}, \tag{9.22}$$

with $S_{kl} = \pi T \sum_n |\omega_n|^{-k}(|\omega_n| + |\Gamma|)^{-l}$. The Matsubara sums here can be easily calculated:

$$S_{01} = \pi T \sum_n{}' \frac{1}{\omega_n + \Gamma} = \ln\frac{2e^C\omega_c}{\pi T} - \mathscr{F}\left(\frac{T}{\Gamma}\right),$$

where $C \simeq 0.577$ is Euler's constant,

$$\mathscr{F}(x) = \Psi\left(\frac{1}{2} + \frac{1}{2\pi x}\right) - \Psi\left(\frac{1}{2}\right), \tag{9.23}$$

and $\Psi(x)$ is the digamma function. Note that the expression (9.23) for the impurity correction to S_{01} is valid if $\Gamma \ll \omega_c$, when it is legitimate to extend the summation in $2\pi T\sum_n[1/(\omega_n+\Gamma) - 1/\omega_n]$ to infinity and express the result in terms of the digamma functions, see Ref. [33]. Similarly, we obtain: $S_{11} = \mathscr{F}(T/\Gamma)/\Gamma$. It is convenient to introduce the following notation for dimensionless coupling constants:

$$g_{\lambda\lambda'} = N_F V_{\lambda\lambda'}\rho_{\lambda'} = V_{\lambda\lambda'}N_{\lambda'} \tag{9.24}$$

(note that the matrix $\hat{g}$ is not symmetric, in general). Then, the superconducting critical temperature T_c is found from the equation $\det(\hat{\tau}_0 + \hat{g}\hat{M}) = 0$, where

$$M_{\lambda\lambda'} = -\delta_{\lambda\lambda'}\ln\frac{2e^C\omega_c}{\pi T_c} + \left(\delta_{\lambda\lambda'} - \frac{\rho_{\lambda'}}{2}\langle\phi_\lambda\rangle\langle\phi_{\lambda'}\rangle\right)\mathscr{F}\left(\frac{T_c}{\Gamma}\right), \tag{9.25}$$

see Ref. [34].

In the absence of impurities, the second term in $M_{\lambda\lambda'}$ vanishes, and we obtain the critical temperature of a clean superconductor:

$$T_{c0} = \frac{2e^C \omega_c}{\pi} e^{-1/g}, \tag{9.26}$$

where

$$g = \frac{g_{++} + g_{--}}{2} + \sqrt{\left(\frac{g_{++} - g_{--}}{2}\right)^2 + g_{+-}g_{-+}} \tag{9.27}$$

is the effective coupling constant. In the presence of impurities, the cases of conventional and unconventional pairing have to be considered separately.

Unconventional pairing. In this case $\langle \phi_\lambda \rangle = 0$, and we obtain the following equation for T_c:

$$\ln \frac{T_{c0}}{T_c} = \mathscr{F}\left(\frac{T_c}{\Gamma}\right). \tag{9.28}$$

The reduction of the critical temperature is described by a universal function, which has the same form as in isotropic centrosymmetric superconductors with magnetic impurities [35], or in anisotropically paired centrosymmetric superconductors with nonmagnetic impurities [29, 36]. In particular, at weak disorder, i.e. in the limit $\Gamma \ll T_{c0}$, we have $T_c = T_{c0} - \pi\Gamma/4$. The superconductivity is completely suppressed at $\Gamma_c = (\pi/2e^C)T_{c0}$.

Conventional pairing. Assuming a completely isotropic pairing with $\phi_\lambda = 1$, we obtain:

$$\ln \frac{T_{c0}}{T_c} = \frac{1 + c_1\mathscr{F}(x)}{c_2 + c_3\mathscr{F}(x) + \sqrt{c_4 + c_5\mathscr{F}(x) + c_6\mathscr{F}^2(x)}} - \frac{1}{g}, \tag{9.29}$$

where $x = T_c/\Gamma$, and

$$c_1 = \frac{\rho_+(g_{--} - g_{+-}) + \rho_-(g_{++} - g_{-+})}{2}, \quad c_2 = \frac{g_{++} + g_{--}}{2}, \quad c_3 = \frac{\det \hat{g}}{2},$$

$$c_4 = \left(\frac{g_{++} - g_{--}}{2}\right)^2 + g_{+-}g_{-+}, \quad c_5 = (c_2 - c_1)\det \hat{g}, \quad c_6 = c_3^2.$$

We see that the critical temperature depends on nonmagnetic disorder, but in contrast to the unconventional case, the effect is not described by a universal Abrikosov-Gor'kov function [34, 37]. At weak disorder the suppression is linear in the scattering rate, but with a non-universal slope:

$$T_c = T_{c0} - \frac{1}{g}\left[c_1 - \frac{\pi}{4}\frac{1}{g}\left(c_3 + \frac{c_5}{2\sqrt{c_4}}\right)\right]\Gamma. \tag{9.30}$$

In the case of strong impurity scattering, $\Gamma \gg T_{c0}$, we use $\mathscr{F}(x) = \ln(1/x) + O(1)$ at $x \to 0$, to find that the critical temperature approaches the limiting value given by

$$T_c^* = T_{c0} \exp\left(\frac{1}{g} - \frac{c_1}{2c_3}\right), \tag{9.31}$$

i.e. superconductivity is not completely destroyed by impurities. The explanation is the same as in the conventional two-gap superconductors, see e.g. Refs. [38–40]: Interband impurity scattering tends to reduce the difference between the gap magnitudes in the two bands, which costs energy and thus suppresses T_c, but only until both gaps become equal. One can show that both the coefficient in front of Γ in Eq. (9.30) and the exponent in Eq. (9.31) are negative, i.e. $T_c^* < T_{c0}$.

In the BCS-like model (9.10) the pairing is isotropic and described by a single coupling constant $V_{\lambda\lambda'} = V/2$, and we have $g = N_F V$. Although the expression (9.26) for the critical temperature in the clean case has the usual BCS form, the analogy is not complete, because the order parameter resides in two nondegenerate bands, and $N_F = (N_+ + N_-)/2$. In the presence of impurities, the right-hand side of Eq. (9.29) vanishes, therefore there is an analog of Anderson's theorem: The nonmagnetic disorder has no effect on the critical temperature.

Another important particular case, possibly relevant to $CePt_3Si$, is the model in which only one band, say $\lambda = +$, is superconducting, while the other band remains normal [24, 34]. This can be described by setting $V_{+-} = V_{--} = 0$. Using Eq. (9.25), we obtain the following equation for the critical temperature:

$$\ln\frac{T_{c0}}{T_c} = c_0 \mathscr{F}\left(\frac{T_c}{\Gamma}\right), \tag{9.32}$$

where $c_0 = 1 - \rho_+ \langle\phi_+\rangle^2/2$. At weak disorder we have $T_c = T_{c0} - c_0(\pi\Gamma/4)$, while at strong disorder $T_c = T_{c0}(\pi T_{c0}/e^C \Gamma)^{1/c_0}$. If the pairing is anisotropic but conventional, then, unlike the unconventional case with $\langle\phi_+\rangle = 0$, the superconductivity is never completely destroyed, even at strong disorder.

9.3.1 Isotropic Model

Finding the superconducting gap at arbitrary temperatures and impurity concentrations from the nonlinear gap equations (9.21) is more difficult than the calculation of T_c. We focus on the case when the pairing is completely isotropic, i.e. $\phi_+(\boldsymbol{k}) = \phi_-(\boldsymbol{k}) = 1$ and $\tilde{\Delta}_\lambda(\boldsymbol{k}) = \eta_\lambda$. The order parameter components can be chosen to be real, and the gap equations take the following form:

$$\sum_{\lambda'} V_{\lambda\lambda'}^{-1} \eta_{\lambda'} = \pi \rho_\lambda N_F T {\sum_n}' \frac{D_\lambda(\omega_n)}{\sqrt{\tilde{\omega}_n^2 + D_\lambda^2(\omega_n)}}. \tag{9.33}$$

We further assume that the difference between ρ_+ and ρ_- can be neglected and the pairing strength, see Eq. (9.9), does not vary between the bands: $V_{++} = V_{--} > 0$. For the interband coupling constants, we have $V_{+-} = V_{-+}$. The gap equations have two solutions: $\eta_+ = \eta_- = \eta$ and $\eta_+ = -\eta_- = \eta$. In the spin representation, the former corresponds to the singlet state, while the latter – to the "protected" triplet state [21].

The impurity responses of these two states turn out to be very different.

$\eta_+ = \eta_- = \eta$. In this case $D_+(\omega_n) = D_-(\omega_n) = D(\omega_n)$, and Eqs. (9.18) and (9.19) take the following form:

$$\tilde{\omega}_n = \omega_n + \Gamma \frac{\tilde{\omega}_n}{\sqrt{\tilde{\omega}_n^2 + D^2}}, \quad D = \eta + \Gamma \frac{D}{\sqrt{\tilde{\omega}_n^2 + D^2}}.$$

Introducing $Z(\omega_n) = 1 + \Gamma/\sqrt{\omega_n^2 + \eta^2}$, the solution of these equations is $D(\omega_n) = Z(\omega_n)\eta$, $\tilde{\omega}_n = Z(\omega_n)\omega_n$. Therefore, the gap equation (9.33) becomes

$$\eta = \pi g_1 T \sum_n{}' \frac{D}{\sqrt{\tilde{\omega}_n^2 + D^2}} = \pi g_1 T \sum_n{}' \frac{\eta}{\sqrt{\omega_n^2 + \eta^2}}, \tag{9.34}$$

where $g_1 = (V_{++} + V_{+-})N_F$. The scattering rate has dropped out, therefore there is an analog of Anderson's theorem: neither the gap magnitude nor the critical temperature are affected by impurities. Namely, we have $T_c(\Gamma) = T_{c0}$, see Eq. (9.26) with $g = g_1$, while the gap magnitude at T=0 is given by the clean BCS expression: $\eta(T = 0) = \eta_0 = (\pi/e^C)T_{c0}$.

$\eta_+ = -\eta_- = \eta$. In this case $D_+(\omega_n) = -D_-(\omega_n) = \eta$, and we obtain from Eqs. (9.18), (9.19), and (9.33):

$$\tilde{\omega}_n = \omega_n + \Gamma \frac{\tilde{\omega}_n}{\sqrt{\tilde{\omega}_n^2 + \eta^2}}, \quad \eta = \pi g_2 T \sum_n{}' \frac{\eta}{\sqrt{\tilde{\omega}_n^2 + \eta^2}}, \tag{9.35}$$

where $g_2 = (V_{++} - V_{+-})N_F$. In the absence of impurities, the critical temperature is given by the BCS expression (9.26) with $g = g_2$. If $V_{+-} > 0$ (attractive interband interaction), then $g_2 < g_1$ and the phase transition occurs into the state $\eta_+ = \eta_-$. However, if $V_{+-} < 0$ (repulsive interband interaction), then $g_2 > g_1$ and the phase transition occurs into the state $\eta_+ = -\eta_-$. In contrast to the previous case, both the critical temperature and the gap magnitude are now suppressed by disorder. The former is determined by the equation (9.29), which takes the same universal form as the Abrikosov-Gor'kov equation (9.28). Superconductivity is completely destroyed if the disorder strength exceeds the critical value $\Gamma_c = (\pi/2e^C)T_{c0}$.

To find the gap magnitude at $T = 0$ as a function of Γ we follow the procedure described in Ref. [41]. Replacing the Matsubara sum by a frequency integral in the second of equations (9.35), we obtain:

$$\ln \frac{\eta_0}{\eta} = \int_0^\infty d\omega \left(\frac{1}{\sqrt{\omega^2 + \eta^2}} - \frac{1}{\sqrt{\tilde{\omega}^2 + \eta^2}} \right), \tag{9.36}$$

where $\eta_0 = 2\Gamma_c$ is the BCS gap magnitude in the clean case, and $\tilde{\omega}$ satisfies the equation $\tilde{\omega} = \omega + \Gamma\tilde{\omega}/\sqrt{\tilde{\omega}^2 + \eta^2}$. Transforming the second term on the right-hand side of Eq. (9.36) into an integral over $\tilde{\omega}$ we arrive at the following equation:

$$\ln\frac{\Gamma_c}{\Gamma}=\frac{\pi x}{4}-\ln(2x)+\theta(x-1)\Bigg[\ln(x+\sqrt{x^2-1})$$
$$-\frac{x}{2}\arctan\sqrt{x^2-1}-\frac{\sqrt{x^2-1}}{2x}\Bigg], \tag{9.37}$$

where $x=\Gamma/\eta$. This equation does not have solutions at $\Gamma>\Gamma_c$, which is consistent with the complete suppression of superconductivity above the critical disorder strength. In the weak disorder limit, $\Gamma\ll\Gamma_c$, the solution is $x\simeq\Gamma/2\Gamma_c$, while at $\Gamma\to\Gamma_c$ we have $x\simeq\sqrt{\Gamma_c/12(\Gamma_c-\Gamma)}$.

9.4 Upper Critical Field at Arbitrary Temperature

In this section, we calculate the upper critical field $H_{c2}(T)$ of a disordered noncentrosymmetric superconductor described by the BCS-like model (9.10). We assume a uniform external field $\boldsymbol{H}$ and neglect the paramagnetic pair breaking. The noninteracting part of the Hamiltonian is given by

$$\hat{h}=\varepsilon_0(\boldsymbol{K})+\boldsymbol{\gamma}(\boldsymbol{K})\hat{\boldsymbol{\sigma}}+U(\boldsymbol{r}), \tag{9.38}$$

where $\boldsymbol{K}=-i\nabla+(e/c)\boldsymbol{A}(\boldsymbol{r})$, $\boldsymbol{A}$ is the vector potential, and e is the absolute value of the electron charge. The superconducting order parameter in the model (9.10) is represented by a single complex function $\eta(\boldsymbol{r})$. According to Sect. 9.3, the zero-field critical temperature is not affected by scalar impurities. The critical temperature at $\boldsymbol{H}\neq 0$, or inversely the upper critical field as a function of temperature, can be found from the condition that the linearized gap equation $[V^{-1}-T\sum_n{}'\hat{X}(\omega_n)]\eta(\boldsymbol{r})=0$ has a nontrivial solution. Here the operator $\hat{X}(\omega_n)$ is defined by the kernel

$$X(\boldsymbol{r},\boldsymbol{r}';\omega_n)=\frac{1}{2}\left\langle \mathrm{tr}\,\hat{g}^\dagger\hat{G}(\boldsymbol{r},\boldsymbol{r}';\omega_n)\hat{g}\hat{G}^T(\boldsymbol{r},\boldsymbol{r}';-\omega_n)\right\rangle_{imp}, \tag{9.39}$$

where $\hat{g}=i\hat{\sigma}_2$. The angular brackets denote the impurity averaging, and $\hat{G}(\boldsymbol{r},\boldsymbol{r}';\omega_n)$ is the Matsubara Green's function of electrons in the normal state, which satisfies the equation $(i\omega_n-\hat{h})\hat{G}(\boldsymbol{r},\boldsymbol{r}';\omega_n)=\delta(\boldsymbol{r}-\boldsymbol{r}')$, with $\hat{h}$ given by expression (9.38).

The impurity average of the product of two Green's functions in Eq. (9.39) can be represented graphically by the ladder diagrams, see Fig. 9.2 (as before, we assume the disorder to be sufficiently weak for the diagrams with crossed impurity lines to be negligible). In order to sum the diagrams, we introduce an impurity-renormalized gap function $\hat{D}(\boldsymbol{r},\omega_n)$, which a matrix in the spin space satisfying the following integral equation:

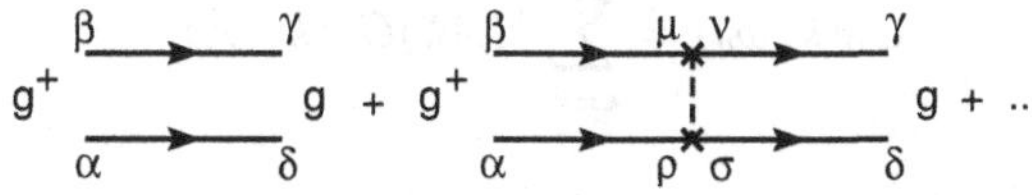

β γ β μ ν γ
g⁺ g + g⁺ g + ...
α δ α ρ σ δ

Fig. 9.2 Impurity ladder diagrams in the Cooper channel. Lines with arrows correspond to the average Green's functions of electrons, $\hat{g} = i\hat{\sigma}_2$, and the impurity (dashed) line is defined in the text, see Eq. (9.41)

$$\begin{aligned}\hat{D}(\boldsymbol{r}, \omega_n) &= \eta(\boldsymbol{r})\hat{g} \\ &+ \frac{1}{2} n_{imp} U_0^2 \hat{g} \int d^3\boldsymbol{r}' \mathrm{tr} \hat{g}^\dagger \langle \hat{G}(\boldsymbol{r}, \boldsymbol{r}'; \omega_n) \rangle_{imp} \hat{D}(\boldsymbol{r}', \omega_n) \langle \hat{G}^T(\boldsymbol{r}, \boldsymbol{r}'; -\omega_n) \rangle_{imp} \\ &+ \frac{1}{2} n_{imp} U_0^2 \hat{\boldsymbol{g}} \int d^3\boldsymbol{r}' \mathrm{tr} \hat{\boldsymbol{g}}^\dagger \langle \hat{G}(\boldsymbol{r}, \boldsymbol{r}'; \omega_n) \rangle_{imp} \hat{D}(\boldsymbol{r}', \omega_n) \langle \hat{G}^T(\boldsymbol{r}, \boldsymbol{r}'; -\omega_n) \rangle_{imp}.\end{aligned} \tag{9.40}$$

This can be easily derived from the ladder diagrams in Fig. 9.2, by representing each "rung" of the ladder as a sum of spin-singlet and spin-triplet terms:

$$n_{imp} U_0^2 \delta_{\mu\nu} \delta_{\rho\sigma} = \frac{1}{2} n_{imp} U_0^2 g_{\mu\rho} g_{\sigma\nu}^\dagger + \frac{1}{2} n_{imp} U_0^2 \boldsymbol{g}_{\mu\rho} \boldsymbol{g}_{\sigma\nu}^\dagger, \tag{9.41}$$

where $\hat{\boldsymbol{g}} = i\hat{\boldsymbol{\sigma}}\hat{\sigma}_2$.

Seeking solution of Eq. (9.40) in the form $\hat{D}(\boldsymbol{r}, \omega_n) = d_0(\boldsymbol{r}, \omega_n)\hat{g} + \boldsymbol{d}(\boldsymbol{r}, \omega_n)\hat{\boldsymbol{g}}$, we obtain a system of four integral equations for d_a, where $a = 0, 1, 2, 3$:

$$\sum_{b=0}^{3} \left[\delta_{ab} - \Gamma \hat{\mathscr{Y}}_{ab}(\omega_n)\right] d_b(\boldsymbol{r}, \omega_n) = \eta(\boldsymbol{r}) \delta_{a0}. \tag{9.42}$$

Here the operators $\hat{\mathscr{Y}}_{ab}(\omega_n)$ are defined by the kernels

$$\mathscr{Y}_{ab}(\boldsymbol{r}, \boldsymbol{r}'; \omega_n) = \frac{1}{2\pi N_F} \mathrm{tr} \hat{g}_a^\dagger \langle \hat{G}(\boldsymbol{r}, \boldsymbol{r}'; \omega_n) \rangle_{imp} \hat{g}_b \langle \hat{G}^T(\boldsymbol{r}, \boldsymbol{r}'; -\omega_n) \rangle_{imp}, \tag{9.43}$$

with $\hat{g}_0 = \hat{g}$, and $\hat{g}_a = \hat{\boldsymbol{g}}_a$ for $a = 1, 2, 3$. We see that, in addition to the spin-singlet component $d_0(\boldsymbol{r}, \omega_n)$, impurity scattering also induces a nonzero spin-triplet component $\boldsymbol{d}(\boldsymbol{r}, \omega_n)$. The linearized gap equation contains only the former. Indeed, using Eq. (9.42) we obtain:

$$\frac{1}{N_F V} \eta(\boldsymbol{r}) - \pi T \sum_n{}' \frac{d_0(\boldsymbol{r}, \omega_n) - \eta(\boldsymbol{r})}{\Gamma} = 0. \tag{9.44}$$

It is easy to see that the triplet component does not appear in the centrosymmetric case. Indeed, in the absence of the Zeeman interaction the spin structure of the Green's function is trivial: $G_{\alpha\beta}(\boldsymbol{r}, \boldsymbol{r}'; \omega_n) = \delta_{\alpha\beta} G(\boldsymbol{r}, \boldsymbol{r}'; \omega_n)$. Then it follows from Eq. (9.43) that $\hat{\mathscr{Y}}_{ab}(\omega_n) = \delta_{ab} \hat{\mathscr{Y}}(\omega_n)$, therefore $d_0 = (1 - \Gamma \hat{\mathscr{Y}})^{-1} \eta$ and $\boldsymbol{d} = 0$.

The next step is to find the spectrum of the operators $\hat{\mathscr{Y}}_{ab}(\omega_n)$. At zero field, the average Green's function has the following form:

$$\hat{G}_0(\boldsymbol{k},\omega_n)=\sum_{\lambda=\pm}\hat{\Pi}_\lambda(\boldsymbol{k})G_\lambda(\boldsymbol{k},\omega_n), \tag{9.45}$$

where $\hat{\Pi}_\lambda=(1+\lambda\hat{\boldsymbol{\gamma}}\hat{\boldsymbol{\sigma}})/2$ are the helicity band projection operators, and $G_\lambda(\boldsymbol{k},\omega_n)$ are the impurity-averaged Green's functions in the band representation, see Eq. (9.7). At $\boldsymbol{H}\neq 0$, we have $\langle\hat{G}(\boldsymbol{r},\boldsymbol{r}';\omega_n)\rangle_{imp}=\hat{G}_0(\boldsymbol{r}-\boldsymbol{r}';\omega_n)\exp[(ie/c)\int_{\boldsymbol{r}}^{\boldsymbol{r}'}\boldsymbol{A}d\boldsymbol{r}]$, where the integration is performed along a straight line connecting $\boldsymbol{r}$ and $\boldsymbol{r}'$ [42]. This approximation is legitimate if the temperature is not very low, so that the Landau level quantization can be neglected. It follows from Eq. (9.43) that $\mathscr{Y}_{ab}(\omega_n)=Y_{ab}(\boldsymbol{q},\omega_n)\big|_{\boldsymbol{q}\to\boldsymbol{D}}$, where $\boldsymbol{D}=-i\nabla+(2e/c)\boldsymbol{A}$ and

$$Y_{ab}(\boldsymbol{q},\omega_n)=\frac{1}{2\pi N_F}\int\frac{d^3k}{(2\pi)^3}\mathrm{tr}\hat{g}_a^\dagger\hat{G}_0(\boldsymbol{k}+\boldsymbol{q},\omega_n)\hat{g}_b\hat{G}_0^T(-\boldsymbol{k},-\omega_n). \tag{9.46}$$

Substituting here expressions (9.45) and calculating the spin traces, we obtain for the singlet-singlet and singlet-triplet contributions:

$$Y_{00}=\frac{1}{2}\sum_\lambda\rho_\lambda\left\langle\frac{1}{|\omega_n|+\Gamma+i\boldsymbol{v}_\lambda(\boldsymbol{k})\boldsymbol{q}\ \mathrm{sign}\,\omega_n/2}\right\rangle, \tag{9.47}$$

$$Y_{0i}=Y_{i0}=\frac{1}{2}\sum_\lambda\lambda\rho_\lambda\left\langle\frac{\hat{\gamma}_i(\boldsymbol{k})}{|\omega_n|+\Gamma+i\boldsymbol{v}_\lambda(\boldsymbol{k})\boldsymbol{q}\ \mathrm{sign}\ \omega_n/2}\right\rangle, \tag{9.48}$$

where $\boldsymbol{v}_\lambda=\partial\xi_\lambda/\partial\boldsymbol{k}$ is the quasiparticle velocity in the λth band. We see that the singlet-triplet mixing occurs due to the SO coupling and vanishes when $\rho_+=\rho_-=1$ and $\boldsymbol{v}_+=\boldsymbol{v}_-=\boldsymbol{v}_F$. The triplet-triplet contributions can be represented as follows: $Y_{ij}=Y_{ij}^{(1)}+Y_{ij}^{(2)}$, where

$$Y_{ij}^{(1)}=\frac{1}{2}\sum_\lambda\rho_\lambda\left\langle\frac{\hat{\gamma}_i(\boldsymbol{k})\hat{\gamma}_j(\boldsymbol{k})}{|\omega_n|+\Gamma+i\boldsymbol{v}_\lambda(\boldsymbol{k})\boldsymbol{q}\ \mathrm{sign}\ \omega_n/2}\right\rangle, \tag{9.49}$$

and

$$Y_{ij}^{(2)}=\frac{1}{2\pi N_F}\sum_\lambda\int\frac{d^3k}{(2\pi)^3}(\delta_{ij}-\hat{\gamma}_i\hat{\gamma}_j-i\lambda e_{ijl}\hat{\gamma}_l)G_\lambda(\boldsymbol{k}+\boldsymbol{q},\omega_n)G_{-\lambda}(-\boldsymbol{k},-\omega_n). \tag{9.50}$$

The singlet impurity scattering channel, which is described by the first term in Eq. (9.41), causes only the scattering of intraband pairs between the bands. In contrast, the triplet impurity scattering can also create interband pairs, which are described by $Y_{ij}^{(2)}$.

It is easy to show that if the SO band splitting exceeds both ω_c and Γ, then the interband term in Y_{ij} is smaller than the intraband one. Let us consider, for example, isotropic bands with $\xi_\pm(\boldsymbol{k})=\varepsilon_0(\boldsymbol{k})\pm\gamma$. Neglecting for simplicity the differences between the densities of states and the Fermi velocities in the two bands and setting $\boldsymbol{q} = 0$, we obtain from Eq. (9.49) and (9.50):

$$Y_{ij}^{(1)}(\mathbf{0}, \omega_n) = \frac{\delta_{ij}}{3(|\omega_n| + \Gamma)} \equiv Y_{intra}(\omega_n)\delta_{ij},$$
$$Y_{ij}^{(2)}(\mathbf{0}, \omega_n) = \frac{2\delta_{ij}}{3(|\omega_n| + \Gamma)(1 + r^2)} \equiv Y_{inter}(\omega_n)\delta_{ij},$$

where $r(\omega_n) = \gamma/(|\omega_n| + \Gamma)$. Due to the BCS cutoff, the maximum value of ω_n in the Cooper ladder is equal to ω_c, therefore $r_{min} \sim E_{SO}/\max(\omega_c, \Gamma)$. We assume that this ratio is large, which is a good assumption for real materials, therefore

$$\max_n \frac{Y_{inter}(\omega_n)}{Y_{intra}(\omega_n)} = \frac{2}{1 + r_{min}^2} \sim \left[\frac{\max(\omega_c, \Gamma)}{E_{SO}}\right]^2 \ll 1, \tag{9.51}$$

at all Matsubara frequencies satisfying $|\omega_n| \leq \omega_c$.

Thus the interband contributions to the Cooper ladder can be neglected, and we obtain:

$$Y_{ab}(\boldsymbol{q}, \omega_n) = \frac{1}{2}\sum_\lambda \rho_\lambda \left\langle \frac{\Lambda_{\lambda,a}(\boldsymbol{k})\Lambda_{\lambda,b}(\boldsymbol{k})}{|\omega_n| + \Gamma + i\boldsymbol{v}_\lambda(\boldsymbol{k})\boldsymbol{q}\,\mathrm{sign}\,\omega_n/2} \right\rangle, \tag{9.52}$$

where $\Lambda_{\lambda,0}(\boldsymbol{k}) = 1$ and $\Lambda_{\lambda,a}(\boldsymbol{k}) = \lambda\hat{\gamma}_a(\boldsymbol{k})$ for $a = 1, 2, 3$. Making the substitution $\boldsymbol{q} \to \boldsymbol{D}$, we represent $\hat{\mathscr{Y}}_{ab}$ as a differential operator of infinite order:

$$\hat{\mathscr{Y}}_{ab}(\omega_n) = \frac{1}{2}\int_0^\infty du\, e^{-u(|\omega_n| + \gamma)} \sum_\lambda \rho_\lambda \left\langle \Lambda_{\lambda,a}(\boldsymbol{k})\Lambda_{\lambda,b}(\boldsymbol{k}) e^{-iu\boldsymbol{v}_\lambda(\boldsymbol{k})\boldsymbol{D}\,\mathrm{sign}\,\omega_n/2} \right\rangle. \tag{9.53}$$

In order to solve Eq. (9.42), with the operators $\hat{\mathscr{Y}}_{ab}(\omega_n)$ given by expressions (9.53), we follow the procedure described in Ref. [43, 44]. Choosing the z-axis along the external field: $\boldsymbol{H} = H\hat{z}$, we introduce the operators $a_\pm = \ell_H(D_x \pm iD_y)/2$, and $a_3 = \ell_H D_z$, where $\ell_H = \sqrt{c/eH}$ is the magnetic length. It is easy to check that $a_+ = a_-^\dagger$ and $[a_-, a_+] = 1$, therefore $a_\pm$ have the meaning of the raising and lowering operators, while $a_3 = a_3^\dagger$ commutes with both of them: $[a_3, a_\pm] = 0$. It is convenient to expand both the order parameter η and the impurity-renormalized gap functions d_a in the basis of Landau levels $|N, p\rangle$, which satisfy $a_+|N, p\rangle = \sqrt{N+1}|N+1, p\rangle$, $a_-|N, p\rangle = \sqrt{N}|N-1, p\rangle$, and $a_3|N, p\rangle = p|N, p\rangle$, where $N = 0, 1, \ldots$, and p is a real number characterizing the variation of the order parameter along the field. We have

$$\eta(\boldsymbol{r}) = \sum_{N,p} \eta_{N,p}\langle \boldsymbol{r}|N, p\rangle, \qquad d_a(\boldsymbol{r}, \omega_n) = \sum_{N,p} d_{N,p}^a(\omega_n)\langle \boldsymbol{r}|N, p\rangle. \tag{9.54}$$

According to Eq. (9.42), the expansion coefficients can be found from the following algebraic equations:

$$\sum_{N',p',b}\Big[\delta_{ab}\delta_{NN'}\delta_{pp'} - \Gamma\langle N,p|\hat{\mathcal{Y}}_{ab}(\omega_n)|N',p'\rangle\Big]d^b_{N',p'}(\omega_n) = \delta_{a0}\eta_{N,p}. \quad (9.55)$$

Substituting the solutions of these equations into

$$\frac{1}{N_F V}\eta_{N,p} - \pi T\sum_n{}' \frac{d^0_{N,p}(\omega_n) - \eta_{N,p}}{\Gamma} = 0, \quad (9.56)$$

see Eq. (9.44), and setting the determinant of the resulting linear equations for $\eta_{N,p}$ to zero, one arrives at an equation for the upper critical field.

9.4.1 $H_{c2}(T)$ in a Cubic Crystal

In the general case, i.e. for arbitrary crystal symmetry and electronic band structure, the procedure outlined above does not yield an equation for $H_{c2}(T)$ in a closed form, since all the Landau levels are coupled. In order to make progress, we focus on the cubic case, $\mathbf{G} = \mathbf{O}$, with a parabolic band and the SO coupling given by $\boldsymbol{\gamma}(\boldsymbol{k}) = \gamma_0\boldsymbol{k}$. As for the parameter δ, which characterizes the difference between the band densities of states, see Eq. (9.20), we assume that

$$\delta_c \ll |\delta| \le 1, \quad (9.57)$$

where $\delta_c = \max(\omega_c, \Gamma)/\varepsilon_F \ll 1$. While the first inequality follows from the condition (9.51), which ensures the smallness of the interband contribution to the Cooper impurity ladder, the second one is always satisfied, with $|\delta| \to 1$ corresponding to the rather unrealistic limit of extremely strong SO coupling.

In order to solve the gap equations, we make a change of variables in the triplet component: $d_\pm = (d_1 \pm id_2)/\sqrt{2}$. Then, Eq. (9.42) take the following form:

$$\begin{pmatrix} 1-\Gamma\hat{\mathcal{Y}}_{00} & -\Gamma\hat{\mathcal{Y}}_{03} & -\Gamma\hat{\mathcal{Y}}_{0-} & -\Gamma\hat{\mathcal{Y}}_{0+} \\ -\Gamma\hat{\mathcal{Y}}_{03} & 1-\Gamma\hat{\mathcal{Y}}_{33} & -\Gamma\hat{\mathcal{Y}}_{3-} & -\Gamma\hat{\mathcal{Y}}_{3+} \\ -\Gamma\hat{\mathcal{Y}}_{0+} & -\Gamma\hat{\mathcal{Y}}_{3+} & 1-\Gamma\hat{\mathcal{Z}} & -\Gamma\hat{\mathcal{Z}}_{+} \\ -\Gamma\hat{\mathcal{Y}}_{0-} & -\Gamma\hat{\mathcal{Y}}_{3-} & -\Gamma\hat{\mathcal{Z}}_{-} & 1-\Gamma\hat{\mathcal{Z}} \end{pmatrix}\begin{pmatrix} d_0 \\ d_3 \\ d_+ \\ d_- \end{pmatrix} = \begin{pmatrix} \eta \\ 0 \\ 0 \\ 0 \end{pmatrix}, \quad (9.58)$$

where

$$\begin{aligned} \hat{\mathcal{Y}}_{0\pm} &= \frac{\hat{\mathcal{Y}}_{01} \pm i\hat{\mathcal{Y}}_{02}}{\sqrt{2}}, \quad & \hat{\mathcal{Y}}_{3\pm} &= \frac{\hat{\mathcal{Y}}_{13} \pm i\hat{\mathcal{Y}}_{23}}{\sqrt{2}}, \\ \hat{\mathcal{Z}} &= \frac{\hat{\mathcal{Y}}_{11} + \hat{\mathcal{Y}}_{22}}{2}, \quad & \hat{\mathcal{Z}}_{\pm} &= \frac{\hat{\mathcal{Y}}_{11} \pm 2i\hat{\mathcal{Y}}_{12} - \hat{\mathcal{Y}}_{22}}{2}, \end{aligned} \quad (9.59)$$

with $\hat{\mathcal{Y}}_{ab} = \hat{\mathcal{Y}}_{ba}$ given by Eq. (9.53).

According to Eq. (9.55), one has to know the matrix elements of the operators $\hat{\mathcal{Y}}_{ab}(\omega_n)$ in the basis of the Landau levels $|N,p\rangle$. After some straightforward algebra, see Ref. [45] for details, we find that $\hat{\mathcal{Y}}_{00}$, $\hat{\mathcal{Y}}_{03}$, $\hat{\mathcal{Y}}_{33}$, and $\hat{\mathcal{Z}}$ are diagonal in the

Landau levels, while for the rest of the operators (9.59) the only nonzero matrix elements are as follows: $\langle N,p|\hat{\mathscr{Y}}_{0-}|N+1,p\rangle = \langle N+1,p|\hat{\mathscr{Y}}_{0+}|N,p\rangle$, $\langle N,p|\hat{\mathscr{Y}}_{3-}|N+1,p\rangle = \langle N+1,p|\hat{\mathscr{Y}}_{3+}|N,p\rangle$, and $\langle N,p|\hat{\mathscr{Z}}_{-}|N+2,p\rangle = \langle N+2,p|\hat{\mathscr{Z}}_{+}|N,p\rangle$. Therefore, the Landau levels are decoupled, and for $\eta(\boldsymbol{r}) = \eta\langle \boldsymbol{r}|N,p\rangle$ (η is a constant) the solution of Eq. (9.58) has the following form:

$$\begin{pmatrix} d_0(\boldsymbol{r},\omega_n) \\ d_3(\boldsymbol{r},\omega_n) \\ d_+(\boldsymbol{r},\omega_n) \\ d_-(\boldsymbol{r},\omega_n) \end{pmatrix} = \begin{pmatrix} d^0_{N,p}(\omega_n)\langle \boldsymbol{r}|N,p\rangle \\ d^3_{N,p}(\omega_n)\langle \boldsymbol{r}|N,p\rangle \\ d^+_{N,p}(\omega_n)\langle \boldsymbol{r}|N+1,p\rangle \\ d^-_{N,p}(\omega_n)\langle \boldsymbol{r}|N-1,p\rangle \end{pmatrix}. \tag{9.60}$$

At arbitrary magnitude of the SO band splitting, the singlet-triplet mixing makes the equation for $H_{c2}(T)$ in noncentrosymmetric superconductors considerably more cumbersome than in the Werthamer-Helfand-Hohenberg (WHH) problem [43, 44], even in our "minimal" isotropic model. It is even possible that, at some values of the parameters, the maximum critical field is achieved for $N>0$ and $p\neq 0$, the latter corresponding to a disorder-induced modulation of the order parameter along the applied field. Leaving these exotic possibilities aside, here we just assume that $N=p=0$. Then the only nonzero components of the impurity-renormalized gap function, see Eq. (9.54), are $d^0_{0,0}$ and $d^+_{0,0}$.

It is convenient to introduce the reduced temperature, magnetic field, and disorder strength:

$$t = \frac{T}{T_{c0}}, \qquad h = \frac{2H}{H_0}, \qquad \zeta = \frac{\Gamma}{\pi T_{c0}},$$

where $H_0 = \Phi_0/\pi\xi_0^2$, $\Phi_0 = \pi c/e$ is the magnetic flux quantum, and $\xi_0 = v_F/2\pi T_{c0}$ is the superconducting coherence length (v_F is the Fermi velocity). Using the expression for the critical temperature to eliminate both the frequency cutoff and the coupling constant from Eq. (9.56), we arrive at the following equation for the upper critical field $h_{c2}(t)$:

$$\ln\frac{1}{t} = 2\sum_{n\geq 0}\left[\frac{1}{2n+1} - t\frac{w_n(1-\zeta p_n) - \zeta\delta^2 q_n^2}{(1-\zeta w_n)(1-\zeta p_n) + \zeta^2\delta^2 q_n^2}\right], \tag{9.61}$$

where

$$\begin{aligned} w_n &= \int_0^\infty d\rho e^{-\theta_n\rho}\int_0^1 ds e^{-h\rho^2(1-s^2)/4}, \\ p_n &= \int_0^\infty d\rho e^{-\theta_n\rho}\int_0^1 ds\frac{1-s^2}{2}\left[1-\frac{h}{2}\rho^2(1-s^2)\right]e^{-h\rho^2(1-s^2)/4}, \\ q_n &= \int_0^\infty d\rho e^{-\theta_n\rho}\int_0^1 ds\sqrt{\frac{h}{4}}\rho(1-s^2)e^{-h\rho^2(1-s^2)/4}, \end{aligned} \tag{9.62}$$

where $\theta_n = (2n+1)t + \zeta$.

In the clean limit, i.e. at $\zeta \to 0$, or if the SO band splitting is negligibly small, i.e. at $\delta \to 0$, the WHH equation for H_{c2} in a centrosymmetric superconductor [43, 44] is recovered. Therefore, the absence of inversion symmetry affects the upper critical field only if disorder is present. One can expect that the effect will be most pronounced in the "dirty" limit, in which Eq. (9.61) takes the form of a universal equation describing the magnetic pair breaking in superconductors [46]:

$$\ln\frac{1}{t} = \Psi\left(\frac{1}{2}+\frac{\sigma}{t}\right) - \Psi\left(\frac{1}{2}\right). \tag{9.63}$$

Here the parameter

$$\sigma = \frac{2+\delta^2}{12\zeta}h \tag{9.64}$$

characterizes the pair breaking strength. Note that the corresponding expression in the centrosymmetric case is different: $\sigma_{CS} = h/6\zeta$. Analytical expressions for the upper critical field can be obtained in the weak-field limit near the critical temperature:

$$h_{c2}|_{t\to 1} = \frac{24\zeta}{(2+\delta^2)\pi^2}(1-t), \tag{9.65}$$

and also at low temperatures:

$$h_{c2}|_{t=0} = \frac{3e^{-C}}{2+\delta^2}\zeta. \tag{9.66}$$

We see that nonmagnetic disorder suppresses the orbital pair breaking and thus enhances the upper critical field.

In the general case, $H_{c2}(T)$ can be calculated analytically only in the vicinity of the critical temperature using the Ginzburg-Landau free energy expansion. The results for the impurity response turn out to be nonuniversal, i.e. dependent on the pairing symmetry, the values of the intra- and interband coupling constants, and the densities of states in the helicity bands [34].

9.5 Spin Susceptibility

In this section we calculate the magnetic response of a noncentrosymmetric superconductor, neglecting the orbital magnetic interaction and taking into account only the Zeeman coupling of the electron spins with a uniform external field $\boldsymbol{H}$:

$$H_{Zeeman} = -\mu_B \boldsymbol{H} \sum_{\boldsymbol{k},\alpha\beta} \boldsymbol{\sigma}_{\alpha\beta} a^{\dagger}_{\boldsymbol{k}\alpha} a_{\boldsymbol{k}\beta} = -\boldsymbol{H} \sum_{\boldsymbol{k},\lambda\lambda'} \boldsymbol{m}_{\lambda\lambda'}(\boldsymbol{k}) c^{\dagger}_{\boldsymbol{k}\lambda} c_{\boldsymbol{k}\lambda'}, \tag{9.67}$$

where μ_B is the Bohr magneton. The components of the spin magnetic moment operator in the band representation have the following form:

$$\hat{m}_x = \mu_B \begin{pmatrix} \hat{\gamma}_x & -(\gamma_x \hat{\gamma}_z + i\gamma_y)/\gamma_\perp \\ -(\gamma_x \hat{\gamma}_z - i\gamma_y)/\gamma_\perp & -\hat{\gamma}_x \end{pmatrix},$$
$$\hat{m}_y = \mu_B \begin{pmatrix} \hat{\gamma}_y & -(\gamma_y \hat{\gamma}_z - i\gamma_x)/\gamma_\perp \\ -(\gamma_y \hat{\gamma}_z + i\gamma_x)/\gamma_\perp & -\hat{\gamma}_y \end{pmatrix},$$
$$\hat{m}_z = \mu_B \begin{pmatrix} \hat{\gamma}_z & \gamma_\perp/\gamma \\ \gamma_\perp/\gamma & -\hat{\gamma}_z \end{pmatrix}, \tag{9.68}$$

where $\gamma_\perp = \sqrt{\gamma_x^2 + \gamma_y^2}$.

The magnetization of the system is expressed in terms of the Green's functions as follows: $\mathscr{M}_i = (1/\mathscr{V})T\sum_n \sum_{\boldsymbol{k},\lambda\lambda'} \mathrm{tr}\, m_{i,\lambda\lambda'}(\boldsymbol{k})\mathscr{G}^{11}_{\lambda'\lambda}(\boldsymbol{k},\omega_n)$, where $\mathscr{G}(\boldsymbol{k},\omega_n)$ is the impurity-averaged 4×4 matrix Green's function in the presence of magnetic field (recall that the upper indices label the Nambu matrix components, see Sect. 9.2.1). In a weak field, we have $\mathscr{M}_i = \sum_j \chi_{ij} H_j$, where χ_{ij} is the spin susceptibility tensor. Treating the Zeeman interaction, Eq. (9.67), as a small perturbation and expanding $\mathscr{G}$ in powers of $\boldsymbol{H}$, we obtain:

$$\chi_{ij} = -T\sum_n \frac{1}{\mathscr{V}}\sum_{\boldsymbol{k},\boldsymbol{k}'} \big\langle \mathrm{Tr} M_i(\boldsymbol{k})\mathscr{G}(\boldsymbol{k},\boldsymbol{k}';\omega_n)M_j(\boldsymbol{k}')\mathscr{G}(\boldsymbol{k}',\boldsymbol{k};\omega_n)\big\rangle_{imp}, \tag{9.69}$$

where $M_i(\boldsymbol{k}) = \mathrm{diag}[\hat{m}_i(\boldsymbol{k}), -\hat{m}_i^T(-\boldsymbol{k})]$. The Green's functions here are unaveraged 4×4 matrix Green's functions at zero field, and the trace is taken in both the electron-hole and helicity indices.

In the clean case, one can evaluate expression (9.69) by summing over the Matsubara frequencies first, followed by the momentum integration. The susceptibility tensor can be represented as $\chi_{ij} = \chi^+_{ij} + \chi^-_{ij} + \tilde{\chi}_{ij}$ (Ref. [22]), where

$$\chi^\lambda_{ij} = -\mu_B^2 T\sum_n \int \frac{d^3\boldsymbol{k}}{(2\pi)^3}\hat{\gamma}_i\hat{\gamma}_j\left(G_\lambda^2 + |\tilde{F}_\lambda|^2\right) = \mu_B^2 N_\lambda \langle \hat{\gamma}_i\hat{\gamma}_j Y_\lambda\rangle \tag{9.70}$$

are the intraband contributions, determined by the thermally excited quasiparticles near the Fermi surfaces. Here $Y_\lambda(\boldsymbol{k},T) = 2\int_0^\infty d\xi(-\partial f/\partial E_\lambda)$ is the angle-resolved Yosida function, $f(\varepsilon) = (e^{\varepsilon/T}+1)^{-1}$ is the Fermi-Dirac distribution function, $E_\lambda(\boldsymbol{k}) = \sqrt{\xi^2 + |\tilde{\Delta}_\lambda(\boldsymbol{k})|^2}$ is the energy of quasiparticle excitations in the λth band, and, as in the previous sections, the subscripts λ in the Fermi-surface averages are omitted for brevity. The interband contribution is given by

$$\tilde{\chi}_{ij} = -2\mu_B^2 T\sum_n \int \frac{d^3\boldsymbol{k}}{(2\pi)^3}(\delta_{ij} - \hat{\gamma}_i\hat{\gamma}_j)\left(G_+G_- + \mathrm{Re}\,\tilde{F}_+^*\tilde{F}_-\right)$$
$$\simeq -\mu_B^2 \int \frac{d^3\boldsymbol{k}}{(2\pi)^3}\frac{\delta_{ij} - \hat{\gamma}_i\hat{\gamma}_j}{|\boldsymbol{\gamma}|}[f(\xi_+) - f(\xi_-)]. \tag{9.71}$$

Since $\tilde{\chi}_{ij}$ is determined by all quasiparticles in the momentum-space shell "sandwiched" between the Fermi surfaces, it is almost unchanged when the system undergoes the superconducting transition, in which only the electrons near the Fermi surface are affected.

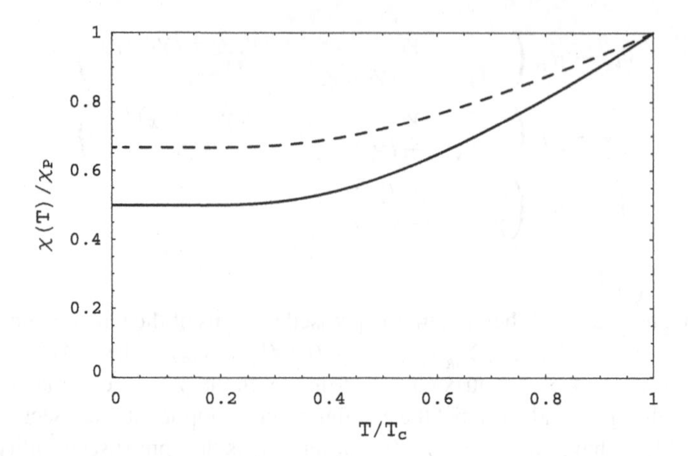

Fig. 9.3 The clean-case temperature dependence of the transverse components of the susceptibility for the 2D model (*solid line*), and of all three components for the 3D model (*dashed line*); $\chi_P = 2\mu_B^2 N_F$ is the Pauli susceptibility

Collecting together the inter- and intraband contributions, we arrive at the following expression for the spin susceptibility of a clean superconductor:

$$\chi_{ij} = \tilde{\chi}_{ij} + \mu_B^2 N_F \sum_\lambda \rho_\lambda \left\langle \hat{\gamma}_i \hat{\gamma}_j Y_\lambda \right\rangle . \tag{9.72}$$

At zero temperature there are no excitations ($Y_\lambda = 0$) and the intraband terms are absent, but the susceptibility still has a nonzero value given by $\tilde{\chi}_{ij}$. The temperature dependence of the susceptibility in the superconducting state at $0 < T \le T_c$ is determined by the intraband terms, with the low-temperature behavior depending crucially on the magnitude of the SO coupling at the gap nodes. While in the fully gapped case the intraband susceptibility is exponentially small in all directions, in the presence of the lines of nodes it is proportional to either T or T^3, depending on whether or not the zeros of $\tilde{\Delta}_\lambda(\boldsymbol{k})$ coincide with those of $\boldsymbol{\gamma}(\boldsymbol{k})$, see Ref. [47]. In Fig. 9.3 the temperature dependence of χ_{ij} is plotted for a Rashba superconductor with $\boldsymbol{\gamma}(\boldsymbol{k}) = \gamma_\perp(\boldsymbol{k} \times \hat{z})$ and a cylindrical Fermi surface (referred to as the 2D model), and also for a cubic superconductor with $\boldsymbol{\gamma}(\boldsymbol{k}) = \gamma_0 \boldsymbol{k}$ and a spherical Fermi surface (the 3D model). In both cases, the gaps in the two helicity bands are assumed to be isotropic and have the same magnitude.

Now let us include the scalar disorder described by Eq. (9.2), or, equivalently, Eq. (9.5). After the impurity averaging, Eq. (9.69) is represented by a sum of the ladder diagrams containing the average Green's functions (9.17). In contrast to the clean case, it is not possible to calculate the Matsubara sums before the momentum integrals. To make progress, one should add to and subtract from Eq. (9.69) the normal-state susceptibility [26]. It is easy to show that the latter is not affected by impurities and is therefore given by $\chi_{N,ij} = \tilde{\chi}_{ij} + \mu_B^2 N_F \sum_\lambda \rho_\lambda \langle \hat{\gamma}_i \hat{\gamma}_j \rangle$, see Eq. (9.72). Then,

$$\chi_{ij}-\chi_{N,ij} = -T\sum_n{}'\frac{1}{\mathscr{V}}\sum_{kk'}\big(\langle \operatorname{Tr} M_i\mathscr{G}M_j\mathscr{G}\rangle_{imp}-\langle \operatorname{Tr} M_i\mathscr{G}_N M_j\mathscr{G}_N\rangle_{imp}\big), \quad (9.73)$$

where $\mathscr{G}_N$ is the unaveraged 4×4 matrix Green's function in the normal state with impurities, and the Matsubara summation is limited to the frequencies $|\omega_n| \leq \omega_c$, at which the gap function is nonzero.

Due to convergence of the expression on the right-hand side of Eq. (9.73), one can now do the momentum integrals first. The second term vanishes, while the calculation of the ladder diagrams in the first term is facilitated by the observation that the integrals of the products of the Green's functions from different bands are small compared with their counterparts containing the Green's functions from the same band. The argument is similar to the one leading to Eq. (9.51). For example, assuming that both the SO band splitting and the pairing are isotropic, i.e. $\xi_\pm(\boldsymbol{k}) = \varepsilon_0(\boldsymbol{k}) \pm \gamma$ and $D_\pm = D$, we have

$$\max_n \frac{\int d\varepsilon_0 G_\lambda G_{-\lambda}}{\int d\varepsilon_0 G_\lambda G_\lambda} = \max_n \frac{1}{1+\gamma^2/\Omega_n^2} \sim \left[\frac{\max(\omega_c,\Gamma)}{E_{SO}}\right]^2 \ll 1,$$

where $\Omega_n = \sqrt{\tilde\omega_n^2 + |D|^2}$. In the same way, one can also obtain estimates for the momentum integrals containing anomalous Green's functions:

$$\max_n \frac{\int d\varepsilon_0 G_\lambda \tilde F_{-\lambda}}{\int d\varepsilon_0 G_\lambda \tilde F_\lambda} = \max_n \frac{1-i\lambda\gamma/\tilde\omega_n}{1+\gamma^2/\Omega_n^2} \sim \frac{\max(\omega_c,\Gamma)}{E_{SO}} \ll 1,$$

$$\max_n \frac{\int d\varepsilon_0 \tilde F_\lambda \tilde F_{-\lambda}}{\int d\varepsilon_0 \tilde F_\lambda \tilde F_\lambda} = \max_n \frac{1}{1+\gamma^2/\Omega_n^2} \sim \left[\frac{\max(\omega_c,\Gamma)}{E_{SO}}\right]^2 \ll 1.$$

Thus we see that it is legitimate to keep only the same-band contributions to the impurity ladder on the right-hand side of Eq. (9.73).

Following Ref. [22], we obtain the following expression for the spin susceptibility:

$$\chi_{ij} = \chi_{N,ij} + \frac{2\pi\mu_B N_F}{\Gamma} T\sum_n \frac{\partial X_i(\omega_n)}{\partial H_j}, \quad (9.74)$$

where X_i are found from the equations

$$\begin{aligned} X_i - \sum_j (A_{1,ij}X_j + A_{2,ij}Y_j + A_{2,ij}^* Y_j^*) &= X_{0,i}, \\ Y_i - \sum_j (2A_{2,ij}^* X_j + A_{3,ij}Y_j + A_{4,ij}Y_j^*) &= Y_{0,i}. \end{aligned} \quad (9.75)$$

The notations here are as follows:

$$A_{1,ij} = \frac{\Gamma}{2}\sum_\lambda \rho_\lambda \left\langle \frac{\hat{\gamma}_i\hat{\gamma}_j|D_\lambda|^2}{\Omega_n^3} \right\rangle, \quad A_{2,ij} = \frac{\Gamma}{4}\sum_\lambda \rho_\lambda \left\langle \frac{i\hat{\gamma}_i\hat{\gamma}_j\tilde{\omega}_n D_\lambda^*}{\Omega_n^3} \right\rangle,$$

$$A_{3,ij} = \frac{\Gamma}{4}\sum_\lambda \rho_\lambda \left\langle \frac{\hat{\gamma}_i\hat{\gamma}_j(2\tilde{\omega}_n^2 + |D_\lambda|^2)}{\Omega_n^3} \right\rangle, \quad A_{4,ij} = \frac{\Gamma}{4}\sum_\lambda \rho_\lambda \left\langle \frac{\hat{\gamma}_i\hat{\gamma}_j D_\lambda^2}{\Omega_n^3} \right\rangle,$$

$$X_{0,i} = -\mu_B \sum_j A_{1,ij} H_j, \quad Y_{0,i} = -2\mu_B \sum_j A_{2,ij}^* H_j,$$

and $\Omega_n = \sqrt{\tilde{\omega}_n^2 + |D_\lambda|^2}$. Due to fast convergence, the Matsubara summation in Eq. (9.74) can be extended to all frequencies.

9.5.1 Residual Susceptibility for Isotropic Pairing

The general expression for the spin susceptibility, Eq. (9.74), is rather cumbersome. On the other hand, application of our results to real noncentrosymmetric materials is complicated by the lack of information about the superconducting gap symmetry and the distribution of the pairing strength between the bands. As an illustration, we focus on the isotropic pairing model introduced in Sect. 9.3.1. In this model, the order parameter magnitudes in the two bands are the same, but the relative phase can be either 0 or π. While in the clean limit the spin susceptibility for both states is given by Eq. (9.72), the effects of impurities have to be analyzed separately.

$\eta_+ = \eta_- = \eta$. Solving equations (9.75) we obtain, in the coordinate system in which $\langle \hat{\gamma}_i\hat{\gamma}_j \rangle$ is diagonal, the following expression for the nonzero components of the susceptibility tensor:

$$\frac{\chi_{ii}(T)}{\chi_P} = 1 - \langle \hat{\gamma}_i^2 \rangle \pi T \sum_n \frac{\eta^2}{\omega_n^2 + \eta^2} \frac{1}{\sqrt{\omega_n^2 + \eta^2} + \Gamma_i}, \tag{9.76}$$

where $\Gamma_i = (1 - \langle \hat{\gamma}_i^2 \rangle)\Gamma$.

We are particularly interested in the effect of disorder on the residual susceptibility at zero temperature. In this limit, the Matsubara sum in Eq. (9.76) can be replaced by a frequency integral, which gives

$$\frac{\chi_{ii}(T=0)}{\chi_P} = 1 - \langle \hat{\gamma}_i^2 \rangle + \langle \hat{\gamma}_i^2 \rangle \Phi_1\left(\frac{\gamma_i}{\eta_0}\right), \tag{9.77}$$

where $\eta_0 = (\pi/e^C)T_c$ is the gap magnitude at $T = 0$, and

$$\Phi_1(x) = 1 - \frac{\pi}{2x}\left(1 - \frac{4}{\pi\sqrt{1-x^2}} \arctan\sqrt{\frac{1-x}{1+x}}\right).$$

While the first two terms on the right-hand side of Eq. (9.77) represent the residual susceptibility in the clean case, the last term describes the impurity effect. In a weakly-disordered superconductor, using the asymptotics $\Phi_1(x) \simeq \pi x/4$, we find that the

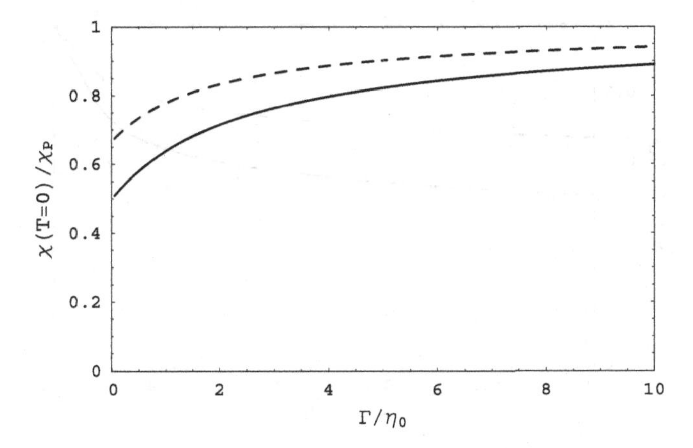

Fig. 9.4 The residual susceptibility at $T = 0$ vs disorder strength for $\eta_+ = \eta_-$. The *solid line* corresponds to the transverse components in the 2D case ($\chi_{zz} = \chi_P$ and is disorder-independent), the *dashed line* to the diagonal components in the 3D case

residual susceptibility increases linearly with disorder. In the dirty limit, $\Gamma \gg \eta_0$, we have $\Phi_1(x) \to 1$, therefore $\chi_{ii}(T = 0)$ approaches the normal-state value χ_P. For the two simple band-structure models (2D and 3D) discussed earlier in this section, the Fermi-surface averages can be calculated analytically, and we obtain the results plotted in Fig. 9.4.

Thus we see that, similarly to spin-orbit impurities in a usual centrosymmetric superconductor [48], scalar impurities in a noncentrosymmetric superconductor lead to an enhancement of the spin susceptibility at $T = 0$. Since the interband contribution is not sensitive to disorder, this effect can be attributed to an increase in the intraband susceptibilities.

$\eta_+ = -\eta_- = \eta$. From Eqs. (9.74) and (9.75) we obtain the nonzero components of the susceptibility tensor:

$$\frac{\chi_{ii}(T)}{\chi_P} = 1 - \langle \hat{\gamma}_i^2 \rangle \pi T \sum_n \frac{\eta^2}{(\tilde{\omega}_n^2 + \eta^2)^{3/2} - \hat{\Gamma} \langle \gamma_i^2 \rangle \eta^2}. \tag{9.78}$$

We note that for a spherical 3D model with $\langle \hat{\gamma}_i^2 \rangle = 1/3$ this expression has exactly the same form as the susceptibility of the superfluid ^{3}He-B in aerogel, see Refs. [41] and [49].

At $T = 0$, the expression (9.78) takes the following form:

$$\frac{\chi_{ii}(T = 0)}{\chi_P} = 1 - \langle \hat{\gamma}_i^2 \rangle + \langle \hat{\gamma}_i^2 \rangle \Phi_2 \left(\frac{\Gamma}{\eta} \right), \tag{9.79}$$

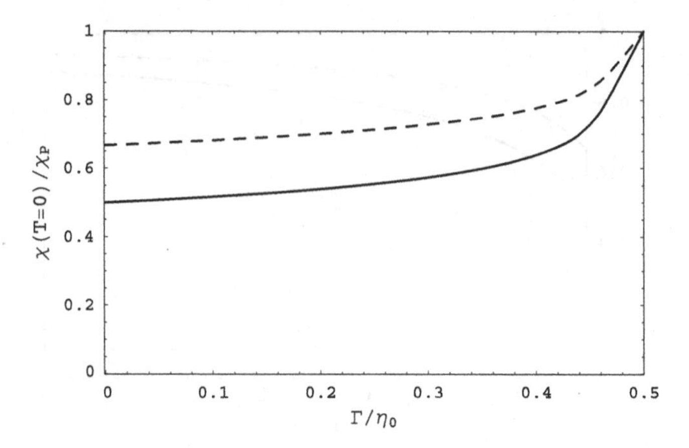

Fig. 9.5 The residual susceptibility at $T = 0$ vs disorder strength for $\eta_+ = -\eta_-$. The *solid line* corresponds to the transverse components in the 2D case ($\chi_{zz} = \chi_P$ and is disorder-independent), the *dashed line* to the diagonal components in the 3D case

where

$$\Phi_2(x) = 1 - \int_{y_{min}}^{\infty} dy \left[1 - \frac{x}{(y^2+1)^{3/2}} \right] \frac{1}{(y^2+1)^{3/2} - x\langle \hat{\gamma}_i^2 \rangle},$$

and $y_{min} = \theta(x-1)\sqrt{x^2-1}$. The last term on the right-hand side of Eq. (9.79) describes the effect of impurities. According to Sect. 9.3.1, superconductivity is suppressed above the critical disorder strength $\Gamma_c = (\pi/2e^C)T_{c0}$. For a given Γ, one should first obtain the gap magnitude from Eq. (9.37) and then calculate $\Phi_2(x)$. In the weak disorder limit, we have $\Phi_2(x) \simeq (3\pi x/16)(1 - \langle \hat{\gamma}_i^2 \rangle)$, i.e. the residual susceptibility increases linearly with disorder. At $\Gamma \to \Gamma_c$, we have $\Phi_2(x) \to 1$ and $\chi_{ii}(T=0) \to \chi_P$. The dependence of $\chi_{ii}(T=0)$ on the disorder strength for the 2D and 3D models is plotted in Fig. 9.5. As in the case $\eta_+ = \eta_-$, the residual susceptibility is enhanced by impurities.

9.6 Conclusions

Scalar disorder in noncentrosymmetric superconductors causes anisotropic mixing of the electron states in the bands split by the SO coupling. The critical temperature is generally suppressed by impurities, but this happens differently for conventional and unconventional pairing. For all types of unconventional pairing (which is defined as corresponding to a non-unity representation of the crystal point group, with vanishing Fermi-surface averages of the gap functions), the impurity effect on T_c is described by the universal Abrikosov-Gor'kov equation. The same is also true for certain types of conventional pairing, in particular the "protected" isotropic triplet state with $\eta_+ = -\eta_-$. Any deviation from the Abrikosov-Gor'kov curve, in particular, an incomplete suppression of superconductivity by strong disorder, is a signature of conventional pairing symmetry.

The impurity-induced mixing of singlet and triplet pairing channels makes the magnetic response of noncentrosymmetric superconductors with the SO coupling different from the centrosymmetric case. In an isotropic BCS-like model, the upper critical field H_{c2} is enhanced by disorder at all temperatures, the magnitude of the effect depending on the SO coupling strength. In general, the effect of impurities on the slope of H_{c2} is sensitive to the pairing symmetry and the band structure.

Concerning the spin susceptibility, we found that scalar impurities in noncentrosymmetric superconductors act similarly to spin-orbit impurities in centrosymmetric superconductors, in the sense that they enhance the residual susceptibility at $T = 0$. The quantitative details again depend on the band structure, the anisotropy of the SO coupling, and the symmetry of the order parameter.

Most of the experimental work on noncentrosymmetric superconductors has been done on $CePt_3Si$. In this compound the Fermi surface is quite complicated and consists of multiple sheets [24]. It is not known which of them are superconducting. The order parameter symmetry is not known either, although there is experimental evidence that there are lines of nodes in the gap [50–52]. The data on the impurity effects are controversial. The experimental samples seem to be rather clean, with the ratio of the elastic mean free path l to the coherence length ξ_0 ranging from 4 (Ref. [1]) to 10–27 (Ref. [53]. There are indications that T_c is indeed suppressed by structural defects and/or impurities in some samples [52]. On the other hand, the values of both the critical temperature and the upper critical field in polycrystalline samples [1] are higher than in single crystals [50]. This is opposite to what has been observed in other unconventional superconductors and also disagrees with the theoretical predictions, assuming that the polycrystals are intrinsically more disordered than the single crystals. In addition, the low-temperature behaviour of the penetration depth in disordered samples is unusual [52] and cannot be explained by existing theoretical models. In order to resolve these issues, more systematic studies of the disorder effects in a wide range of impurity concentrations are needed.

Acknowledgements The author is grateful to D. Agterberg, I. Bonalde, S. Fujimoto, V. Mineev, B. Mitrovic, and M. Sigrist for useful discussions. This work was supported by a Discovery Grant from the Natural Sciences and Engineering Research Council of Canada.

References

1. Bauer, E., Hilscher, G., Michor, H., Paul, Ch., Scheidt, E.W., Gribanov, A., Seropegin, Yu., Noël, H., Sigrist, M., Rogl, P.: Phys. Rev. Lett. **92**, 027003 (2004)
2. Akazawa, T., Hidaka, H., Fujiwara, T., Kobayashi, T.C., Yamamoto, E., Haga, Y., Settai, R., Onuki, Y.: J. Phys. Condens. Matter **16**, L29 (2004)
3. Kimura, N., Ito, K., Saitoh, K., Umeda, Y., Aoki, H., Terashima, T.: Phys. Rev. Lett. **95**, 247004 (2005)

4. Sugitani, I., Okuda, Y., Shishido, H., Yamada, T., Thamizhavel, A., Yamamoto, E., Matsuda, T.D., Haga, Y., Takeuchi, T., Settai, R., Onuki, Y.: J. Phys. Soc. Jpn. **75**, 043703 (2006)
5. Amano, G., Akutagawa, S., Muranaka, T., Zenitani, Y., Akimitsu, J.: J. Phys. Soc. Jpn. **73**, 530 (2004)
6. Togano, K., Badica, P., Nakamori, Y., Orimo, S., Takeya, H., Hirata, K.: Phys. Rev. Lett. **93**, 247004 (2004)
7. Badica, P., Kondo, T., Togano, K.: J. Phys. Soc. Jpn. **74**, 1014 (2005)
8. Samokhin, K.V.: Ann. Phys. (N. Y.) **324**, 2385 (2009)
9. Mineev, V.P. Samokhin, K.V.: Sov. Phys. JETP **78**, 401 (1994)
10. Agterberg, D.F.: Physica C **387**, 13 (2003)
11. Dimitrova, O.V., Feigel'man, M.V.: JETP Lett. **78**, 637 (2003)
12. Samokhin, K.V.: Phys. Rev. B **70**, 104521 (2004)
13. Kaur, R.P., Agterberg, D.F., Sigrist, M.: Phys. Rev. Lett. **94**, 137002 (2005)
14. Mineev, V.P., Samokhin, K.V.: Phys. Rev. B **78**, 144503 (2008)
15. Levitov, L.S., Nazarov, Yu.V., Eliashberg, G.M.: JETP Lett. **41**, 445 (1985)
16. Edelstein, V.M.: Sov. Phys. JETP **68**, 1244 (1989)
17. Yip, S.K.: Phys. Rev. B **65**, 144508 (2002)
18. Fujimoto, S.: Phys. Rev. B **72**, 024515 (2005)
19. Edelstein, V.M.: Phys. Rev. B **72**, 172501 (2005)
20. Gor'kov, L.P., Rashba, E.I.: Phys. Rev. Lett. **87**, 037004 (2001)
21. Frigeri, P.A., Agterberg, D.F., Koga, A., Sigrist, M.: M Phys. Rev. Lett. **92**, 097001 (2004) Erratum 93:099903(E) (2004)
22. Samokhin, K.V.: Phys. Rev. B **76**, 094516 (2007)
23. Rashba, E.I.: Sov. Phys. Solid State **2**, 1109 (1960)
24. Samokhin, K.V., Zijlstra, E.S., Bose, S.K.: Phys. Rev. B **69**, 094514 (2004) Erratum: 70:069902(E) (2004)
25. Lee, K-.W., Pickett, W.E.: Phys. Rev. B **72**, 174505 (2005)
26. Abrikosov, A.A., Gor'kov, L.P., Dzyaloshinski, I.E.: Methods of Quantum Field Theory in Statistical Physics. Dover, New York (1975)
27. Sergienko, I.A., Curnoe, S.H.: Phys. Rev. B **70**, 214510 (2004)
28. Samokhin, K.V., Mineev, V.P.: Phys. Rev. B **77**, 104520 (2008)
29. Mineev, V.P., Samokhin, K.V.: Introduction to Unconventional Superconductivity. Gordon and Breach, London (1999)
30. Suhl, H., Matthias, B.T., Walker, L.R.: Phys. Rev. Lett. **3**, 552 (1959)
31. Moskalenko, V.A.: Sov. Phys. Met. Metallogr. **8**, 25 (1959)
32. Liu, B., Eremin, I.: Phys. Rev. B **78**, 014518 (2008)
33. Allen, P.B., Mitrović, B.: Solid state physics. In: Ehrenreich, H., Seitz, F., Turnbull, D. (eds.) vol. 37, p. 1. Academic Press, New York (1982)
34. Mineev, V.P., Samokhin, K.V.: Phys. Rev. **75**, 184529 (2007)
35. Abrikosov, A.A., Gor'kov, L.P.: Sov. Phys. JETP **12**, 1243 (1960)
36. Larkin, A.I.: JETP Lett. **2**, 130 (1965)
37. Frigeri, P.A., Agterberg, D.F., Milat, I., Sigrist, M.: Eur. Phys. J. B **54**, 435 (2006)
38. Moskalenko, V.A., Palistrant, M.E.: Sov. Phys. JETP **22**, 536 (1966)
39. Kusakabe, T.: Progr. Theor. Phys. **43**, 907 (1970)
40. Golubov, A.A., Mazin, I.I.: Phys. Rev. B **55**, 15146 (1997)
41. Mineev, V.P., Krotkov, P.L.: Phys. Rev. B **65**, 024501 (2001)
42. Gor'kov, L.P.: Sov. Phys. JETP **91**, 364 (1959)
43. Helfand, E., Werthamer, N.R.: Phys. Rev. **147**, 288 (1966)
44. Werthamer, N.R., Helfand, E., Hohenberg, P.C.: Phys. Rev. **147**, 295 (1966)
45. Samokhin, K.V.: Phys. Rev. B **78**, 144511 (2008)
46. Tinkham, M.: Introduction to Superconductivity, Ch 10.2, McGraw-Hill, New York (1996)
47. Samokhin, K.V.: Phys. Rev. Lett. **94**, 027004 (2005)
48. Abrikosov, A.A., Gor'kov, L.P.: Sov. Phys. JETP **15**, 752 (1962)

49. Sharma, P., Sauls, J.A.: J. Low Temp. Phys. **125**, 115 (2001)
50. Yasuda, T., Shishido, H., Ueda, T., Hashimoto, S., Settai, R., Takeuchi, T., Matsuda, T.D., Haga, Y., Onuki, Y.: J. Phys. Soc. Jpn. **73**, 1657 (2004)
51. Izawa, K., Kasahara, Y., Matsuda, Y., Behnia, K., Yasuda, T., Settai, R., Onuki, Y.: Phys. Rev. Lett. **94**, 197002 (2005)
52. Bonalde, I., Ribeiro, R.L., Brämer-Escamilla, W., Rojas, C., Bauer, E., Prokofiev, A., Haga, Y., Yasuda, T., Onuki, Y.: New J. Phys. **11**, 055054 (2009)
53. Yogi, M., Mukuda, H., Kitaoka, Y., Hashimoto, S., Yasuda, T., Settai, R., Matsuda, T.D., Haga, Y., Onuki, Y., Rogl, P., Bauer, E.: J. Phys. Soc. Jpn. **75**, 013709 (2006)

Chapter 10
Vortex Dynamics in Superconductors Without Inversion Symmetry

Corneliu F. Miclea, Ana-Celia Mota and Manfred Sigrist

Abstract In this chapter we give an overview on some recent experimental results on vortex dynamics in the non-centrosymmetric superconductors $CePt_3Si$ and Li_2Pt_3B. In both compounds the flux creep from a metastable vortex configuration is anomalously slow—slower than in any other superconductor. Additionally, Li_2Pt_3B shows very strong avalanche-like flux release after waiting times of several hours at millikelvin temperatures. Since critical currents are also low, the origin of these properties cannot be simply attributed to conventional flux pinning by defects. We speculate that both properties might be connected with crystalline twinning of the samples. We show that twin boundaries in non-centrosymmetric superconductors can host states with broken time reversal symmetry which can carry fractionally quantized flux lines. These flux lines are strongly pinned to the twin boundaries such that they impede the usual flux motion without affecting the critical current.

10.1 Introduction

Since a few years non-centrosymmetric superconductors that lack spatial inversion symmetry have revealed several new properties which are discussed in the various chapters of this book. In this chapter we will focus on the vortex dynamics and

C. F. Miclea (✉)
Max-Planck-Institute for Chemical Physics of Solids, Dresden, Germany
e-mail: corneliufm@gmail.com

Ana-Celia Mota
Laboratory for Solid State Physics, ETH Zurich, Zurich, Switzerland
e-mail: mota@phys.ethz.ch

M. Sigrist
Institute for Theoretical Physics, ETH Zurich, Zurich, Switzerland
e-mail: sigrist@itp.phys.ethz.ch

E. Bauer and M. Sigrist (eds.), *Non-centrosymmetric Superconductors*,
Lecture Notes in Physics 847, DOI: 10.1007/978-3-642-24624-1_10,

pinning properties in such materials, that exhibit unusual behaviors not observed in other superconductors.

It is commonly observed in superconductors that a vortex phase in non-equilibrium configurations moves towards equilibrium, unless vortices are prevented to move by pinning due to defects in the material. If the pinning is strong, the critical current of the superconductor, that is the minimal current able to move vortices, is high, and the creep rate of vortices is low. For this reason high-field applications of superconductivity depend crucially on strong pinning. However, here we discuss the unusual combination of extremely weak flux creep along with very weak critical currents in the non-centrosymmetric superconductors $CePt_3Si$ and Li_2Pt_3B.

We study here the flux dynamics in these superconductors by observing the relaxation of the remnant magnetization trapped in the material in the critical state after a magnetization cycle. In $CePt_3Si$ [1] and Li_2Pt_3B [2] the creep of flux lines from the critical state, is weaker than in any other superconductor known up to date, and their critical currents are also very low. This apparently contradictory fact indicates that a novel pinning mechanism different from the well-known pinning by defects, such as impurities or dislocations, may be at work in these non-centrosymmetric superconductors. This unusual behavior cannot simply be attributed to the antiferromagnetic phase which coexists with superconductivity in the heavy fermion compound $CePt_3Si$, since the same kind of vortex behavior is found in the non-magnetic Li_2Pt_3B, considered as a weakly correlated electron system.

10.2 Measuring Setup

Susceptibility and relaxation measurements were performed in a custom-built mixing chamber of a dilution refrigerator. In this arrangement, the sample is placed inside the mixing chamber in contact with the $^3He-^4He$ dilute solution, and stays stationary in the detection and field coils built in the walls of the mixing chamber. The schematic structure of this arrangement is depicted in Fig. 10.1. Samples are introduced in the measuring towers and sealed with a mixture of soap and glycerine. Residual fields at the sample position are kept at less than 2 mOe by means of several cryoperm shields in the helium bath. The field coils are designed to be used only in driven mode (not persistent) and no superconducting shielding is used around them. Thus inhomogeneities in either applied or residual fields in the cryostat have no influence on the measurement.

For ac susceptibility measurements, an ac impedance bridge with a SQUID as a null detector is used. The amplitude of the field can be varied in nine fixed steps from 0.07 to 33 mOe, and the excitation frequency can be chosen in four steps, from 16 to 160 Hz. For dc relaxation measurements a digital flux counter is used to record the signal from the SQUID unit. Typically creep measurements are taken from 10 s after removing the dc magnetic field to 2×10^4–5×10^4 s. Both flux dynamics and ac-susceptibility are investigated in the same arrangement, in sequence, without altering the sample position between measurements.

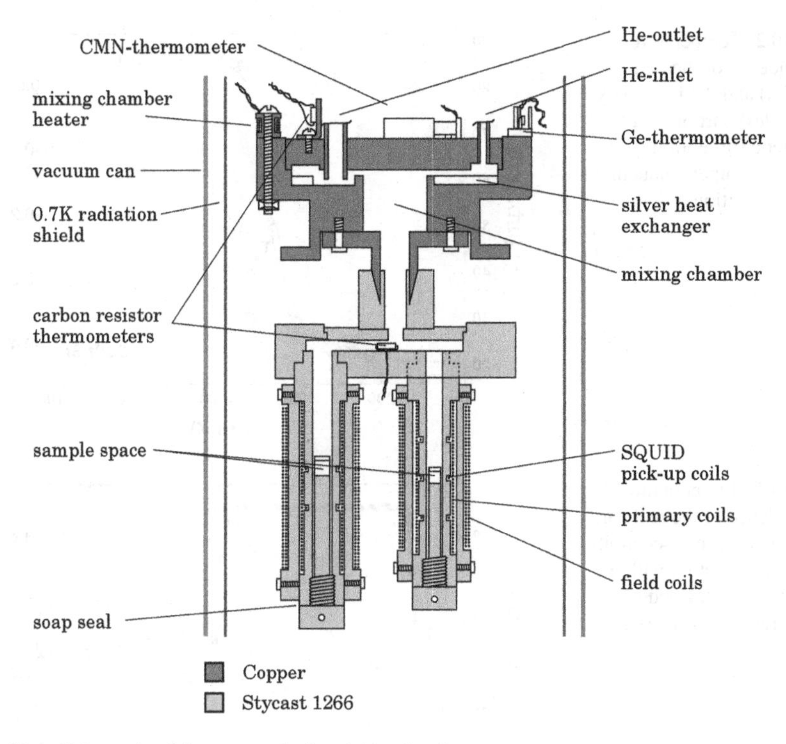

Fig. 10.1 Schematic of the custom built mixing chamber

Specific heat measurements were done on a custom made platform built from Ag foil suspended on nylon wires. The platform is installed in a dilution refrigerator and the measurements are done using a quasi-adiabatic method in the temperature range $0.050\,\mathrm{K} \leq T \leq 5\,\mathrm{K}$ and magnetic fields up to $H = 12\,\mathrm{T}$.

10.3 $CePt_3Si$

The sample investigated was cut and polished from a single crystal prepared using the Bridgman technique in the group of E. Bauer at TU Wien. It has a mass of 140 mg and is $4.6 \times 2.65 \times 1.05\,\mathrm{mm}^3$ in size, with the *c*-axis pointing along the 2.65-mm direction. The magnetic field *H* was applied perpendicular to the *c*-axis, and in the direction of the longest dimension. A refinement of the crystal structure of $CePt_3Si$ from x-ray intensity data collected on a small piece cut from the same starting crystal, shows twinning with a contribution of 87% of the main inversion twin component [1]. In Fig. 10.2 we show the in-phase and out-of-phase magnetic susceptibility as a function of temperature. It was measured with a field of 1.3 mOe and a low frequency $f = 80$ Hz. The mid-point of the superconducting transition occurs at $T_c = 0.45\,\mathrm{K}$ with a width of $\Delta T_c = 0.1\,\mathrm{K}$. This value of the transition temperature agrees well

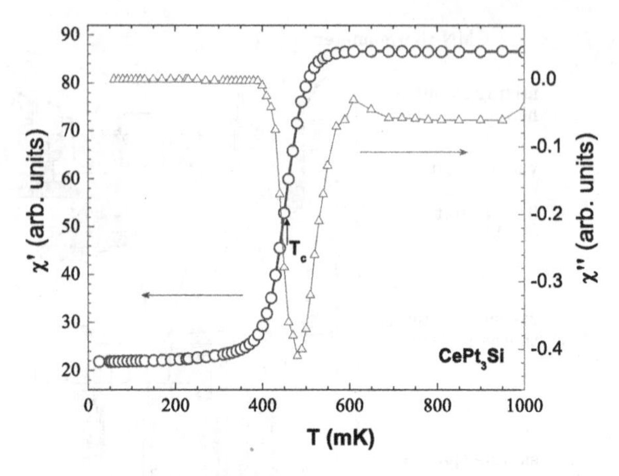

Fig. 10.2 Temperature dependence of the real (*circles*) and the imaginary (*triangles*) part of the ac magnetic susceptibility across the superconducting phase transition

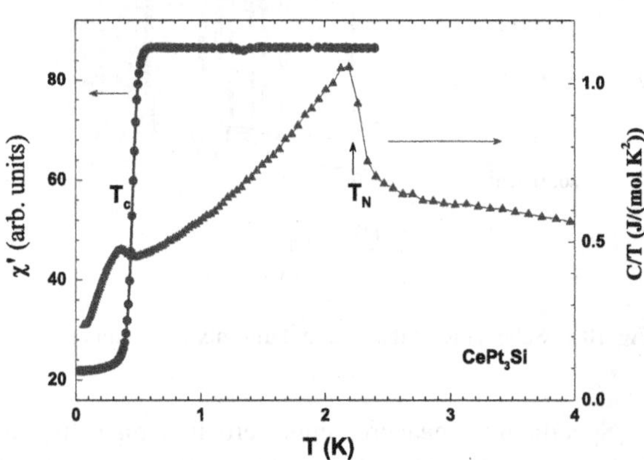

Fig. 10.3 Temperature dependence of the real part of the magnetic susceptibility (*left axis*) together with the specific heat devided by temperature (*right axis*)

with the one reported by Takeuchi et al. [3] from specific heat measurements on a high-quality single crystal. However, this value is substantially lower than the $T_c = 0.75$ K reported for polycrystalline $CePt_3Si$ by Bauer et al. [4]. The origin of the difference in the superconducting critical temperature between single crystal and polycrystalline samples is unclear so far.

In Fig. 10.3 we show specific heat data taken on another piece cut from the same starting single crystal. The antiferromagnetic transition at $T_N = 2.2$ K, is clearly visible. We also notice that across the antiferromagnetic transition, the ac susceptibility, taken with the magnetic field perpendicular to the *c*-axis does not change, since the antiferromagnetic moments in $CePt_3Si$ are aligned along the *c*-direction of the tetragonal lattice [5].

Isothermal relaxation curves of the remnant magnetization, M_{rem} and magnetic susceptibility were measured on the same single crystal in the same experimental configuration. For each curve, the specimen is first zero-field-cooled to the desired temperature, then the magnetic field is raised to a value high enough to drive the sample to the Bean critical state. At this point, the field is removed and after it

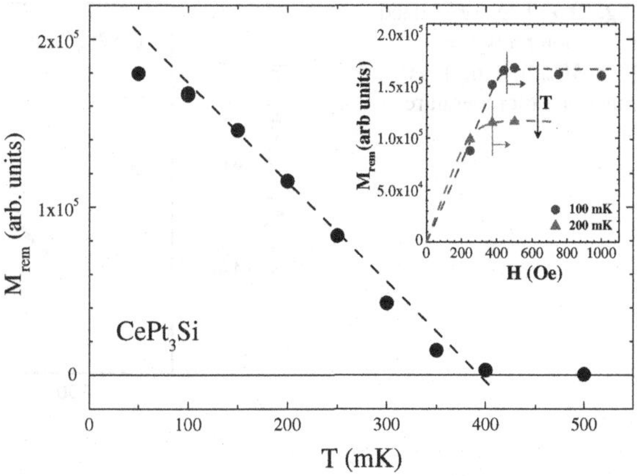

Fig. 10.4 Temperature dependence of the total remnant magnetization. The *dashed line* is a linear fit to the data. *Inset* The total remnant magnetization at $T = 0.1$ K and $T = 0.2$ K as function of the external magnetic field applied H

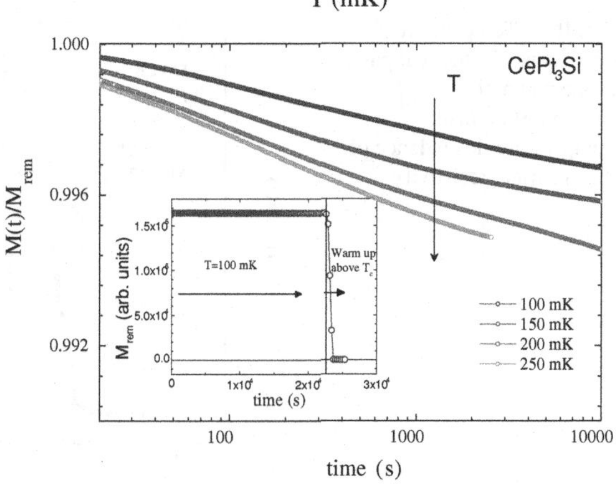

Fig. 10.5 The normalized remnant magnetization as function of time at different constant temperatures. *Inset* the remnant magnetization as function of time at $T = 0.100$ K. After 2.25×10^4 s the sample is warmed up above T_c and all the trapped magnetic flux is expelled

reaches zero, the decay of the remnant magnetization is recorded with a quantum flux counter for several hours. The time involved in rising and lowering the field is controlled in each case, in order to keep eddy current heating at a minimum.

The Bean critical field H_s is determined at the lowest temperature of our investigation by measuring the total remnant magnetization as function of the external magnetic field applied as shown in the inset of Fig. 10.4. In the Bean model, H_s is the maximum external field that can be completely screened from the mid-plane of a superconducting slab. It the applied field is twice the value of H_s or higher, the remnant magnetization reached immediately after removing the field is constant. This means that the whole sample is penetrated by vortices.

After each decay measurement, the specimen is heated above its critical temperature T_c, and the expelled flux is measured in order to determine M_{rem} at the beginning of the decay, as the sum of the decayed flux plus the flux expelled during heating. Values of M_{rem} as function of temperature are given in Fig. 10.4. The critical current can be estimated from values of the Bean critical field H_s at each temperature and the thickness of the sample perpendicular to the applied field. In the Bean model,

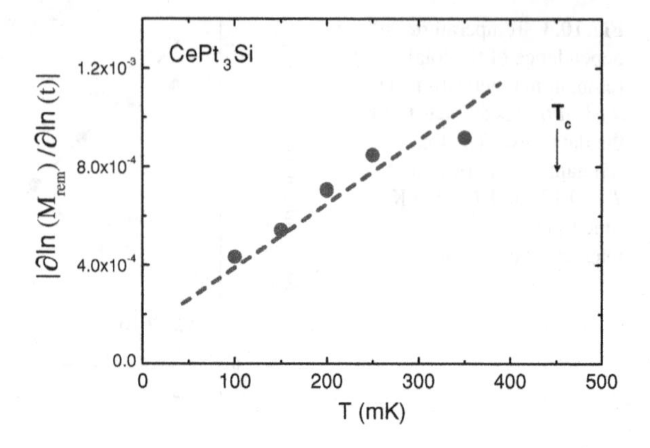

Fig. 10.6 The normalized relaxation rates $S = |\partial \ln(M_{rem})/\partial \ln(t)|$ as function of temperature

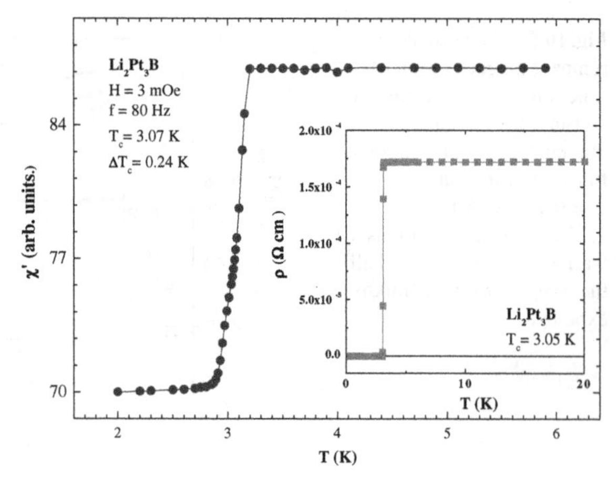

Fig. 10.7 Temperature dependence of the real part of the magnetic susceptibility. *Inset* Temperature dependence of the electrical resistivity

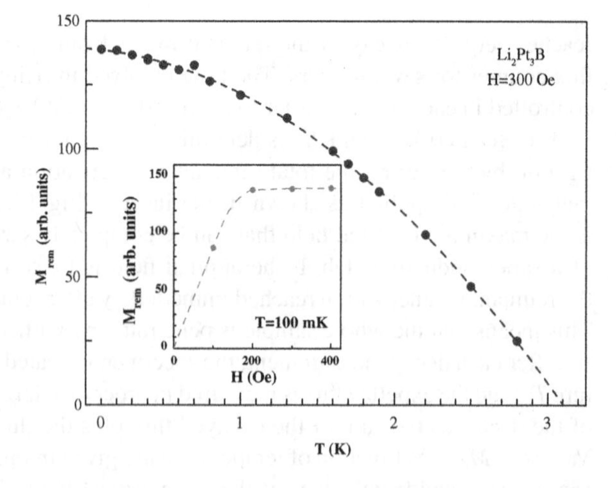

Fig. 10.8 Temperature dependence of the remnant magnetization M_{rem}. The *dotted line* is a parabolic fit to the data. *Inset* M_{rem} as function of the external magnetic field at constant temperature, $T = 100$ mK. The *dotted line* is a guide to the eye

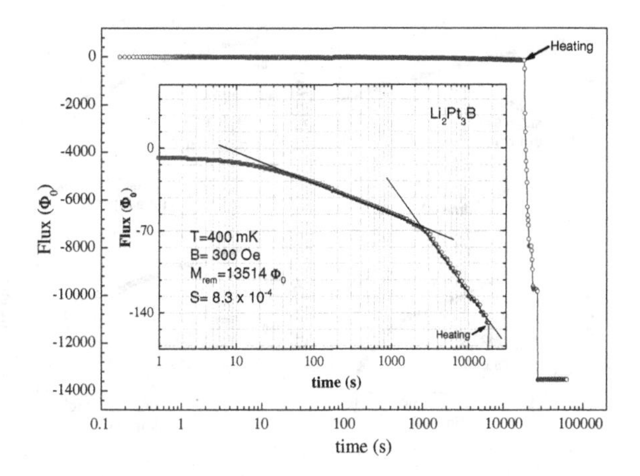

Fig. 10.9 A typical relaxation curve exemplified for $T = 400$ mK. The magnetic flux is measured as it is expelled from the sample at constant temperature. After a certain time (marked by *arrows* in the figure), the sample is gradually heated and driven into the normal state. *Inset*: Two relaxation regimes are visible in the expanded scale

flux density profiles are roughly linear in space with a slope proportional to j_c [6]. We apply at all temperatures the field H_s which is determined at the lowest T of our investigation, since H_s is reduced upon increasing temperature.

Figure 10.5 shows typical decays of the remnant magnetization of the $CePt_3Si$ specimen at different temperatures. In the inset, a decay at 100 mK is shown in another scale to illustrate the flux expulsion on heating above T_c. The logarithmic decays are extremely weak. For example, at T = 200 mK, only 0.6% of the trapped flux has decayed after 10^4 s. Common knowledge would suggest that this low creep rate implies a very high critical current resulting from strong vortex pinning. However, the critical current in $CePt_3Si$ is very low too, as will be discussed later on. In Fig. 10.6 the normalized creep rates, $S = |\partial \ln(M_{rem})/\partial \ln(t)|$ for $CePt_3Si$ are given in linear scale. We notice that they do not extrapolate to zero as expected for thermally activated creep according to the Kim-Anderson theory. This could be a sign that a small contribution originates from quantum tunneling of vortices.

10.4 Li_2Pt_3B

The polycrystalline specimen investigated was synthesized in an arc furnace utilizing a two-step process similar to that outlined by Badica et al. [7, 8]. For the magnetic measurements, a 0.225-mm thin slice, was cut and polished. The magnetic field was applied perpendicular to its smallest dimension. From ac susceptibility data taken with a field of 3 mOe at a frequency $f = 80$ Hz (Fig. 10.7) we determined the midpoint of the superconducting transition of the Li_2Pt_3B sample at $T_c = 3.07$ K with a transition width of $\Delta T_c = 0.240$ K. In the inset of Fig. 10.7 we show electrical resistivity, ρ data around the superconducting transition measured using a standard four-wire configuration in a ^{3}He cryostat. $\rho(T)$ reaches zero at $T = 3.05$ K with a transition width of $\Delta T_c = 55$ mK in close agreement with the susceptibility data.

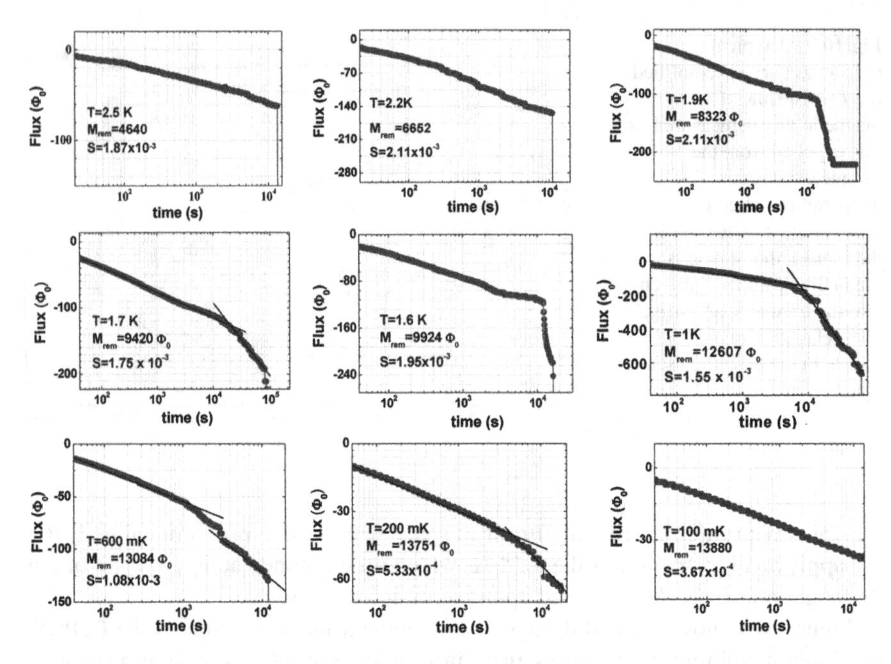

Fig. 10.10 Relaxation decays at different temperatures

From $T = 60\,\mathrm{K}$ to just above T_c, the resistivity revealed typical metallic behavior with a resistance ratio $RRR = 2$.

Vortex dynamics in Li_2Pt_3B was investigated in the temperature range $0.1\,\mathrm{K} \leq T \leq 2.8\,\mathrm{K}$, in a similar way as described previously for $CePt_3Si$. In Fig. 10.8 we show data of M_{rem} as function of T and in the inset, M_{rem} at $T = 0.100\,\mathrm{K}$ as a function of cycling fields. For this specimen, the Bean critical state at $T = 0.100\,\mathrm{K}$ is established at a magnetic field $H_s = 100\,\mathrm{Oe}$. In this sample M_{rem} decreases monotonically upon increasing temperature following roughly a parabola which reaches zero at $T = 3.1\,\mathrm{K}$, in agreement with the susceptibility, specific heat and resistivity measurements.

Figure 10.9 shows a relaxation decay of the remnant magnetization from the Bean critical state at $T = 0.4\mathrm{K}$. The decay was measured for about 1.8×10^4 s. At that time the sample was heated above T_c in order to account for the remaining flux which had not decayed in the first 1.8×10^4 s. In the inset we show the same data in an expanded scale. In the first 2.4×10^3 s the creep has a clearly defined logarithmic time dependence with only about 0.5% of the total flux creeping out of the sample. Around 2.4×10^3 s, an abrupt change in the logarithmic creep rate occurs, with a logarithmic slope which is about four times higher. Indeed it seems as if at that time vortices had overcome some barrier allowing flux lines to escape at a faster rate. We observed clear avalanche-like behavior at all temperatures, except for relaxations much below $T = 0.400$ K and above $T = 2$ K as can be seen in Fig. 10.10. In some cases the logarithmic creep rate caused by the avalanche is up to 10 times higher than the initial creep rate at the same temperature. These increased creep rates are

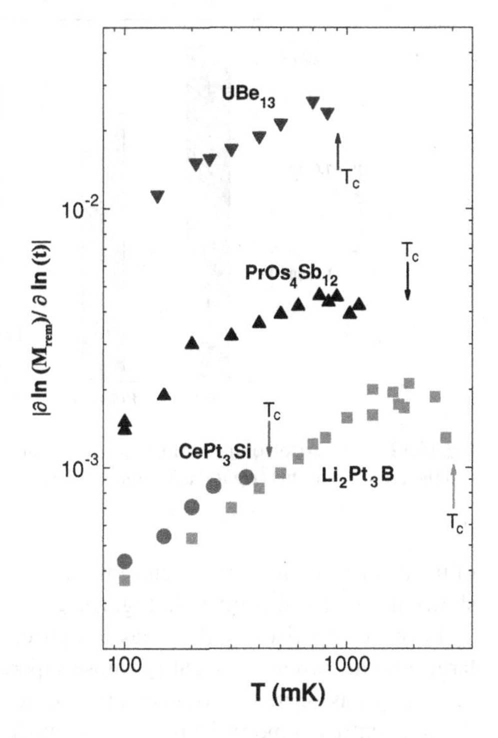

Fig. 10.11 Comparison of the normalized relaxation rates $S = |\partial \ln(M_{rem})/\partial \ln(t)|$ as function of temperature for different compounds in a log–log representation

also much higher than initial creep rates at temperatures close to T_c. This fact rules out an increase of the temperature of the sample as the origin of the avalanches. Concerning the lack of avalanches at $T = 2$ K and above, we notice that on increasing the temperature the flux vortex density is reduced causing the different behavior. At temperatures much below $T \approx 0.4$ K, vortices move so slowly that our time of observation might not be sufficient to detect an avalanche.

A comparison of initial relaxation rates of the two non-centrosymmetric superconductors $CePt_3Si$ and Li_2Pt_3B with two heavy fermion superconductors, UBe_{13} (time reversal invariant phase) and $PrOs_4Sb_{12}$ (time reversal symmetry breaking phase) [9] is shown in Fig. 10.11. In this figure we notice that the initial relaxation rates of Li_2Pt_3B and of $CePt_3Si$ are very similar. However, they are by more than a factor of 10 lower than the creep rates of UBe_{13} and about a factor of 4 lower than the rates of $PrOs_4Sb_{12}$. It has been suggested [10] that the slow creep rate of $PrOs_4Sb_{12}$ might be connected with the apparent violation of time reversal symmetry in the superconducting state inferred from zero-field μSR experiments [11].

It is interesting to relate the creep rates to the critical current. We can obtain a rough estimate of the critical current from the values of the Bean critical field H_s and the smallest dimension, d of the sample perpendicular to the field. In this model, the critical current is related to H_s by the expression $H_s = (1/2) j_c/d$. A comparison

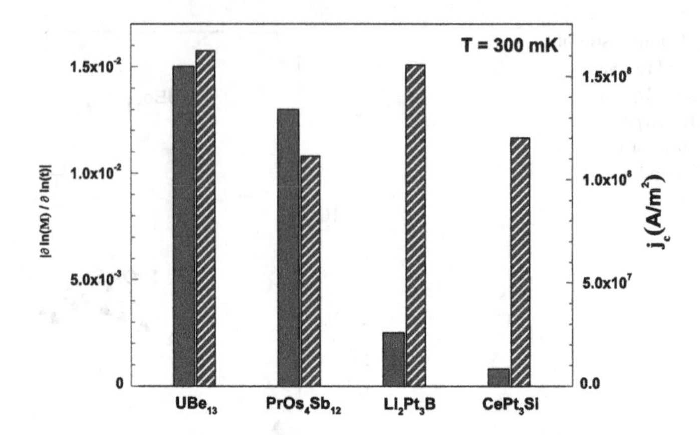

Fig. 10.12 Comparison of the normalized relaxation rates $S = |\partial \ln(M_{rem})/\partial \ln(t)|$ (*left solid bar*) and the critical currents (*right dashed bar*) at $T = 0.3$ K for different compounds

of the creep rates and critical currents at $T = 300$ mK for the same superconductors shown in Fig. 10.11 is given in Fig. 10.12.

In our comparison, UBe_{13} has a high creep rate accompanied by a relatively large critical current. $PrOs_4Sb_{12}$ whose superconducting phase breaks time reversal symmetry, has a critical current comparable to UBe_{13} but a low creep rate. In contrast, the non-centrosymmetric heavy fermion superconductor $CePt_3Si$ looks highly anomalous with a very weak creep rate as well as very small critical current. The critical current in Li_2Pt_3B is not as low as in $CePt_3B$ but still low considering its very low creep rate. One possible reason for the relatively enhanced critical current in Li_2Pt_3B might be the fact that the samples of UBe_{13}, $PrOs_4Sb_{12}$ and $CePt_3B$ were single crystals, while samples of $LiPt_3B$ are here polycrystalline.

10.5 Are Twin Boundaries the Origin of the Anomalous Vortex Dynamics?

In this section we would like to consider a particular aspect of non-centrosymmetric superconductors which is possibly relevant for the understanding of the experimental findings on their vortex dynamics. A crystal lattice without inversion center can be reached from a centrosymmetric crystal by displacing certain atoms, e.g. moving certain atoms out of a symmetry plane and so destroying the mirror symmetry of this plane, as depicted in Fig. 10.13.

Since there are in this case two ways of displacing these atoms, up or down, there are two degenerate forms of the same non-centrosymmetric crystal, related by the mirror operation at the plane (Fig. 10.13). It is, thus, possible to encounter twinning in such a crystal, i.e. both "twin domains" are present separated by twin boundaries.

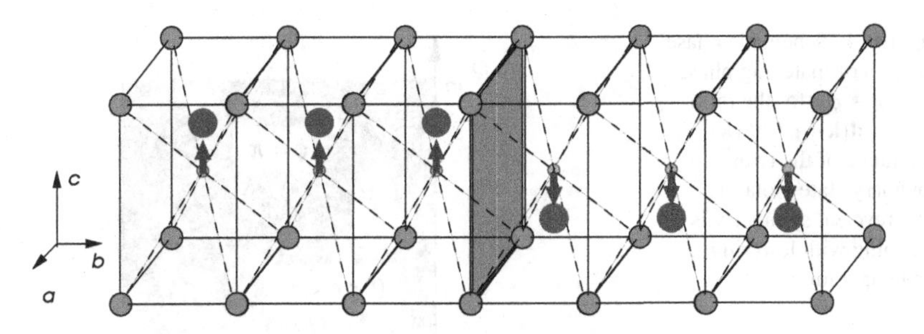

Fig. 10.13 Degenerate twins of non-centrosymmetric tetragonal crystal. The red atom can be moved up or down along the c-axis to remove the mirror symmetry with respect of the basal plane. The blue shaded plane marks the twin boundary between domains of the two twin crystals

10.5.1 Twin Boundary States

The superconducting phase is strongly influenced by the antisymmetric spin-orbit coupling induced by the non-centrosymmetricity of the crystal. We may now ask how the superconducting state is changed at a twin boundary (see Chapter by Mineev and Sigrist in this book). The pairing state in the non-centrosymmetric superconductor has mixed parity and may be expressed through a gap function which, in the simplest case, can be given by

$$\hat{\Delta}_{\mathbf{k}} = \left\{\psi(\mathbf{k}) + \mathbf{d}(\mathbf{k}) \cdot \hat{\sigma}\right\} i\hat{\sigma}_y \tag{10.1}$$

where $\psi(\mathbf{k})$ is an even function of $\mathbf{k}$, the even-parity component, and $\mathbf{d}(\mathbf{k}) \propto \psi(\mathbf{k})\gamma(\mathbf{k})$, as an odd function of $\mathbf{k}$, is the odd-parity component. Here $\gamma(\mathbf{k})$ is the vector entering the anisotropic spin-orbit coupling in the Hamiltonian: $\sum_{\mathbf{k},s,s'} \gamma(\mathbf{k}) \cdot \sigma_{ss'} c^{\dagger}_{\mathbf{k}s} c_{\mathbf{k}s'}$. We may assume a situation like in $CePt_3Si$ with the point group C_{4v} which has two types of domains, characterized by a Rashba-type spin-orbit coupling of opposite sign. In the two twin domains, A and B, the mixed-parity state is then different,

$$\hat{\Delta}_{\mathbf{k},A} = e^{i\phi_A} \left\{\psi(\mathbf{k}) + \mathbf{d}(\mathbf{k}) \cdot \hat{\sigma}\right\} i\hat{\sigma}_y \tag{10.2}$$

and

$$\hat{\Delta}_{\mathbf{k},B} = e^{i\phi_B} \left\{\psi(\mathbf{k}) - \mathbf{d}(\mathbf{k}) \cdot \hat{\sigma}\right\} i\hat{\sigma}_y. \tag{10.3}$$

The relative sign between the even- and odd-parity components of Δ_A and Δ_B is opposite. Through the twin boundary the two gap functions have to change smoothly into each other over a length scale comparable to the coherence length. The phase $\phi_{A,B}$ plays an important role when the two gap functions are matched at a twin boundary. The matching conditions are influenced by the relative magnitude of the even- and odd-parity components. The phase difference $\phi = \phi_B - \phi_A$ between two twin domains is $\phi = 0$, if the even-parity component is dominating $(\langle|\psi(\mathbf{k})|\rangle_{\mathbf{k}} \gg$

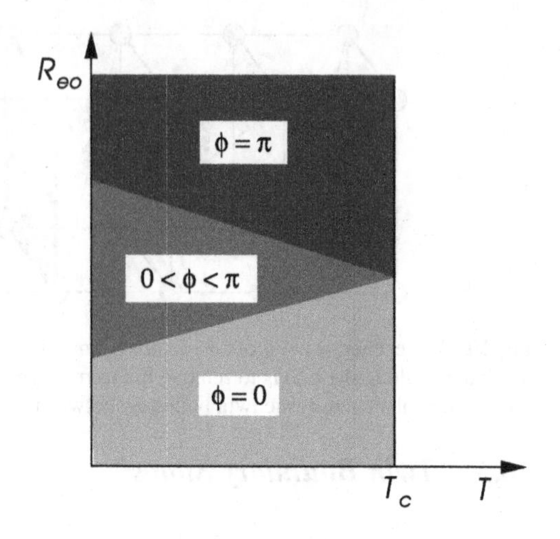

Fig. 10.14 Schematic phase diagram of matching phase $\phi = \phi_B - \phi_A$ for the ratio $R_{eo} = \langle|\mathbf{d}(\mathbf{k})|\rangle_{\mathbf{k}} / \langle|\psi(\mathbf{k})|\rangle_{\mathbf{k}}$. The range of the twin boundary state violating time reversal symmetry is widening with lowering the temperature

$\langle|\mathbf{d}(\mathbf{k})|\rangle_{\mathbf{k}}$), and $\phi = \pi$, in the opposite limit. The question arises how the phase ϕ evolves, if we change continuously between the two limits. Iniotakis et al. have shown [12] that in a certain range ϕ changes continuously between 0 and π, as shown in the phase diagram in Fig. 10.14. This leads to a special twin boundary state with $\phi \neq 0, \pi$, which breaks time reversal symmetry. Therefore we find two degenerate states in this case, connected by the time reversal operation, $\phi = \gamma$ and $\phi = -\gamma (0 < \gamma < \pi)$.

On such a twin boundary it is possible to create a line defect separating the two types of twin-boundary states. It can be shown that such a line defect carries a magnetic flux which is different from a multiple of a standard flux quantum $\Phi_0 = hc/2e$:

$$\frac{\Phi}{\Phi_0} = n \pm \frac{\gamma}{\pi} \tag{10.4}$$

where the sign depends on which of the two line defects we are looking at ($\gamma \to -\gamma$ or $-\gamma \to \gamma$). It is obvious that two neighboring line defects are of opposite type such that the total flux

$$\Phi = \Phi_1 + \Phi_2 = \Phi_0 \left\{ n_1 + \frac{\gamma}{\pi} + n_2 - \frac{\gamma}{\pi} \right\} = \Phi_0 (n_1 + n_2), \tag{10.5}$$

adds up to an integer multiple of Φ_0 [10]. Thus, conventional vortices could decay on such a twin boundary into two fractionally quantized vortices with $0 < \Phi_1, \Phi_2 < \Phi_0$ and $\Phi_1 + \Phi_2 = \Phi_0$.

10.5.2 Influence on Vortex Dynamics

The described fractional vortices can only exist at twin boundaries and can only be removed by recombining into conventional vortices. A twin boundary decorated with many fractional vortices would then act as a fence preventing vortices from easy passage. It has been argued that such a configuration could be a severe impediment to flux motion [12]. Flux lines encircled by such twin boundaries would be prevented from escaping. Thus the flux creep should be small in samples with a large density of twin boundaries. On the other hand, the critical current density j_c which is connected with the slope in the Bean profile of the vortex distribution is governed by the standard vortex pinning by defects and is, thus, unrelated to the barrier effect of the twin boundaries. Estimates of the critical current based on the value of the remnant magnetization may overestimate j_c, since twin boundaries would keep more flux inside the superconductor than predicted based on the simple Bean profile.

These arguments would essentially be sufficient to explain the basic experimental findings in $CePt_3Si$ which has low flux creep and small critical current, two features only seemingly in conflict. In Li_2Pt_3B there is an additional effect, i.e. the avalanche of flux decay after a certain longer period of time. Does the fence of fractional flux lines on twin boundaries allow for such an effect? Actually, we can argue that the barrier effect of the twin boundary depends on the density of vortices piling up on one side. This density also determines the density of the fractional flux quanta. For a sufficiently high density, the fractional vortices approach each other close enough to recombine and so loose their strong pinning. Then the barrier opens and allows a large amount of flux to escape suddenly.

Such a condition may not be realized immediately after the external field has been turned off and a Bean critical state is reached which decays slowly in time. Only if the encircled flux lines have rearranged themselves after some time in a way that the density of vortices at the barriers has increased sufficiently, then the flux lines will suddenly be able to escape through the barrier. Obviously this effect needs a high density of vortices as it can be reached more easily at low temperatures with a high remnant magnetization. This may be the reason that the avalanche feature disappears at higher temperatures in Li_2Pt_3B.

A similar type of scenario was in the past suggested for superconductors with broken time reversal symmetry, where fractional vortices can appear on domain walls separating superconducting phases of opposite chirality and leading to a similar quench of the flux creep rate without affecting the critical current [10]. Such effects have been reported for $U_{1-x}Th_xBe_{13}$ $(0.2 < x0.45)$ and UPt_3 which both show a superconducting double transition where it is assumed that time reversal symmetry is broken at the second transition coinciding with a drastic decrease of creep rate [13, 14]. In contrast, in UBe_{13}, which does neither have a second transition nor signs of broken time reversal symmetry, no unusual drop of the flux creep rate has been observed.

10.6 Conclusion

The anomalously low flux creep rate in Li_2Pt_3B and $CePt_3Si$ which, counterintuitively, is not correlated with a large critical current, is an extraordinary feature observed in the two non-centrosymmetric superconductors $CePt_3Si$ and Li_2Pt_3B. Note that in $CePt_3Si$ superconductivity coexists with antiferromagnetic order whose influence, if any, on vortex pinning is not clear so far. On the other hand, the novel flux dynamics in Li_2Pt_3B cannot be attributed to any magnetic order coexisting with superconductivity. In this chapter we addressed the question whether the origin of this extraordinary behavior is specific to non-centrosymmetric superconductors. Obviously the traditional pinning of individual vortices at impurities and other lattice defects would not be sufficient to account for the whole set of properties. Twin boundaries which should in principle be present in non-centrosymmetric crystals may provide a possible consistent explanation through novel flux line defects sitting on twin boundaries. This scenario is speculative and has not been tested directly so far. One type of test could be the direct observation of fractional vortices on twin boundaries. Another test could address the sample "quality" in the sense that one would influence the presence or the density of twin boundaries. This is more difficult in the case of non-centrosymmetric crystals, as twin domains do not directly couple to uniform uniaxial strain. The definite understanding of the anomalous vortex dynamics indeed needs additional experiments.

Acknowledgements The authors would like to thank E. Bauer, R. Cardoso, T. Cichorek, C.A. McElroy, M.B. Maple, M. Nicklas, A. Prokofiev, T.A. Sayles, F. Steglich and B.J. Taylor for contributions and discussions. This work was financially supported by the Swiss Nationalfonds, the NCCR MaNEP and the German Research Foundation (DFG).

References

1. Miclea, C.F., Mota, A.C., Nicklas, M., Cardoso, R., Steglich, F., Sigrist, M., Bauer, E.: Phys Rev B **81**, 014527 (2010)
2. Miclea C.F., Mota A.C., Sigrist M. Steglich F, Sayles T.A., Taylor B.J., McElroy C., Maple M.B. Phys Rev B **80**, 132502 (2009)
3. Takeuchi, T., Tsujin, M., Yasuda, T., Hashimoto, S., Settai, R., Onuki, Y.: J Magn Magn Mater **310**, 557 (2007)
4. Bauer, E., Hilscher, G., Michor, H., Paul, Ch., Scheidt, E.W., Gribanov, A., Seropegin, Yu., Noel, H., Sigrist, M., Rogl, P.: Phys Rev Lett **92**, 027003 (2004)
5. Bauer, E., Hilscher, G., Michor, H., Paul, Ch., Scheidt, E.W., Gribanov, A., Seropegin, Yu., Noel, H., Sigrist, M., Rogl, P.: Phys Rev Lett **92**, 027003 (2004)
6. Tinkham, M.: Introduction to Superconductivity. McGraw-Hill, Inc., New York (1996)
7. Badica, P., Kondo, T., Togano, K.: J Phys Soc Jpn **74**, 1014 (2005)
8. Togano, K., Badica, P., Nakamori, Y., Orimo, S., Takeya, H., Hirata, K.: Phys Rev Lett **93**, 247004 (2004)
9. Cichorek T. et al., to be published.
10. Sigrist, M., Agterberg, D.: Prog Theor Phys **102**, 965 (1999)

11. Aoki, Y., Tayama, T., Sakakibara, T., Kuwahara, K., Iwasa, K., Kohgi, M., Higemoto, W., Maclaughlin, D.E., Sugawara, H., Sato, H.: J Phys Soc Jpn **76**, 051006 (2007)
12. Iniotakis, C., Fujimoto, S., Sigrist, M.: J Phys Soc Jpn **77**, 083701 (2008)
13. Amann, A., Mota, A.C., Maple, M.B., Löhneysen, H.v.: Phys Rev B **57**, 3640 (1998)
14. Dumont E.M.M. Phd. Thesis ETH13938 (2000)

Chapter 11
Properties of Interfaces and Surfaces in Non-centrosymmetric Superconductors

Matthias Eschrig, Christian Iniotakis and Yukio Tanaka

Abstract Tunneling spectroscopy at surfaces of unconventional superconductors has proven an invaluable tool for obtaining information about the pairing symmetry. It is known that mid-gap Andreev bound states manifest themselves as zero-bias conductance peaks in tunneling spectroscopy. The zero-bias conductance peak is a signature for a non-trivial pair potential that exhibits different signs on different regions of the Fermi surface. Here, we review recent theoretical results on the spectrum of Andreev bound states near interfaces and surfaces in non-centrosymmetric superconductors. We introduce a theoretical scheme to calculate the energy spectrum of a non-centrosymmetric superconductor. Then, we discuss the interplay between the spin-orbit vector field on the Fermi surface and the order parameter symmetry. The Andreev states carry a spin supercurrent and represent a helical edge mode along the interface. We study the topological nature of the resulting edge currents. If the triplet component of the order parameter dominates, then the helical edge mode exists. If, on the other hand, the singlet component dominates, the helical edge mode is absent.

M. Eschrig (✉)
SEPnet and Hubbard Theory Consortium, Department of Physics,
Royal Holloway, University of London, Egham, Surrey TW20 0EX, United Kingdom
e-mail: matthias.eschrig@rhul.ac.uk

Fachbereich Physik, Universität Konstanz, D-78464 Konstanz, Germany

Institut für Theoretische Festkörperphysik and DFG-Center for Functional Nanostructures,
Karlsruhe Institute of Technology, D-76128 Karlsruhe, Germany

C. Iniotakis
Institute for Theoretical Physics, ETH Zurich, 8093 Zurich, Switzerland
e-mail: iniotaki@phys.ethz.ch

Y. Tanaka
Department of Applied Physics, Nagoya University, Nagoya 464-8603, Japan
e-mail: ytanaka@nuap.nagoya-u.ac.jp

E. Bauer and M. Sigrist (eds.), *Non-centrosymmetric Superconductors*,
Lecture Notes in Physics 847, DOI: 10.1007/978-3-642-24624-1_11,

A quantum phase transition occurs for equal spin singlet and triplet order parameter components. We discuss the tunneling conductance and the Andreev point-contact conductance between a normal metal and a non-centrosymmetric superconductor.

11.1 Introduction

In this chapter, we will discuss the surface and interface properties of non-centrosymmetric superconductors [47] focusing on the Andreev conductance. Since the early sixties tunneling spectroscopy has played an important role in gathering information about the gap function of conventional superconductors [1]. In the context of unconventional superconductivity tunneling spectroscopy appeared as an important tool to probe the internal phase structure of the Cooper pair wave functions [2, 3]. Surface states with sub-gap energy, known as Andreev bound states (ABS) [4–8] provide channels for resonant tunneling leading to so-called zero-bias anomalies in dI/dV. Zero-bias anomalies observed in high-temperature superconductors showed the presence of zero-energy bound states at the surface, giving strong evidence for *d*-wave pairing [2–6]. Similarly the tunneling spectrum observed in Sr_2RuO_4 is consistent with the existence of chiral surface states as expected for a chiral *p*-wave superconductor [9–12]. Zero bias conductance peaks due to Andreev bound states have been observed in numerous experiments, e.g. in high-T_c cuprates [13–17], Sr_2RuO_4 [18–20], UBe_{13} [21], $CeCoIn_5$ [22], the two dimensional organic superconductor κ-(BEDT-TTF)$_2$Cu[N(CN)$_2$]Br [23] and $PrOs_4Sb_{12}$ [24]. Andreev bound states have also been observed in the Balian-Werthammer phase of superfluid ^{3}He [25]. The study of Andreev bound states in unconventional superconductors and superfluids has emerged as an important phase sensitive probe.

In Sect. 11.2 we present the theory for Andreev spectroscopy using Bogoliubov wave function technique in Andreev approximation. Starting with superconductors exhibiting *d*-wave or *p*-wave pairing, we proceed with non-centrosymmetric superconductors. In Sect. 11.3, we develop the theoretical tools for describing Andreev spectroscopy in non-centrosymmetric superconductors in the framework of Nambu-Gor'kov Green's functions within the quasiclassical theory of superconductivity.

11.2 Andreev Spectroscopy in Unconventional Superconductors

11.2.1 Andreev Conductance in *s*- and *d*-Wave Superconductors

We discuss first the example of zero-bias resonant states at the interface of a normal metal/spin-singlet *d*-wave superconductor junction. In general, the pair potential can be expressed in terms of two coordinates, $\boldsymbol{x}$ and $\boldsymbol{x}'$, as $\Delta(\boldsymbol{x}, \boldsymbol{x}')$. In uniform systems it only depends on the relative coordinate $\boldsymbol{x} - \boldsymbol{x}'$, and a Fourier transform with respect to this relative coordinate yields $\Delta(\boldsymbol{k})$ with relative momentum $\boldsymbol{k}$. For illustrative purposes, we assume in the following a cylindrical Fermi surface and concentrate on two-dimensional systems. The pair potential for spin-singlet d-wave pairing is $\Delta(\theta) = \Delta_0 \cos(2\theta)$, with $e^{i\theta} = (k_x + ik_y)/\mid \boldsymbol{k} \mid$, while the corresponding spin-singlet s-wave pair potential is isotropic, $\Delta(\theta) = \Delta_0$. The bulk quasiparticle density of states normalized by its value in the normal state is given by

$$\rho_{\mathrm{B}}(E) = \frac{1}{\pi} \int_0^{\pi} d\theta \rho_0\Big(E, \Delta(\theta)\Big), \quad \rho_0\Big(E, \Delta(\theta)\Big) = \frac{E}{\sqrt{E^2 - \Delta_0^2 \cos^2(2\theta)}}. \tag{11.1}$$

For a spin-singlet d-wave superconductor this quantity behaves linearly at low energies, $\rho(E) \propto |E|$. As shown below if the angle between the interface normal and the lobe direction of the d-wave pair potential has a nonzero value α with $0 < \alpha < \pi/2$, then the resulting tunneling conductance has a zero bias conductance peak.

The Andreev conductance, $\sigma_{\mathrm{T}}(E)$, for a normal metal/insulator/spin singlet s-wave superconductor junction is described by the model of Blonder, Tinkham, and Klapwijk (BTK) [26]. Within this model, $\sigma_{\mathrm{T}}(E)$ at zero temperature is given by

$$\sigma_{\mathrm{T}}(E) \propto \sum_{\theta} \Big(1 + \mid a(E, \theta) \mid^2 - \mid b(E, \theta) \mid^2\Big) \tag{11.2}$$

where $a(E, \theta)$, and $b(E, \theta)$ are probability amplitude coefficients for Andreev reflection and for normal reflection, respectively. We apply the BTK model in the following to the case of spin-singlet d-wave pairing.

We assume that the Fermi energy E_F is much larger than $\mid \Delta(\theta) \mid$, such that the Andreev approximation can be applied to the Bogoliubov wave functions. For simplicity, we also assume equal effective masses and Fermi momenta in the normal metal and in the superconductor. The spatial dependence of the pair potential is chosen to be $\Delta(\theta)\Theta(x)$ (with the Heaviside step function Θ). The insulating barrier at the atomically clean interface is modeled by a δ-function potential, $V(x) = H\delta(x)$. Since the momentum parallel to the interface is conserved, the two component Bogoliubov wave function is given in Andreev approximation by

$$\Psi(\theta, x) = \begin{pmatrix} u_+(\theta, x) \\ v_+(\theta, x) \end{pmatrix} \exp(ik_{\mathrm{F}} x \cos\theta) + \begin{pmatrix} u_-(\theta, x) \\ v_-(\theta, x) \end{pmatrix} \exp(-ik_{\mathrm{F}} x \cos\theta) \tag{11.3}$$

where $u_j(\theta, x)$ and $v_j(\theta, x)$ with $j \in \{+, -\}$ obey the Andreev equations

$$\begin{aligned} Eu_j(\theta, x) &= -\Big[\frac{i\hbar^2 \sigma_j k_{\mathrm{F}} \cos\theta}{m} \frac{d}{dx} - H\delta(x)\Big] u_j(\theta, x) + \Delta(\theta_j)\Theta(x) v_j(\theta, x), \\ Ev_j(\theta, x) &= \Big[\frac{i\hbar^2 \sigma_j k_{\mathrm{F}} \cos\theta}{m} \frac{d}{dx} - H\delta(x)\Big] v_j(\theta, x) + \Delta^*(\theta_j)\Theta(x) u_j(\theta, x), \end{aligned} \tag{11.4}$$

with $\sigma_+ = 1, \theta_+ = \theta$, and $\sigma_- = -1, \theta_- = \pi - \theta$. For a d-wave superconductor the corresponding effective pair potentials $\Delta(\theta_\pm)$ are given by

$$\Delta(\theta_+) = \Delta_0 \cos(2\theta - 2\alpha), \quad \Delta(\theta_-) = \Delta_0 \cos(2\theta + 2\alpha), \tag{11.5}$$

where the angle between the interface normal and the lobe direction of the d-wave pair is α. The wave functions $u_\pm(\theta, x)$ and $v_\pm(\theta, x)$ resulting from Eqs. (11.4) are obtained from the ansatz

$$\begin{pmatrix} u_+(\theta,x) \\ v_+(\theta,x) \end{pmatrix} = \begin{cases} \begin{pmatrix} 1 \\ 0 \end{pmatrix} \exp(i\delta x) + a(E,\theta) \begin{pmatrix} 0 \\ 1 \end{pmatrix} \exp(-i\delta x) & x < 0, \\ c(E,\theta) \begin{pmatrix} \sqrt{(E+\Omega_+)/2E} \\ \exp(-i\phi_+)\sqrt{(E-\Omega_+)/2E} \end{pmatrix} \exp(i\gamma_+ x) & x > 0, \end{cases} \tag{11.6}$$

$$\begin{pmatrix} u_-(\theta,x) \\ v_-(\theta,x) \end{pmatrix} = \begin{cases} b(E,\theta) \begin{pmatrix} 1 \\ 0 \end{pmatrix} \exp(-i\delta x) & x < 0, \\ d(E,\theta) \begin{pmatrix} \exp(i\phi_-)\sqrt{(E-\Omega_-)/2E} \\ \sqrt{(E+\Omega_-)/2E} \end{pmatrix} \exp(i\gamma_- x) & x > 0, \end{cases} \tag{11.7}$$

where we used the abbreviations

$$\delta = \frac{Em}{\hbar k_F \cos\theta}, \quad \gamma_\pm = \frac{\Omega_\pm m}{\hbar k_F \cos\theta}, \quad \Omega_\pm = \sqrt{E^2 - \Delta^2(\theta_\pm)}, \quad \exp(i\phi_\pm) = \frac{\Delta(\theta_\pm)}{|\Delta(\theta_\pm)|}. \tag{11.8}$$

With the help of appropriate boundary conditions,

$$\Psi(\theta, x)|_{x=0_-} = \Psi(\theta, x)|_{x=0_+}$$

$$\left.\frac{d}{dx}\Psi(\theta,x)\right|_{x=0_+} - \left.\frac{d}{dx}\Psi(\theta,x)\right|_{x=0_-} = \left.\frac{2mH}{\hbar^2}\Psi(\theta,x)\right|_{x=0_+} \tag{11.9}$$

one obtains $a(E,\theta)$, $b(E,\theta)$, $c(E,\theta)$, and $d(E,\theta)$. The resulting conductance is

$$\sigma_T(E) = \frac{\left(\int_{-\pi/2}^{\pi/2} d\theta\, D(\theta)\sigma_R(E,\theta)\cos\theta\right)}{\left(\int_{-\pi/2}^{\pi/2} d\theta\, D(\theta)\cos\theta\right)},$$

with the angular resolved Andreev conductance

$$\sigma_R(E,\theta) = \frac{1 + D(\theta)\,|\,\Gamma_+\,|^2 - R(\theta)\,|\,\Gamma_+\Gamma_-\,|^2}{|\,1 - R(\theta)\Gamma_+\Gamma_- \exp[i(\phi_- - \phi_+)]\,|^2}, \tag{11.10}$$

where $\Gamma_\pm = (E - \Omega_\pm)/|\Delta(\theta_\pm)|$. The quantities $D(\theta)$ and $R(\theta)$ above are transmission and reflection probabilities given by

$$D(\theta) = 4\cos^2\theta/(4\cos^2\theta + Z^2), \quad R(\theta) = 1 - D(\theta),$$

with injection angle θ and $Z = 2mH/\hbar^2 k_F$ [2]. Choosing $\Delta(\theta_\pm) = \Delta_0$ reproduces the BTK formula for an s-wave superconductor. Typical line shapes of $\sigma_T(eV)$ with

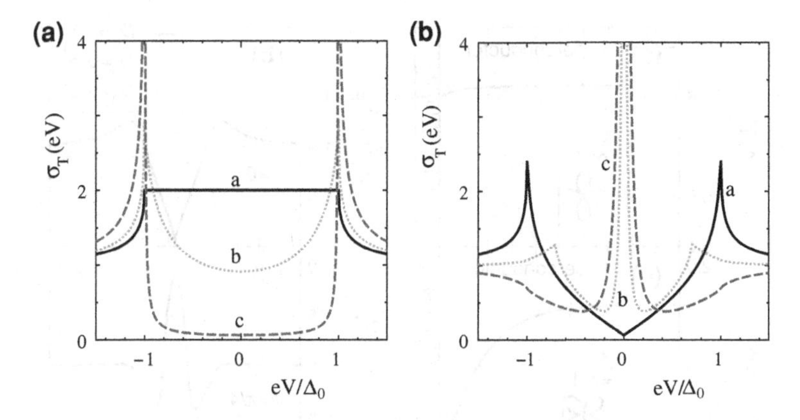

Fig. 11.1 **a** Andreev conductance for an *s*-wave superconductor for various barrier heights, a $Z = 0$, b $Z = 1$ and c $Z = 5$. **b** Andreev conductance for a *d*-wave superconductor for $Z = 5$, and various surface alignment angles, a $\alpha = 0$, b $\alpha = 0.125\pi$ and c $\alpha = 0.25\pi$

$eV = E$ for *s*-wave and *d*-wave superconductors are shown in Fig. 11.1. The *d*-wave case is shown in Fig. 11.1(b). As can be seen there, if the angle α deviates from 0, the resulting $\mathrm{d}I/\mathrm{d}V$ has a zero bias conductance peak (ZBCP) (curves *b* and *c*); the only exceptional case is $\alpha = 0$, as shown in curve *a*. The width of the ZBCP is proportional to D, while its height is proportional to the inverse of D. The origin of this peak are mid-gap Andreev bound states (MABS). The condition for the formation of Andreev bound states at the surface of an isolated *d*-wave superconductor ($D \to 0$) is expressed by

$$1 = \Gamma_+\Gamma_- \exp[i(\phi_- - \phi_+)]. \tag{11.11}$$

At zero energy $\Gamma_+\Gamma_- = -1$ is satisfied, and consequently a MABS appears, provided $\exp[i(\phi_+ - \phi_-)] = -1$. For this case, on the superconducting side of the interface the injected electron and the reflected hole experience a different sign of the pair potential. For $\alpha = \pi/4$, there is a MABS independent of the injection angle. In this case, the energy dispersion of the resulting ABS, E_b, is given by

$$E_b = 0. \tag{11.12}$$

Finally, we comment on the effects of order-parameter suppression near a surface or interface in a *d*-wave superconductor. In Fig. 11.2 we reproduce a self-consistent solution for a layered *d*-wave superconductor, showing that a strong order-parameter suppression is always present for $\alpha = \pi/4$, whereas for $\alpha = 0$ in the tunneling limit the order-parameter suppression can be neglected. The corresponding local density of states at the interface is shown in Fig. 11.2(b) and (d). The interface is modeled by a δ-potential as above, with a transmission $D(\theta) = D_0 \cos^2\theta/(1 - D_0 \sin^2\theta)$, and the parameter D_0 is related to Z via $D_0 = 1/[1 + (Z/2)^2]$.

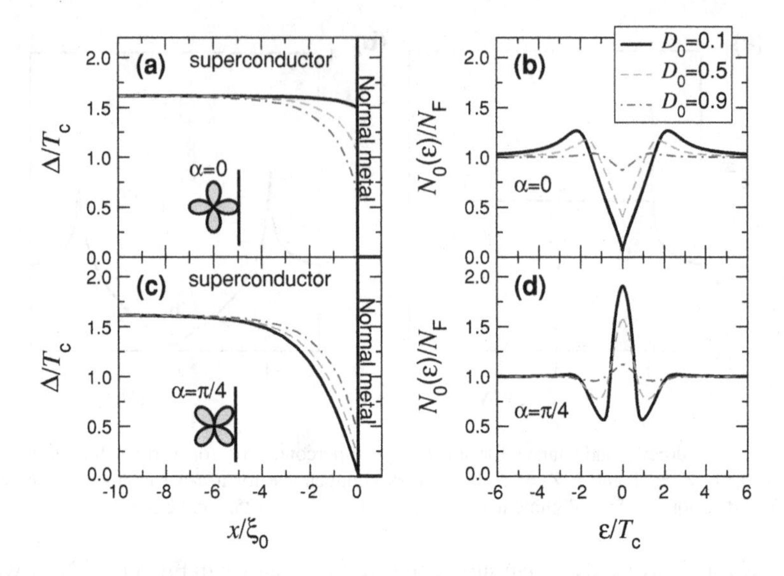

Fig. 11.2 **a, c** order parameter amplitude and **b, d** local density of states at the interface for a layered d-wave-superconductor/normal-metal junction. **a, b** $\alpha = 0$, **c, d** $\alpha = \pi/4$. The interface is at $x = 0$. The curves are for the indicated transmission coefficients D_0. The temperature is $T = 0.3T_c$, and the mean free path $\ell = 10\xi_0$. After Ref. [27]

11.2.2 Andreev Conductance in Chiral *p*-Wave Superconductor

In this section, we discuss the Andreev conductance of a normal metal/chiral p-wave superconductor junction. There is evidence supporting the realization of spin-triplet pairing with broken time reversal symmetry in the superconducting state of Sr_2RuO_4 [28, 29, 30, 31, 32, 33, 34, 35, 36]. A possible pairing symmetry is given by two-dimensional (in the $k_x - k_y$ plane) chiral p-wave pairing, where the pair potentials are given by $\Delta_{\uparrow,\uparrow} = \Delta_{\downarrow,\downarrow} = 0$, $\Delta_{\uparrow,\downarrow} = \Delta_{\downarrow,\uparrow} = \Delta_0 \exp(i\theta)$. In the following, θ is measured from the interface normal. In the actual sample, the presence of chirality may produce chiral domain structures. A recent experiment is consistent with the presence of chiral domains [37]. Also, there are several theoretical proposals to detect chiral domain structures [38, 39]. Here, for simplicity, we consider a single domain chiral p-wave superconductor.

Since the z-component of the Cooper pair spin is zero, we can also use Eq. (11.10) to obtain the Andreev conductance for normal metal/chiral p-wave superconductor junctions. Before discussing the Andreev conductance, we first consider the bulk local density of states (LDOS) of a chiral p-wave superconductor. In contrast to the spin-singlet d-wave pairing case, $\rho_0(E, \Delta(\theta))$ in Eq. (11.1) is given by

$$\rho_0(E, \Delta(\theta)) = E/\sqrt{E^2 - \Delta_0^2}.$$

It has a fully gapped density of states as in the spin-singlet s-wave case.

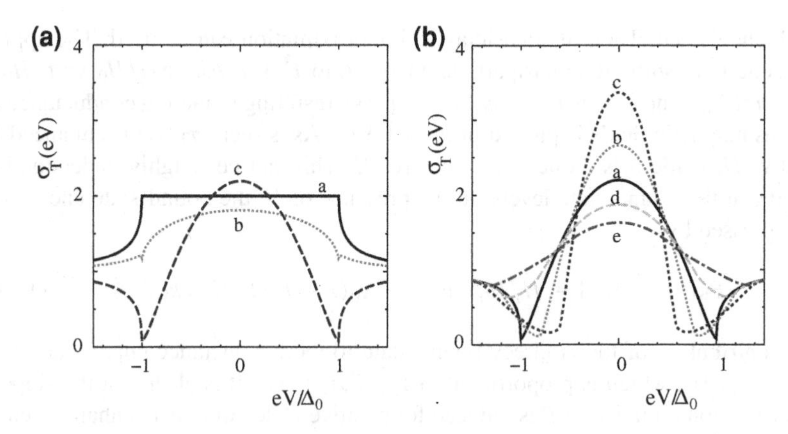

Fig. 11.3 **a** Andreev conductance σ_T for a chiral p-wave superconductor, for various barrier heights, a $Z = 0$, b $Z = 1$ and c $Z = 5$. **b** Andreev conductance for a chiral p-wave superconductor in the presence of a magnetic field for $Z = 5$. The magnetic fields for the various curves are a $H = 0$, b $H = 0.2H_0$ and c $H = 0.4H_0$, d $H = -0.2H_0$ and *e* $H = -0.4H_0$

We now discuss the condition when an ABS is formed at the surface of an isolated chiral p-wave superconductor. The bound-state condition is given by [9, 10, 11]

$$E + \sqrt{E^2 - \Delta_0^2} = -\left(E - \sqrt{E^2 - \Delta_0^2}\right) \exp(-2i\theta), \qquad (11.13)$$

showing that the bound-state level E_b satisfies

$$E_b(\theta) = \Delta_0 \sin\theta. \qquad (11.14)$$

Note that the ABS has a dispersion different from that in the d-wave case with $\alpha = \pi/4$. The presence of the edge state with a dispersion induces a spontaneous dissipationless current.

As in the previous section we consider the Andreev conductance $\sigma_T(E)$ in a normal metal/chiral p-wave superconductor junction, which is shown in Fig. 11.3(a). As can be seen, for $Z = 0$, the line shape of the conductance is identical to that of a spin-singlet s-wave superconductor (see curve a), whereas with increasing Z a zero bias conductance peak emerges (curves b and c). The resulting ZBCP is broad in contrast to the spin-singlet d-wave case due to the fact that the position of the ABS depends on the injection angle θ according to Eq. (11.14) [9, 10]. The presence of the ABS has been confirmed by tunneling experiments [18–20].

Next we consider the situation where a magnetic field H is applied in z-direction, perpendicular to the two-dimensional superconducting planes, which induces a shielding current along the interface (we consider the interface normal in x-direction). When the penetration depth for the chiral p-wave superconducting material is much longer than the coherence length, the vector potential can be approximated as $\boldsymbol{A}(\boldsymbol{r}) = [0, A_y(x), 0]$ with $A_y(x) = -\lambda_m H \exp(-x/\lambda_m)$, where λ_m is the penetration depth. In the following we consider the situation where Landau level quantization can be neglected. Then the quasiclassical approximation can be used. The applied magnetic field shifts the quasiparticle energy E to $E + H\Delta_0 \sin\phi/H_0$ with $H_0 = h/(2e\pi^2\xi\lambda_m)$ and $\xi = \hbar^2 k_F/(\pi m \Delta_0)$ [40]. The resulting tunneling conductance for various magnetic fields is plotted in Fig. 11.3(b). As is seen, $\sigma_T(E)$ is enhanced for positive H, while it is reduced for negative H. This can be roughly understood by looking at the bound-state levels. In the presence of H, the bound-state energy can be expressed by

$$E_b(\theta) = \Delta_0(1 - H/H_0)\sin\theta \sim \Delta_0(1 - H/H_0)k_y/k_F. \tag{11.15}$$

The contribution of the Andreev bound state to the conductance enters via a term $\delta(E - E_b(\theta))$, which is proportional to $1/|dE_b(\theta)/d\theta|$. It is clear that the slope of the dispersion around $\theta = 0$ is reduced for positive H, leading to an enhancement of the numerator in Eq. (11.10) around $\theta = 0$, where the bound states are close to zero energy. On the other hand, for negative H, the height of the ZBCP is reduced since the slope of the curve of E_b around zero energy becomes steeper [12].

In p-wave superconductors self-consistency of the order parameter and impurity effects can be of importance. In Fig. 11.4 we show self-consistent order parameters and Andreev spectra at a surface of a layered p-wave superconductor. In addition to the bulk $k_x + ik_y$ component a subdominant $k_x - ik_y$ component is stabilized within a few coherence lengths ($\xi_0 = v_F/2\pi k_B T_c$) near the surface. The full lines are results assuming a mean free path of $\ell = 10\xi_0$ everywhere. When replacing the mean free path in a surface layer (gray shaded region in Fig. 11.4) by $\ell = 0.3\xi_0$, we obtain the results shown as dashed lines. In contrast to the first case, for the second case both order parameter components are strongly suppressed near the surface. The presence of an increased scattering in a surface layer also modifies the form of point contact spectra and the tunneling conductance as seen in the insets of Fig. 11.4. In contrast to the surface density of states which for a clean surface is constant in energy, the tunneling conductance shows a broad peak similar as in Fig. 11.3, which is however reduced in height for a self consistent order parameter [42].

Finally, we would like to comment on the observation that the above edge state is topologically equivalent to that of a quantum Hall system. In a quantum Hall system it is established that the edge channel supports the accurate quantization of the Hall conductance σ_H, which is related to a topological integer [43, 44]. In the edge state of a chiral p-wave superconductor, such a topological number can be also defined [45, 46]. For this case, the edge state is topologically protected by the bulk energy gap Δ_0. The topological properties of the electronic states have been attracting intensive interest in condensed matter physics. In Sect. 11.2.3.3 we will return to this question in connection with non-centrosymmetric superconductors. Before that, we discuss in the following section theoretical predictions for the Andreev conductance spectra for non-centrosymmetric superconductors.

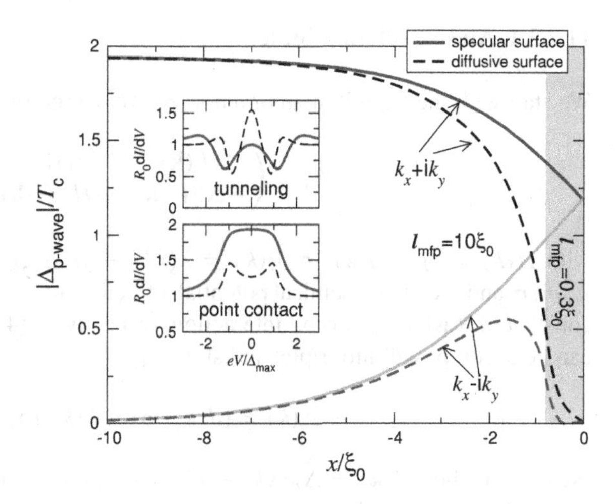

Fig. 11.4 Self-consistent order parameter near a surface of a layered p-wave superconductor. Full lines are for mean free path $\ell_{mfp} = 10\xi_0$ everywhere; *dashed lines* are for a shortened $\ell_{mfp} = 0.3\xi_0$ in the gray shaded region, and $\ell_{mfp} = 10\xi_0$ else. The calculations are for $T = 0.1T_c$. *Insets*: point contact spectra for fully transparent interface (*bottom*) and tunneling conductance ($D_0 = 0.05$) (*top*). After Ref. [41]

11.2.3 Andreev Conductance in Non-centrosymmetric Superconductors

Non-centrosymmetric superconductors such as $CePt_3Si$ are a central topic of current research [47, 48]. Two-dimensional non-centrosymmetric superconductors are expected, e.g., at interfaces and/or surfaces due to a strong potential gradients. An interesting example is superconductivity at a $LaAlO_3/SrTiO_3$ interface [49, 50]. In non-centrosymmetric materials spin-orbit interaction becomes very important. Frigeri *et al.* [48] have shown that the $(p_x \pm ip_y)$-pairing state has the highest T_c within the triplet channel in $CePt_3Si$. It has been shown that singlet (s-wave) and triplet (p-wave) pairing is mixed, and several novel properties related to that mixing, such as a large upper critical field beyond the Pauli limit, have been focused on [48]. On the other hand, a pure $(p_x \pm ip_y)$-pairing state has been studied as a superconducting analogue of a quantum spin Hall system [51–53]. Therefore, it is an important and urgent issue to study the spin transport properties of the NCS superconductors from a topological viewpoint.

In this section, we discuss charge and spin transport in non-centrosymmetric superconductors [54]. We concentrate on non-centrosymmetric superconductors with time-reversal symmetry, where a spin-triplet $(p_x \pm ip_y)$-wave and a spin-singlet s-wave pair potential can mix with each other, similar as discussed in the last section. We show that when the amplitude of the $(p_x \pm ip_y)$-wave component is larger than that of the s-wave component, then the superconducting state belongs to a topologically nontrivial class analogous to a quantum spin Hall system, and the resulting helical edge modes are spin-current-carrying Andreev bound states that are topologically protected. Below, we study Andreev reflection [55] at low energy, which is determined mostly by the helical edge modes, and find the spin polarized current flowing through an interface as a function of incident angle. When a magnetic field is applied, even the angle-integrated current is spin polarized.

11.2.3.1 Andreev Bound States

We start with the Hamiltonian of a non-centrosymmetric superconductor

$$\hat{\mathscr{H}}_{\mathrm{S}} = \begin{pmatrix} H(\mathbf{k}) & \Delta(\mathbf{k}) \\ -\Delta^*(-\mathbf{k}) & -H^*(-\mathbf{k}) \end{pmatrix}$$

with $H(\boldsymbol{k}) = \xi_{\boldsymbol{k}} + \boldsymbol{g}(\boldsymbol{k}) \cdot \boldsymbol{\sigma}$, $\boldsymbol{g}(\boldsymbol{k}) = \lambda(\hat{\boldsymbol{x}} k_y - \hat{\boldsymbol{y}} k_x)$, $\xi_{\boldsymbol{k}} = \hbar^2 \boldsymbol{k}^2/(2m) - \mu$. Here, $\mu, m, \boldsymbol{\sigma}$ and λ denote chemical potential, effective mass, Pauli matrices and coupling constant of Rashba spin-orbit interaction, respectively [48]. The pair potential $\Delta(\boldsymbol{k})$ can be decomposed into triplet and singlet parts,

$$\Delta(\boldsymbol{k}) = [\boldsymbol{d}(\boldsymbol{k}) \cdot \boldsymbol{\sigma} + \psi(\boldsymbol{k})] i\sigma_y. \tag{11.16}$$

We consider here $\boldsymbol{d}(\boldsymbol{k}) = \Delta_p(\hat{\boldsymbol{x}} k_y - \hat{\boldsymbol{y}} k_x)/ \mid \boldsymbol{k} \mid$ for the spin-triplet component [48], and $\psi(\boldsymbol{k}) = \Delta_s$ for the spin-singlet component, with $\Delta_p \geq 0$ and $\Delta_s \geq 0$. The superconducting gaps $\Delta_1 = \Delta_p + \Delta_s$ and $\Delta_2 = \mid \Delta_p - \Delta_s \mid$ open for the two spin-split energy bands, respectively, in the homogeneous state [56]. As we show below, surface states are crucially influenced by the relative magnitude between Δ_p and Δ_s.

Let us consider the wave functions, focusing on those for ABS localized at the surface. Consider a two-dimensional semi-infinite superconductor for $x > 0$ where the surface is located at $x = 0$. The corresponding wave function is given by [57]

$$\Psi_{\mathrm{S}}(x) = e^{ik_y y} \left[c_1 \psi_1 e^{iq_{1x}^+ x} + c_2 \psi_2 e^{-iq_{1x}^- x} + c_3 \psi_3 e^{iq_{2x}^+ x} + c_4 \psi_4 e^{-iq_{2x}^- x} \right],$$
$$q_{jx}^{\pm} = k_{jx}^{\pm} \pm (k_j/k_{jx}^{\pm}) \sqrt{(E^2 - \Delta_j^2)/(\lambda^2 + 2\hbar^2\mu/m)}, \tag{11.17}$$

with $j \in \{1, 2\}$, and $k_{jx}^+ = k_{jx}^- = k_{jx}$ for $|k_y| \leq k_j$ and $k_{jx}^+ = -k_{jx}^- = k_{jx}$ for $|k_y| > k_j$. Here,

$$k_{1(2)} = \mp m\lambda/\hbar^2 + \sqrt{(m\lambda/\hbar^2)^2 + 2m\mu/\hbar^2} \tag{11.18}$$

are the Fermi momenta of the small (large) Fermi surface (the upper sign holds for k_1), and $k_{jx} = (k_j^2 - k_y^2)^{1/2}$ denotes the x component of the Fermi momentum $\boldsymbol{k}_j$. The wave functions are given by

$$\psi_1 = \begin{pmatrix} u_1 \\ -i\alpha_1^{-1} u_1 \\ i\alpha_1^{-1} v_1 \\ v_1 \end{pmatrix}, \psi_2 = \begin{pmatrix} v_1 \\ -i\tilde{\alpha}_1^{-1} v_1 \\ i\tilde{\alpha}_1^{-1} u_1 \\ u_1 \end{pmatrix}, \psi_3 = \begin{pmatrix} u_2 \\ i\alpha_2^{-1} u_2 \\ i\gamma\alpha_2^{-1} v_2 \\ -\gamma v_2 \end{pmatrix}, \psi_4 = \begin{pmatrix} v_2 \\ i\tilde{\alpha}_2^{-1} v_2 \\ i\gamma\tilde{\alpha}_2^{-1} u_2 \\ -\gamma u_2 \end{pmatrix}$$

with $\gamma = \mathrm{sgn}(\Delta_p - \Delta_s)$. In the above,

$$u_j = \sqrt{\left(E + \sqrt{E^2 - \Delta_j^2}\right)/2E}, \quad v_j = \sqrt{\left(E - \sqrt{E^2 - \Delta_j^2}\right)/2E}, \tag{11.19}$$

and $\alpha_1 = (k_{1x}^+ - ik_y)/k_1, \alpha_2 = (k_{2x}^+ - ik_y)/k_2, \tilde{\alpha}_1 = (-k_{1x}^- - ik_y)/k_1$, and $\tilde{\alpha}_2 = (-k_{2x}^- - ik_y)/k_2$. Finally, E is the quasiparticle energy measured from the Fermi energy. By postulating $\Psi_S(x) = 0$ at $x = 0$, we can determine the ABS.

The bound-state condition can be expressed by

$$\sqrt{(\Delta_1^2 - E^2)(\Delta_2^2 - E^2)} = \frac{1-\zeta}{1+\zeta}(E^2 + \gamma\Delta_1\Delta_2), \tag{11.20}$$

$$\zeta = \begin{cases} \frac{\sin^2[\frac{1}{2}(\theta_1+\theta_2)]}{\cos^2[\frac{1}{2}(\theta_1-\theta_2)]} & \text{for } \mid \theta_2 \mid \le \theta c \\ 1 & \text{for } \theta_c < \mid \theta_2 \mid \le \pi/2, \end{cases} \tag{11.21}$$

with $\zeta \le 1, \cos\theta_1 = k_{1x}/k_1$ and $\cos\theta_2 = k_{2x}/k_2$. The critical angle θ_c is defined as $\arcsin(k_1/k_2)$. For $\lambda = 0$, Eq. (11.20) reproduces the previous results [56]. As seen from Eq. (11.20), a zero energy ABS is only possible for $\mid \theta_2 \mid \le \theta_c$ and $\gamma = 1$, i.e. $\Delta_p > \Delta_s$. This ABS corresponds to a state in which a localized quasiparticle can move along the edge. The energy level of this edge state depends crucially on the direction of the motion of the quasiparticle. The inner gap edge modes are absent for large magnitude of k_y, i.e. large θ_2. In this case, k_{1x} becomes a purely imaginary number due to the conservation of the Fermi momentum component parallel to the surface. The parameter regime where the edge modes survive is reduced with increasing λ. However, as far as we concentrate on normal injection, the edge modes survive as midgap ABS [2, 5] irrespective of the strength of λ. If we focus on the low-energy limit, the ABS energy can be written as

$$E = \pm\Delta_p\left(1 - \frac{\Delta_s^2}{\Delta_p^2}\right)\frac{k_1 + k_2}{2k_1k_2}k_y, \tag{11.22}$$

with $\Delta_s < \Delta_p$ for any λ with small magnitude of k_y. For $\Delta_s \ge \Delta_p$, the ABS vanishes since the value of right hand side of Eq. (11.20) becomes negative due to the negative sign of γ for $\mid E \mid < \Delta_1$ and $\mid E \mid < \Delta_2$.

It should be remarked that the ABS under consideration does not break time reversal symmetry, since the edge currents carried by the two partners of the Kramers doublet flow in opposite directions. Thus they can be regarded as helical edge modes, with the two modes related to each other by a time reversal operation.

11.2.3.2 Charge and Spin Conductance

Now we turn to transport properties governed by the ABS in NCS superconductors [58–61]. First, we point out that the spin Hall effect, i.e., the appearance of the spin Hall voltage perpendicular to the superconducting current, is suppressed by the compressive nature of the superconducting state by the factor of $(k_F\lambda_m)^{-2}$ (k_F: Fermi momentum, λ_m: penetration depth) [45, 46]. Instead, we will show below that spin transport through the junction between a ballistic normal metal at $x < 0$ and a NCS superconductor, i.e., through a N/NCS junction, can be enhanced by the Doppler effect during Andreev reflection. The Hamiltonian $\hat{\mathscr{H}}_{\rm N}$ of N is given by putting $\Delta(\boldsymbol{k}) = 0$ and $\lambda = 0$ in $\hat{\mathscr{H}}_{\rm S}$. We assume an insulating barrier at $x = 0$, expressed by a delta-function potential $U\delta(x)$.

The quantities of interest are the angle-resolved spin conductance $f_{\rm S}(\theta)$ and charge conductance $f_{\rm C}(\theta)$ defined by [62]

$$f_{\rm S}(\theta) = \frac{1}{2}\Big[\sum_{\sigma,\rho} s_\rho(\mid a_{\sigma,\rho}\mid^2 - \mid b_{\sigma,\rho}\mid^2)\Big]\cos\theta, \tag{11.23}$$

$$f_{\rm C}(\theta) = \Big[1 + \frac{1}{2}\sum_{\sigma,\rho}(\mid a_{\sigma,\rho}\mid^2 - \mid b_{\sigma,\rho}\mid^2)\Big]\cos\theta, \tag{11.24}$$

where $s_\rho = +(-)1$ for $\rho = \uparrow(\downarrow)$, and θ denotes the injection angle measured from the normal to the interface. Here, $b_{\sigma,\rho}$ and $a_{\sigma,\rho}$ with $\sigma, \rho \in \{\uparrow, \downarrow\}$ are spin-dependent reflection and Andreev reflection coefficients, respectively. These coefficients are determined as follows. The wave function for spin σ in the normal metal $\Psi_{\rm N}(x)$ is given by

$$\Psi_{\rm N}(x) = \exp(ik_{{\rm F}y}y)[(\psi_{i\sigma} + \sum_{\rho=\uparrow,\downarrow} a_{\sigma,\rho}\psi_{a\rho})\exp(ik_{{\rm F}x}x) + \sum_{\rho=\uparrow,\downarrow} b_{\sigma,\rho}\psi_{b\rho}\exp(-ik_{{\rm F}x}x)]$$

with ${}^T\psi_{i\uparrow} = {}^T\psi_{b\uparrow} = (1,0,0,0)$, ${}^T\psi_{i\downarrow} = {}^T\psi_{b\downarrow} = (0,1,0,0)$, ${}^T\psi_{a\uparrow} = (0,0,1,0)$, and ${}^T\psi_{a\downarrow} = (0,0,0,1)$. The corresponding $\Psi_{\rm S}(x)$ is given by Eq. (11.17). The coefficients $a_{\sigma,\rho}$ and $b_{\sigma,\rho}$ are determined by postulating the boundary condition $\Psi_{\rm N}(0) = \Psi_{\rm S}(0)$, and

$$\hbar\hat{v}_{{\rm S}x}\Psi_{\rm S}(0) - \hbar\hat{v}_{{\rm N}x}\Psi_{\rm N}(0) = -2iU\hat{\tau}_3\Psi_{\rm S}(0)$$

with $\hbar\hat{v}_{{\rm S(N)}x} = \partial\hat{H}_{\rm S(N)}/\partial k_x$, and the diagonal matrix $\hat{\tau}_3$ given by $\hat{\tau}_3 = {\rm diag}(1, 1, -1, -1)$.

The resulting angle-averaged charge conductance (Andreev conductance) is given by

$$\sigma_{\rm C} \equiv \sigma_{\rm T} = \left(\int_{-\pi/2}^{\pi/2} f_{\rm C}(\theta)d\theta\right) / \left(\int_{-\pi/2}^{\pi/2} f_{\rm NC}(\theta)d\theta\right). \tag{11.25}$$

We plot in Fig. 11.5 the charge conductance for various ratios Δ_s/Δ_p in the presence of a splitting of the Fermi surface [54]. For $\Delta_s < \Delta_p$, $\sigma_{\rm T}(eV)$ has a zero-bias conductance peak (ZBCP) due to the presence of the helical edge modes (curves a and b in Fig. 11.5). For $\Delta_s = \Delta_p$, due to the closing of the bulk energy gap, the resulting $\sigma_{\rm T}(eV)$ is almost constant. For $\Delta_s > \Delta_p$, $\sigma_{\rm T}(eV)$ has a gap like structure similar to a spin-singlet s-wave superconductor.

Next, we focus on the spin conductance. First we consider a pure spin-triplet $(p_x \pm ip_y)$-wave state. In Fig. 11.6, the angle-resolved spin conductance is plotted as

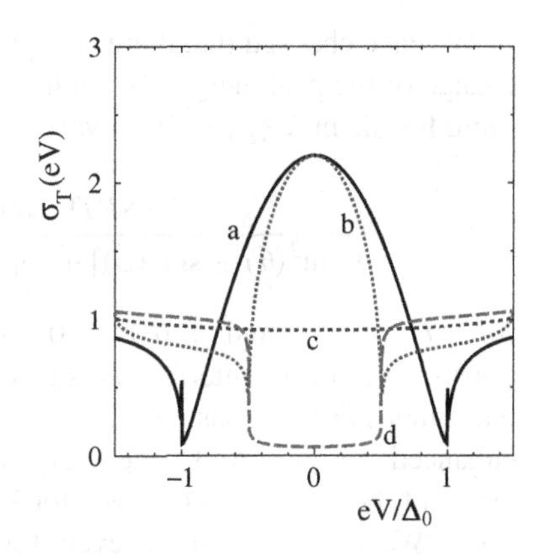

Fig. 11.5 The Andreev conductance σ_T for an NCS superconductor with $2m\lambda/k_F\hbar^2 = 0.1$ and $Z = 5$. The *curves* correspond to various ratios of Δ_s/Δ_p, with a $\Delta_s = 0$, b $\Delta_s = 0.5\Delta_p$, c $\Delta_s = \Delta_p$ and *d* $\Delta_s = 1.5\Delta_p$

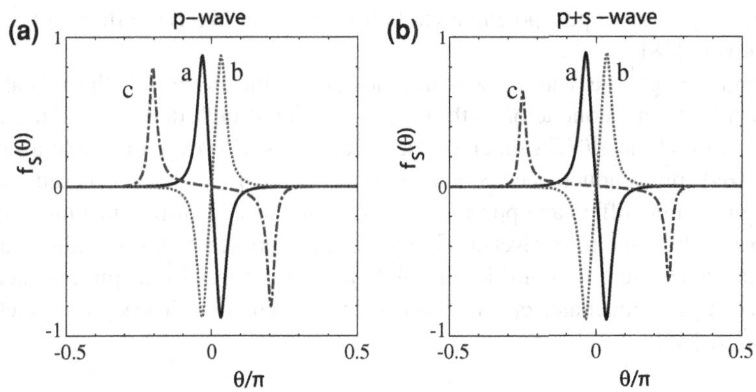

Fig. 11.6 Angle-resolved spin conductance for an NCS superconductor for $Z = 5$. **a** pure $(p_x \pm ip_y)$-wave case with $\Delta_s = 0$; **b** $\Delta_s = 0.3\Delta_p$. The *curves* are for $\lambda k_F = 0.1\mu$ and various voltages, a $eV = 0.1\Delta_p$, b $eV = -0.1\Delta_p$ and c $eV = 0.6\Delta_p$. From Fig. 11.2 of Ref. [54]

a function of injection angle θ and bias voltage V with $E = eV$. Note here that the k_y is related to θ as $k_y = k_F \sin\theta$. It is remarkable that the spin conductance has a non-zero value although the NCS superconductor does not break time reversal symmetry. The quantity $f_S(\theta)$ has a peak whenever the angle θ or the momentum component k_y corresponds to an Andreev bound-state energy E in the energy dispersion. With this condition, the spin-dependent Andreev reflection results in a spin current. Besides this property, we can show that $f_S(\theta) = -f_S(-\theta)$ is satisfied. By changing the sign of eV, $f_S(\theta)$ changes sign as seen in Fig. 11.6(a). Next, we look at the case where an *s*-wave component coexists. We calculate the spin conductance similar to that for the pure $(p_x \pm ip_y)$-wave case. For $\Delta_s < \Delta_p$, where helical edge modes exist, $f_S(\theta)$ shows a sharp peak and $f_S(\theta) = -f_S(-\theta)$ is satisfied [see Fig. 11.6(b)]. These features are similar to those of the pure $(p_x \pm ip_y)$-wave case. On the other hand, for $\Delta_s > \Delta_p$, where the helical edge modes are absent, sharp peaks of $f_S(\theta)$ as shown in Fig. 11.6 are absent.

We have checked that there is negligible quantitative change, i.e., less than 0.5% change of the peak height, by taking the $\lambda = 0$ limit compared to Fig. 11.6. In this limit, for the pure $(p_x \pm ip_y)$-wave state, $f_S(\theta)$ is given as follows

$$\frac{-8RD^2 \sin(2\theta)\sin(2\varphi)\cos(\theta)}{\mid 4[\sin^2(\theta) - \sin^2(\varphi)] + D[2\cos(2\theta) - (1+R)\exp(-2i\varphi)] \mid^2}$$

for $\mid E \mid < \Delta_p$ and $f_S(\theta) = 0$ for $\mid E \mid > \Delta_p$ with $\sin\varphi = E/\Delta_p$. The transparency of the interface D is given as before by $4\cos^2\theta/(4\cos^2\theta + Z^2)$, with the dimensionless constant $Z = 2mU/\hbar^2 k_F$. The magnitude of $f_S(\theta)$ is largely enhanced at $E = \pm\Delta_p \sin\theta$ corresponding to the energy dispersion of the ABS. The origin of the nonzero $f_S(\theta)$ even for $\lambda = 0$ is due to spin-dependent Andreev bound states. We have checked that even if we take into account the spatial dependence of the $(p_x \pm ip_y)$-wave pair potential explicitly, the resulting $f_S(\theta)$ does not change qualitatively [58].

Summarizing these features, we can conclude that the presence of the helical edge modes in NCS superconductors is the origin of the large angle dependent spin current through normal-metal/NCS superconductor junctions. However, the angle-averaged normalized spin conductance becomes zero since $f_S(\theta) = -f_S(-\theta)$ is satisfied.

Magnetic field offers an opportunity to observe the spin current in a more accessible way, where the time reversal (T) symmetry is broken by the shielding current at the interface. Here we consider the angle-averaged normalized spin conductance σ_S and charge conductance σ_C as a function of magnetic field. The spin conductance is given by [62]

$$\sigma_S = \left(\int_{-\pi/2}^{\pi/2} f_S(\theta)\, d\theta\right) / \left(\int_{-\pi/2}^{\pi/2} f_{NC}(\theta)\, d\theta\right), \tag{11.26}$$

where $f_{NC}(\theta)$ denotes the angle-resolved charge conductance in the normal state with $\Delta_p = \Delta_s = 0$. We consider a magnetic field H applied perpendicular to the two-dimensional plane, which induces a shielding current along the normal-metal/NCS superconductor interface. When the penetration depth of the NCS superconductor is much longer than the coherence length, the vector potential can be approximated as described in Sect. 11.2. As in the case of a chiral p-wave superconductor, the applied magnetic field shifts the quasiparticle energy E to $E + H\Delta_p \sin\theta/H_0$. For typical values of $\xi \sim 10$ nm, $\lambda_m \sim 100$ nm, the magnitude of H_0 is of the order of 0.2 Tesla. The order of magnitude of the Doppler shift is given by $H\Delta_p/H_0$. Since the Zeeman energy is given by $\mu_B H$, the energy shift due to the Doppler effect is by a factor $k_F\lambda_m$ larger than that due to the Zeeman effect. Thus, we can neglect the Zeeman effect in the present analysis. This is in sharp contrast to quantum spin Hall systems where the Zeeman effect is the main effect of a magnetic field, which opens a gap in the helical edge modes and modulates the transport properties [63]. The enhanced spin current due to Doppler shifts is specific to the superconducting state, and is not realized in quantum spin Hall systems.

11.2.3.3 Topological Aspects

We now focus on the topological aspect of non-centrosymmetric superconductors. Recently, the concept of the quantum Hall system has been generalized to time-reversal (T) symmetric systems, i.e., quantum spin Hall systems [64–68]. A quantum spin Hall system could be regarded as two copies of a quantum Hall system, for up and down spins, that are characterized by opposite chiralities. In the generic case, however, a mixture of up and down spins occurs due to spin-orbit interaction, which necessitates a new topological number to characterize a quantum spin Hall system [64, 65, 67, 68]. In quantum spin Hall systems, there exist helical edge modes, i.e., time-reversed partners of right- and left-going one-dimensional modes. This has been experimentally demonstrated for the quantum well of the HgTe system by measurements of the charge conductance [63].

As shown in Fig. 11.6, to discuss the topological nature of the helical edge modes, it is sufficient to consider the pure $(p_x \pm ip_y)$-wave state. Here, we give an argument from the viewpoint of the Z_2 (topological) class [64, 65], why the superconducting state with $\Delta_p > \Delta_s$ has an Andreev bound state. We commence with a pure $(p_x \pm ip_y)$-wave state without spin-orbit interaction, i.e. $\lambda = 0$. The spin Chern number [67, 68] for the corresponding Bogoliubov-de Gennes Hamiltonian is 2. Turning on λ adiabatically leaves the time reversal T-symmetry intact and keeps the gap open. Upon this adiabatic change of λ, the number of the helical edge mode pairs does not change. The reason is that this number is a topological number and consequently can only change by integer values. We now increase the magnitude of Δ_s from zero. As far as $\Delta_p > \Delta_s$ is satisfied, the number of helical edge modes does not change. However, if Δ_s exceeds Δ_p, the helical mode disappears. In this regime, the topological nature of the superconducting state belongs to a pure s-wave state with $\lambda = 0$. It is remarkable that just at $\Delta_s = \Delta_p$ one of the two energy gaps for quasiparticles in the bulk closes. At precisely this point a quantum phase transition occurs.

In the following, we discuss the pure $(p_x \pm ip_y)$-wave case in more detail. In Fig. 11.7, the spin conductance σ_S and charge conductance σ_C normalized by the charge conductance in the normal state are plotted. It should be noted that σ_S becomes nonzero in the presence of a magnetic field H (see curves b, c and d), since $f_S(\theta)$ is no more an odd function of θ due to the imbalance of the helical edge modes. For $\lambda = 0$, the corresponding helical edge modes are given by $E = \Delta_p(1 - H/H_0)\sin\theta$ and $E = -\Delta_p(1 + H/H_0)\sin\theta$. As seen from the curves b and c, the sign of σ_S is reversed when changing the direction of the applied magnetic field. On the other hand, the corresponding charge conductance has different features. For $H = 0$, the resulting line shape of σ_C is the same as that for a chiral p-wave superconductor (see curve a of right panel) [56, 57, 59]. As seen from curves b and c in the right panel, σ_C does not change with the direction of the magnetic field.

In summary, we have clarified the charge and spin transport properties of non-centrosymmetric superconductors from the viewpoint of topology and Andreev bound states. We have found spin-polarized current flowing through the interface that depends on the incident angle. When a weak magnetic field is applied, even the

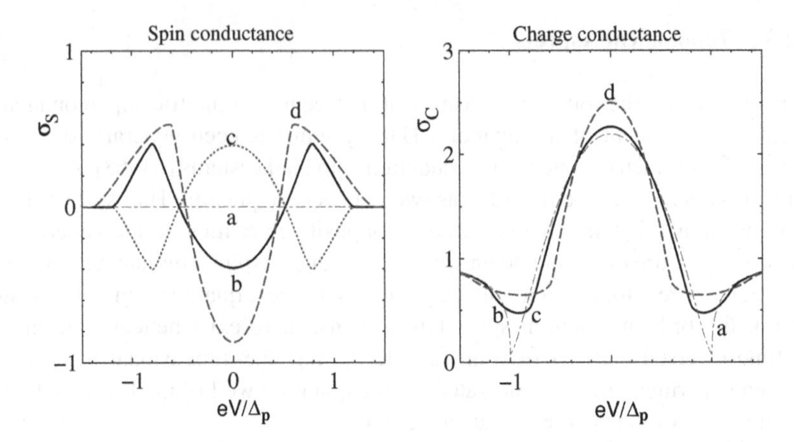

Fig. 11.7 Angle-averaged spin conductance and charge conductance as a function of eV for $\lambda k_F = 0.1\mu$. The various magnetic field values are a $H = 0$, b $H = -0.2H_0$, c $H = 0.2H_0$, and d $H = -0.4H_0$. Curves b and c of the right panel are identical. From Fig. 11.2 of Ref. [54]

angle-integrated current is largely spin polarized. In analogy to quantum spin Hall systems, the Andreev bound states in non-centrosymmetric superconductors correspond to helical edge modes. Andreev reflection via helical edge modes produces the enhanced spin current specific to non-centrosymmetric superconductors.

11.3 Quasiclassical Theory of Superconductivity for Non-centrosymmetric Superconductors

11.3.1 Quasiparticle Propagator

Electronic quasiparticles in normal Landau Fermi liquids are restricted in phase space to a region that comprises only a small part of the entire electronic phase space [69, 70]. It consists of a narrow (compared to the Fermi momentum $\boldsymbol{p}_F$) shell around the Fermi surface, and a small (compared to the Fermi energy E_F) region around the chemical potential. Quasiparticles are characterized by their spin and charge, and their group velocity is the Fermi velocity, $\boldsymbol{v}_F(\boldsymbol{p}_F)$. Quasiclassical theory is the appropriate framework to describe such a system. It consists of a *systematic* classification of all interaction processes according to their relevance, i.e. their smallness with respect to an expansion parameter SMALL [71–75]. This expansion parameter assumes the existence of a well defined scale separation between a *low-energy scale* and a *high-energy scale*.

Superconducting phenomena are governed by the low-energy scale. That means that the energy scales determined by the energy gap Δ and the transition temperature T_c are small. In contrast the energy scales determined by the Fermi energy E_F or the Coulomb repulsion U_C are large energies. Disorder can be described within the quasiclassical approximation as long as the energy associated with the scattering rate, $\hbar/\tau$, is classified as a small energy. A systematic classification shows that a consistent treatment of disorder requires the t-matrix approximation. Localization effects due to disorder are beyond the leading-order precision of quasiclassical theory. Associated with the energy scales are small and large length scales. For example the superconducting coherence length $\xi_0 = \hbar v_F / 2\pi k_B T_c$, and the elastic mean free path $\ell = v_F \tau$ are large compared to the lattice constant a and the Fermi wave length $\lambda_F = \hbar / p_F$.

This separation in energy and length scales is associated with a low-energy region in phase space, that includes low quasiparticle energies $\sim\Delta$, $k_B T$, and a momentum shell around the quasiparticle Fermi momentum $\boldsymbol{p}_F$ of extent $\delta p \sim \Delta/|\boldsymbol{v}_F(\boldsymbol{p}_F)|$. The phase-space volume of this low-energy region, divided by the entire phase-space volume, is employed for a systematic diagrammatic expansion of a Dyson series within a path-ordered Green's function technique (e.g. Matsubara technique for the Matsubara path, Keldysh-Nambu-Gor'kov technique for the Schwinger-Keldysh path). Within the framework of Green's function technique, all diagrams in a Feynman diagrammatic expansion can be classified according to their order in this expansion parameter, which is denoted as SMALL. The leading-order theory in this expansion parameter is called the "Quasiclassical Theory of Metals and Superconductors" [71–78].

The possibility to define a quasiparticle Fermi surface around which all quasiparticle excitations reside is a requirement for the quasiclassical theory to work. Its presence ensures that the Pauli principle is still effective in placing stringent kinetic restrictions on the possible scattering events. It is essential to note that such a definition need not be sharp, i.e. the theory is not restricted to normal Fermi liquids with a jump in the momentum distribution at zero temperature. Thus, the theory includes superconducting phenomena as well as strong-coupling metals. It is convenient to introduce a local coordinate system at each momentum point of the Fermi surface $\boldsymbol{p}_F$, with a variation along the surface normal, i.e. in direction of the Fermi velocity $\boldsymbol{v}_F(\boldsymbol{p}_F)$, that is determined by a variable ξ_p (this variable is zero at the Fermi momentum), and a tangential variation along the Fermi surface at constant ξ_p. A consistent approximation requires to consider the Fermi velocity constant across the low-energy momentum shell, and thus the local coordinate system stays an orthogonal system as long as ξ_p varies within this momentum shell, and furthermore, ξ_p stays small within this momentum shell. The coordinate ξ_p around each Fermi surface point $\boldsymbol{p}_F$ varies then approximately as $\xi_p \approx \boldsymbol{v}_F(\boldsymbol{p} - \boldsymbol{p}_F)$.

The quasiclassical theory is obtained by defining quasiparticle propagators for the low-energy regions of the phase space, and in combining all diagrams involving Green's functions with their variables residing in the high-energy regions into new effective *high-energy interaction vertices*. This process of integrating out high-energy degrees of freedom is highly non-trivial and must be solved by microscopic theories. In the spirit of Fermi-liquid theory it is, however, possible to regard all high-energy interaction vertices as phenomenological parameters of the theory. In the quasiclassical approximation they do not depend on any low-energy variables as temperature or superconducting gap, and they do not vary as function of ξ_p as long as ξ_p stays within the momentum shell that harbors the quasiparticle excitations. However they do depend in general on the position of the Fermi momentum on the Fermi surface.

In addition to introducing new effective interaction vertices the above procedure also introduces a quasiparticle renormalization factor $a^2(\boldsymbol{p}) \sim 1/Z(\boldsymbol{p})$ that is due to the self energies of the low-energy quasiparticles moving in the background of the high energy electrons. This renormalization leads to a modification of the quasiparticle Fermi velocity compared to the bare Fermi velocity of the system, and to a deformation of the quasiparticle Fermi surface compared to the bare Fermi surface. It also determines quasiparticle weight as the residua of the quasiparticle poles in the complex energy plane.

One has to keep these remarks in consideration when including additional interaction, like spin-orbit interaction or exchange interaction, in a quasiclassical theory. First, it is important to decide if this interaction is going to be treated among the low-energy terms or among the high-energy terms. Depending on this issue, one obtains two different quasiclassical theories that cannot in general adiabatically be connected with each other. Going from one limit to the other includes the un-dressing of all effective interaction vertices and of the quasiparticles, and re-dressing with new types of effective interaction vertices and self energies. Importantly, this dressing leads to strongly spin-dependent effective interactions and quasiparticle renormalizations in one limit, and to leading order spin-symmetric interactions and quasiparticle renormalizations in the other limit. The former case, when spin-dependent interactions are included in the high-energy scale, leads to a complete reorganization of the Fermi surface geometry, with in general new spin-dependent quasiparticle energy bands. In this case, it is not sensible anymore to keep the spin as a good quantum number, but it is necessary to deal directly with the representation that diagonalizes the energy bands including the spin-dependent interaction. In the case of a strong exchange energy this leads to exchange-split energy bands, and in the case of strong spin-orbit interaction this leads to helicity bands.

The basic quantities in the theory are the quasiparticle Fermi surface, the quasiparticle velocity, and quasiparticle interactions. Here we give a short sketch of how they enter the theory. The bare propagator (without inclusion of exchange interaction or spin-orbit coupling) in the quasiparticle region of the phase space has the general structure

$$G^{(0)}_{\alpha\beta}(\boldsymbol{p}, \varepsilon) = \frac{\delta_{\alpha\beta}}{\varepsilon - \xi^{(0)}(\boldsymbol{p})} \tag{11.27}$$

where $\xi^{(0)}(\boldsymbol{p})$ is the bare energy dispersion of the energy band (measured from the electrochemical potential of the electrons). It does not include electron-electron interaction effects yet, and thus determines a *bare* Fermi surface that does not coincide with the quasiparticle Fermi surface defined below. The quantum number α labels the spin. The leading-order self energy is solely due to coupling of low-energy electrons (superscript L) to high-energy electrons (superscript H), and consequently the corresponding self energy, $\Sigma^{(\mathrm{H})}$, must be classified as a pure high-energy quantity.

In general, when either exchange interaction or spin-orbit coupling are large energy scales, this self-energy contribution will be spin-dependent (and will ultimately lead to new, spin-split energy bands as explained below). The self energy $\Sigma^{(\mathrm{H})}$ is, however, slowly varying in energy on the low-energy scale and thus can be expanded around the chemical potential,

$$\Sigma_{\alpha\beta}^{(\mathrm{H})}(\boldsymbol{p},\varepsilon)=\Sigma_{\alpha\beta}^{(\mathrm{H})}(\boldsymbol{p},0)+\varepsilon\partial_{\varepsilon}\Sigma_{\alpha\beta}^{(\mathrm{H})}(\boldsymbol{p},\varepsilon)\Big|_{\varepsilon=0}+\mathscr{O}(\varepsilon^{2}). \tag{11.28}$$

The second term can be combined with energy ε into a renormalization function

$$\varepsilon-\varepsilon\partial_{\varepsilon}\Sigma_{\alpha\beta}^{(\mathrm{H})}(\boldsymbol{p},0)=Z_{\alpha\beta}^{(\mathrm{H})}(\boldsymbol{p})\varepsilon, \tag{11.29}$$

such that to leading order the Dyson equation for the low-energy propagator $G_{\beta\gamma}^{(\mathrm{L})}$ reads

$$\left\{Z_{\alpha\beta}^{(\mathrm{H})}(\boldsymbol{p})\varepsilon-[\xi^{(0)}(\boldsymbol{p})\delta_{\alpha\beta}+\Sigma_{\alpha\beta}^{(\mathrm{H})}(\boldsymbol{p},0]-\Sigma_{\alpha\beta}^{(\mathrm{L})}(\boldsymbol{p},\varepsilon)\right\}\otimes G_{\beta\gamma}^{(\mathrm{L})}(\boldsymbol{p},\varepsilon)=\delta_{\alpha\gamma}, \tag{11.30}$$

where $\Sigma^{(\mathrm{L})}$ includes all self-energy terms of order SMALL . Here, and in the following, summation over repeated indices is implied. The $\otimes$ sign accounts for possible spatial or temporal inhomogeneities, where it has the form of a convolution product in Wigner representation (see Ref. [71] for details). The equation holds in this form either in Matsubara or in Keldysh representation (in which case all quantities are 2×2 matrices in Keldysh space [88, 89]). Low-energy excitations reside in momentum regions differing considerably from that for the bare propagators Eq. (11.27).

The quantities $Z_{\alpha\beta}^{(\mathrm{H})}(\boldsymbol{p})$ and $\Sigma_{\alpha\beta}^{(\mathrm{H})}(\boldsymbol{p},0)$ can be defined such that they have real eigenvalues. The next step is to eliminate the high-energy renormalization factor $Z_{\alpha\beta}^{(\mathrm{H})}(\boldsymbol{p})$ from the low-energy theory. This is done with the help of quasiparticle weight factors $a_{\alpha\beta}(\boldsymbol{p})$, that are the solution of

$$a_{\alpha\gamma}(\boldsymbol{p})Z_{\gamma\gamma'}^{(\mathrm{H})}(\boldsymbol{p})a_{\gamma'\beta}(\boldsymbol{p})=\delta_{\alpha\beta}. \tag{11.31}$$

They exist as long as the matrix $Z^{(\mathrm{H})}(\boldsymbol{p})$ has non-zero eigenvalues. Then we can define the *quasiparticle Green's function* $G_{\alpha\beta}^{(\mathrm{QP})}$ as the solution of

$$a_{\alpha\gamma}(\boldsymbol{p})G_{\gamma\gamma'}^{(\mathrm{QP})}a_{\gamma'\beta}(\boldsymbol{p})=G_{\alpha\beta}^{(\mathrm{L})}(\boldsymbol{p},\varepsilon), \tag{11.32}$$

which exists under the condition that the matrix $a(\boldsymbol{p})$ has non-zero eigenvalues (i.e. the quasiparticle weights are non-zero; otherwise the quasiparticle approximation breaks down). It fulfills the Dyson equation

$$\left[\varepsilon-\xi^{(\mathrm{QP})}(\boldsymbol{p})-\Sigma^{(\mathrm{QP})}(\boldsymbol{p},\varepsilon)\right]_{\alpha\beta}\otimes G_{\beta\gamma}^{(\mathrm{QP})}(\boldsymbol{p},\varepsilon)=\delta_{\alpha\gamma} \tag{11.33}$$

with the quasiparticle dispersion

$$\xi^{(QP)}_{\alpha\beta}(\boldsymbol{p}) = a_{\alpha\gamma}(\boldsymbol{p})\left(\xi^{(0)}(\boldsymbol{p})\delta_{\gamma\gamma'} + \Sigma^{(H)}_{\gamma\gamma'}(\boldsymbol{p},0)\right)a_{\gamma'\beta}(\boldsymbol{p}), \tag{11.34}$$

and the quasiparticle self energies

$$\Sigma^{(QP)}_{\alpha\beta}(\boldsymbol{p},\varepsilon) = a_{\alpha\gamma}(\boldsymbol{p})\Sigma^{(L)}_{\gamma\gamma'}(\boldsymbol{p},\varepsilon)a_{\gamma'\beta}(\boldsymbol{p}). \tag{11.35}$$

The effective (renormalized by high-energy processes) interactions vertices for the low-energy propagators, $G^{(L)}_{\alpha\beta}$, which enter the diagrammatic expressions for the quasiparticle self energy, have the general structure $V_{\beta_1\ldots\beta_n}(\varepsilon_1,\boldsymbol{p}_1;\ldots;\varepsilon_n,\boldsymbol{p}_n)$. In leading order the energy dependence of these vertices can be neglected near the chemical potential, i.e. the arguments can be restricted to the chemical potential. Furthermore, instead of working with $G^{(L)}_{\alpha\beta}$ and $V_{\beta_1\ldots\beta_n}$ the common and completely equivalent description in terms of the quasiparticle propagators, $G^{(QP)}_{\alpha\beta}$ defined above, and renormalized quasiparticle interactions, $V^{(QP)}_{\beta_1\ldots\beta_n}$, given by

$$V^{(QP)}_{\beta_1\ldots\beta_n}(\boldsymbol{p}_1\ldots\boldsymbol{p}_n) = a_{\beta_1\beta_1'}(\boldsymbol{p}_1)\ldots a_{\beta_n\beta_n'}(\boldsymbol{p}_n)V_{\beta_1'\ldots\beta_n'}(0,\boldsymbol{p}_1;\ldots;0,\boldsymbol{p}_n) \tag{11.36}$$

can be used.

It is important to note that the quasiparticle self energies can be written as functionals of the quasiparticle Green's functions only in leading order in the expansion in SMALL, which is the order relevant for the quasiclassical approximation. In this case, the quasiparticle weights have disappeared from the theory and cannot in principle be determined from low-energy processes that only involve quasiparticle dynamics. They must be obtained from a microscopic theory by considering high energy scattering processes, which is beyond the quasiclassical approximation.

It is obvious, that the appearance of the quasiparticle renormalization factors renders all self energies and interactions non-diagonal in spin unless spin-dependent interactions are small enough the be omitted from the high-energy quantities. From the above expressions one obtains the quasiparticle Fermi surfaces by diagonalizing the quasiparticle dispersion (here we make the summation explicit)

$$\sum_\alpha U_{\lambda\alpha}(\boldsymbol{p})\xi^{(QP)}_{\alpha\beta}(\boldsymbol{p}) = \xi^{(QP)}_\lambda(\boldsymbol{p})U_{\lambda\beta}(\boldsymbol{p}) \tag{11.37}$$

with band index λ, and solving the equation

$$\xi^{(QP)}_\lambda(\boldsymbol{p}) = 0 \rightarrow \boldsymbol{p} = \boldsymbol{p}^\lambda_F. \tag{11.38}$$

The corresponding quasiparticle Fermi velocity is then given by

$$\boldsymbol{v}^\lambda_F = \frac{\partial}{\partial \boldsymbol{p}}\xi^{(QP)}_\lambda(\boldsymbol{p})\Big|_{\boldsymbol{p}=\boldsymbol{p}^\lambda_F}. \tag{11.39}$$

In the band-diagonal frame, the quasiparticle propagator is given by

$$\left\{[\varepsilon - \xi_\lambda^{(QP)}(\boldsymbol{p})]\delta_{\lambda\lambda_1} - \Sigma_{\lambda\lambda_1}^{(QP)}(\boldsymbol{p}, \varepsilon)\right\} \otimes G_{\lambda_1\lambda'}^{(QP)}(\boldsymbol{p}, \varepsilon) = \delta_{\lambda\lambda'}, \tag{11.40}$$

where the self energy (and all interactions in the self-energy expressions) must be transformed accordingly, e.g.

$$\Sigma_{\lambda\lambda'}^{(QP)}(\boldsymbol{p}, \varepsilon) = U_{\lambda\alpha}(\boldsymbol{p})\Sigma_{\alpha\beta}^{(QP)}(\boldsymbol{p}, \varepsilon)U_{\lambda'\beta}(\boldsymbol{p})^*. \tag{11.41}$$

In the next section this procedure is carried out for the case of strong spin-orbit interaction, e.g. appropriate for some non-centrosymmetric materials.

11.3.2 Spin-Orbit Interaction and Helicity Representation

As discussed in Chap. 4 by V.P. Mineev and M. Sigrist, for treating a non-centrosymmetric material it is convenient to perform a canonical transformation from a spin basis with fermion annihilation operators $a_{\boldsymbol{k}\alpha}$ for spin $\alpha = \uparrow, \downarrow$ to the so-called helicity basis with fermion annihilation operators $c_{\boldsymbol{k}\lambda}$ for helicity $\lambda = \pm$. This canonical transformation diagonalizes the kinetic part of the Hamiltonian,

$$\mathscr{H}_{kin} = \sum_{\boldsymbol{k}} \sum_{\alpha\beta=\uparrow,\downarrow} \left[\xi(\boldsymbol{k}) + \boldsymbol{g}(\boldsymbol{k}) \cdot \boldsymbol{\sigma})_{\alpha\beta}\right] a_{\boldsymbol{k}\alpha}^\dagger a_{\boldsymbol{k}\beta} = \sum_{\boldsymbol{k}} \sum_{\lambda=\pm} \xi_\lambda(\boldsymbol{k}) c_{\boldsymbol{k}\lambda}^\dagger c_{\boldsymbol{k}\lambda}. \tag{11.42}$$

Here, $\xi(\boldsymbol{k})$ is the band dispersion relative to the chemical potential in the absence of spin-orbit interaction, $\boldsymbol{g}(\boldsymbol{k})$ is the spin-orbit pseudovector, which is odd in momentum, $\boldsymbol{g}(-\boldsymbol{k}) = -\boldsymbol{g}(\boldsymbol{k})$, and $\boldsymbol{\sigma}$ is the vector of Pauli matrices. The resulting helicity band dispersion is

$$\xi_\pm(\boldsymbol{k}) = \xi(\boldsymbol{k}) \pm |\boldsymbol{g}(\boldsymbol{k})|. \tag{11.43}$$

As is easily seen, spin-orbit interaction locks the orientation of the quasiparticle spin with respect to its momentum in each helicity band. The Hamiltonian, Eq. (11.42), is time-reversal invariant, however lifts the spin degeneracy.

It is convenient to introduce polar and azimuthal angles for the vector $\boldsymbol{g}$, defined by $\{g_x, g_y, g_z\} = |\boldsymbol{g}|\{\sin(\theta_g)\cos(\varphi_g), \sin(\theta_g)\sin(\varphi_g), \cos(\theta_g)\}$ (where $0 \le \theta_g \le \pi$). In terms of those, the transformation from spin to helicity basis, $U_{\boldsymbol{k}\lambda\alpha}$, is defined by [48]

$$U_{\boldsymbol{k}\lambda\alpha} = \begin{pmatrix} \cos(\theta_g/2) & \sin(\theta_g/2)\mathrm{e}^{-i\varphi_g} \\ -\sin(\theta_g/2)\mathrm{e}^{i\varphi_g} & \cos(\theta_g/2) \end{pmatrix}, \quad c_{\boldsymbol{k}\lambda} = \sum_\alpha U_{\boldsymbol{k}\lambda\alpha} a_{\boldsymbol{k}\alpha}. \tag{11.44}$$

Obviously, $\sum_{\alpha\beta} U_{\boldsymbol{k}\lambda\alpha}[\boldsymbol{g}(\boldsymbol{k}) \cdot \boldsymbol{\sigma}_{\alpha\beta}]U_{\boldsymbol{k}\lambda'\beta}^* = |\boldsymbol{g}(\boldsymbol{k})|\sigma_{\lambda\lambda'}^{(3)}$.

For the superconducting state the Nambu-Gor'kov formalism is appropriate [84–87]. The Nambu spinor, $\hat{A}_{\boldsymbol{k}} = (a_{\boldsymbol{k}\uparrow}, a_{\boldsymbol{k}\downarrow}, a^\dagger_{-\boldsymbol{k}\uparrow}, a^\dagger_{-\boldsymbol{k}\downarrow})^T$ transforms under the above canonical transformation into the helical object $\hat{C}_{\boldsymbol{k}} = (c_{\boldsymbol{k}+}, c_{\boldsymbol{k}-}, c^\dagger_{-\boldsymbol{k}+}, c^\dagger_{-\boldsymbol{k}-})^T$, where

$$\hat{C}_{\boldsymbol{k}} = \hat{U}_{\boldsymbol{k}} \hat{A}_{\boldsymbol{k}}, \quad \hat{U}_{\boldsymbol{k}} = \begin{pmatrix} U_{\boldsymbol{k}} & 0 \\ 0 & U^*_{-\boldsymbol{k}} \end{pmatrix}. \tag{11.45}$$

Correspondingly, one can construct 4×4 retarded Green's functions in spin basis,

$$\hat{G}^{(s)}_{\boldsymbol{k}_1\boldsymbol{k}_2}(t_1, t_2) = -i\theta(t_1 - t_2)\langle\{\hat{A}_{\boldsymbol{k}_1}(t_1), \hat{A}^\dagger_{\boldsymbol{k}_2}(t_2)\}\rangle_{\mathscr{H}}, \tag{11.46}$$

and in helicity basis,

$$\hat{G}_{\boldsymbol{k}_1\boldsymbol{k}_2}(t_1, t_2) = -i\theta(t_1 - t_2)\langle\{\hat{C}_{\boldsymbol{k}_1}(t_1), \hat{C}^\dagger_{\boldsymbol{k}_2}(t_2)\}\rangle_{\mathscr{H}}, = \hat{U}_{\boldsymbol{k}_1}\hat{G}^{(s)}_{\boldsymbol{k}_1\boldsymbol{k}_2}(t_1, t_2)\hat{U}^\dagger_{\boldsymbol{k}_2}, \tag{11.47}$$

where $\hat{A}(t)$ and $\hat{C}(t)$ are Heisenberg operators, the braces denote an anticommutator, $\langle\ldots\rangle_{\mathscr{H}}$ is a grand canonical average, and θ is the usual Heaviside step function. Analogously, advanced, Keldysh, and Matsubara propagators can be defined in helicity representation. For dealing with superconducting phenomena it is often convenient to introduce Wigner coordinates,

$$\hat{G}(\boldsymbol{k}, \boldsymbol{R}, \varepsilon, t) = \int (d\boldsymbol{q})(d\tau) e^{i(\boldsymbol{q}\boldsymbol{R}+\varepsilon\tau)} \hat{G}_{\boldsymbol{k}+\frac{\boldsymbol{q}}{2}, \boldsymbol{k}-\frac{\boldsymbol{q}}{2}}(t + \frac{\tau}{2}, t - \frac{\tau}{2}). \tag{11.48}$$

From here, one can proceed along different lines. Either, the Dyson equation for the full Gor'kov Green's functions is solved, which is equivalent to the Bogoliubov-de Gennes description in wave-function techniques. Or, the quasiclassical approximation is employed, which is equivalent to the Andreev approximation in wave-function language. In the following section we will adopt the second line.

11.3.3 Quasiclassical Propagator

In the following, the quasiclassical theory of superconductivity [71–83] will be employed to calculate electronic transport properties across interfaces with non-centrosymmetric superconductors. This method is based on the observation that, in most situations, the superconducting state varies on the length scale of the superconducting coherence length $\xi_0 = \hbar v_{\mathrm{F}}/2\pi k_{\mathrm{B}} T_{\mathrm{c}}$. The appropriate many-body Green's function for describing the superconducting state has been introduced by Gor'kov [84–87], and the Gor'kov Green's function can then be decomposed in a fast oscillating component, varying on the scale of $1/k_{\mathrm{F}}$, and an envelope function varying on the scale of ξ_0. The quasiclassical approximation consists of integrating out the fast oscillating component for each quasiparticle band separately:

$$\check{g}(\boldsymbol{p}_{\mathrm{F}}^{\lambda}, \boldsymbol{R}, \varepsilon, t) = \int d\xi_{\boldsymbol{p}}^{\lambda}\, \hat{\tau}_3\, \check{G}^{(\mathrm{QP})}(\boldsymbol{p}, \boldsymbol{R}, \varepsilon, t) \tag{11.49}$$

where a "check" denotes a matrix in Keldysh-Nambu-Gor'kov space,[88, 89] a "hat" denotes a matrix in Nambu-Gor'kov particle-hole space, $\xi_{\boldsymbol{p}}^{\lambda} = v_{\mathrm{F}}^{\lambda}(\boldsymbol{p} - \boldsymbol{p}_{\mathrm{F}}^{\lambda})$, and $\hat{\tau}_3$ is the third Pauli matrix in particle-hole space.

The quasiclassical Green's function obeys the transport equation [76, 77, 78]

$$i\hbar \boldsymbol{v}_{\mathrm{F}} \cdot \nabla_{\boldsymbol{R}} \check{g} + [\varepsilon \hat{\tau}_3 - \check{\Delta} - \check{h}, \check{g}]_{\circ} = \check{0}. \tag{11.50}$$

Here, ε is the quasiparticle energy, $\check{\Delta}$ is the superconducting order parameter and $\check{h}$ contains all other self-energies and external perturbations, related to external fields, impurities etc. The notation $\circ$ combines a time convolution with matrix multiplication, and $[\check{x}, \check{y}]_{\circ}$ denotes the commutator of $\check{x}$ and $\check{y}$ with respect to the $\circ$-product. Equation (11.50) must be supplemented by a normalization condition that must be obtained from an explicit calculation in the normal state [76, 77, 90],

$$\check{g} \circ \check{g} = -\check{1}\pi^2. \tag{11.51}$$

From the knowledge of $\check{g}$ one can calculate measurable quantities, e.g. the current density is related to the Keldysh component of the Green's function via

$$\boldsymbol{j}(\boldsymbol{R}, t) = qN_{\mathrm{F}} \int \frac{d\varepsilon}{8\pi i} \mathrm{Tr} \langle \boldsymbol{v}_{\mathrm{F}} \hat{\tau}_3 \hat{g}^{\mathrm{K}}(\boldsymbol{p}_{\mathrm{F}}^{\lambda}, \boldsymbol{R}, \varepsilon, t) \rangle, \tag{11.52}$$

where $q = -|e|$ is the electron charge, and $\langle \cdots \rangle$ denotes a Fermi surface average, which is defined by

$$\langle \cdots \rangle = \frac{1}{N_{\mathrm{F}}} \sum_{\lambda} \int \frac{d^3 p_{\mathrm{F}}^{\lambda}}{(2\pi\hbar)^3 |\boldsymbol{v}_{\mathrm{F}}^{\lambda}|} \cdots \quad N_{\mathrm{F}} = \sum_{\lambda} \int \frac{d^3 p_{\mathrm{F}}^{\lambda}}{(2\pi\hbar)^3 |\boldsymbol{v}_{\mathrm{F}}^{\lambda}|}, \tag{11.53}$$

and Tr denotes a trace over the Nambu-Gor'kov matrix.

11.3.3.1 Case of Weak Spin-Orbit Splitting

In the case of weak spin-orbit splitting the quasiclassical propagator can be obtained in either spin or helicity representation. It is possible then to define a common Fermi surface $\boldsymbol{p}_{\mathrm{F}}$ for both spin bands or, equivalently, both helicity bands. This case applies when $|\boldsymbol{g}(\boldsymbol{p}_{\mathrm{F}})| \ll E_{\mathrm{F}}$ for any Fermi momentum $\boldsymbol{p}_{\mathrm{F}}$, where E_{F} is the Fermi energy (in addition to the condition that the superconducting energy scales ($k_{\mathrm{B}} T_{\mathrm{c}}$ and the gap Δ are much smaller than E_{F}). Under these circumstances quasiparticles with different helicity but with the same $\hat{\boldsymbol{k}} \equiv \boldsymbol{k}/|\boldsymbol{k}|$ propagate coherently along a common classical trajectory over distances much longer than the Fermi wavelength. The transport equation is the usual Eilenberger equation modified by a spin-orbit interaction term [71, 91]

$$i\hbar \mathbf{v}_{\mathrm{F}} \cdot \nabla_{\mathbf{R}} \check{g} + [\varepsilon \hat{\tau}_3 - \check{\Delta} - \check{v}_{\mathrm{SO}}, \check{g}]_{\circ} = \check{0} \qquad (11.54)$$

with normalization $\hat{g} \circ \hat{g} = -\pi^2 \hat{1}$. Here, in helicity basis $\hat{v}_{SO} = |\mathbf{g}_{\mathbf{k}_{\mathrm{F}}}| \sigma^{(3)}$, and in spin basis $\hat{v}_{SO} = \mathbf{g}_{\mathbf{k}_{\mathrm{F}}} \cdot \hat{\boldsymbol{\sigma}} \hat{\tau}_3$, with

$$\hat{\boldsymbol{\sigma}} = \begin{pmatrix} \boldsymbol{\sigma} & 0 \\ 0 & \boldsymbol{\sigma}^* \end{pmatrix} = \begin{pmatrix} \boldsymbol{\sigma} & 0 \\ 0 & -\sigma^{(2)} \boldsymbol{\sigma} \sigma^{(2)} \end{pmatrix}. \qquad (11.55)$$

The velocity renormalization of order $|g|/E_{\mathrm{F}} \ll 1$ can safely be neglected. The quasiparticle trajectories are doubly degenerate in either spin or helicity space, and coherent mixing between spin states or between helicity states can take place.

11.3.3.2 Case of Strong Spin-Orbit Splitting

In the case of strong spin-orbit splitting the only possible representation for quasiclassical theory is the helicity representation. In this case, the spin-orbit interaction does not appear anymore as a source term in the transport equations, however does so explicitly as the presence of well-defined helicity bands. The transport equation takes the form

$$i\hbar \mathbf{v}_{\mathrm{F}}^{\lambda} \cdot \nabla_{\mathbf{R}} \check{g} + [\varepsilon \hat{\tau}_3 - \check{\Delta}^{\lambda}, \check{g}]_{\circ} = \check{0} \qquad (11.56)$$

with normalization condition $\hat{g} \circ \hat{g} = -\pi^2 \hat{1}$. Here, the velocity is strongly renormalized due to spin-orbit interaction. The quasiparticle trajectories are different for different helicity, and no coherence exists between the different helicity states. The matrix dimension can be reduced by a factor 2 compared to the case of weak spin-orbit splitting, and instead the number of Fermi surface sheets is increased by a factor of 2. If measurements are made that are spin-selective, the corresponding vector of Pauli spin matrices must be transformed according to

$$\boldsymbol{\sigma}_{\lambda\lambda'} = (U_{\mathbf{k}} \boldsymbol{\sigma} U_{\mathbf{k}}^{\dagger})_{\lambda\lambda'} = U_{\mathbf{k}\lambda\alpha} \boldsymbol{\sigma}_{\alpha\beta} U^*_{\mathbf{k}\lambda'\beta}. \qquad (11.57)$$

11.3.4 Riccati Parameterization

One of the main obstacles of the quasiclassical theory has been the non-linearity that is introduced by the normalization condition. A powerful way to deal with this problem is the choice of a parameter representation that ensures the normalization condition by definition. In this representation, the Keldysh quasiclassical Green's function is determined by six parameters in particle-hole space, $\gamma^{\mathrm{R,A}}, \tilde{\gamma}^{\mathrm{R,A}}, x^{\mathrm{K}}, \tilde{x}^{\mathrm{K}}$, where $\gamma^{\mathrm{R,A}}, \tilde{\gamma}^{\mathrm{R,A}}$ are the coherence functions, describing the coherence between particle-like and hole-like states, whereas $x^{\mathrm{K}}, \tilde{x}^{\mathrm{K}}$ are distribution functions, describing the occupation of quasiparticle states [27, 92]. The coherence functions are a generalization of the so-called Riccati amplitudes [92, 93, 94, 95, 96, 97, 98, 99] to non-equilibrium situations. All six parameters are matrix functions with the dimension determined by the degeneracy of the quasiparticle trajectories, and depend on Fermi momentum, position, energy, and time. The parameterization is simplified by the fact that, due to symmetry relations, only two functions of the six are independent. The particle-hole symmetry is expressed by the operation $\tilde{X}$ which is defined for any function X of the phase-space variables by

$$\tilde{Q}(\boldsymbol{p}_{\mathrm{F}}, \boldsymbol{R}, z, t) = Q(-\boldsymbol{p}_{\mathrm{F}}, R, -z^*, t)^*. \tag{11.58}$$

Here, $z = \varepsilon$ is real for the Keldysh components and z is situated in the upper (lower) complex energy half plane for retarded (advanced) quantities. Furthermore, the symmetry relations

$$\gamma^{\mathrm{A}} = (\tilde{\gamma}^{\mathrm{R}})^{\dagger}, \quad \tilde{\gamma}^{\mathrm{A}} = (\gamma^{\mathrm{R}})^{\dagger}, \quad x^{\mathrm{K}} = (x^{\mathrm{K}})^{\dagger} \tag{11.59}$$

hold. As a consequence, it suffices to determine fully the parameters γ^{R} and x^{K}.

The quasiclassical Green's function is related to these amplitudes in the following way [here the upper (lower) sign corresponds to retarded (advanced)]:

$$\hat{g}^{\mathrm{R,A}} = \mp i\pi \begin{pmatrix} (1-\gamma\circ\tilde{\gamma})^{-1}\circ(1+\gamma\circ\tilde{\gamma}) & 2(1-\gamma\circ\tilde{\gamma})^{-1}\circ\gamma \\ -2(1-\tilde{\gamma}\circ\gamma)^{-1}\circ\tilde{\gamma} & -(1-\tilde{\gamma}\circ\gamma)^{-1}\circ(1+\tilde{\gamma}\circ\gamma) \end{pmatrix}^{\mathrm{R,A}}, \tag{11.60}$$

which can be written in more compact form as [111]

$$\hat{g}^{\mathrm{R,A}} = \mp 2\pi i \begin{pmatrix} \mathscr{G} & \mathscr{F} \\ -\tilde{\mathscr{F}} & -\tilde{\mathscr{G}} \end{pmatrix}^{\mathrm{R,A}} \pm i\pi\hat{\tau}_3, \tag{11.61}$$

with the abbreviations $\mathscr{G} = (1-\gamma\circ\tilde{\gamma})^{-1}$ and $\mathscr{F} = \mathscr{G}\circ\gamma$. For the Keldysh component one can write [111]

$$\hat{g}^{\mathrm{K}} = -2\pi i \begin{pmatrix} \mathscr{G} & \mathscr{F} \\ -\tilde{\mathscr{F}} & -\tilde{\mathscr{G}} \end{pmatrix}^{\mathrm{R}} \circ \begin{pmatrix} x^{\mathrm{K}} & 0 \\ 0 & \tilde{x}^{\mathrm{K}} \end{pmatrix} \circ \begin{pmatrix} \mathscr{G} & \mathscr{F} \\ -\tilde{\mathscr{F}} & -\tilde{\mathscr{G}} \end{pmatrix}^{\mathrm{A}}. \tag{11.62}$$

Here, the $\circ$-symbol includes a time convolution as well as matrix multiplication; the inversion is defined with respect to the $\circ$-operation [111].

From the transport equation for the quasiclassical Green's functions one obtains a set of matrix equations of motion for the six parameters above [27, 92]. For the coherence amplitudes this leads to Riccati differential equations [96, 97], hence the name Riccati parameterization.

11.3.5 Transport Equations

The central equations that govern the transport phenomena have been derived in Ref. [27, 92]. The transport equation for the coherence functions $\gamma(\boldsymbol{p}_F, \boldsymbol{R}, \varepsilon, t)$ are given by

$$(i\hbar v_F \cdot \nabla_R + 2\varepsilon)\gamma^{R,A} = [\gamma \circ \tilde{\Delta} \circ \gamma + \Sigma \circ \gamma - \gamma \circ \tilde{\Sigma} - \Delta]^{R,A}. \tag{11.63}$$

For the distribution functions $x(\boldsymbol{p}_F, \boldsymbol{R}, \varepsilon, t)$ the transport equations read

$$(i\hbar v_F \cdot \nabla_R + i\hbar\partial_t)x^K - [\gamma \circ \tilde{\Delta} + \Sigma]^R \circ x^K - x^K \circ [\Delta \circ \tilde{\gamma} - \Sigma]^A$$
$$= -\gamma^R \circ \tilde{\Sigma}^K \circ \tilde{\gamma}^A + \Delta^K \circ \tilde{\gamma}^A + \gamma^R \circ \tilde{\Delta}^K - \Sigma^K. \tag{11.64}$$

The equations for the remaining components are obtained by the symmetry relation Eq. (11.58).

11.3.6 Boundary Conditions

The transport equations must be complemented with boundary conditions for the coherence amplitudes and distribution functions at interfaces and surfaces [100, 101, 102, 103]. For spin-active scattering such conditions were obtained in Ref. [104]. Explicit formulations in terms of special parameterizations were given in Refs. [27, 105–110]. Further developments include strongly spin-polarized systems [111–116], diffusive interface scattering [117] or multi-band systems [118]. We adopt the notation [27] that incoming amplitudes are denoted by small case letters and outgoing ones by capital case letters, see Fig. 11.8. Note that the velocity direction of trajectories is opposite for holelike and particlelike amplitudes as well as advanced and retarded ones. The boundary conditions express outgoing amplitudes as a function of incoming ones and as a function of the parameters of the normal-state scattering matrix.

11.3.6.1 Coherence Amplitudes

The boundary conditions for the coherence amplitudes are formulated in terms of the solution of the equation [111]

$$[\gamma'_{kk'}]^R = \sum_p S^R_{kp} \circ \gamma^R_p \circ \tilde{S}^R_{pk'} \tag{11.65}$$

$$[\Gamma_{k\leftarrow k'}]^R = \left[\gamma'_{kk'} + \sum_{k_1 \neq k} \Gamma_{k\leftarrow k_1} \circ \tilde{\gamma}_{k_1} \circ \gamma'_{k_1 k'}\right]^R, \tag{11.66}$$

(the trajectory index p runs over all incoming trajectories) for $[\Gamma_{k\leftarrow k'}]^R$, where the trajectory indices k, k', k_1 run over outgoing trajectories involved in the interface scattering process, and the scattering matrix parameters enter only via the "elementary scattering event" $[\gamma'_{kk'}]^{R,A}$. The quasiclassical coherence amplitude is given by the forward scattering contribution of $[\Gamma_{k\leftarrow k'}]^R$,

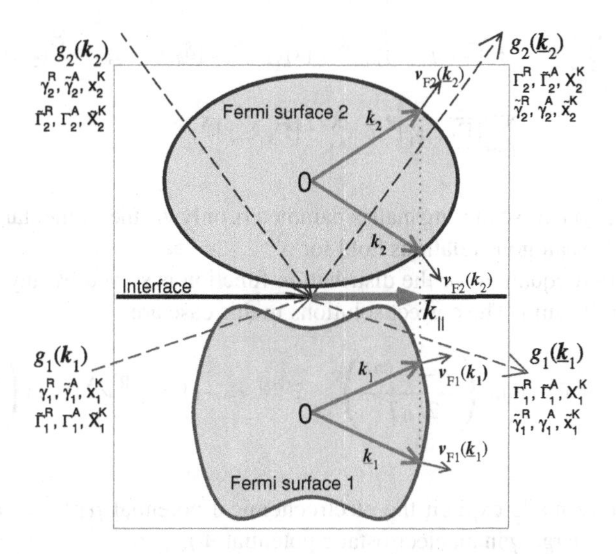

Fig. 11.8 Notation for the coherence amplitudes and distribution functions at an interface. Indices 1 and 2 refer to the sides of the interface. The arrows for the Fermi velocities are for particle like excitations. The Fermi velocity directions are given by the directions perpendicular to the Fermi surface at the corresponding Fermi momentum. Quasiparticles move along the Fermi velocity directions (*dashed lines*). The components of the Fermi momenta parallel to the surface are conserved (indicated by the thin *dotted line*). For each trajectory, small case letters denote coherence functions and distribution functions with initial conditions from the bulk, and capital case letters denote functions with initial conditions at the interface. The interface boundary conditions must express all capital case quantities in terms of the small case quantities. Here, the simplest case, that involves only one Fermi surface sheet on either side ('two-trajectory scattering'), is shown. After Ref. [27]

$$\Gamma_k^{\mathrm{R}} = \Gamma_{k\leftarrow k}^{\mathrm{R}}. \tag{11.67}$$

Analogous equations [111] hold for the advanced and particle-hole conjugated components, $[\tilde{\Gamma}_{p\leftarrow p'}]^{\mathrm{R}}$, $[\Gamma_{p'\rightarrow p}]^{\mathrm{A}}$, and $[\tilde{\Gamma}_{k'\rightarrow k}]^{\mathrm{A}}$.

11.3.6.2 Distribution Functions

For the Keldysh component not only the forward scattering contribution of $[\Gamma_{k\leftarrow k'}]^{\mathrm{R}}$ is required, but also the off-scattering part

$$[\overline{\Gamma}_{k\leftarrow k'}]^{\mathrm{R}} = [\Gamma_{k\leftarrow k'} - \Gamma_k \delta_{kk'}]^{\mathrm{R}}. \tag{11.68}$$

The boundary conditions for the distribution functions read [111]

$$[x'_{kk'}]^{\mathrm{K}} = \sum_p S_{kp}^{\mathrm{R}} \circ x_p^{\mathrm{K}} \circ S_{pk'}^{\mathrm{A}}. \tag{11.69}$$

$$X_k^{\mathrm{K}} = \sum_{k_1,k_2} [\delta_{kk_1} + \overline{\Gamma}_{k\leftarrow k_1} \circ \tilde{\gamma}_{k_1}]^{\mathrm{R}} \circ [x'_{k_1k_2}]^{\mathrm{K}} \circ [\delta_{k_2k} + \gamma_{k_2} \circ \overline{\tilde{\Gamma}}_{k_2\rightarrow k}]^{\mathrm{A}}$$
$$- \sum_{k_1} [\overline{\Gamma}_{k\leftarrow k_1}]^{\mathrm{R}} \circ \tilde{x}_{k_1}^{\mathrm{K}} \circ [\overline{\tilde{\Gamma}}_{k_1\rightarrow k}]^{\mathrm{A}}, \qquad (11.70)$$

which depends on the scattering matrix parameters only via the elementary scattering event $[x'_{kk'}]^{\mathrm{K}}$. Analogous relations hold for $\tilde{X}_p^{\mathrm{K}}$.

The transport equation for the distribution function is solved by any function of energy in equilibrium. The correct solutions in this case are

$$x^{(\mathrm{eq})} = (1 - \gamma^{\mathrm{R}}\tilde{\gamma}^{\mathrm{A}}) \tanh\left(\frac{\varepsilon - \mu^{(\mathrm{el})}}{2k_{\mathrm{B}}T}\right), \quad \tilde{x}^{(\mathrm{eq})} = -(1 - \tilde{\gamma}^{\mathrm{R}}\gamma^{\mathrm{A}}) \tanh\left(\frac{\varepsilon + \mu^{(\mathrm{el})}}{2k_{\mathrm{B}}T}\right) \qquad (11.71)$$

(where we have made explicit the electrochemical potential $\mu^{(\mathrm{el})} = \mu + q\Phi$ for excitations of charge q in an electrostatic potential Φ).

11.3.6.3 Case 1: One-Trajectory Scattering

In this case for each given value of parallel component of the momentum, only one incoming and one outgoing trajectory are coupled via the boundary conditions. The corresponding normal state scattering matrix is denoted by S and is a scalar in trajectory space. The boundary conditions read in this case simply

$$[\gamma']^{\mathrm{R}} = S^{\mathrm{R}} \circ \gamma^{\mathrm{R}} \circ \tilde{S}^{\mathrm{R}}, \quad \Gamma^{\mathrm{R}} = [\gamma']^{\mathrm{R}}, \qquad (11.72)$$

and

$$[x']^{\mathrm{K}} = S^{\mathrm{R}} \circ x^{\mathrm{K}} \circ S^{\mathrm{A}}, \quad X_k^{\mathrm{K}} = [x']^{\mathrm{K}}. \qquad (11.73)$$

11.3.6.4 Case 2: Two-Trajectory Scattering

This is the case of scattering from two incoming trajectories into two outgoing trajectories. Examples are reflection and transmission at an interface, or reflection from a surface in a two-band system. The scattering matrix and the elementary scattering events have in this case the form

$$S = \begin{pmatrix} S_{11} & S_{12} \\ S_{21} & S_{22} \end{pmatrix}, \quad [\gamma'_{ij}]^{\mathrm{R}} = \sum_{l=1,2} S_{il}^{\mathrm{R}} \circ \gamma_l^{\mathrm{R}} \circ \tilde{S}_{lj}^{\mathrm{R}}, \quad [x'_{ij}]^{\mathrm{K}} = \sum_{l=1,2} S_{il}^{\mathrm{R}} \circ x_l^{\mathrm{K}} \circ S_{lj}^{\mathrm{A}}. \qquad (11.74)$$

We give the solutions for trajectory 1, the remaining solution can be obtained by interchanging the indices 1 and 2. The boundary conditions read for $i, j = 1, 2$

$$\Gamma^{R}_{1\leftarrow 1} = \left[\gamma'_{11} + \Gamma_{1\leftarrow 2} \circ \tilde{\gamma}_2 \circ \gamma'_{21}\right]^{R}, \quad \Gamma^{R}_{1\leftarrow 2} = \left[\gamma'_{12} + \Gamma_{1\leftarrow 2} \circ \tilde{\gamma}_2 \circ \gamma'_{22}\right]^{R}. \tag{11.75}$$

The equation for the $\Gamma_{1\leftarrow 2}$ can be solved by simple inversion,

$$\Gamma^{R}_{1\leftarrow 2} = \left[\gamma'_{12} \circ (1 - \tilde{\gamma}_2 \circ \gamma'_{22})^{-1}\right]^{R}, \tag{11.76}$$

and the such obtained solution introduced into the equation for $\Gamma^{R}_{1\leftarrow 1} = \Gamma^{R}_{1}$,

$$\Gamma^{R}_{1} = \left[\gamma'_{11} + \gamma'_{12} \circ (1 - \tilde{\gamma}_2 \circ \gamma'_{22})^{-1} \circ \tilde{\gamma}_2 \circ \gamma'_{21}\right]^{R}. \tag{11.77}$$

For the distribution function one needs the components $\overline{\Gamma}^{R}_{1\leftarrow 2} = \Gamma^{R}_{1\leftarrow 2}$ and obtains

$$\begin{aligned} X^{K}_{1} &= [x'_{11}]^{K} + \Gamma^{R}_{1\leftarrow 2} \circ \tilde{\gamma}^{R}_{2} \circ [x'_{21}]^{K} + [x'_{12}]^{K} \circ \gamma^{A}_{2} \circ \tilde{\Gamma}^{A}_{2\to 1} \\ &\quad + \Gamma^{R}_{1\leftarrow 2} \circ \left(\tilde{\gamma}^{R}_{2} \circ [x'_{22}]^{K} \circ \gamma^{A}_{2} - \tilde{x}^{K}_{2}\right) \circ \tilde{\Gamma}^{A}_{2\to 1}. \end{aligned} \tag{11.78}$$

We present here formulas for the special case of the zero temperature conductance when a single band system is contacted by a normal metal. We assign the index 1 to the normal metal side of the interface and the index 2 to the superconducting side. The momentum for incoming trajectories on the superconducting side of the interface is denoted by $\boldsymbol{k}_2$, and that for the outgoing trajectory on the superconducting side by $\underline{\boldsymbol{k}}_2$. For the normal side the corresponding momenta are $\boldsymbol{k}_1$ and $\underline{\boldsymbol{k}}_1$ (see Fig. 11.8 for the scattering geometry). The projection on the interface of all four momenta is equal. The corresponding incoming coherence functions in the superconductor are $\gamma_2(\varepsilon) \equiv \gamma^{R}_{2}(\boldsymbol{k}_2, \varepsilon)$ and $\tilde{\gamma}_2(\varepsilon) \equiv \tilde{\gamma}^{R}_{2}(\underline{\boldsymbol{k}}_2, \varepsilon)$. Furthermore, $S_{12} \equiv S^{R}_{12}(\underline{\boldsymbol{k}}_1, \boldsymbol{k}_2)$, $S_{22} \equiv S^{R}_{22}(\underline{\boldsymbol{k}}_2, \boldsymbol{k}_2)$, and $\tilde{S}_{22} = \tilde{S}^{R}_{22}(\boldsymbol{k}_2, \underline{\boldsymbol{k}}_2)$. The Fermi velocity for outgoing directions on the normal side will be denoted by $\boldsymbol{v}_{F1} \equiv \boldsymbol{v}_{F1}(\underline{\boldsymbol{k}}_1)$. Having thus specified all momentum dependencies, we will suppress in the formulas below the momentum variables. In the case under consideration, after introducing Eqs. (11.76), (11.77), and (11.78) into Eq. (11.52), we obtain after some algebra (we omit hereafter the $\circ$ sign)

$$\begin{aligned} \frac{G(eV)}{G_{\rm N}} &= \left\langle \hat{\boldsymbol{n}}\boldsymbol{v}_{F1} \left\{ \left|\left| S_{12}\left[1 + A_2(\varepsilon) S_{22}\right] \right|\right|^2 - \left|\left| S_{12} A_2(\varepsilon) \right|\right|^2 \right\} \right\rangle^{+}_{\varepsilon = eV} \\ &\quad + \left\langle \hat{\boldsymbol{n}}\boldsymbol{v}_{F1} \left|\left| S_{12}\left[1 + A_2(\varepsilon) S_{22}\right] \gamma_2(\varepsilon) \tilde{S}_{21} \right|\right|^2 \right\rangle^{+}_{\varepsilon = -eV} \end{aligned} \tag{11.79}$$

where

$$A_2(\varepsilon) = \left(1 - \gamma_2(\varepsilon) \tilde{S}_{22} \tilde{\gamma}_2(\varepsilon) S_{22}\right)^{-1} \gamma_2(\varepsilon) \tilde{S}_{22} \tilde{\gamma}_2(\varepsilon) \tag{11.80}$$

and we used the notation $||A||^2 = \frac{1}{2} Tr(AA^{\dagger})$ for any 2×2 matrix A. The symbol $\langle \ldots \rangle^{+}_{\varepsilon = eV}$ denotes Fermi surface average only over outgoing directions, and the argument is to be taken at energy eV. For $S_{22} = \tilde{S}_{22} = -\sqrt{R(\theta)}$, $S_{12} = \tilde{S}_{21} = \sqrt{D(\theta)}$ (with impact angle θ), Eq. (11.79) reduces to Eq. (11.10). For the tunneling limit we can neglect the second line in Eq. (11.79), and using the relation

$$1 + A_2(\varepsilon)S_{22} = \frac{1}{2}\{\mathcal{N}_2(\varepsilon) + 1\} \tag{11.81}$$

with the complex quantity

$$\mathcal{N}_2(\varepsilon) = \left\{\left(1 - \gamma_2(\varepsilon)\tilde{S}_{22}\tilde{\gamma}_2(\varepsilon)S_{22}\right)^{-1}\left(1 + \gamma_2(\varepsilon)\tilde{S}_{22}\tilde{\gamma}_2(\varepsilon)S_{22}\right)\right\}$$

the conductance simplifies after some re-arrangements to

$$\frac{G(eV)}{G_{\mathrm{N}}} = \frac{1}{2}\mathrm{ReTr}\left\langle \hat{\boldsymbol{n}}\boldsymbol{v}_{\mathrm{F}1}\left\{S_{12}\mathcal{N}_2(eV)(S_{12})^{\dagger}\right\}\right\rangle^{+}. \tag{11.82}$$

For the tunneling limit, in $\mathcal{N}_2$ the surface scattering matrix (i.e. for $S_{12} = S_{21} = 0$) can be used, for which the local density of states at an impenetrable surface is

$$\frac{N_2(\varepsilon)}{N_{2,\mathrm{F}}} = \frac{1}{2}\mathrm{ReTr}\left\langle \mathcal{N}_2(\varepsilon) + S_{22}\mathcal{N}_2(\varepsilon)S_{22}^{\dagger}\right\rangle^{+}. \tag{11.83}$$

11.3.6.5 Scattering Matrix for Non-centrosymmetric/Normal-Metal Junction

For the case that a non-centrosymmetric material with small spin-orbit splitting is brought in contact with a normal metal, we can use the formulas of the last subsection. The scattering matrix for scattering between the two helicity bands in the non-centrosymmetric metal (index 2) and the two spin bands in the normal metal (index 1) can be expressed in terms of the scattering matrix for scattering between spin states on both sides of the interface. The corresponding transformation is

$$\begin{pmatrix} S'_{11} & S'_{12} \\ S'_{21} & S'_{22} \end{pmatrix} = \begin{pmatrix} 1 & 0 \\ 0 & U_{\underline{k}} \end{pmatrix} \cdot \begin{pmatrix} S_{11} & S_{12} \\ S_{21} & S_{22} \end{pmatrix} \cdot \begin{pmatrix} 1 & 0 \\ 0 & U_{\underline{k}}^{\dagger} \end{pmatrix}, \tag{11.84}$$

$$\begin{pmatrix} \tilde{S}'_{11} & \tilde{S}'_{12} \\ \tilde{S}'_{21} & \tilde{S}'_{22} \end{pmatrix} = \begin{pmatrix} 1 & 0 \\ 0 & U^{*}_{-\underline{k}} \end{pmatrix} \cdot \begin{pmatrix} \tilde{S}_{11} & \tilde{S}_{12} \\ \tilde{S}_{21} & \tilde{S}_{22} \end{pmatrix} \cdot \begin{pmatrix} 1 & 0 \\ 0 & U^{T}_{-\underline{k}} \end{pmatrix}. \tag{11.85}$$

For the simple case of a spin-conserving scattering in the spin/spin representation, the spin/helicity representation of the scattering matrix takes the form

$$\begin{pmatrix} S'_{11} & S'_{12} \\ S'_{21} & S'_{22} \end{pmatrix} = \begin{pmatrix} r & t\, U_{\underline{k}}^{\dagger} \\ t^{*} U_{\underline{k}} & -r\, U_{\underline{k}} U_{\underline{k}}^{\dagger} \end{pmatrix}, \quad \begin{pmatrix} \tilde{S}'_{11} & \tilde{S}'_{12} \\ \tilde{S}'_{21} & \tilde{S}'_{22} \end{pmatrix} = \begin{pmatrix} r & t^{*}\, U^{T}_{-\underline{k}} \\ t\, U^{*}_{-\underline{k}} & -r\, U^{*}_{-\underline{k}} U^{T}_{-\underline{k}} \end{pmatrix}. \tag{11.86}$$

where $r \equiv r_{\underline{k}k}$ and $t = t_{\underline{k}k}$ with $r^2 + |t|^2 = 1$ are reflection and transmission coefficients that depend on the (conserved) momentum projection on the interface. We have chosen r real, as in quasiclassical approximation possible reflection phases do not affect the results. The case $t = 0$ can be used to describe scattering at a surface.

In the case of a contact with a non-centrosymmetric metal with strong spin-orbit split bands the scattering matrix has a more complicated structure. It connects in this case three incoming with three outgoing trajectories, and the scattering at the interface will not be spin-conserving. For this case, it does then not make sense anymore do use a spin/spin representation, but a spin/helicity representation must be used consistently. The scattering matrix must be obtained in agreement with the symmetry group of the interface, and it cannot in general be related anymore to the U_k matrices in a simple way.

11.3.7 Superconducting Order Parameter

For the case of weak spin-orbit splitting one expects that to leading order in the small expansion parameters either a singlet or a triplet component nucleates. On the other hand, any finite spin-orbit interaction leads to a mixture of spin singlet (Δ_s) and triplet (Δ_t) components [119, 120]. Consequently, the singlet or triplet states are never pure, but they are mixed. This mixing becomes in particular prominent when the spin-orbit interaction is strong. In this case, it does not make sense anymore to speak about singlet or triplet components, but it is necessary to start from the helicity basis.

In the following, we concentrate on the case of weak splitting. In this case the triplet component is expected to be induced directly by the structure of the spin-orbit interaction, and the spin triplet component aligns with $\boldsymbol{g}(\boldsymbol{k})$. The order parameter matrix is in this case in spin representation given by,

$$\Delta^{\text{spin}} = (\Delta_{\boldsymbol{k}} + D_{\boldsymbol{k}} \boldsymbol{g}(\boldsymbol{k}) \cdot \boldsymbol{\sigma}) i\sigma^{(2)} \tag{11.87}$$

which transforms in helicity basis into

$$\begin{aligned}\Delta &= U_{\boldsymbol{k}}(\Delta_{\boldsymbol{k}} + D_{\boldsymbol{k}} \boldsymbol{g}(\boldsymbol{k}) \cdot \boldsymbol{\sigma}) i\sigma^{(2)} U_{-\boldsymbol{k}}^T \\ &= U_{\boldsymbol{k}}(\Delta_{\boldsymbol{k}} + D_{\boldsymbol{k}} \boldsymbol{g}(\boldsymbol{k}) \cdot \boldsymbol{\sigma}) U_{\boldsymbol{k}}^\dagger U_{\boldsymbol{k}} i\sigma^{(2)} U_{-\boldsymbol{k}}^T \\ &= (\Delta_{\boldsymbol{k}} + D_{\boldsymbol{k}} |\boldsymbol{g}(\boldsymbol{k})| \sigma^{(3)}) U_{\boldsymbol{k}} U_{-\boldsymbol{k}}^\dagger i\sigma^{(2)}.\end{aligned} \tag{11.88}$$

For the following, we introduce the notation

$$(U_{\boldsymbol{k}} U_{-\boldsymbol{k}}^\dagger)_{\lambda\lambda'} = \begin{pmatrix} 0 & e^{-i\varphi_g} \\ -e^{i\varphi_g} & 0 \end{pmatrix} \equiv -i\sigma_{\lambda\lambda'}^{(g)}. \tag{11.89}$$

Note that the identities $(\sigma^{(g)})^2 = 1, \sigma^{(-g)} = -\sigma^{(g)}$, and $\sigma^{(2)}\sigma^{(g)}\sigma^{(2)} = -\sigma^{(g)*}$, hold. With this notation, we can obtain the Nambu-Gor'kov space structure of the order parameter as

$$\hat{\Delta} = \begin{pmatrix} 0 & \Delta \\ \tilde{\Delta} & 0 \end{pmatrix} = \begin{pmatrix} 0 & (\Delta_k + D_k|g|\sigma^3)\sigma^{(g)}\sigma^{(2)} \\ (\Delta_{-k} + D_{-k}|g|\sigma^{(3)})^*\sigma^{(g)*}\sigma^{(2)} & 0 \end{pmatrix}$$
$$= \begin{pmatrix} 0 & (\Delta_k + D_k|g|\sigma^{(3)})i\sigma^{(2)} \\ (\Delta_k^* + D_k^*|g|\sigma^{(3)})i\sigma^{(2)} & 0 \end{pmatrix} \begin{pmatrix} i\sigma^{(g)} & 0 \\ 0 & -i\sigma^{(-g)*} \end{pmatrix}, \tag{11.90}$$

where $\Delta_{-k} = \Delta_k$, and $D_{-k} = D_k$, and we have used $\sigma^{(g)*}\sigma^{(3)}\sigma^{(g)*} = -\sigma^{(3)}$. With $\Delta_\pm(\boldsymbol{k}) = \Delta_k \pm D_k|g|$ the order parameter can be cast in the form

$$\Delta(\boldsymbol{k}) = \begin{pmatrix} \Delta_+(\boldsymbol{k})t_+(\boldsymbol{k}) & 0 \\ 0 & \Delta_-(\boldsymbol{k})t_-(\boldsymbol{k}) \end{pmatrix} \tag{11.91}$$

$$\tilde{\Delta}(\boldsymbol{k}) = \begin{pmatrix} \Delta_+(\boldsymbol{k})^*t_+(-\boldsymbol{k})^* & 0 \\ 0 & \Delta_-(\boldsymbol{k})^*t_-(-\boldsymbol{k})^* \end{pmatrix}. \tag{11.92}$$

with phase factors $t_\lambda(\boldsymbol{k}) = -e^{-i\lambda\varphi_g}$. Note that $t_\lambda(-\boldsymbol{k}) = -t_\lambda(\boldsymbol{k})$, and $|t_\lambda(\boldsymbol{k})| = 1$, and $\Delta_\pm(-\boldsymbol{k}) = \Delta_\pm(\boldsymbol{k})$.

We note in passing that other possibilities to define the canonical transformation that diagonalizes the kinetic part of the Hamiltonian exist, which differ by the relation between particle and hole components. Using these alternative definitions (e.g. in Refs. [48, 58]), the order parameter is purely off-diagonal instead of diagonal in the band representation, and the symmetry relation Eq. (11.58) becomes non-trivial (see e.g. Ref. [56]). Here, we prefer a transformation that preserves the symmetry (11.58), and renders the order parameter above diagonal. This is a natural choice when treating strongly spin-orbit split systems, where the order parameter should be band diagonal.

The coherence amplitudes in a bulk system with order parameter Eq. (11.91) are of a similar form,

$$\gamma(\boldsymbol{k}, \varepsilon) = \begin{pmatrix} \gamma_+(\boldsymbol{k}, \varepsilon)t_+(\boldsymbol{k}) & 0 \\ 0 & \gamma_-(\boldsymbol{k}, \varepsilon)t_-(\boldsymbol{k}) \end{pmatrix} \tag{11.93}$$

$$\tilde{\gamma}(\boldsymbol{k}, \varepsilon) = \begin{pmatrix} \tilde{\gamma}_+(\boldsymbol{k}, \varepsilon)t_+(-\boldsymbol{k})^* & 0 \\ 0 & \tilde{\gamma}_-(\boldsymbol{k}, \varepsilon)t_-(-\boldsymbol{k})^* \end{pmatrix} \tag{11.94}$$

with $\tilde{\gamma}_\pm(\boldsymbol{k}, \varepsilon) = \gamma_\pm(-\boldsymbol{k}, -\varepsilon)^*$. In inhomogeneous systems helicity mixing can take place. If this happens, the form of the coherence functions is the same band-diagonal form as above for the case of strong spin-orbit splitting, however has the full matrix structure for the case of weak spin-orbit splitting.

11.3.8 Results

11.3.8.1 Andreev Bound States Near the Surface

The surface bound states are determined by the poles of the Green's function. Following Refs. [56, 58], we consider specular reflection, whereby the component of $\boldsymbol{k}$ normal to surface changes sign, $\boldsymbol{k} \to \underline{\boldsymbol{k}}$, whereas the component parallel to the surface is conserved. We find the amplitudes $\gamma(\boldsymbol{k}, \varepsilon)$ by integrating forward along the incoming, $\boldsymbol{k}$, trajectory starting from the values in the bulk, and the amplitudes $\tilde{\gamma}(\underline{\boldsymbol{k}}, \varepsilon)$ by integrating backward along the outgoing, $\underline{\boldsymbol{k}}$, trajectory, again starting from the values in the bulk [27]. For the homogeneous solutions one obtains

$$\gamma^0_{\pm}(\boldsymbol{k}, \varepsilon) = \frac{-\Delta_{\pm}(\boldsymbol{k})}{\varepsilon + i\sqrt{|\Delta_{\pm}(\boldsymbol{k})|^2 - \varepsilon^2}}, \quad \tilde{\gamma}^0_{\pm}(\underline{\boldsymbol{k}}, \varepsilon) = \frac{\Delta_{\pm}(\underline{\boldsymbol{k}})^*}{\varepsilon + i\sqrt{|\Delta_{\pm}(\underline{\boldsymbol{k}})|^2 - \varepsilon^2}}, \tag{11.95}$$

Note that the spin-orbit interaction in the helicity basis enters as a term proportional to $\sigma^{(3)}$, see Eq. (11.54). Consequently, this term commutes with any term diagonal in the helicity basis, and thus drops out of the homogeneous solutions in Eq. (11.95) (see Ref. [121] for the case of a Rashba-type spin-orbit coupling). However, this is not in general the case for non-homogeneous solutions: when helicity mixing takes place due to impurities or surfaces and interfaces, and a fully self-consistent solution is obtained, then the spin-orbit coupling term in Eq. (11.54) enters through the transport equation.

The amplitudes $\Gamma_{\underline{\boldsymbol{k}}}$ and $\tilde{\Gamma}_{\boldsymbol{k}}$, are determined from the boundary conditions at the surface. We consider here a simple model of a non-magnetic surface, that conserves the spin under reflection (this assumption only holds for a *small* spin-orbit interaction in the bulk material). In this case the components of $\hat{g}$ in the spin basis,

$$\hat{g}^{\text{spin}}(\boldsymbol{k}, \varepsilon) = \hat{U}^{\dagger}_{\boldsymbol{k}}\, \hat{g}(\boldsymbol{k}, \varepsilon)\hat{U}_{\boldsymbol{k}}, \tag{11.96}$$

are continuous at the surface. This leads to a surface-induced mixing of the helicity bands according to

$$U^{\dagger}_{\underline{\boldsymbol{k}}}\Gamma(\underline{\boldsymbol{k}}, \varepsilon)U^*_{-\underline{\boldsymbol{k}}} = \Gamma^{\text{spin}}(\underline{\boldsymbol{k}}, \varepsilon) = \gamma^{\text{spin}}(\boldsymbol{k}, \varepsilon) = U^{\dagger}_{\boldsymbol{k}}\gamma(\boldsymbol{k}, \varepsilon)U^*_{-\boldsymbol{k}}, \tag{11.97}$$

$$U^T_{-\boldsymbol{k}}\tilde{\Gamma}(\boldsymbol{k}, \varepsilon)U_{\boldsymbol{k}} = \tilde{\Gamma}^{\text{spin}}(\boldsymbol{k}, \varepsilon) = \tilde{\gamma}^{\text{spin}}(\underline{\boldsymbol{k}}, \varepsilon) = U^T_{-\underline{\boldsymbol{k}}}\tilde{\gamma}(\underline{\boldsymbol{k}}, \varepsilon)U_{\underline{\boldsymbol{k}}}. \tag{11.98}$$

Note that these boundary conditions correspond to Eq. (11.72) with S^{R} and $\tilde{S}^{\text{R}}$ given by the (11.28)-components of Eq. (11.86).

We proceed with discussing the local density of states at the surface, $N(\varepsilon)$, that is defined in terms of the momentum resolved density of states, $N(\boldsymbol{k}, \varepsilon)$ by

$$N(\boldsymbol{k}, \varepsilon)/N_{\text{F}} = -(2\pi)^{-1}\text{ImTr}_{\lambda}\,\{g(\boldsymbol{k}, \varepsilon)\}\,, \quad N(\varepsilon) = \langle N(\boldsymbol{k}, \varepsilon)\rangle, \tag{11.99}$$

which can be expressed in terms of the coherence amplitudes in the following way (here $\boldsymbol{k}$ points towards the surface and $\underline{\boldsymbol{k}}$ away from it),

$$N(\boldsymbol{k},\varepsilon)/N_{\rm F} = {\rm ReTr}_\lambda \left\{ \left[1-\gamma(\boldsymbol{k},\varepsilon)\tilde{\Gamma}(\boldsymbol{k},\varepsilon) \right]^{-1} - 1/2 \right\}$$
$$N(\underline{\boldsymbol{k}},\varepsilon)/N_{\rm F} = {\rm ReTr}_\lambda \left\{ \left[1-\Gamma(\underline{\boldsymbol{k}},\varepsilon)\tilde{\gamma}(\underline{\boldsymbol{k}},\varepsilon) \right]^{-1} - 1/2 \right\} . \qquad (11.100)$$

We obtain $\Gamma(\underline{\boldsymbol{k}},\varepsilon)$ and $\tilde{\Gamma}(\boldsymbol{k},\varepsilon)$ from Eqs. (11.97) and (11.98), with $\gamma(\boldsymbol{k},\varepsilon)$ and $\tilde{\gamma}(\underline{\boldsymbol{k}},\varepsilon)$ from Eqs. (11.93), (11.94), (11.95), and (11.104).

The bound states in the surface density of states correspond to the zero eigenvalues of the matrix

$$1-\gamma(\boldsymbol{k},\varepsilon)\tilde{\Gamma}(\boldsymbol{k},\varepsilon) = 1-\gamma(\boldsymbol{k},\varepsilon)(U^*_{-\boldsymbol{k}}U^T_{-\underline{\boldsymbol{k}}})\tilde{\gamma}(\underline{\boldsymbol{k}},\varepsilon)(U_{\underline{\boldsymbol{k}}}U^\dagger_{\boldsymbol{k}}) \qquad (11.101)$$

at the surface. An explicite calculation results in an equation for the Andreev bound-state energy in terms of the surface coherence amplitudes in the helicity basis [58],

$$\frac{\{1+\gamma_+\tilde{\gamma}_+\}\{1+\gamma_-\tilde{\gamma}_-\}}{\{1+\gamma_+\tilde{\gamma}_-\}\{1+\gamma_-\tilde{\gamma}_+\}} = -\mathcal{M}, \qquad (11.102)$$

where we used the abbreviations $\gamma_\pm \equiv \gamma_\pm(\boldsymbol{k},\varepsilon)$ and $\tilde{\gamma}_\pm \equiv \tilde{\gamma}_\pm(\underline{\boldsymbol{k}},\varepsilon)$. The "mixing" factor $\mathcal{M}$ is determined by the change of $\boldsymbol{g}(\boldsymbol{k}) \to \boldsymbol{g}(\underline{\boldsymbol{k}})$ under reflection $\boldsymbol{k} \to \underline{\boldsymbol{k}}$ at the surface,

$$\mathcal{M} = \frac{\sin^2\frac{\theta_g-\theta_{\underline{g}}}{2} + \sin^2\frac{\theta_g+\theta_{\underline{g}}}{2}\tan^2\frac{\varphi_g-\varphi_{\underline{g}}}{2}}{\cos^2\frac{\theta_g-\theta_{\underline{g}}}{2} + \cos^2\frac{\theta_g+\theta_{\underline{g}}}{2}\tan^2\frac{\varphi_g-\varphi_{\underline{g}}}{2}}, \qquad (11.103)$$

where θ_g, φ_g and $\theta_{\underline{g}}, \varphi_{\underline{g}}$ are the polar and azimuthal angles of $\boldsymbol{g}(\boldsymbol{k})$ and $\boldsymbol{g}(\underline{\boldsymbol{k}})$, respectively.

In general, the order parameter must be obtained self-consistently at the surface. Helicity mixing at the surface will lead necessarily to a suppression of the order parameter. To gain insight in the role of the order parameter suppression it is useful to model it by a normal layer of width W next to the interface. Trajectories incident at an angle $\alpha_{\boldsymbol{k}}$ from the surface normal travel through a normal region of an effective width $2W_{\boldsymbol{k}} = 2W/\cos(\alpha_{\boldsymbol{k}})$. Thus, the surface coherence amplitudes gain a phase factor,

$$\gamma_\pm(\boldsymbol{k},\varepsilon) = \gamma^0_\pm(\boldsymbol{k},\varepsilon)e^{2\imath\varepsilon W/v_{\rm F}\cos(\alpha_{\boldsymbol{k}})}, \quad \tilde{\gamma}_\pm(\underline{\boldsymbol{k}},\varepsilon) = \tilde{\gamma}^0_\pm(\underline{\boldsymbol{k}},\varepsilon)e^{2\imath\varepsilon W/v_{\rm F}\cos(\alpha_{\boldsymbol{k}})}. \qquad (11.104)$$

Similar as for $\boldsymbol{g}$, we will use in the following polar and azimuthal angles for the vector $\boldsymbol{k}$, defined by $\{k_x, k_y, k_z\} = |\boldsymbol{k}|\{\sin(\theta_{\boldsymbol{k}})\cos(\varphi_{\boldsymbol{k}}), \sin(\theta_{\boldsymbol{k}})\sin(\varphi_{\boldsymbol{k}}), \cos(\theta_{\boldsymbol{k}})\}$ (where $0 \le \theta_{\boldsymbol{k}} \le \pi$). We also introduce the notation $\hat{\boldsymbol{g}} = \boldsymbol{g}/\max(|\boldsymbol{g}|)$. For the order parameter, we assume isotropic $\Delta_{\boldsymbol{k}} = \Delta$ and $D_{\boldsymbol{k}} = D$ in Eq. (11.87), and introduce the parameter $q = \Delta/D'$ where $D' = D \cdot \max(|\boldsymbol{g}|)$ [56]. In this case $\Delta_\pm = D'(q \pm |\hat{\boldsymbol{g}}|)$ with maximal gap amplitudes $\Delta_0 = D'(q+1)$.

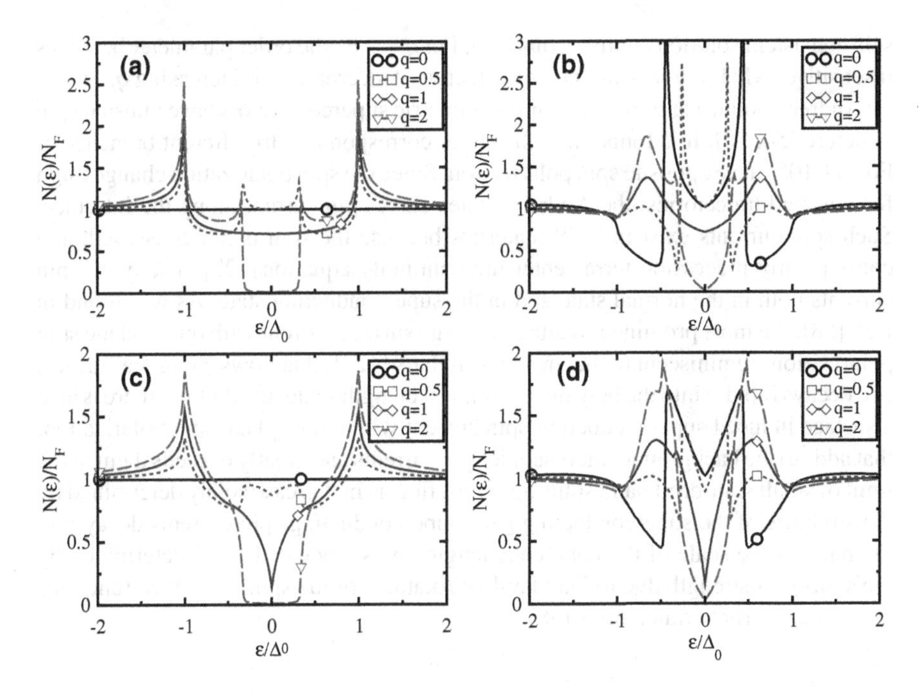

Fig. 11.9 Local surface density of states $N(\varepsilon)/N_{\mathrm{F}}$ for a Rashba superconductor, $\boldsymbol{g}(\boldsymbol{k}) = \alpha_R \boldsymbol{k} \times \hat{z}$. The surface is parallel to $\hat{z}$. The curves are for $\Delta_\pm = \Delta_0(q \pm |\hat{\boldsymbol{g}}(\boldsymbol{k})|)/(q+1)$, with q ranging from 0 to 2. In **a** and **c** the order parameter is assumed constant up to the surface, and in **b** and **d** a suppression of the order parameter to zero in a surface layer of thickness $W = 2\xi_0$ with $\xi_0 = \hbar v_{\mathrm{F}}/2\pi k_{\mathrm{B}} T_c$ is assumed. **a** and **b** is for a cylindrical Fermi surface, $\boldsymbol{v}_{\mathrm{F}} = (v_x, v_y, 0)$, and **c** and **d** is for a spherical Fermi surface. (The symbols are labels for the curves only)

In Fig. 11.9 we show results for a Rashba-type spin-orbit coupling, $\boldsymbol{g}(\boldsymbol{k}) = \alpha_R \boldsymbol{k} \times \hat{z}$. The surface is aligned with the $\hat{z}$ direction. In (a) and (b) we use a cylindrical Fermi surface, for which $|\hat{\boldsymbol{g}}(\boldsymbol{k})| = 1$. In (c) and (d) the results for a spherical Fermi surface are shown, for which $|\hat{\boldsymbol{g}}(\boldsymbol{k})| = \sin(\theta_{\boldsymbol{k}})$. The effect of a surface layer with suppressed order parameter is illustrated in Fig. 11.9 (b) and (d), where Eq. (11.104) with $W = 2\xi_0$ is used, where ξ_0 is the coherence length $\xi_0 = \hbar v_{\mathrm{F}}/2\pi k_{\mathrm{B}} T_c$.

For the special case $q = 0$ we have $\Delta_+ = -\Delta_- = \Delta_0 \sin(\theta_{\boldsymbol{k}})$ (we use a real gauge). For this case, $\theta_g = \theta_{\underline{g}} = \pi/2$, and consequently, $\mathscr{M} = \tan^2 \varphi_g = \cot^2 \varphi_{\boldsymbol{k}}$. The bound states are then given by [56, 58]

$$\frac{\varepsilon}{\Delta_0} = -\sin\left(\frac{2W\varepsilon}{v_{\mathrm{F}} \cos \varphi_{\boldsymbol{k}}} \pm \varphi_{\boldsymbol{k}}\right) \sin(\theta_{\boldsymbol{k}}). \tag{11.105}$$

Numerical solution of the problem shows that the "principal" bound-state branches $\varepsilon(\varphi_{\boldsymbol{k}})$ with energies away from the continuum edge contribute the most to the subgap DOS. For $W \neq 0$ the main branch $\varepsilon_{\mathrm{bs}}(\varphi_{\boldsymbol{k}})$ develops a maximum at $\varepsilon^\star < \Delta_0$, which gives rise to a peak in the surface DOS near $\varepsilon^\star$, see Fig. 11.9 (b) and (d). Fully self-consistent solution confirms this [58]. For $q \to \infty$ the order parameter becomes insensitive to helicity mixing, i.e. the effective W decreases for increasing q.

Andreev bound states in non-centrosymmetric superconductors have unusual spin structure [54, 58]. It is found, that the states corresponding to different branches of Eq. (11.105) have opposite spin polarization. Since the spin polarization changes sign for reversed trajectories, the Andreev states carry spin current along the interface. Such spin currents exist in NCS materials because the spin is not conserved, and consequently precession terms enter the continuity equation [122]. There are spin currents both in the normal state and in the superconducting state. As was found in Ref. [58], the most prominent feature is a large surface current with out of plane spin polarization (reminiscent to that in spin Hall bars [123]) that flows along the surface, and decays rapidly into the bulk on a Fermi wavelength scale. In addition, there is also a surface induced superconducting spin current with out of plane spin polarization, that adds to the background microscopic spin currents and greatly exceeds them in the limit of small spin-orbit band splitting. This effect is in this case solely determined by the structure of the superconducting gap. Superconducting spin currents decay into the bulk on the scale of the coherence length and show oscillations determined by the spin-orbit strength due to Faraday-like rotations of the spin coherence functions along quasiparticle trajectories [58].

11.3.8.2 Tunneling Conductance

For a three-dimensional model, which for the Rashba-type spin-orbit coupling was discussed in Ref. [56], we present in the following tunneling conductances in various geometries. We will discuss several types of spin-orbit interaction:

$$\begin{aligned} &\mathsf{C}_{4v}: \quad \boldsymbol{g} = \eta \begin{pmatrix} \hat{k}_y \\ -\hat{k}_x \\ 0 \end{pmatrix} + \eta' \begin{pmatrix} \hat{0} \\ 0 \\ \hat{k}_x\hat{k}_y\hat{k}_z(\hat{k}_x^2 - \hat{k}_y^2) \end{pmatrix}, \\ &\mathsf{T}_d: \quad \boldsymbol{g} = \eta \begin{pmatrix} \hat{k}_x(\hat{k}_y^2 - \hat{k}_z^2) \\ \hat{k}_y(\hat{k}_z^2 - \hat{k}_x^2) \\ \hat{k}_z(\hat{k}_x^2 - \hat{k}_y^2) \end{pmatrix}, \quad \mathsf{O}: \quad \boldsymbol{g} = \eta \begin{pmatrix} \hat{k}_x \\ \hat{k}_y \\ \hat{k}_z \end{pmatrix}, \end{aligned} \tag{11.106}$$

For the symmetry C_{4v}, corresponding to the tetragonal point group, the two parameters η and η' can both be non-zero. We will discuss below the special cases $\eta = 0$ and $\eta' = 0$. The case $\eta' = 0$ corresponds to a Rashba spin-orbit coupling. The type of spin-orbit coupling we consider for the full tetrahedral point group, T_d, is also known as Dresselhaus coupling. Finally, for the cubic point group, O, the simplest form for $\boldsymbol{g}$ is considered here, which is fully isotropic. All the cases above are relevant for non-centrosymmetric superconductors: C_{4v} for $CePt_3Si$, $CeRhSi_3$, and $CeIrSi_3$, T_d for Y_2C_3 and possibly KOs_2O_6, and O for $Li_2(Pd_{1-x}Pt_x)_3B$.

The zero temperature tunneling conductance is obtained according to the formula Eq. (11.82), which leads for a spin-inactive δ-function barrier to

$$\frac{G(eV)}{G_{\mathrm{N}}} = \frac{\langle \cos(\alpha_{\boldsymbol{k}}) D(\alpha_{\boldsymbol{k}}) N(\boldsymbol{k}, eV) \rangle}{\langle \cos(\alpha_{\boldsymbol{k}}) D(\alpha_{\boldsymbol{k}}) \rangle}, \quad D(\alpha_{\boldsymbol{k}}) = \frac{D_0 \cos^2(\alpha_{\boldsymbol{k}})}{1 - D_0 \sin^2(\alpha_{\boldsymbol{k}})} \tag{11.107}$$

where $\alpha_{\boldsymbol{k}}$ is the angle between the surface normal and $\boldsymbol{k}$. A remark is in place here. In principle, the interface barrier will be spin-dependent once a spin-orbit split material is brought in contact with a normal metal. However, for the limit of small spin-orbit splitting we can neglect the spin-dependence of the interface potential consistent with the quasiclassical approximation. The corrections are of the same order as the corrections for the quasiparticle velocity in this case, and are of higher order in the parameter SMALL.

In Fig. 11.10 we show the tunneling conductance $G(eV)/G_{\mathrm{N}}$ obtained from Eq. (11.107) with Eqs. (11.99)–(11.100) for various types of spin-orbit coupling, corresponding to that in Eq. (11.106), and for various alignments of the surface normal with respect to the crystal symmetry directions. The tunneling parameter in this figure is $D_0 = 0.1$. We show curves for an order parameter $\Delta_\pm = \Delta_0(q \pm |\hat{\boldsymbol{g}}(\boldsymbol{k})|)/(q+1)$, with q ranging from 0 to 2. For simplicity, we concentrate here on the assumption that the order parameter is constant up to the surface, i.e. we use the bulk solutions Eqs. (11.93)–(11.95). For a detailed quantitative description, a self-consistent determination of the order parameter suppression near the surface must be obtained. We also use the simplifying assumptions of a spherical Fermi surface with isotropic Fermi velocity.

As seen from Fig. 11.10, a rich structure of Andreev bound states below the bulk gap energy develops, which depends strongly on the alignment of the surface with the crystal symmetry axes. In (a) and (b) a pure Rashba spin-orbit coupling $\boldsymbol{k} = [\hat{k}_y, -\hat{k}_x, 0]$ on a Fermi sphere is assumed. Below the critical value $q = 1$, a zero-bias peak appears for tunneling in the direction perpendicular to the $\hat{z}$ direction, i.e. the (1,0,0) or (0,1,0) direction, however a dependence quadratic in energy appears for tunneling parallel to the $\hat{z}$ direction, i.e. the (0,0,1) direction [56]. For $q > 1$ the tunneling density of states acquires a gap, as then the singlet character of the order parameter dominates.

In Fig. 11.10 (b) and (c), we show results for a hypothetical spin-orbit coupling of the form $\boldsymbol{g} = \eta'[0, 0, \hat{k}_x \hat{k}_y \hat{k}_z (\hat{k}_x^2 - \hat{k}_y^2)]$, that is consistent with the same point group symmetry D_{4v} as the Rashba spin-orbit coupling. For this case, a sharp zero-bias conductance peak exists for $q < 1$ in all tunneling directions. In contrast, for $q > 1$, the zero bias conductance peak only exists when tunneling perpendicular to the z-direction, however, not when tunneling parallel to the z-direction.

In Fig. 11.10 (d) we consider the cubic point group symmetry, and assume the simplest form of a fully isotropic spin-orbit interaction of the form $\hat{\boldsymbol{g}} = [\hat{k}_x, \hat{k}_y, \hat{k}_z]$. Here, for $q < 1$ the tunneling conductance is zero at zero bias, but raises sharply away from zero bias, showing side peaks due to Andreev bound states. At $q = 1$ this structure disappears with only a pseudogap remaining. For $q < 1$ a gap opens.

Finally, in Fig. 11.10 (e)–(g) we show results for the full tetrahedral point group T_d. We compare tunneling in (1,0,0), (1,1,0), and (1,1,1) directions. Note that in this case, the relation (11.106) between $\boldsymbol{g}$ and $\boldsymbol{k}$ is not invariant under a rotation of both vectors by 90 degree around the $\hat{k}_x$-, $\hat{k}_y$-, or $\hat{k}_z$-axis, but an overall sign change

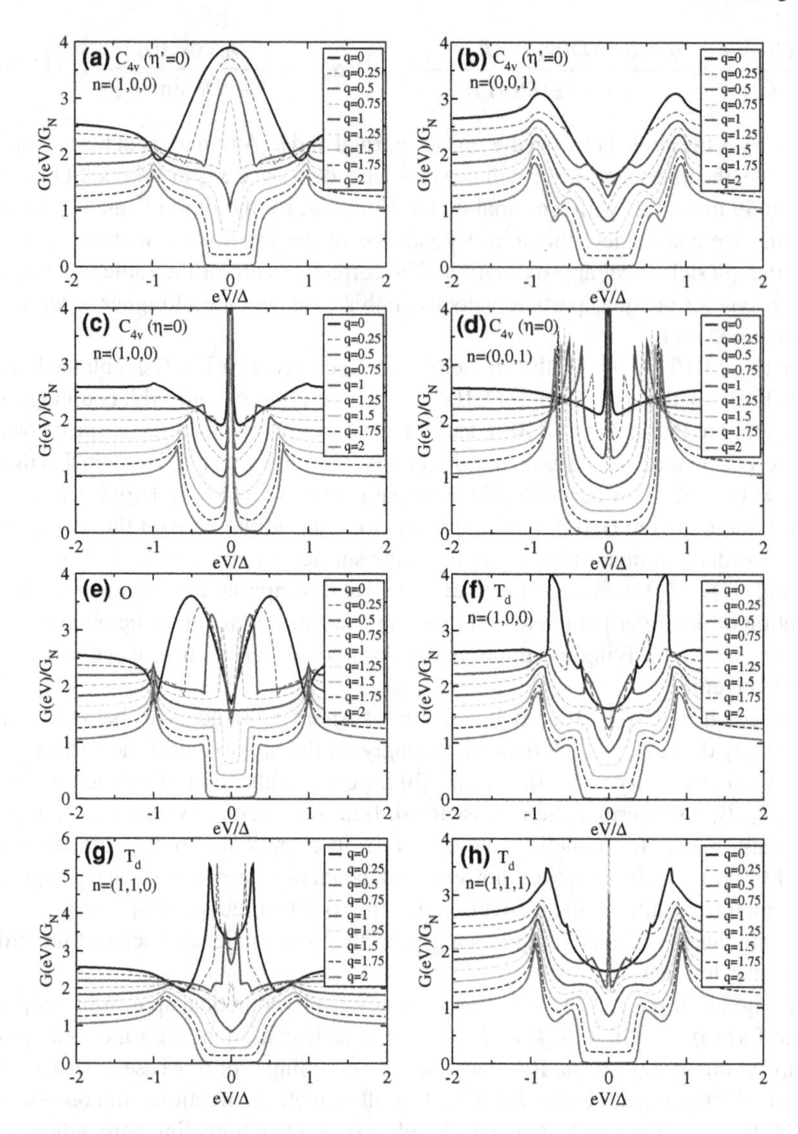

Fig. 11.10 Tunneling conductance $G(eV)/G_{\mathrm{N}}$ for various types of spin-orbit coupling corresponding to the indicated symmetry groups, and for various alignments of the surface normal $\hat{\boldsymbol{n}}$ as indicated. The spin-orbit vector is of the form C_{4v}: $\boldsymbol{g} = \eta[\hat{k}_y, -\hat{k}_x, 0] + \eta'[0, 0, \hat{k}_x\hat{k}_y\hat{k}_z(\hat{k}_x^2 - \hat{k}_y^2)]$; O: $\hat{\boldsymbol{g}} = [\hat{k}_x, \hat{k}_y, \hat{k}_z]$ (this case is fully isotropic); T_d: $\hat{\boldsymbol{g}} = 2[\hat{k}_x(\hat{k}_y^2 - \hat{k}_z^2), \hat{k}_y(\hat{k}_z^2 - \hat{k}_x^2), \hat{k}_z(\hat{k}_x^2 - \hat{k}_y^2)]$. The curves are for $\Delta_\pm = \Delta_0(q \pm |\hat{\boldsymbol{g}}(\boldsymbol{k})|)/(q+1)$, with q ranging from 0 to 2. The order parameter is assumed constant up to the surface, and a spherical Fermi surface with isotropic Fermi velocity is assumed. The tunneling parameter is $D_0 = 0.1$. Curves are vertically shifted by multiples of 0.2

appears; however, the conductance spectra are insensitive to this sign change. For $q < 1$ there is a vanishing zero-bias conductance for tunneling in (1,0,0) direction, and a low-energy dispersive Andreev bound-state branch for tunneling in (1,1,0) direction. For tunneling in (1,1,1) direction, the zero-bias conductance vanishes for $q = 0$, and shows a sharp zero-bias peak for $0 < q < 1$. For $q > 1$ the tunneling conductance becomes gapped for all directions.

As can be seen from these results, studying directional resolved tunneling in non-centrosymmetric superconductors gives important clues about the order parameter symmetry and the type of spin-orbit interaction.

11.3.8.3 Andreev Point-Contact Spectra

Here we present results for the case of a point contact between a normal metal and a non-centrosymmetric superconductor. We use Eq. (11.79) to calculate the spectra, with a scattering matrix that has the form shown in Eq. (11.86). We assume isotropic Fermi surfaces in the materials on both sides of the interface, and for simplicity use equal magnitudes for Fermi momenta and velocities. The transmission amplitude is modeled by that for a δ-function barrier,

$$t(\alpha_{\boldsymbol{k}}) = \frac{t_0 \cos(\alpha_{\boldsymbol{k}})}{\sqrt{1 - t_0^2 \sin^2(\alpha_{\boldsymbol{k}})}}, \tag{11.108}$$

and the component of the Fermi velocity along the interface normal in direction of current transport is $\hat{\boldsymbol{n}} \boldsymbol{v}_{\mathrm{F}1} = v_{\mathrm{F}} \cos(\alpha_{\boldsymbol{k}})$.

In Fig. 11.11, the Andreev conductance $G(eV)/G_{\mathrm{N}}$ for various types of spin-orbit coupling and for various alignments of the surface normal $\hat{\boldsymbol{n}}$ are shown. Here, the transmission probability $D_0 = t_0^2$ is varied from zero to one. We restrict here to the case $q = 0$, i.e. an order parameter of the form $\Delta_{\pm} = \pm\Delta_0 |\hat{\boldsymbol{g}}(\boldsymbol{k})|$). Again, a spherical Fermi surface with isotropic Fermi velocity is assumed. We also compare the case of a surface layer with suppressed order parameter (dashed lines) with that of an order parameter constant up to the surface (full lines). To model the order parameter suppression, we assume a layer of thickness $W = 2\xi_0$ with $\xi_0 = \hbar v_{\mathrm{F}}/2\pi k_{\mathrm{B}} T_c$ (dotted lines) in which the order parameter vanishes. Thus, we use Eq. (11.104) as incoming solutions for the coherence amplitudes. Note that for $D_0 = 1$ the surface layer with zero order parameter does not affect the Andreev conductance. This is due to the fact that for perfect transmission the normal region simply extends slightly further towards the superconductor, and within our approximation we neglect the spin-orbit effects in interface potential.

For a larger spin-orbit coupling the interface between a normal metal and a normal conducting non-centrosymmetric metal with strong spin-orbit interaction becomes necessarily spin-active, as the interface potential term in the Hamiltonian must be hermitian. Thus, a perfect transmission is not realistic in such a case. For weak spin-orbit splitting these effects are also present, however modify the results only to order $\hat{v}_{\mathrm{SO}}/E_{\mathrm{F}}$, or on energy scales $\hat{v}_{\mathrm{SO}}^2/E_{\mathrm{F}}$.

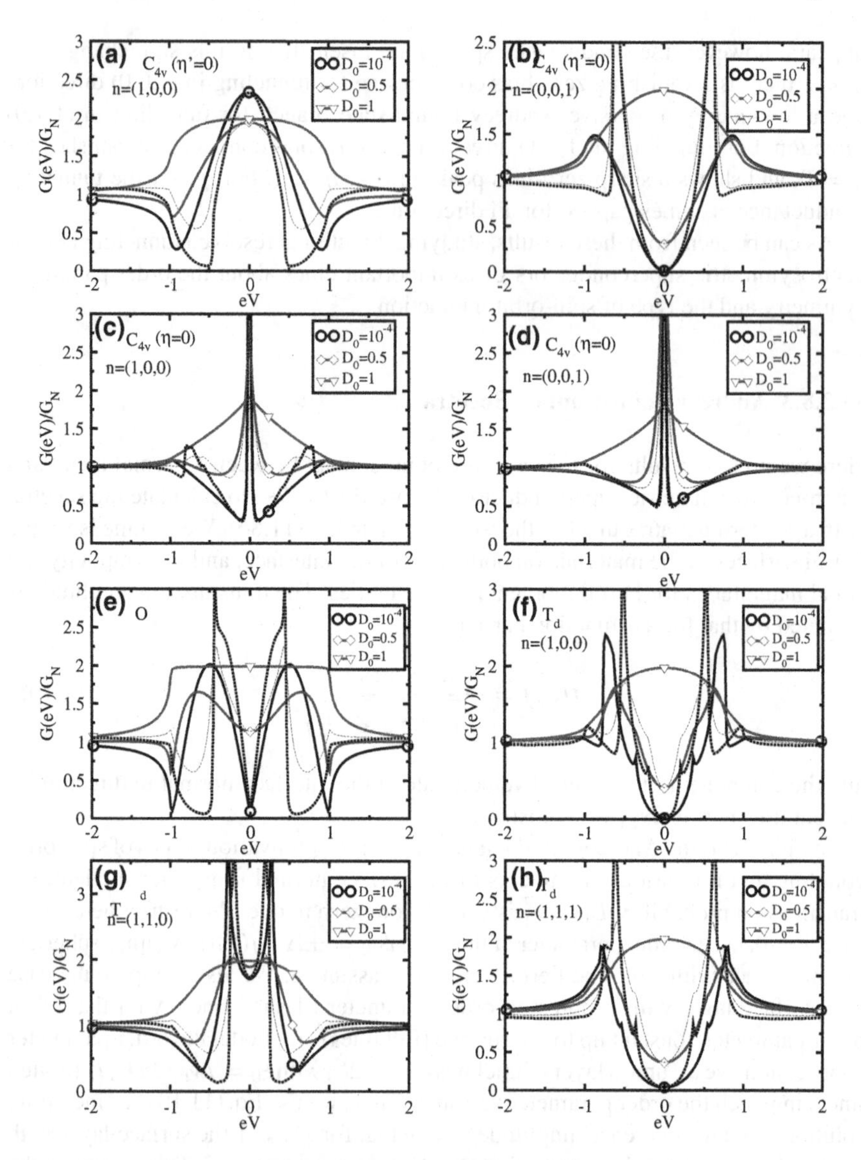

Fig. 11.11 Andreev conductance $G(eV)/G_{\mathrm{N}}$ for various types of spin-orbit coupling corresponding to the indicated symmetry groups, and for the indicated alignments of the surface normal $\hat{\boldsymbol{n}}$. The transmission probability $D_0 = t_0^2$ is varied. The spin-orbit vector is of the form C_{4v}: $\boldsymbol{g} = \eta[\hat{k}_y, -\hat{k}_x, 0] + \eta'[0, 0, \hat{k}_x\hat{k}_y\hat{k}_z(\hat{k}_x^2 - \hat{k}_y^2)]$; O: $\hat{\boldsymbol{g}} = [\hat{k}_x, \hat{k}_y, \hat{k}_z]$ (this case is fully isotropic); T_d: $\hat{\boldsymbol{g}} = 2[\hat{k}_x(\hat{k}_y^2 - \hat{k}_z^2), \hat{k}_y(\hat{k}_z^2 - \hat{k}_x^2), \hat{k}_z(\hat{k}_x^2 - \hat{k}_y^2)]$. The curves are for $\Delta_\pm = \pm\Delta_0|\hat{\boldsymbol{g}}(\boldsymbol{k})|)$. The order parameter is assumed constant up to the surface (full lines) or suppressed to zero in a surface layer of thickness $W = 2\xi_0$ with $\xi_0 = \hbar v_{\mathrm{F}}/2\pi k_{\mathrm{B}} T_c$ (dotted lines). For $D_0 = 1$ these two cases give identical results. A spherical Fermi surface with isotropic Fermi velocity is assumed

For lower transmission, we remark as an overall observation that the suppression of the order parameter does not affect the value of the Andreev conductance at zero bias. This is simply due to the fact that in the clean limit the coherence amplitudes become effectively spatially constant for $\varepsilon = 0$. For higher bias, deviations can be observed, that in general lead to a shift of Andreev bound states to lower bias.

We turn now to the Andreev point contact spectra for $t_0 = 1$. As can be seen, the form of the spectrum is sensitive to the type of spin-orbit coupling, and the associated order-parameter symmetry. For a Rashba spin-orbit coupling, Fig. 11.11 (a) and (b), the Andreev conductance is enhanced to twice the normal conductance at zero bias, however to a smaller value for finite bias. There is a pronounced anisotropy in the shape of the Andreev conductance spectra. In (c) and (d) the Andreev conductance shows a sharp kink feature at zero bias, associated with the complex nodal structure of the spin-orbit vector. For cubic symmetry, (e), we observe an Andreev conductance resembling that of an *s*-wave spin-singlet superconductor. And, finally, for a Dresselhaus spin-orbit coupling, (f)–(h), the Andreev conductance shows a behavior similar to the case of a Rashba spin-orbit interaction, however with a less pronounced anisotropy.

11.4 Conclusions

We have given an overview over the current status of the theoretical understanding of Andreev bound states at the surface of a non-centrosymmetric superconducting material, and have presented results for tunneling conductance, point-contact spectra, and spin polarized Andreev bound-state spectra.

The new feature in non-centrosymmetric superconductors is the possible appearance of spin polarized Andreev states, that carry a spin-current along the interface or surface. The presence of such Andreev bound states that cross the chemical potential as a function of incident angle to the surface, is a topologically stable superconducting property. Such bound states exist as long as triplet order parameter components (in spin representation) dominate singlet components of the order parameter. When both components are equal, the bound state at the chemical potential disappears, and a topologically new ground state is established. The transition between the two states is a quantum phase transition.

The spectrum of Andreev states at the surface provides valuable information about both the structure of the superconducting order parameter and the vector field of spin-orbit vectors on the Fermi surface. In this chapter we have concentrated on the rich structure that appears for the limiting case of a small spin-orbit splitting of the energy bands in the non-centrosymmetric material. In this limit, the spin quantum number is approximately conserved during scattering from surfaces and interfaces with normal metals, which leads to strong mixing between the helicity bands in the non-centrosymmetric material. The opposite limit of strong spin-orbit splitting is still largely unexplored. We have provided a theoretical basis in this chapter that allows to treat this case as well.

Finally, we would like to mention that interesting effects, like e.g. effects related to the spin Hall effect, or to Berry phases associated with the change of the spin-orbit vector along closed paths, are interesting subjects left for future studies.

Acknowledgements The authors would like to thank A. Balatsky, W. Belzig, J. Inoue, S. Kashiwaya, K. Kuroki, N. Nagaosa, J.A. Sauls, G. Schön, M. Sigrist, Y. Tanuma, I. Vekhter, A. Vorontsov, and T. Yokoyama for valuable discussions or contributions in connection with the topic of this chapter.

References

1. Giaever, I.: Phys. Rev. Lett. **5**, 147 (1960)
2. Tanaka, Y., Kashiwaya, S.: Phys. Rev. Lett. **74**, 3451 (1995)
3. Kashiwaya, S., Tanaka, Y.: Rep. Prog. Phys. **63**, 1641 (2000)
4. Bruder, C.: Phys. Rev. B **41**, 4017 (1990)
5. Hu, C.R.: Phys. Rev. Lett. **72**, 1526 (1994)
6. Buchholtz, L.J., Palumbo, M., Rainer, D., Sauls, J.A.: J. Low Temp. Phys. **101**, 1099 (1995)
7. Buchholtz, L.J., Zwicknagl, G.: Phys. Rev. B **23**, 5788 (1981)
8. Hara, J., Nagai, K.: Prog. Theor. Phys. **74**, 1237 (1986)
9. Honerkamp, C., Sigrist, M.: J. Low Temp. Phys. **111**, 895 (1998)
10. Yamashiro, M., Tanaka, Y., Kashiwaya, S.: Phys. Rev. B **56**, 7847 (1997)
11. Matsumoto, M., Sigrist, M.: J. Phys. Soc. Jpn. **68**, 994 (1999)
12. Tanaka, Y., Tanuma, T., Kuroki, K., Kashiwaya, S.: J. Phys. Soc. Jpn. **71**, 2102 (2002)
13. Geerk, J., Xi, X.X., Linker, G.: Z. Phys. B. **73**, 329 (1988)
14. Kashiwaya, S., Tanaka, Y., Koyanagi, M., Takashima, H., Kajimura, K.: Phys. Rev. B **51**, 1350 (1995)
15. Alff, L., Takashima, H., Kashiwaya, S., Terada, N., Ihara, H., Tanaka, Y., Koyanagi, M., Kajimura, K.: Phys. Rev. B **55**, R14757 (1997)
16. Covington, M., Aprili, M., Paraoanu, E., Greene, L.H., Xu, F., Zhu, J., Mirkin, C.A.: Phys. Rev. Lett. **79**, 277 (1997)
17. Wei, J.Y.T., Yeh, N.-C., Garrigus, D.F., Strasik, M.: Phys. Rev. Lett. **81**, 2542 (1998)
18. Laube, F., Goll, G., Löhneysen, H.v., Fogelström, M., Lichtenberg, F.: Phys. Rev. Lett. **84**, 1595 (2000)
19. Mao, Z.Q., Nelson, K.D., Jin, R., Liu, Y., Maeno, Y.: Phys. Rev. Lett. **87**, 037003 (2001)
20. Kawamura, M., Yaguchi, H., Kikugawa, N., Maeno, Y., Takayanagi, H.: J. Phys. Soc. Jpn. **74**, 531 (2005)
21. Wälti, Ch., Ott, H.R., Fisk, Z., Smith, J.L.: Phys. Rev. Lett. **84**, 5616 (2000)
22. Rourke, P.M.C., Tanatar, M.A., Turel, C.S., Berdeklis, J., Petrovic, C., Wei, J.Y.T.: Phys. Rev. Lett. **94**, 107005 (2005)
23. Ichimura, K., Higashi, S., Nomura, K., Kawamoto, A.: Synth. Met. **153**, 409 (2005)
24. Turel, C.S., Wei, J.Y.T., Yuhasz, W.M., Maple, M.B.: Physica C **32**, 463–465 (2007)
25. Aoki, Y., Wada, Y., Saitoh, M., Nomura, R., Okuda, Y., Nagato, Y., Yamamoto, M., Higashitani, S., Nagai, K.: Phys. Rev. Lett. **95**, 075301 (2005)
26. Blonder, G.E., Tinkham, M., Klapwijk, T.M.: Phys. Rev. B **25**, 4515 (1982)
27. Eschrig, M.: Phys. Rev. B **61**, 9061 (2000)
28. Mackenzie, A.P., Maeno, Y.: Rev. Mod. Phys. **75**, 657 (2003)
29. Maeno, Y. et al.: Nature **394**, 532 (1994)
30. Maeno, Y., Hashimoto, H., Yoshida, K., Nishizaki, S., Fujita, T., Bednorz, J.G., Lichtenberg, F.: Nature (London) **372**, 532 (1994)

31. Ishida, K., Mukuda, H., Kitaoka, Y., Asayama, K., Mao, Z.Q., Mori, Y., Maeno, Y.: Nature (London) **396**, 658 (1998)
32. Luke, G.M., Fudamoto, Y., Kojima, K.M., Larkin, M.I., Merrin, J., Nachumi, B., Uemura, Y.J., Maeno, Y., Mao, Z.Q., Mori, Y., Nakamura, H., Sigrist, M.: Nature (London) **394**, 558 (1998)
33. Mackenzie, A.P., Maeno, Y.: Rev. Mod. Phys. **75**, 657 (2003)
34. Nelson, K.D., Mao, Z.Q., Maeno, Y., Liu, Y.: Science **306**, 1151 (2004)
35. Asano, Y., Tanaka, Y., Sigrist, M., Kashiwaya, S.: Phys. Rev. B **67**, 184505 (2003)
36. Asano, Y., Tanaka, Y., Sigrist, M., Kashiwaya, S.: Phys. Rev. B **71**, 214501 (2005)
37. Kambara, H., Kashiwaya, S., Yaguchi, H., Asano, Y., Tanaka, Y., Maeno, Y.: Phys. Rev. Lett. **101**, 267003 (2008)
38. Tanuma, Y., Hayashi, N., Tanaka, Y., Golubov, A.A.: Phys. Rev. Lett. **102**, 117003 (2009)
39. Yokoyama, T., Iniotakis, C., Tanaka, Y., Sigrist, M.: Phys. Rev. Lett. **100**, 177002 (2008)
40. Fogelström, M., Rainer, D., Sauls, J.A.: Phys. Rev. Lett. **79**, 281 (1997)
41. Laube, F., Goll, G., Eschrig, M., Fogelström, M., Werner, R.: Phys. Rev. B **69**, 014516 (2004)
42. Yamashiro, M., Tanaka, Y., Yoshida, N., Kashiwaya, S.: J. Phys. Soc. Jpn. **68**, 2019 (1999)
43. Prange, R.E., Girvin, S.M. (eds.) The Quantum Hall Effect. Springer, New York (1987)
44. Thouless, D. J., Kohmoto, M., Nightingale, M.P., den Nijs, M.: Phys. Rev. Lett. **49**, 405 (1982)
45. Goryo, J., Ishikawa, K.: J. Phys. Soc. Jpn. **67**, 3006 (1998)
46. Furusaki, A., Matsumoto, M., Sigrist, M.: Phys. Rev. B **64** 054514 (2001)
47. Bauer, E., Hilscher, G., Michor, H., Paul, Ch., Scheidt, E.W., Gribanov, A., Seropegin, Yu., Noël, H., Sigrist, M., Rogl, P.: Phys. Rev. Lett. **92**, 027003 (2004)
48. Frigeri, P.A., Agterberg, D.F., Koga, A., Sigrist, M.: Phys. Rev. Lett. **92**, 097001 (2004)
49. Reyren, N. et al.: Science **317**, 1196 (2007)
50. Yada, K., Onari, S., Tanaka, Y., Inoue, J.: Phys. Rev. B **80**, 140509 (2009)
51. Qi, X.L., Hughes, T.L., Raghu, S., Zhang, S.C.: Phys. Rev. Lett. **102**, 187001 (2009)
52. Sato, M., Fujimoto, S.: Phys. Rev. B **79**, 094504 (2009)
53. Roy, R.: arXiv:cond-mat/0608064; Lu C.K., Yip S.-K.: Phys. Rev. B **78**, 132502 (2008) Yip, S.-K.: arXiv:0910.0696
54. Tanaka, Y., Yokoyama, T., Balatsky, A.V., Nagaosa, N.: Phys. Rev. B **79**, 060505(R) (2009)
55. Andreev, A.F.: Sov. Phys. JETP **19**, 1228 (1964)
56. Iniotakis, C., Hayashi, N., Sawa, Y., Yokoyama, T., May, U., Tanaka, Y., Sigrist, M.: Phys. Rev. B **76**, 012501 (2007)
57. Yokoyama, T., Tanaka, Y., Inoue, J.: Phys. Rev. B **72**, 220504(R) (2005)
58. Vorontsov, A.B., Vekhter, I., Eschrig, M.: Phys. Rev. Lett. **101**, 127003 (2008)
59. Linder, J., Sudbø, A.: Phys. Rev. B **76**, 054511 (2007)
60. Børkje, K., Sudbø, A.: Phys. Rev. B **74**, 054506 (2006)
61. Børkje, K.: Phys. Rev. B **76**, 184513 (2007)
62. Kashiwaya, S., Tanaka, Y., Yoshida, N., Beasley, M.R.: Phys. Rev. B **60**, 3572 (1999)
63. König, M., Wiedmann, S., Brüne, C., Roth, A., Buhmann, H., Molenkamp, L., Qi, X.-L., Zhang, S.-C.: Science **318**, 766 (2007)
64. Kane, C.L., Mele, E.J.: Phys. Rev. Lett. **95**, 146802 (2005)
65. Kane, C.L., Mele, E.J.: Phys. Rev. Lett. **95**, 226801 (2005)
66. Bernevig, B.A., Zhang, S.C.: Phys. Rev. Lett. **96**, 106802 (2006)
67. Fu, L., Kane, C.L.: Phys. Rev. B **74**, 195312 (2006)
68. Fu, L., Kane, C.L.: Phys. Rev. B **76**, 045302 (2007)
69. Landau, L.D.: Zh. Eksp. Teor. Fiz. **32**, 59 (1957), [Sov. Phys. JETP **5**, 101 (1957)]
70. Landau, L.D.: Sov. Phys. JETP **8**, 70 (1959)
71. Serene, J.W., Rainer, D.: Phys. Rep. **101**, 221 (1983)
72. Rainer, D.: In: Brewer D.F. (ed.) Progress in Low Temperature Physics X, p. 371. Elsevier, Amsterdam (1986)
73. Eschrig, M., Heym, J., Rainer, D.: J. Low Temp. Phys. **95**, 323 (1994)

74. Rainer, D., Sauls, J.A.: In: Butcher, P.N., Lu, Y. (eds.) Superconductivity: From Basic Physics to New Developments, pp. 45–78. World Scientific, Singapore, (1995)
75. Eschrig, M., Rainer, D., Sauls, J.A.: Phys. Rev. B **59**, 12095 (1999); Appendix C
76. Larkin, A.I., Ovchinnikov, Y.N.: Zh. Eksp. Teor. Fiz. **55**, 2262 (1968)
77. Larkin, A.I., Ovchinnikov, Y.N.: Sov. Phys. JETP **28**, 1200 (1969)
78. Eilenberger, G.: Z. Phys. **214**, 195 (1968)
79. Schmid, A., Schön, G.: J. Low. Temp. Phys. **20**, 207 (1975)
80. Schmid, A.: In: Gray, K.E. (ed.) Nonequilibrium Superconductivity, Phonons and Kapitza Boundaries, Proceedings of NATO Advanced Study Institute. Plenum Press, New York (1981), Chapter 14
81. Rammer, J., Smith, H.: Rev. Mod. Phys. **58**, 323 (1986)
82. Larkin, A.I., Ovchinnikov, Y.N.: In: Langenberg, D.N., Larkin, A.I. (eds.) Nonequilibrium Superconductivity, p. 493. Elsevier, Amsterdam (1986)
83. Eschrig, M., Sauls, J.A., Burkhardt, H., Rainer, D.: In: Drechsler, S.-L., Mishonov, T. (eds.) High-Tc Superconductors and Related Materials, Fundamental Properties, and Some Future Electronic Applications, Proceedings of the NATO Advanced Study Institute, pp. 413–446. Kluwer Academic, Norwell (2001)
84. Gor'kov, L.P.: Zh. Eksp. Teor. Fiz. **34**, 735 (1958)
85. Gor'kov, L.P.: Sov. Phys. JETP **7**, 505 (1958)
86. Gor'kov, L.P.: Zh. Eksp. Teor. Fiz. **36**, 1918 (1959)
87. Gor'kov, L.P.: Sov. Phys. JETP **9**, 1364 (1959)
88. Keldysh, L.V.: Zh. Eksp. Teor. Fiz. **47**, 1515 (1964)
89. Keldysh, L.V.: Sov. Phys. JETP **20**, 1018 (1965)
90. Shelankov, A.L.: J. Low Temp. Phys. **60**, 29 (1985)
91. Alexander, J.A., Orlando, T.P., Rainer, D., Tedrow, P.M.: Phys. Rev. B **31**, 5811 (1985)
92. Eschrig, M., Sauls, J.A., Rainer, D.: Phys. Rev. B **60**, 10447 (1999)
93. Nagato, Y., Nagai, K., Hara, J.: J. Low Temp. Phys. **93**, 33 (1993)
94. Higashitani, S., Nagai, K.: J. Phys. Soc. Jpn. **64**, 549 (1995)
95. Nagato, Y., Higashitani, S., Yamada, K., Nagai, K.: J. Low Temp. Phys. **103**, 1 (1996)
96. Schopohl, N., Maki, K.: Phys. Rev. B **52**, 490 (1995)
97. Schopohl, N.: cond-mat/9804064 (unpublished, 1998)
98. Eschrig, M., Kopu, J., Konstandin, A., Cuevas, J.C., Fogelström, M., Schön, G.: Singlet-triplet mixing in superconductor-ferromagnet hybrid devices. In: Bernhard Kramer (ed.) Advances in Solid State Physics, vol. **44**, pp. 533–546. Springer, Berlin (2004)
99. Cuevas, J.C., Hammer, J., Kopu, J., Viljas, J.K., Eschrig, M.: Phys. Rev. B **73**, 184505 (2006)
100. Shelankov, A.L.: Sov. Phys. Solid State **26**, 981 (1984)
101. Shelankov, A.L.: Fiz. Tved. Tela **26**, 1615 (1984)
102. Zaitsev, A.V.: Zh. Eksp. Teor. Fiz. **86**, 1742 (1984)
103. Zaitsev, A.V.: Sov. Phys. JETP **59**, 1015 (1984)
104. Millis, A., Rainer, D., Sauls, J.A.: Phys. Rev. B **38**, 4504 (1988)
105. Yip, S.-K.: J. Low Temp. Phys. **109**, 547 (1997)
106. Lu, C.-K., Yip, S.-K.: Phys. Rev. B **80**, 024504 (2009)
107. Fogelström, M.: Phys. Rev. B **62**, 11812 (2000)
108. Shelankov, A., Ozana, M.: Phys. Rev. B **61**, 7077 (2000)
109. Shelankov, A., Ozana, M.: J. Low Temp. Phys. **124**, 223 (2001)
110. Zhao, E., Löfwander, T., Sauls, J.A.: Phys. Rev. B **70**, 134510 (2004)
111. Eschrig, M.: Phys. Rev. B **80**, 134511 (2009)
112. Grein, R., Eschrig, M., Metalidis, G., Schön, G.: Phys. Rev. Lett. **102**, 227005 (2009)
113. Grein, R., Löfwander, T., Metalidis, G., Eschrig, M.: Phys. Rev. B **81**, 094508 (2010)
114. Eschrig, M., Kopu, J., Cuevas, J.C, Schön, G.: Phys. Rev. Lett. **90**, 137003 (2003)
115. Kopu, J., Eschrig, M., Cuevas, J.C., Fogelström, M.: Phys. Rev. B **69**, 094501 (2004)
116. Eschrig, M., Löfwander, T.: Nat. Phys. **4**, 138 (2008)
117. Lück, T., Schwab, P., Eckern, U., Shelankov, A.: Phys. Rev. B **68**, 174524 (2003)

118. Graser, S., Dahm, T.: Phys. Rev. B **75**, 014507 (2007)
119. Frigeri P.A. et al.: Eur. Phys. J. B **54**, 435 (2006)
120. Sergienko, I.A., Curnoe, S.H.: Phys. Rev. B **70**, 213410 (2004)
121. Hayashi, N., Wakabayashi, K., Frigeri, P.A., Sigrist, M.: Phys. Rev. B **73**, 024504 (2006)
122. Rashba, E.I.: Phys. Rev. B **68**, 241315 (2003)
123. Mishchenko, E.G., Shytov, A.V., Halperin, B.I.: Phys. Rev. Lett. **93**, 226602 (2004)